Optoelectronic Nanodevices

Optoelectronic Nanodevices

Special Issue Editor

Minas M. Stylianakis

MDPI • Basel • Beijing • Wuhan • Barcelona • Belgrade • Manchester • Tokyo • Cluj • Tianjin

Special Issue Editor
Minas M. Stylianakis
Department of Electrical &
Computer Engineering,
Hellenic Mediterranean
University (HMU)
Greece

Editorial Office
MDPI
St. Alban-Anlage 66
4052 Basel, Switzerland

This is a reprint of articles from the Special Issue published online in the open access journal *Nanomaterials* (ISSN 2079-4991) (available at: https://www.mdpi.com/journal/nanomaterials/special_issues/optoelectronic_nanodevices).

For citation purposes, cite each article independently as indicated on the article page online and as indicated below:

LastName, A.A.; LastName, B.B.; LastName, C.C. Article Title. *Journal Name* **Year**, *Article Number*, Page Range.

ISBN 978-3-03928-696-6 (Pbk)
ISBN 978-3-03928-697-3 (PDF)

Cover image courtesy of Minas M. Stylianakis.

Contents

About the Special Issue Editor

Minas M. Stylianakis received his Ph.D. degree in July 2015 from the Chemistry Department of the University of Crete, Greece. He is an expert in synthesis, solution processing, and characterization of novel universal porous and non-porous carbon- and graphene-based metal oxides and other graphene related materials (GRMs), through chemical functionalization processes. He is highly experienced in 2D materials treatment with polymers, small molecules, biomolecules, metal atoms, and heteroatoms, through simple and facile mechanical and chemical routes for their properties tuning regarding high specific area, porosity, electrical, and improved catalytic, as well as biological, properties. In this context, his research activities include various optoelectronics, energy generation and storage applications, such as organic and hybrid photovoltaic (PV) devices, LEDs, batteries, capacitors, and catalytic applications, such as catalysts of reduced noble metals content (Pt, Au, Pd) and high corrosion resistance for fuel cells fabrication. Recently, water treatment technologies (desalination, wastewater treatment) and biomedical applications concerning graphene-based drug delivery systems for eye diseases treatment have been integrated into his research activities. He has published more than 45 articles in SCIE journals, 3 book chapters, 3 in peer-reviewed conference proceedings, and 2 laboratory manuals. He has received more than 2100 citations with an h-index of 28 (March 2020).

 *nanomaterials*

Editorial

Optoelectronic Nanodevices

Minas M. Stylianakis

Department of Electrical & Computer Engineering, Hellenic Mediterranean University (HMU), Estavromenos, 71410 Heraklion, Greece; stylianakis@hmu.gr; Tel.: +30-2810-379775

Received: 24 February 2020; Accepted: 11 March 2020; Published: 13 March 2020

Over the last decade, novel materials such as graphene derivatives, transition metal dichalcogenides (TMDs), other two-dimensional (2D) layered materials, perovskites, as well as metal oxides and other metal nanostructures have centralized the interest of the scientific community [1–4]. Due to their extraordinary physical, optical, thermal, and electrical properties, which are correlated with their 2D ultrathin atomic layer structure, large interlayer distance, ease of functionalization, as well as bandgap tunability, these nanomaterials have been applied towards the development or the improvement of innovative optoelectronic applications, as well as the expansion of theoretical studies and simulations in very fast-growing fields of energy (photovoltaics, energy storage, fuel cells, hydrogen storage, catalysis, etc.), electronics, photonics, spintronics and sensing devices [5–8]. The continuous nanostructure-based applications development provides the confidence to significantly improve existing products and to explore the design of materials and devices with novel functionalities [9–11].

This Special Issue reports some of the most recent trends and advances in the interdisciplinary field of optoelectronics, including 24 original research articles and two review papers. Most articles focus on light emitting diodes (LEDs) and solar cells (SCs), including organic, inorganic and hybrid configurations, while the rest concern some excellent studies on photodetectors, transistors, as well as on other well-known dynamic optoelectronic devices, as depicted in Figure 1. In this context, this exceptional collection of articles is directed at a broad scientific audience of chemists, materials scientists, physicists, and engineers, with the goals of not only pointing out the potential of innovative optoelectronic applications incorporating nanostructures but also inspiring their realization.

The first section of the special issue includes five original articles from the broad field of solar cells technology. First, the article by Deng at al. [22], reports on the preparation of a Ti porous film supported $NiCo_2S_4$ nanotube and its role as a fluorine doped tin oxide (FTO)/Pt alternative counter electrode (CE) in CdS/CdSe quantum-dot-sensitized solar cells (QDSSCs). The novel porous counter electrode, prepared by acid etching and a two-step hydrothermal method, exhibited better electrocatalytic properties, improved loading, as well as higher stability, resulting in a photovoltaic performance improvement of the fabricated QDSSC by 240%, compared to the conventional FTO/Pt based one. Next, Ho et al. [23] demonstrate their efforts to boost the performance and the output power of textured silicon solar cells through plasmonic forward scattering. They individually inserted a single and a double layer of two-dimensional indium NPs within a SiN_x/SiO_2 double-layer antireflective coating. As a result, the conversion efficiency, for both approaches, was improved compared to the reference cell up to 5%. They also investigated the effect of In NPs layers within the double layer of ARC on light-trapping performance of the devices at different slopes. Finally, in a same manner, the electrical output power was evaluated, exhibiting an enhancement over 50%.

In another study, Shi et al. [24] present the growth of high performance ultrathin MoO_3/Ag transparent electrodes via thermal evaporation at different deposition rates, by assessing their optical, electrical, and morphological characteristics. They proved that the synergy of MoO_3 with silver, even as a very thin nucleation layer, was beneficial for the fabrication of uniform, semitransparent, and highly conductive porous films. The said ultrathin electrode was used in inverted organic solar cells as top semitransparent electrode, exhibiting a significant PV performance (2.76%) when the device was illuminated from the top side. The article by Svrcek et al. [19] deals with the development of low roughness InN thin films through the pulsed metalorganic vapor-phase epitaxy (MOVPE) process. The optimized InN/p-GaN photoelectric heterojunction exhibited excellent electron extraction efficiency compared to the literature for similar studies based on pulsed MOVPE.

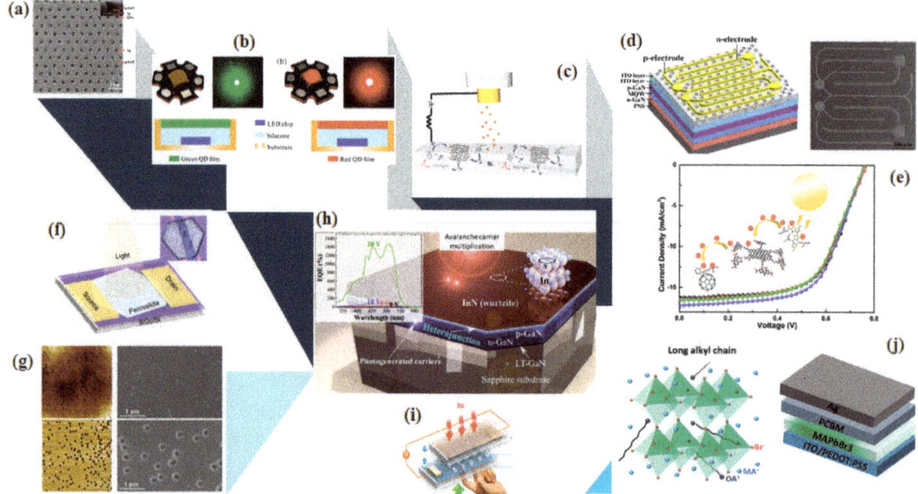

Figure 1. Various configurations of the optoelectronic devices reported in the Special Issue. (**a**) Ag–photonic crystal (PhC) SEM image. Reproduced with permission from [12]; (**b**) real representations and schematic diagrams of a green and a red quantum-dot light emitting diode (LED). Reproduced with permission from [13]; (**c**) representation of a field emitter cold cathode based on graphene oxide r(GO):organic charge transfer materials composites. Reproduced with permission from [14]; (**d**) schematic representation of a patterned double-layer indium tin oxide (ITO) ultraviolet (UV) LED. Reproduced with permission from [15]; (**e**) typical *J-V* curves of a ternary organic solar cells (OSC) device and schematic representation of charge transfer between the active layer materials. Reproduced with permission from [16]; (**f**) schematic representation of a two-dimensional (2D) perovskite platelet phototransistor. Reproduced with permission from [17]; (**g**) atomic force microscopy (AFM) and SEM images of green LEDs with and without InGaN/GaN superlattice. Reproduced with permission from [18]; (**h**) schematic representation of an InN/p-GaN photoelectric heterojunction. Reproduced with permission from [19]; (**i**) a CdTe microdots array photodetector configuration. Reproduced with permission from [20]; (**j**) configuration of a light emitting diode (PeLED) based on octylammonium substituted perovskite. Reproduced with permission from [21]. Copyright of all figures MDPI publisher, 2020.

The ternary configuration has shown a great potential in the field of organic solar cells. In this context, Stylianakis et al. [16] synthesized a novel graphene derivative with favorable energy levels to the binary blend materials of the active layer of a typical inverted organic solar cell. The new graphene-based molecule was incorporated in ink form as the third component within the active layer, in various ratios. In that way, highly efficient inverted ternary organic solar cells (OSC) devices were realized, exhibiting a performance improvement by 13%, compared to the control device, leading to a record PCE value of 8.71%. Power conversion efficiency (PCE) enhancement was attributed to the cascade effect, facilitating electron transport from the active layer to ITO bottom electrode, as well as the morphology improvement between the interfaces of the binary blend components. The last article of this section is by Chen et al. [25], who examined the effect of solvent polarity, as well as the concentration of the precursor methylammonium iodide (MAI) on the morphology of MAPbI$_3$

perovskite thin films. They found that better coverage and compactness were achieved according to the longer alkyl chain of the solvent (lower polarity), although the film's roughness was higher. Upon the parameters' optimization, highly efficient perovskite solar cells were prepared with a champion PCE of 16.66%.

The second section consists of ten original contributions concerning the field of LEDs. Feng et al. [12] present their study on the combination of localized surface plasmon (LSP) and quantum wells (QWs), deriving from the incorporation of silver NPs into the holes of a photonic crystal within the *p*-GaN layer of a green LED. In order to demarcate the impact of light and e-beam induced excitations on LSP–QW coupling mechanism, they adopted a 3D finite difference time domain (FDTD) numerical simulation model during the photophysics measurements, while they suggest a beneficial way towards the energy dissipation reduction in Ag NPs. Next, Yan et al. [13] evaluated the performance of green and red CdSe quantum dots (QDs) based LED devices, at various excitation wavelengths, by assessing the photoluminescence (PL) of the respective thin films. In addition, the indefinite role of a V-pits-embedded InGaN/GaN superlattice (SL) in green LEDs performance improvement was explored by Liu et al. [18], who achieved to improve the device's external quantum efficiency (EQE) by ~30%, upon the development of the V-pits-embedded InGaN/GaN SL layer. They used scanning electron microscopy (SEM) synergistically to a room temperature cathodoluminescence (CL) to validate light emission properties of InGaN/GaN multiple quantum wells (MQWs), proving that V-pits may act as a barrier for carriers' diffusion into nonrecombination centers. Moreover, Chen et al. [21] prepared highly fluorescent and uniform MAPbBr$_3$ thin films by tuning the content of octylammonium bromide (OAB) additive within the perovskite precursors. In this manner, high performance perovskite-based LEDs were fabricated, exhibiting very high luminance and luminous current efficiency values, due to high exciton binding energy of the nanocrystals grain size, resulting in nonradiative recombination mitigation, as well as emission efficiency augmentation.

Li et al. [26] report on their efforts towards the improvement of color-conversion efficiency (CCE) and stability of quantum-dot-based light-emitting diodes upon the incorporation of a blue anti-transmission film (BATF). They proved that under the optimum tradeoff between the BATF thickness and QDs concentration, both CCE and luminous efficacy can be significantly improved by over 42% and 24%, respectively. The study by Zhou et al. [27] proposes the design of two novel p-type layers a) a gradually reduced indium content p-InGaN and b) a *p*-GaN. The first one acted as an ideal replacement of *p*-AlGaN, while the second boosted the light output power of a GaN-based green LED. Indeed, the champion green LED device exhibited an improvement by ~14% compared to the reference one in light intensity while the indium content was gradually reduced from 10% to 0%, as confirmed by experimental and simulation studies. In another theoretical report on GaN-based LEDs, Jin et al. [28] state an error-grating simulation model and recommend some nano-grating strategies, including the use of alternative materials, as well as the adaption of different fabrication structural parameters towards the improvement of light extraction efficiency of LEDs. They compare the results from the literature regarding the use of several materials as patterned bottom reflection layer and suggest that the incorporation of SiO$_2$ nanorod array (NR) is the optimum approach.

The fabrication of a multilayered transparent conductive electrode (TCE) for AlGaN-based UV LEDs, with improved optoelectronic properties, of the structure ITO/Ga$_2$O$_3$/Ag/Ga$_2$O$_3$ is suggested by Wang et al. [29]. In that way, the complex TCE exhibited very high transmittance values at 365 nm and extremely low specific contact resistance, upon annealing. The combination of two single ITO layers is demonstrated by Zhao et al. [15] for GaN-based UV LEDs, in order to improve light extraction efficiency (LEE) and to reach low-resistance ohmic contact with the layer of p-GaN. The concept was implemented using laser direct writing and was further supported by numerical simulations. In the last article of the section, Tang et al. [30] describe their great efforts on LEE enhancement in flip-chip (mini) GaN-based LEDs, through the management of the total internal reflection at sapphire/air interface, using a tetramethylammonium hydroxide (TMAH) based etching.

In this manner, hierarchical prism-structured sidewalls were formed in the device, to boost light output power, due to light trapping, and thus scattering phenomena exploitation.

Three novel studies and a review article in the field of photodetectors are presented in the third section of this SI. More specifically, Shih et al. [31] fabricated p- and n-type Si heterojunction photodetectors incorporating graphene oxide (GO) of different oxidation degrees. They proved that the oxidation degree, adjusted by tuning the total of hydrogen peroxide during the oxidation process, increased the photoresponse in case of *n*-type Si heterojunction devices and vice versa in case of *p*-type ones. In addition, Xue et al. [17] demonstrate their dual approach regarding a) the controllable synthesis of high quality 2D perovskite platelets, compatible to conventional substrates with a melting point over 100 °C and b) the following fabrication of high-performance photodetectors. On the other hand, Lee et al. [20] report on the preparation of an ultrahigh sensitivity CdTe microdots-based photodetector onto a bottom bismuth coated ITO/glass substrate. The devices exhibited excellent durability and efficiency under stress conditions mainly due to piezo-phototronic effect, which highly affected the height of the Schottky barrier. A review article, by Shi et al. [32] is the last contribution of the section, summarizing the current insights and achievements in the field of organic photomultiplication photodetectors, due to trap assisted carrier tunneling effect, while alternative operational mechanisms, future perspectives, and challenges are also considered.

Novel contributions related to other dynamic devices are also included in the fourth section, which consists of three research articles and a review one. Guo et al. [33] successfully tuned the geometry of GaN-based nanobricks to be incorporated as efficient dielectric metasurface towards the manipulation of orthogonal linear polarizations simultaneously in the visible light. Moreover, they conceived and constructed a polarization beam splitter (PBS), as well as the required focusing lenses at the wavelength of 530 nm, featuring the great options of visible light optical devices. In another simulation study, Ma et al. [34] suggest a metamaterial structure composed of a gold split-ring and a graphene one, in order to reinforce the electromagnetically induced transparency (EIT) effect in the mid infrared (MIR) region and to create a tunable transparency window, by adjusting the coupling distance and/or the Fermi level of the graphene-based split-ring. Their results could be exploited towards the development of various light management nanodevices. Wu et al. [35] demonstrate the development of a reusable and flexible tunable filter through an advanced electrowetting fluid-manipulation technology. They observed that upon the simulated adjustment of the period of the grating structure filter, the reflection of CMY (cyan, magenta, yellow) primary colors was accomplished. On the other hand, the significantly shorter device's response time was attributed in the fast speed of electrowetting fluid manipulation process. The section closes with an extensive summary regarding the most recent accomplishments in the field of dynamically tunable metasurface-based applications incorporating liquid crystals (LCs) by Ma et al. [36]. They provide various rational architectures, as well as the most common factors to adjust the optical properties of LCs. In this context, they list various combinations between LCs and a range of metasurfaces that could be accomplished, aiming to increase their compatibility with high performance functional dynamic nanodevices.

The last part of the Special Issue incorporates original articles demonstrating the realization of other efficient optoelectronic nanodevices, such as field emission (FE) devices and transistors, respectively. First, the article by Stylianakis et al. [14] compares the FE performance of four solution-processed approaches including the use of polymeric and fullerene derivatives' composites towards the development of rGO-based in different ratios cold cathodes for FE devices. The prepared devices displayed excellent stability, higher field enhancement factor, and much lower turn-on field compared to the reference n+-Si/rGO FE devices. Celebrano et al. [37] close the Special Issue demonstrating their investigation on the photocurrent behavior of an erbium-doped and co-doped with oxygen silicon-based transistor at room temperature (RT), upon alteration of the laser source wavelength. They proved that the photocurrent strongly depends both on the power, as well as the frequency of the laser source, while the potential incorporation of Er-doped silicon in RT single photon resonators is highlighted.

To sum up, the present Special Issue entitled "Optoelectronic Nanodevices" assembles original research contributions from various subfields of optoelectronics. This collection showcases outstanding experimental and simulation studies pointing out the potential of novel organic, inorganic, hybrid composites, as well as 2D nanomaterials to be incorporated in order to enhance the performance and to extend the lifetime of conventional optoelectronic applications.

Funding: This research is co-financed by Greece and the European Union (European Social Fund-ESF) through the Operational Programme «Human Resources Development, Education and Lifelong Learning» in the context of the project "Reinforcement of Postdoctoral Researchers" (MIS-5001552), implemented by the State Scholarships Foundation (IKY), Grant No. 13992.

Acknowledgments: The guest editor would like to thank all authors, reviewers, and the publishing team. Their contribution towards the completion of this successful Special Issue is highly appreciated. Finally, special thanks to Ms. Alisa Zhai for her valuable and continuous assistance.

Conflicts of Interest: The author declares no conflict of interest.

References

1. Cheng, J.; Wang, C.; Zou, X.; Liao, L. Recent Advances in Optoelectronic Devices Based on 2D Materials and Their Heterostructures. *Adv. Opt. Mater.* **2019**, *7*, 1800441. [CrossRef]
2. Stylianakis, M.M.; Konios, D.; Kakavelakis, G.; Charalambidis, G.; Stratakis, E.; Coutsolelos, A.G.; Kymakis, E.; Anastasiadis, S.H. Efficient ternary organic photovoltaics incorporating a graphene-based porphyrin molecule as a universal electron cascade material. *Nanoscale* **2015**, *7*, 17827–17835. [CrossRef]
3. Viskadouros, G.; Zak, A.; Stylianakis, M.; Kymakis, E.; Tenne, R.; Stratakis, E. Enhanced Field Emission of WS2 Nanotubes. *Small* **2014**, *10*, 2398–2403. [CrossRef]
4. Stylianakis, M.M.; Maksudov, T.; Panagiotopoulos, A.; Kakavelakis, G.; Petridis, K. Inorganic and Hybrid Perovskite Based Laser Devices: A Review. *Materials* **2019**, *12*, 859. [CrossRef]
5. Bao, Q.; Hoh, H.Y. *2D Materials for Photonic and Optoelectronic Applications*, 1st ed.; Elsevier: Amsterdam, The Netherlands, 2019; pp. 1–336.
6. Aslam, U.; Rao, V.G.; Chavez, S.; Linic, S. Catalytic conversion of solar to chemical energy on plasmonic metal nanostructures. *Nat. Catal.* **2018**, *1*, 656–665. [CrossRef]
7. Noori, K.; Konios, D.; Stylianakis, M.M.; Kymakis, E.; Giustino, F. Energy-level alignment and open-circuit voltage at graphene/polymer interfaces: theory and experiment. *2D Mater.* **2016**, *3*, 015003. [CrossRef]
8. Toma, A.; Zaccaria, P.R.; Krahne, R.; Alabastri, A.; Coluccio, M.L.; Das, G.; Liberale, C.; De Angelis, F.; Francardi, M.; Mecarini, F.; et al. Nanostructures for Photonics. In *Encyclopedia of Nanotechnology*; Bhushan, B., Ed.; Springer: Dordrecht, The Netherlands, 2016; pp. 2827–2843.
9. Wen, Y.; He, P.; Yao, Y.; Zhang, Y.; Cheng, R.; Yin, L.; Li, N.; Li, J.; Wang, J.; Wang, Z.; et al. Bridging the van der Waals Interface for Advanced Optoelectronic Devices. *Adv. Mater.* **2019**, 1906874. [CrossRef]
10. Yan, W.; Dong, C.; Xiang, Y.; Jiang, S.; Leber, A.; Loke, G.; Xu, W.; Hou, C.; Zhou, S.; Chen, M.; et al. Thermally drawn advanced functional fibers: New frontier of flexible electronics. *Mater. Today* **2020**. [CrossRef]
11. Ou, Q.; Bao, X.; Zhang, Y.; Shao, H.; Xing, G.; Li, X.; Shao, L.; Bao, Q. Band structure engineering in metal halide perovskite nanostructures for optoelectronic applications. *Nano Mater. Sci.* **2019**, *1*, 268–287. [CrossRef]
12. Feng, Y.; Chen, Z.; Jiang, S.; Li, C.; Chen, Y.; Zhan, J.; Chen, Y.; Nie, J.; Jiao, F.; Kang, X.; et al. Study on the Coupling Mechanism of the Orthogonal Dipoles with Surface Plasmon in Green LED by Cathodoluminescence. *Nanomaterials* **2018**, *8*, 244. [CrossRef]
13. Yan, C.; Du, X.; Li, J.; Ding, X.; Li, Z.; Tang, Y. Effect of Excitation Wavelength on Optical Performances of Quantum-Dot-Converted Light-Emitting Diode. *Nanomaterials* **2019**, *9*, 1100. [CrossRef] [PubMed]
14. Stylianakis, M.M.; Viskadouros, G.; Polyzoidis, C.; Veisakis, G.; Kenanakis, G.; Kornilios, N.; Petridis, K.; Kymakis, E. Updating the Role of Reduced Graphene Oxide Ink on Field Emission Devices in Synergy with Charge Transfer Materials. *Nanomaterials* **2019**, *9*, 137. [CrossRef]
15. Zhao, J.; Ding, X.; Miao, J.; Hu, J.; Wan, H.; Zhou, S. Improvement in Light Output of Ultraviolet Light-Emitting Diodes with Patterned Double-Layer ITO by Laser Direct Writing. *Nanomaterials* **2019**, *9*, 203. [CrossRef] [PubMed]

16. Stylianakis, M.M.; Kosmidis, D.M.; Anagnostou, K.; Polyzoidis, C.; Krassas, M.; Kenanakis, G.; Viskadouros, G.; Kornilios, N.; Petridis, K.; Kymakis, E. Emphasizing the Operational Role of a Novel Graphene-Based Ink into High Performance Ternary Organic Solar Cells. *Nanomaterials* **2020**, *10*, 89. [CrossRef] [PubMed]

17. Xue, Y.; Yuan, J.; Liu, J.; Li, S. Controllable Synthesis of 2D Perovskite on Different Substrates and Its Application as Photodetector. *Nanomaterials* **2018**, *8*, 591. [CrossRef] [PubMed]

18. Liu, M.; Zhao, J.; Zhou, S.; Gao, Y.; Hu, J.; Liu, X.; Ding, X. An InGaN/GaN Superlattice to Enhance the Performance of Green LEDs: Exploring the Role of V-Pits. *Nanomaterials* **2018**, *8*, 450. [CrossRef]

19. Svrcek, V.; Kolenda, M.; Kadys, A.; Reklaitis, I.; Dobrovolskas, D.; Malinauskas, T.; Lozach, M.; Mariotti, D.; Strassburg, M.; Tomašiūnas, R. Significant Carrier Extraction Enhancement at the Interface of an InN/p-GaN Heterojunction under Reverse Bias Voltage. *Nanomaterials* **2018**, *8*, 1039. [CrossRef]

20. Lee, D.J.; Mohan Kumar, G.; Ilanchezhiyan, P.; Xiao, F.; Yuldashev, S.; Woo, Y.D.; Kim, D.Y.; Kang, T.W. Arrayed CdTe Microdots and Their Enhanced Photodetectivity via Piezo-Phototronic Effect. *Nanomaterials* **2019**, *9*, 178. [CrossRef]

21. Chen, L.-C.; Tseng, Z.-L.; Lin, D.-W.; Lin, Y.-S.; Chen, S.-H. Improved Performance of Perovskite Light-Emitting Diodes by Quantum Confinement Effect in Perovskite Nanocrystals. *Nanomaterials* **2018**, *8*, 459. [CrossRef]

22. Deng, J.; Wang, M.; Song, X.; Yang, Z.; Yuan, Z. Ti Porous Film-Supported NiCo2S4 Nanotubes Counter Electrode for Quantum-Dot-Sensitized Solar Cells. *Nanomaterials* **2018**, *8*, 251. [CrossRef]

23. Ho, W.-J.; Liu, J.-J.; Yang, Y.-C.; Ho, C.-H. Enhancing Output Power of Textured Silicon Solar Cells by Embedding Indium Plasmonic Nanoparticles in Layers within Antireflective Coating. *Nanomaterials* **2018**, *8*, 1003. [CrossRef] [PubMed]

24. Shi, L.; Cui, Y.; Gao, Y.; Wang, W.; Zhang, Y.; Zhu, F.; Hao, Y. High Performance Ultrathin MoO3/Ag Transparent Electrode and Its Application in Semitransparent Organic Solar Cells. *Nanomaterials* **2018**, *8*, 473. [CrossRef] [PubMed]

25. Chen, L.; Zhang, H.; Zhang, J.; Zhou, Y. A Compact and Smooth CH3NH3PbI3 Film: Investigation of Solvent Sorts and Concentrations of CH3NH3I towards Highly Efficient Perovskite Solar Cells. *Nanomaterials* **2018**, *8*, 897. [CrossRef] [PubMed]

26. Li, J.; Tang, Y.; Li, Z.; Ding, X.; Yu, S.; Yu, B. Improvement in Color-Conversion Efficiency and Stability for Quantum-Dot-Based Light-Emitting Diodes Using a Blue Anti-Transmission Film. *Nanomaterials* **2018**, *8*, 508. [CrossRef]

27. Zhou, Q.; Wang, H.; Xu, M.; Zhang, X.-C. Quantum Efficiency Enhancement of a GaN-Based Green Light-Emitting Diode by a Graded Indium Composition p-Type InGaN Layer. *Nanomaterials* **2018**, *8*, 512. [CrossRef]

28. Jin, X.; Trieu, S.; Chavoor, G.J.; Halpin, G.M. Enhancing GaN LED Efficiency through Nano-Gratings and Standing Wave Analysis. *Nanomaterials* **2018**, *8*, 1045. [CrossRef]

29. Wang, H.; Zhou, Q.; Liang, S.; Wen, R. Fabrication and Characterization of AlGaN-Based UV LEDs with a ITO/Ga2O3/Ag/Ga2O3 Transparent Conductive Electrode. *Nanomaterials* **2019**, *9*, 66. [CrossRef]

30. Tang, B.; Miao, J.; Liu, Y.; Wan, H.; Li, N.; Zhou, S.; Gui, C. Enhanced Light Extraction of Flip-Chip Mini-LEDs with Prism-Structured Sidewall. *Nanomaterials* **2019**, *9*, 319. [CrossRef]

31. Shih, C.-K.; Ciou, Y.-T.; Chiu, C.-W.; Li, Y.-R.; Jheng, J.-S.; Chen, Y.-C.; Lin, C.-H. Effects of Different Oxidation Degrees of Graphene Oxide on P-Type and N-Type Si Heterojunction Photodetectors. *Nanomaterials* **2018**, *8*, 491. [CrossRef]

32. Shi, L.; Liang, Q.; Wang, W.; Zhang, Y.; Li, G.; Ji, T.; Hao, Y.; Cui, Y. Research Progress in Organic Photomultiplication Photodetectors. *Nanomaterials* **2018**, *8*, 713. [CrossRef]

33. Guo, Z.; Xu, H.; Guo, K.; Shen, F.; Zhou, H.; Zhou, Q.; Gao, J.; Yin, Z. High-Efficiency Visible Transmitting Polarizations Devices Based on the GaN Metasurface. *Nanomaterials* **2018**, *8*, 333. [CrossRef] [PubMed]

34. Ma, Q.; Zhan, Y.; Hong, W. Tunable Metamaterial with Gold and Graphene Split-Ring Resonators and Plasmonically Induced Transparency. *Nanomaterials* **2019**, *9*, 7. [CrossRef] [PubMed]

35. Wu, J.; Du, Y.; Xia, J.; Zhang, T.; Lei, W.; Wang, B. Dynamically Tunable Light Absorbers as Color Filters Based on Electrowetting Technology. *Nanomaterials* **2019**, *9*, 70. [CrossRef] [PubMed]

36. Ma, Z.; Meng, X.; Liu, X.; Si, G.; Liu, Y.J. Liquid Crystal Enabled Dynamic Nanodevices. *Nanomaterials* **2018**, *8*, 871. [CrossRef]
37. Celebrano, M.; Ghirardini, L.; Finazzi, M.; Ferrari, G.; Chiba, Y.; Abdelghafar, A.; Yano, M.; Shinada, T.; Tanii, T.; Prati, E. Room Temperature Resonant Photocurrent in an Erbium Low-Doped Silicon Transistor at Telecom Wavelength. *Nanomaterials* **2019**, *9*, 416. [CrossRef]

 *nanomaterials*

Article

Ti Porous Film-Supported NiCo$_2$S$_4$ Nanotubes Counter Electrode for Quantum-Dot-Sensitized Solar Cells

Jianping Deng [1,*], Minqiang Wang [2,*], Xiaohui Song [3], Zhi Yang [2] and Zhaolin Yuan [1]

[1] Shaanxi Key Laboratory of Industrial Automation, School of Physics and Telecommunication Engineering, Shaanxi University of Technology, Hanzhong 723001, China; yuanzhaolin98@126.com

[2] Electronic Materials Research Laboratory, Key Laboratory of the Ministry of Education & International Center for Dielectric Research, School of Electronic and Information Engineering, Xi'an Jiaotong University, Xi'an 710049, China; yangzhi029@xjtu.edu.cn

[3] Henan Key Laboratory of Photovoltaic Materials, College of Physics and Materials Science, Henan Normal University, Xinxiang 453000, China; songsxh@126.com

* Correspondence: jpdeng0214@163.com (J.D.); mqwang@mail.xjtu.edu.cn (M.W.); Tel.: +86-(0)91-6264-1715 (J.D.); +86-(0)29-8266-8794 (M.W.)

Received: 22 March 2018; Accepted: 13 April 2018; Published: 17 April 2018

Abstract: In this paper, a novel Ti porous film-supported NiCo$_2$S$_4$ nanotube was fabricated by the acid etching and two-step hydrothermal method and then used as a counter electrode in a CdS/CdSe quantum-dot-sensitized solar cell. Measurements of the cyclic voltammetry, Tafel polarization curves, and electrochemical impedance spectroscopy of the symmetric cells revealed that compared with the conventional FTO (fluorine doped tin oxide)/Pt counter electrode, Ti porous film-supported NiCo$_2$S$_4$ nanotubes counter electrode exhibited greater electrocatalytic activity toward polysulfide electrolyte and lower charge-transfer resistance at the interface between electrolyte and counter electrode, which remarkably improved the fill factor, short-circuit current density, and power conversion efficiency of the quantum-dot-sensitized solar cell. Under illumination of one sun (100 mW/cm^2), the quantum-dot-sensitized solar cell based on Ti porous film-supported NiCo$_2$S$_4$ nanotubes counter electrode achieved a power conversion efficiency of 3.14%, which is superior to the cell based on FTO/Pt counter electrode (1.3%).

Keywords: NiCo$_2$S$_4$ nanotubes; Ti porous film; quantum dot; solar cells; counter electrode

1. Introduction

In recent years, the quantum-dot-sensitized solar cell (QDSSC) has aroused a widespread attention due to the large absorption coefficient, multiple exciton generation, and the tunable absorption spectrum based on quantum confinement effect. QDSSC is composed of three parts: QD-sensitized TiO$_2$ or ZnO photoanode, electrolyte (S$_n^{2-}$/S^{2-}), and counter electrode (CE) [1]. As one of the most important parts in QDSSC, CE is used as a catalyst to reduce S$_n^{2-}$/S^{2-} after the electron injection from external circuit so that QD can be regenerated. For achieving this function, CE materials should provide superior catalytic activity and high chemical stability against the corrosive polysulfide electrolyte.

As is well known, Pt is a poor electrocatalyst for reducing S$_n^{2-}$/S^{2-} due to its strong chemisorption with S^{2-} ions, resulting in the serious corrosion and much higher overpotentials for electrolyte regeneration. Therefore, the QDSSC with Pt CE show a low fill factor (FF) and power conversion efficiency (PCE) [2], and the high cost is another disadvantage. Recently, some Pt-free CE materials with low cost, such as carbon materials [3–5], conductive polymers [6], and inorganic compound [7], have been widely developed and demonstrated to have attractive performances. Of these Pt-free CE materials, the transition-metal sulfides, such as CuS, FeS, CoS, NiS, and $NiCo_2S_4$ [8–12], have attracted tremendous interest. Especially, $NiCo_2S_4$ contains higher electrochemical characteristics compared with binary NiS and CoS. $NiCo_2S_4$ has been regarded as one of the most potential electrode materials for a super-capacitor for are several main reasons, as follows: (1) $NiCo_2S_4$ has a high electric conductivity, which is approximately 100 times higher than that of $NiCo_2O_4$ and higher than that of NiS and CoS [13]; (2) $NiCo_2S_4$ has good mechanical and thermal stability and two different metal cations (Co and Ni) supplying richer redox reactions, leading to better electrochemical performance [14–17]. Up until now, $NiCo_2S_4$ as an efficient CE has also been widely used in dye-sensitized solar cell (DSSC). Shi et al. reported that $NiCo_2S_4$ nanosheet films were used as a CE of DSSC, the photocurrent density is increased by 3 mA/cm^2 [18]. Huo et al. fabricated the flower-like $NiCo_2S_4/NiS$ micro-spheres, then the $NiCo_2S_4/NiS$ was coated on FTO (fluorine doped tin oxide) conductive glass as a CE for DSSC and the PCE of DSSC increased by 8.24% compared with that of the DSSC based on Pt CE [19]. A compact $NiCo_2S_4$ film with a thickness of 40 nm and a cross-linked network of $NiCo_2S_4$ nanosheet film coated FTO conductive glass were used as CEs for DSSC [20,21], and the DSSC with $NiCo_2S_4$ CE exhibited higher PCE compared with that of Pt CE-based DSSC. In addition, one-dimensional (1D) nanomaterials (e.g., nanorod, nanowire, nanotube) with direct electrical pathways show excellent application prospects in nanoscale electronic devices. One-dimensional $NiCo_2S_4$ nanotube arrays were used for supercapacitors [22] and 1D Co_9S_8 hollow nanoneedle arrays were used as CE for QDSSC [11].

The substrate supporting CE materials is also very important for the performance of DSSC; it should have large surface area, excellent conductivity, and good corrosion resistance to the electrolyte. Owing to the poor conductivity of metal sulfides, improving the catalytic activity of CE by increasing the thickness of CE is limited. In order to solve this problem, the porous microstructure with a large surface is used to load CE materials. For example, porous SnO_2, ZnO, TiO_2, and carbon and nickel foam were used as catalyst support [7,9,23,24]. The porous Cu_2S and FeS CEs were directly prepared on Cu and Fe substrates by in situ corrosion method [25], but they can easily peeled off from the substrates because of the reaction of Cu and Fe substrates with S^{2-} in electrolyte. In addition, it is important that the electrons from the external circuit quickly transfer to CE materials and reduce the electrolyte. So far, various substrates have been used, such as FTO and ITO (indium doped tin oxide) conductive glass [3–6], C fiber cloth and C paper [16], Ti mesh [26], and abovementioned Fe and Cu. Generally, two methods are available to prepare CE: one is the in situ growth method, the other is the ex situ coating method. For the former, there is a good adhesion between the substrate and CE materials, but the load of CE materials is limited. For the latter, although CE materials can be increased by increasing the coating several times, the adhesion is poor. In order to solve the adhesion problem, the adhesive was added into the CE materials [27], but it increased the electron transfer resistance.

In this paper, we have designed $NiCo_2S_4$ nanotubes supported on Ti porous film (Ti-PF) as CE for QDSSC. Firstly, Ti-PF was prepared by acid etching, then $NiCo_2S_4$ nanotubes were synthesized on Ti-PF by two-step hydrothermal method. $NiCo_2S_4$ nanotubes not only provide the effective path for electron transport but also have more electroactive sites for reducing polysulfide electrolyte. In addition, Ti-PF/$NiCo_2S_4$ CE exhibits lower charge-transfer resistance compared with FTO/Pt CE owing to the high conductivity of Ti and the porous structure increases the load of CE materials and improves the stability via the pore-wall. As a result, the PCE (3.14%) of QDSSC based on Ti-PF/$NiCo_2S_4$ CE is higher than that (1.3%) of QDSSC based on Pt CE.

2. Experimental Section

In this work, $NiCo_2S_4$ nanotubes supported on Ti-PF were prepared by the following three steps (Figure 1): (1) Ti-PF was prepared by acid etching as the substrates (Step 1); (2) Ni—Co precursor $((Ni,Co)_2(CO_3)(OH)_2)$ nanorods were hydrothermally grown (Step 2); and (3) $(Ni,Co)_2(CO_3)(OH)_2$ nanorods were converted into $NiCo_2S_4$ nanotubes in Na_2S solution via an anion-exchange reaction (Step 3).

2.1. Preparation of Ti-PF

Firstly, Ti sheets with high purity (TA1, 99.9%) and 0.2 mm thickness were washed in the acetone and ethanol using an ultrasonic bath for 30 min, respectively, and rinsed with deionized water. Then, the cleaned Ti sheets were immersed in 90 mL of HCl solution (25 wt·%) for 24–72 h at room temperature. Next, Ti-PF sheets were washed thoroughly with deionized water until the pH was close to 7 and further dried in air.

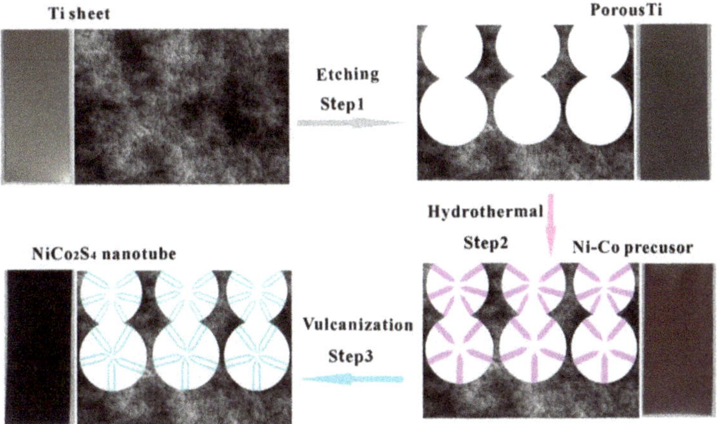

Figure 1. Schematic diagram to illustrate the preparation process of $NiCo_2S_4$ nanotubes on Ti porous film (Ti-PF).

2.2. Fabrication of NiCo₂S₄ Nanotubes

$NiCo_2S_4$ nanotubes were prepared by two-step hydrothermal method according to the literature [14,28]. All the reagents were of analytical grade in this experiment and purchased from Sinopharm (Beijing, China). Firstly, 4 mmol $CoCl_2 \cdot 6H_2O$, 2 mmol $NiCl_2 \cdot 6H_2O$, and 12 mmol urea were dissolved in 35 mL deionized water and stirred to form a pink homogeneous solution. Subsequently, the mixed-solution and Ti-PF sheets were transferred into 50 mL Teflon-lined stainless-steel autoclave and then heated at 120 °C for 10 h. After being cooled to room temperature, Ti-PF sheets with pink product were washed with deionized water and ethanol and then dried at 60 °C in air for 10 h. The $(Ni,Co)_2(CO_3)(OH)_2$ nanorods were obtained. In the next step, the $(Ni,Co)_2(CO_3)(OH)_2$ was transformed into $NiCo_2S_4$ by a hydrothermal process in 0.1 M $Na_2S \cdot 9H_2O$ solution at 120 °C for 10 h. After being cooled to room temperature, Ti-PF sheets with black products were washed with deionized water and ethanol and then dried in air at 60 °C. The $NiCo_2S_4$ nanotube CE was obtained.

2.3. Assembly of QDSSCs

The QDSSC was fabricated using a screen printing technique with the home-made TiO_2/ZnO paste [29]. Firstly, the TiO_2 compact layer was prepared via a spin-coating method and followed by calcination at 400 °C for 0.5 h. Subsequently, the mesoscopic photoanodes were prepared through four circulars screen printing of TiO_2/ZnO paste on FTO conductive glass with TiO_2 compact layer and sintered at 450 °C for 0.5 h. The active area of the film is 0.25 cm^2. Next, the growth of ZnO nanowires was performed in a procedure similar to that in our previous paper. Lastly, CdS/CdSe/ZnS QDs were deposited by successive ionic layer adsorption and reaction (SILAR) method [1,29]. A polysulfide electrolyte used in the QDSSCs and the symmetric cells was prepared by dissolving 1 M Na_2S, 1 M S and 0.2 M KCl in a methanol/water solution (7:3, v/v).

2.4. Characterization and Measurements

Field emission scanning electron microscopy (FE-SEM, S-4800, Hitachi, Tokyo, Japan) and transmission electron microscope (TEM, Tecnai G2F20, FEI, Columbus, OH, USA) were carried out to investigate the morphology and composition. The X-ray diffraction (XRD) patterns were obtained by D/max-2400 X-ray diffraction spectrometer (Rigaku, Akishima-Shi, Japan) with Cu Ka radiation at 40 kV and 100 mA. The current-voltage (I–V) characterization was performed under AM 1.5 G simulated sunlight (100 mW/cm^2) and recorded by a Keithley 2400 Source Meter (Keithley Instruments, Inc., Cleveland, OH, USA). The cyclic voltammetry (CV), Tafel polarization curves, and electrochemical impedance spectroscopy (EIS) were performed in the symmetric cells on the workstation (CHI660E, CH Instruments Ins., Shanghai, China). These tests were used to investigate the electrocatalytic ability of CE towards the reduction of S_n^{2-}/S^{2-} electrolyte and the electronic transport properties of CE. In CV tests, the scanning potential range is from -1.2 V to 1 V with a scan rate of 100 mV/s. EIS curves were recorded at bias voltage of 0 V over a frequency range of 0.1 Hz to 1 MHz with AC amplitude of 10 mV, all EIS spectra were analyzed by ZsimpWin software. Polarization Tafel curves were recorded from -0.6V to 0.6 V at the scan rate of 10 mV/s.

3. Results and Discussion

3.1. Morphology of Ti-PF

Figure 2 shows the SEM images for Ti-PF with different etching time in HCl. Generally, metal Ti is stable in low concentration of HCl at room temperature. However, Ti is slowly etched when HCl concentration is greater than 20%. In our experiment, HCl solution with concentration of 25% is used to etch Ti sheets at room temperature and the morphology of Ti-PF is controlled by adjusting the etching time. Figure 2a–f show the morphologies of Ti-PF with etching 24 h, 48 h, and 72 h, respectively. It can be seen clearly that with the increase of etching time, the porous structure has changed significantly. When the etching time is 24 h, the holes with the size range of about 5–10 μm exist only in very few places (Figure 2a,b). When the etching time increases to 48 h, the holes with the size range of 10–20 μm are uniformly formed on the surface of Ti sheet (Figure 2c,d). With an increase in etching time (up to 72 h), the holes disappeared completely (Figure 2e) and the shallow pits with the size of below 5 μm were formed (Figure 2f). In this experiment, Ti-PF plays three roles to improve the catalytic activity of CE to polysulfide electrolyte: (1) metal Ti provides a fast electronic transmission channel, (2) the porous structure gives a large surface area and thus increases the load of CE materials, and (3) the deeper holes in which $NiCo_2S_4$ nanotubes are limited by the wall increase the stability of CE. Therefore, the Ti-PF etched for 48 h is most suitable for using as the CE substrate.

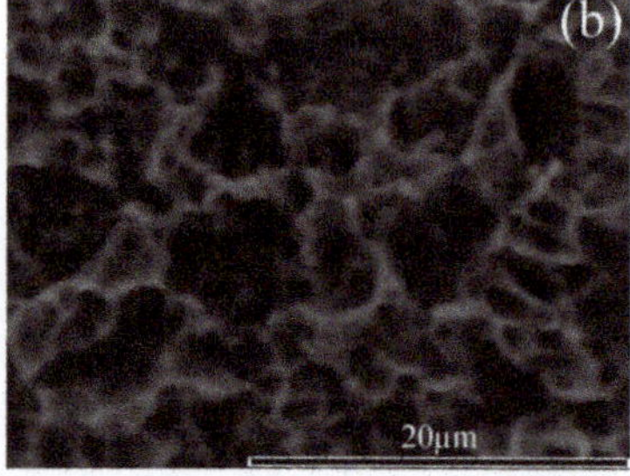

Figure 2. *Cont.*

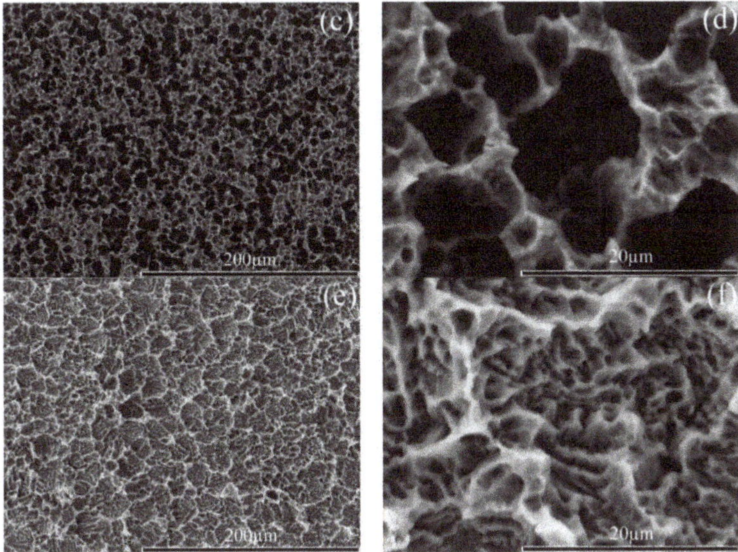

Figure 2. SEM images for Ti-PF from different etching time. (**a,b**)24 h, (**c,d**)48 h, and (**e,f**)72 h.

3.2. Morphology of NiCo₂S₄ Nanotubes

The conversion of $(Ni,Co)_2(CO_3)(OH)_2$ nanorods into $NiCo_2S_4$ nanotubes can be explained by the anion-exchange reaction mechanism [22,30,31]. Firstly, S^{2-} in the Na_2S solution exchanges with CO_3^{2-} and OH^- on the surface of $(Ni, Co)_2(CO_3)(OH)_2$ nanorods to form $NiCo_2S_4$, CO_3^{2-}, and OH^- react with H^+ in the solution to produce CO_2 and H_2O. At the same time, the internal $(Ni,Co)_2(CO_3)(OH)_2$ diffuse spontaneously to the surface of the nanorod, where it provides a source of $(Ni,Co)_2(CO_3)(OH)_2$ for further anion exchange. The continuous outward diffusion results in the generation of void space inside the nanorod. When $(Ni,Co)_2(CO_3)(OH)_2$ has been completely converted into $NiCo_2S_4$, nanorods become nanotubes.

To illustrate the morphology of as-synthesized samples, SEM measurements were performed. Figure 3a,b,e show the representative low-magnification and high-magnification SEM images of $(Ni,Co)_2(CO_3)(OH)_2$ nanorods, respectively. Figure 3c,b,f correspond to the low-magnification and high-magnification SEM images of $NiCo_2S_4$ nanotubes, respectively. In Figure 3a–d, $(Ni,Co)_2(CO_3)(OH)_2$ nanorods and $NiCo_2S_4$ nanotubes were homogeneously deposited on Ti-PF, suggesting that the two-hydrothermal method is favorable in forming this structure. Moreover, the $NiCo_2S_4$ nanotubes were well preserved during sulfurization process. It can be seen that the $(Ni,Co)_2(CO_3)(OH)_2$ array film is composed by many multi-directional nanorods due to the nucleation sites from the wall of holes and there is considerable inter-nanorod space, it will help electrolyte full contact with CE materials at the bottom, improving the utilization rate of the CE materials. A rough comparison between Figure 3b,d the diameters of $NiCo_2S_4$ nanotubes are larger than that of $(Ni,Co)_2(CO_3)(OH)_2$ nanorods, owing to the Ni and Co ions diffusion from inside to outside of the nanorod during an anion-exchange process. As shown in Figure 3e,f $(Ni,Co)_2(CO_3)(OH)_2$ nanorods with a smooth surface and solid structure and $NiCo_2S_4$ nanotubes with a rough surface are clearly seen from the damaged film.

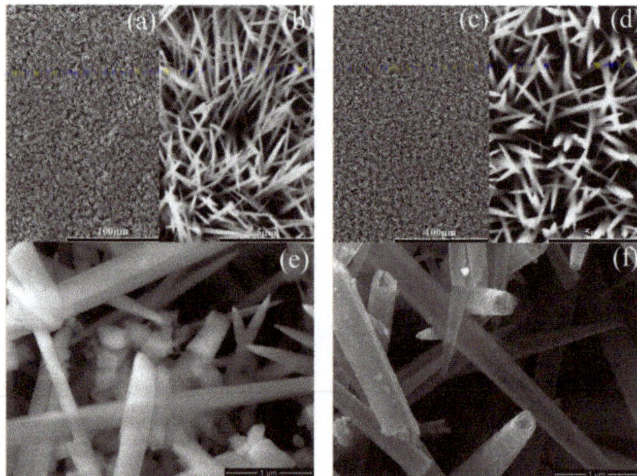

Figure 3. SEM images for $(Ni,Co)_2(CO_3)(OH)_2$ nanorods (**a**,**b**,**e**) and $NiCo_2S_4$ nanotubes (**c**,**d**,**f**) on Ti-PF.

The detailed structure of $NiCo_2S_4$ nanotube scraped from Ti-PF sheet was further confirmed by TEM, as shown in Figure 4. From Figure 4a, the nanotube structure and the porous wall can be evidently seen, indicating the successful preparation of $NiCo_2S_4$ nanotubes on Ti-PF. By a closer examination of the wall in Figure 4b, it is found that $NiCo_2S_4$ nanotube is composed of many nanoparticles with a size of about 25 nm (marked with red line) and numerous pores locate at the nanotube. The $NiCo_2S_4$ nanotube with a rough surface (Figure 4a–c) and a thin wall of about 25 nm (Figure 4c) effectively increases the electroactive sites. The nanotube structure can greatly enhance the electrolyte penetration and improve the performance of cells. In Figure 4c, the corresponding selected area electron diffraction (SAED) pattern indicates the polycrystalline nature of $NiCo_2S_4$ nanotubes and the diffraction rings can be readily indexed to the (111), (220), (311), (400), (511), and (440) planes of $NiCo_2S_4$ phase. In addition, Figure 4d reveals that the lattice spacings are about 0.51 nm, 0.284 nm, and 0.234 nm, which can be assigned to the (111), (311), and (400) crystal planes of the cubic $NiCo_2S_4$ phase, respectively, indicating the successful formation of crystalline $NiCo_2S_4$.

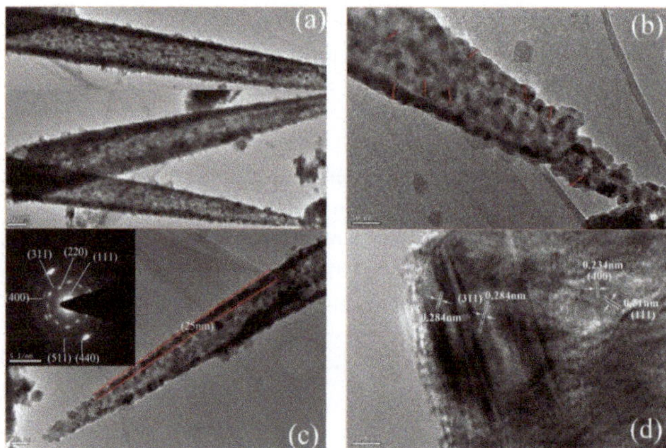

Figure 4. (**a**–**c**) TEM and (inset) SAED and (**d**) HRTEM images of $NiCo_2S_4$ nanotube.

The as-synthesized $NiCo_2S_4$ nanotubes were further confirmed by the XRD and electron energy loss spectroscopy (EELS) elemental mapping. In order to investigate the effect of annealing treatment on the catalytic activity of $NiCo_2S_4$ CE and the performance of cells, $NiCo_2S_4$ CE was annealed at 400 °C for 30 min in the nitrogen atmosphere ($NiCo_2S_4$-an). Figure 5a shows XRD images for Ti-PF, $NiCo_2S_4$, and $NiCo_2S_4$-an on Ti-PF. The diffraction peaks located at 31.6°, 38.3°, 50.5°, and 55.3° can be indexed to the (311), (400), (511), and (440) planes of the cubic phase $NiCo_2S_4$ (JCPDS 20-0782), which is consistent with SAED analysis. There are also two strong peaks at 29.84° and 52.0°, which may correspond to the (311) and (440) planes of the cubic phase Co_9S_8 (JCPDS no.19-0364). The existence of Co_9S_8 phase is because that $(Ni,Co)_2(CO_3)(OH)_2$ nanorods were incompletely sulfurized in Na_2S solution, which was verified by many reports [14,32]. It can also be seen that the intensity of (311) and (440) diffraction peaks of $NiCo_2S_4$ increased after annealing, indicating an increase in the crystallinity. Moreover, the TEM and EELS mapping images (Figure 5b) indicate that the elements (Ni, Co, and S) are uniformly distributed in the $NiCo_2S_4$ nanotube.

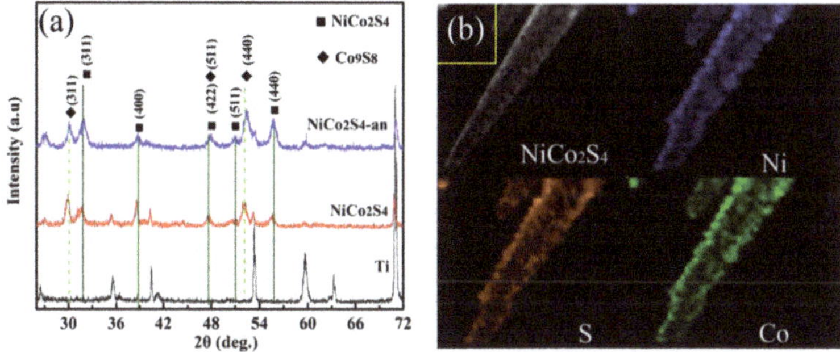

Figure 5. (a) XRD images for Ti-PF, $NiCo_2S_4$, and $NiCo_2S_4$-an on Ti-PF; (b) TEM image and the corresponding EELS elemental mapping images of a single $NiCo_2S_4$ nanotube.

3.3. Electrochemical Properties of CEs

To investigate the electrochemical properties of Pt, $NiCo_2S_4$, and $NiCo_2S_4$-an CEs, CV test of a symmetrical cell was carried out, as shown in Figure 6a. The peaks explain the catalytic reaction at the interface between CE and electrolyte as follows $S_n^{2-} + e^- \rightarrow nS^{2-}$. From the CV curves, $NiCo_2S_4$ and $NiCo_2S_4$-an CEs present a similar shape with two typical pairs of redox peaks [16,21]. As a matter of fact, the reduction peak of the left pair is assigned to the reaction $S_n^{2-} + e^- \rightarrow nS^{2-}$ and the right one is assigned as $S + 2ne^- \rightarrow S_n^{2-}$ [19,20,33]. The role of the CE in a QDSSC is to catalyze the reduction of S_n^{2-} to S^{2-} ions in the polysulfide electrolyte, so the left pair of redox peaks is directly related to the catalytic activity of CE, the positive and negative peaks correspond to the oxidation of S^{2-} and the reduction of Sn^{2-}, respectively [34,35]. However, because of high over-potential, Pt CE has only one pair of redox peaks, which correspond to the oxidation of S^{2-} and the reduction of S_n^{2-} [36]. The higher current density of left cathodic peak indicates that the CE has an excellent electrocatalytic activity for the reduction of S_n^{2-} to nS^{2-} [17]. It can be seen that $NiCo_2S_4$ and $NiCo_2S_4$-an CEs show higher current than Pt CE and the reduction current density of $NiCo_2S_4$-an CE is the biggest. This results indicate that the $NiCo_2S_4$ CE is expected to enhance QD regeneration and photoelectron generation, thus beneficial for improving QDSSC's photocurrent, and the annealing treatment enhanced the crystallinity of $NiCo_2S_4$ CE, which increases the reduction current.

Tafel polarization technique is an important method to evaluate the catalytic activity of CEs. Theoretically, the Tafel curve includes the diffusion, Tafel and polarization zones at the high-, middle-, and low-potential areas, respectively. In Tafel analysis, the exchange current density (J_0) (Tafel zone) and the limiting diffusion current density (J_{lim}) (diffusion zone) are two key parameters to evaluate the electrocatalytic activity of CEs. Tafel polarization curves of Pt, $NiCo_2S_4$, and $NiCo_2S_4$-an CEs are shown in Figure 6b. It can be seen that, in the Tafel zone, the slopes of the anodic or cathodic branches are in the order of $NiCo_2S_4$-an > $NiCo_2S_4$ > Pt. A larger slope indicates a higher J_0. According to the following equation:

$$J_0 = RT/nFR_{ct},\tag{1}$$

where R, T, F and n are the gas constant, the temperature, Faraday's constant, and the electron number involved in Sn^{2-}/S^{2-} redox couple, respectively [35]; the charge-transfer resistance (R_{ct}) can be calculated by J_0 values. The change trends of the R_{ct} is $NiCo_2S_4$-an < $NiCo_2S_4$ < Pt, which is consistent with the EIS results. In addition, J_{lim} derived from the horizontal part of the curve at high potential is also closely related to the catalytic activity of CEs, which is given by equation:

$$D = L\,J_{lim}/2nFC,\tag{2}$$

where D, L, F, C, and n are the diffusion coefficient of the polysulfide, the electrolyte thickness, the Faraday constant, the polysulfide concentration, and the number of electrons involved in the reduction of disulphide at the counter electrode, respectively [37]. It can be noticed that the change trend of J_{lim} is consistent with CV results, suggesting the D of redox couple in the electrolyte increases with enhanced electrocatalytic activity of CEs.

To further understand the reason for the good performance of the as-prepared CEs, EIS was carried out using the symmetrical cells, as shown in Figure 6c, and the corresponding parameters are shown in Table 1. Figure 6c shows Nyquist plots of Pt, $NiCo_2S_4$, and $NiCo_2S_4$4-an CEs and the insets of Figure 6c are the equivalent circuit and the magnified Nyquist impedance. In the equivalent circuit, R_s represents the series resistance including the sheet resistance of the substrates (FTO and Ti sheet) and the contact resistance of the symmetrical cell, which is mainly correlated to electron transfer rates to the interface of CE/electrolyte and it can be estimated from the intercept on the real axis at the high frequency. The intercept of the middle frequency semicircle on the real axis represents R_{ct} at the interface between CE and electrolyte. The R_{ct}, which is closely related to the electrocatalytic activity and the reaction kinetics of the CE, is an important parameter to determine the FF of cell [36,38,39]. Generally, because of the symmetrical structure, R_{ct1} and R_{ct2} at the two CE/electrolyte interfaces are equal ($R_{ct1} = R_{ct2}$), so every Nyquist plot has one semicircle [35,40]. The obtained impedance spectra are fitted by Z-View software, as shown in Table 1. The R_s values of Pt, $NiCo_2S_4$ and $NiCo_2S_4$-an CEs are 8.639 Ω, 3.139 Ωand 3.01 Ω, respectively. Among them, the R_s values of $NiCo_2S_4$, and $NiCo_2S_4$-an CEs are close, which may be ascribed to the same Ti-PF substrates. The R_s of Pt CE is much higher than that of $NiCo_2S_4$ and $NiCo_2S_4$-an CEs, which is attributed to the strong chemisorption of S^{2-} ions on Pt. R_{ct} directly reflects the electrochemical reaction at CE/electrolyte interface, R_{ct} values of Pt, $NiCo_2S_4$, and $NiCo_2S_4$-an CEs are 6860 Ω, 67.47 Ω, and 33.31 Ω, respectively; this means that it is easier for charges transfer through the $NiCo_2S_4$-an/electrolyte interface than Pt/electrolyte and $NiCo_2S_4$/electrolyte interfaces. Thus, it is anticipated that the QDSSC with Ti-PF-supported $NiCo_2S_4$ nanotube CE will show better photovoltaic performance. Furthermore, the proper annealing treatment reduced the R_{ct} of $NiCo_2S_4$/electrolyte interface and improved the short-circuit photocurrent density (J_{sc}) and PCE of QDSSC.

Table 1. Photovoltaic parameters obtained from J–V curves of quantum-dot-sensitized solar cells (QDSSCs) and EIS parameters of symmetric cells.

Samples	V_{oc} (V)	J_{sc} (mA/cm²)	FF (%)	PCE (%)	R_s (Ω)	R_{ct} (Ω)
Pt	0.489 (0.483)	11.76 (10.29)	22.56 (25.19)	1.30 (1.21)	8.639	6860
$NiCo_2S_4$	0.456 (0.45)	13.72 (13.56)	40.60 (40.46)	2.54 (2.51)	3.139	67.47
$NiCo_2S_4$-an	0.489 (0.478)	16.68 (15.32)	38.52 (38.61)	3.14 (2.82)	3.010	33.31

Complete photovoltaic cells based on $NiCo_2S_4$ and $NiCo_2S_4$–an CEs were fabricated and the cell based on Pt CE is used as the reference. In this study, Pt, $NiCo_2S_4$, and $NiCo_2S_4$-a CEs were soaked in S_n^{2-}/S^{2-} electrolyte for 24 h and then were used in the complete photovoltaic cells. The complete photovoltaic cells and symmetric cells were fixed by clip and spacer with 90 μm thickness. The photovoltaic curves and the photovoltaic parameters (open-circuit voltage (V_{oc}), J_{sc}, FF, and PCE) are shown in Figure 6d and Table 1, respectively. From Figure 6d, obviously, the performance of QDSSCs based on $NiCo_2S_4$ and $NiCo_2S_4$-an CEs are better than that of QDSSC with Pt CE.

The champion QDSSC based on Pt CE has an V_{oc} of 0.489V, a J_{sc} of 11.76 mA/cm^2, a FF of 22.56%, and a PCE of 1.3%. The champion QDSSC with $NiCo_2S_4$ CE has a V_{oc} of 0.456V, a J_{sc} of 13.72 mA/cm^2, a FF of 40.6%, and a PCE of 2.54%. The champion QDSSC employing $NiCo_2S_4$-an CE has a V_{oc} of 0.489V, a J_{sc} of 16.68 mA/cm^2, a FF of 38.52%, and a PCE of 3.14%. Notably, the PCE increased from 1.3% to 3.14% when Pt CE was replaced with $NiCo_2S_4$-an CE. In addition, the average values obtained of the three best cells (up to nine) based on an optimal photoanode and three CEs are given in brackets, as shown in Table 1, and the change trends of the average values of V_{oc}, J_{sc}, FF, and PCE are consistent with the champion QDSSCs. This improvement in the cell performance originates from the significant increases in both J_{sc} and FF, which closely related to the higher electrocatalytic ability of CE. Furthermore, the QDSSC with $NiCo_2S_4$-an CE shows a higher J_{sc} and V_{oc} than that with $NiCo_2S_4$ CE and thus obtains a higher PCE. V_{oc} of QDSSC depends upon the difference between the quasi Fermi level of the photoanode and the redox potential of the electrolyte. The annealing treatment improved the crystallinity and the conductivity of $NiCo_2S_4$, so the fast charge transfer at CE/electrolyte interface can cause a change in the concentration gradient in the electrolyte solution, which influences the recombination rates at the photoanode/electrolyte interface and consequently the conduction band position of the photoanode. Meanwhile, the high conductivity of $NiCo_2S_4$-an CE also increased the photocurrent of cell [41]. J–V parameters are in line with the electrocatalytic ability of CEs discussed in the CV, Tafel polarization, and EIS.

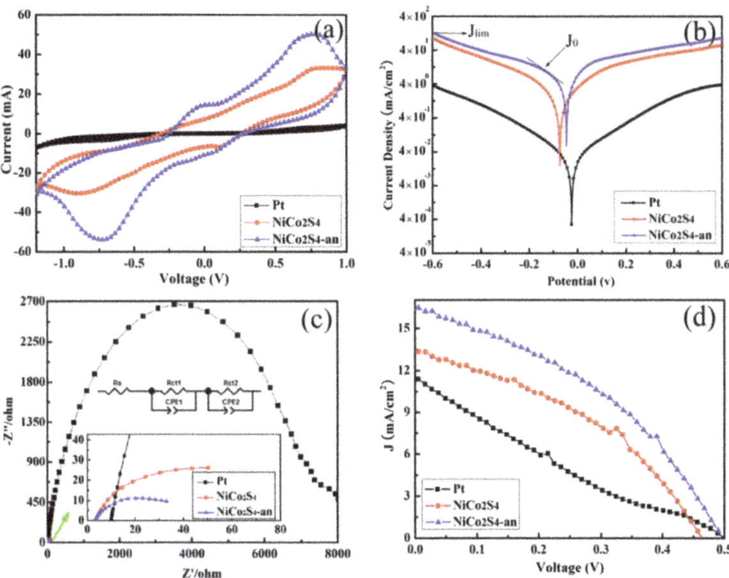

Figure 6. (**a**) Cyclic voltammetry (CV); (**b**) Tafel curves; and (**c**) EIS of the symmetric cells with Pt, $NiCo_2S_4$, and $NiCo_2S_4$-an CEs and (**d**) J–V characteristics for QDSSCs based on Pt, $NiCo_2S_4$, and $NiCo_2S_4$-an CEs, respectively.

Nanomaterials **2018**, *8*, 251

4. Conclusions

In summary, we have prepared Ti-PF by the acid etching technique and Ti-PF supporting $NiCo_2S_4$ nanotubes via two-step hydrothermal method; furthermore, Ti-PF supporting $NiCo_2S_4$ nanotubes are used as CE in QDSSCs. The morphology of Ti-PF is affected with the etching time. When etching time is 48 h in hydrochloric acid with a weight concentration of 25% at room temperature, the holes are uniformly formed on the surface of Ti sheet, which is most suitable for use as the substrate to support CE materials. SEM, TEM, and XRD results show that the as-synthesized $NiCo_2S_4$ nanotube with porous surface is the cubic phase. Using a polysulfide electrolyte in the symmetric cells, Ti-PF/$NiCo_2S_4$ CE provided greater electrocatalytic activity (a higher reduction current density, a higher J_0 and J_{lim}) and lower internal resistance (R_s and R_{ct}). Also, Ti-PF/$NiCo_2S_4$ was used to fabricate QDSSC, it has a higher performance (J_{sc} = 16.68 mA/cm^2, V_{oc} = 0.489 V, FF = 38.52%, and PCE = 3.14%) than that based on FTO/Pt CE (J_{sc} = 11.76 mA/cm^2, V_{oc} = 0.489 V, FF = 22.56%, and PCE = 1.3%).

Acknowledgments: Scientific Research Project of Education Department of Shaanxi Provincial Government (Grant No. 16JS016), Scientific Research Project of Science and Technology Department of Shaanxi Provincial Government (Grant No. 2014SZS16-K02), Natural Science Foundation of Shaanxi Province (Grant No. 2017JM6090). The authors gratefully acknowledge financial support from Natural Science Foundation of China (Grant Nos. 61176056 and 51572216), the industrial science and technology research project in Shaanxi province (2015GY005), the Key Scientific and Technological Project of Henan Province, China (No. 172102210344), and the Key Program of the Higher Education Institutions of Henan Province in China (grant No. 17A140008).

Author Contributions: Jianping Deng and Minqiang Wang designed the experiments. Jianping Deng and Zhaolin Yuan performed the experiments. Xiaohui Song and Zhi Yang analyzed the data and wrote the paper.

References

1. Deng, J.; Wang, M.; Yang, Z.; Yang, Y.; Zhang, P. Preparation of TiO_2 Nanoparticles Two-Dimensional Photonic-Crystals: A Novel Scattering Layer of Quantum Dot-Sensitized Solar Cells. *Mater. Lett.* **2016**, *183*, 307–310. [CrossRef]

2. Radich, J.G.; Dwyer, R.; Kamat, P.V. Cu_2S Reduced Graphene Oxide Composite for High-Efficiency Quantum Dot Solar Cells. Overcoming the Redox Limitations of S^{2-}/Sn^{2-} at the Counter Electrode. *J. Phys. Chem. Lett.* **2011**, *2*, 2453–2460. [CrossRef]

3. Wang, H.; Hu, Y.H. Graphene as a Counter Electrode Material for Dye-Sensitized Solar Cells. *Energy Environ. Sci.* **2012**, *5*, 8182–8188. [CrossRef]

4. Batmunkh, M.; Biggs, M.J.; Shapter, J.G. Carbon Nanotubes for Dye-Sensitized Solar Cells. *Small* **2015**, *11*, 2963–2989. [CrossRef] [PubMed]

5. Hwang, S.; Batmunkh, M.; Nine, M.J.; Chung, H.; Jeong, H. ChemPhysChem Dye-Sensitized Solar Cell Counter Electrodes Based on Carbon Nanotubes. *ChemPhysChem* **2015**, *16*, 53–65. [CrossRef] [PubMed]

6. Jafari, F.; Behjat, A.; Khoshro, A.R.; Ghoshani, M. A Dye-Sensitized Solar Cell Based on Natural Photosensitizers and a PEDOT:PSS/TiO_2 Film as a Counter Electrode. *Eur. Phys. J. Appl. Phys.* **2015**, *69*, 20502. [CrossRef]

7. Song, X.; Wang, M.; Deng, J.; Ju, Y.; Xing, T.; Ding, J.; Yang, Z.; Shao, J. ZnO/PbS Core/Shell Nanorod Arrays as Efficient Counter Electrode for Quantum Dot-Sensitized Solar Cells. *J. Power Sources* **2014**, *269*, 661–670. [CrossRef]

8. Yang, Z.; Chen, C.-Y.; Liu, C.-W.; Li, C.-L.; Chang, H.-T. Quantum Dot–Sensitized Solar Cells Featuring CuS/CoS Electrodes Provide 4.1% Efficiency. *Adv. Energy Mater.* **2011**, *1*, 259–264. [CrossRef]

9. Geng, H.; Zhu, L.; Li, W.; Liu, H.; Quan, L.; Xi, F.; Su, X. FeS/Nickel Foam as Stable and Efficient Counter Electrode Material for Quantum Dot Sensitized Solar Cells. *J. Power Sources* **2015**, *281*, 204–210. [CrossRef]

10. Faber, M.S.; Park, K.; Caban-Acevedo, M.; Santra, P.K.; Jin, S. Earth-Abundant Cobalt Pyrite (CoS_2) Thin Film on Glass as a Robust, High-Performance Counter Electrode for Quantum Dot-Sensitized Solar Cells. *J. Phys. Chem. Lett.* **2013**, *4*, 1843–1849. [CrossRef] [PubMed]

11. Chen, C.; Ye, M.; Zhang, N.; Wen, X.; Zheng, D.; Lin, C. Preparation of Hollow Co_9S_8 Nanoneedle Arrays as Effective Counter Electrodes for Quantum Dot-Sensitized Solar Cells. *J. Mater. Chem. A* **2015**, *3*, 6311–6314. [CrossRef]

12. Manjceevan, A.; Bandara, J. Optimization of Performance and Stability of Quantum Dot Sensitized Solar Cells by Manipulating the Electrical Properties of Different Metal Sulfide Counter Electrodes. *Electrochim. Acta* **2017**, *235*, 390–398. [CrossRef]

13. Xiao, J.; Wan, L.; Yang, S.; Xiao, F.; Wang, S. Design Hierarchical Electrodes with Highly Conductive $NiCo_2S_4$ Nanotube Arrays Grown on Carbon Fiber Paper for High-Performance Pseudocapacitors. *Nano Lett.* **2014**, *14*, 831–838. [CrossRef] [PubMed]

14. Yan, M.; Yao, Y.; Wen, J.; Long, L.; Kong, M.; Zhang, G.; Liao, X.; Yin, G.; Huang, Z. Construction of a Hierarchical $NiCo_2S_4$@PPy Core–Shell Heterostructure Nanotube Array on Ni Foam for a High-Performance Asymmetric Supercapacitor. *ACS Appl. Mater. Interfaces* **2016**, *8*, 24525–24535. [CrossRef] [PubMed]

15. Peng, S.; Li, L.; Li, C.; Tan, H.; Cai, R.; Yu, H.; Mhaisalkar, S.; Srinivasan, M.; Ramakrishna, S.; Yan, Q. In Situ Growth of $NiCo_2S_4$ Nanosheets on Graphene for High-Performance Supercapacitors. *Chem. Commun.* **2013**, *49*, 10178–10180. [CrossRef] [PubMed]

16. Chen, L.; Zhou, Y.; Dai, H.; Yu, T.; Liu, J.; Zou, Z. One-Step Growth of $CoNi_2S_4$ Nanoribbons on Carbon Fibers as Platinum-Free Counter Electrodes for Fiber-Shaped Dye-Sensitized Solar Cells with High Performance: Polymorph-Dependent Conversion Efficiency. *Nano Energy* **2015**, *11*, 697–703. [CrossRef]

17. Shi, Z.; Deng, K.; Li, L. Pt-Free and Efficient Counter Electrode with Nanostructured CoNi2S4 for Dye-Sensitized Solar Cells. *Sci. Rep.* **2015**, *5*, 9317. [CrossRef] [PubMed]

18. Shi, Z.; Lu, H.; Liu, Q.; Cao, F.; Guo, J.; Deng, K.; Li, L. Efficient p-Type Dye-Sensitized Solar Cells with All-Nano-Electrodes:$NiCo_2S_4$ Mesoporous Nanosheet Counter Electrodes Directly Converted from $NiCo_2O_4$ Photocathodes. *Nanoscale Res. Lett.* **2014**, *9*, 608. [CrossRef] [PubMed]

19. Huo, J.; Wu, J.; Zheng, M.; Tu, Y.; Lan, Z. Flower-Like Nickel Cobalt Sulfide Microspheres Modified with Nickel Sulfide as Pt-Free Counter Electrode for Dye-Sensitized Solar Cells. *J. Power Sources* **2016**, *304*, 266–272. [CrossRef]

20. Huang, N.; Zhang, S.; Huang, H.; Liu, J.; Sun, Y.; Sun, P.; Bao, C.; Zheng, L.; Sun, X.; Zhao, X. Pt-Sputtering-Like $NiCo_2S_4$ Counter Electrode for Efficient Dye-Sensitized Solar Cells. *Electrochim. Acta* **2016**, *192*, 521–528. [CrossRef]

21. Khoo, S.Y.; Miao, J.; Yang, H.; He, Z.; Leong, K.C.; Liu, B.; Thatt, T.; Tan, Y. One-Step Hydrothermal Tailoring of $NiCo_2S_4$ Nanostructures on Conducting Oxide Substrates as an Efficient Counter Electrode in Dye-Sensitized Solar Cells. *Adv. Mater. Interfaces* **2015**, *2*, 1500384. [CrossRef]

22. Chen, H.; Jiang, J.; Zhang, L.; Xia, D.; Zhao, Y.; Guo, D.; Qi, T.; Wan, H. In Situ Growth of $NiCo_2S_4$ Nanotube Arrays on Ni Foam for Supercapacitors: Maximizing Utilization Efficiency at High Mass Loading to Achieve Ultrahigh Areal Pseudocapacitance. *J. Power Sources* **2014**, *254*, 249–257. [CrossRef]

23. Park, J.T.; Lee, C.S.; Kim, J.H. High Performance Electrocatalyst Consisting of CoS Nanoparticles on an Organized Mesoporous SnO_2 Film: Use as Counter Electrodes for Pt-Free, Dye-Sensitized Solar Cells. *Nanoscale* **2015**, *7*, 670–678. [CrossRef] [PubMed]

24. Xu, S.; Luo, Y.; Zhong, W.; Xiao, Z.; Luo, Y.; Ou, H. Nanoporous TiO_2/SnO_2/Poly(3,4-ethylene-dioxythiophene): Polystyrenesulfonate Composites as Efficient Counter Electrode for Dye Sensitized Solar Cells. *J. Nanosci. Nanotechnol.* **2016**, *16*, 392–399. [CrossRef] [PubMed]

25. Chen, H.; Zhu, L.; Liu, H.; Li, W. Efficient Iron Sulfide Counter Electrode for Quantum Dots Sensitized Solar Cells. *J. Power Sources* **2014**, *245*, 406–410. [CrossRef]

26. Du, J.; Du, Z.; Hu, J.-S.; Pan, Z.; Shen, Q.; Sun, J.; Long, D.; Dong, H.; Sun, L.; Zhong, X.; et al. Zn–Cu–In–Se Quantum Dot Solar Cells with a Certified Power Conversion Efficiency of 11.6%. *J. Am. Chem. Soc.* **2016**, *138*, 4201–4209. [CrossRef] [PubMed]

27. Yang, Y.; Zhu, L.; Sun, H.; Huang, X.; Luo, Y.; Li, D.; Meng, Q. Composite Counter Electrode Based on Nanoparticulate PbS and Carbon Black: Towards Quantum Dot-Sensitized Solar Cells with Both High Efficiency and Stability. *ACS Appl. Mater. Interfaces* **2012**, *4*, 6162–6168. [CrossRef] [PubMed]

28. Fu, W.; Zhao, C.; Han, W.; Liu, Y.; Zhao, H.; Ma, Y.; Xie, E. Cobalt Sulfide Nanosheets Coated on $NiCo_2S_4$ Nanotube Arrays as Electrode Materials for High-performance Supercapacitors. *J. Mater. Chem. A* **2015**, *3*, 10492–10497. [CrossRef]

29. Deng, J.; Wang, M.; Zhang, P.; Ye, W. Preparing ZnO Nanowires in Mesoporous TiO_2 Photoanode by an in-Situ Hydrothermal Growth for Enhanced Light-Trapping in Quantum Dots-Sensitized Solar Cells. *Electrochim. Acta* **2016**, *200*, 12 20. [CrossRef]

30. Li, R.; Wang, S.; Huang, Z.; Lu, F.; He, T. $NiCo_2S_4$@$Co(OH)_2$ Core-Shell Nanotube Arrays in Situ Grown on Ni Foam for High Performances Asymmetric Supercapcitors. *J. Power Sources* **2016**, *312*, 156–164. [CrossRef]

31. Park, J.; Zheng, H.; Jun, Y.; Alivisatos, A.P. Hetero-Epitaxial Anion Exchange Yields Single-Crystalline Hollow Nanoparticles. *J. Am. Chem. Soc.* **2009**, *131*, 13943–13945. [CrossRef] [PubMed]

32. Lu, F.; Zhou, M.; Li, W.; Weng, Q.; Li, C.; Xue, Y.; Jiang, X.; Zeng, X.; Bando, Y.; Golberg, D. Engineering Sulfur Vacancies and Impurities in $NiCo_2S_4$ Nanostructures Toward Optimal Supercapacitive Performance. *Nano Energy* **2016**, *26*, 313–323. [CrossRef]

33. Su, A.-L.; Lu, M.-N.; Chang, C.-Y.; Wei, T.-C.; Lin, J.-Y. Scalable Fabrication of Efficient $NiCo_2S_4$ Counter Electrodes for Dye-sensitized Solar Cells Using a Facile Solution Approach. *Electrochim. Acta* **2016**, *222*, 1410–1416. [CrossRef]

34. Du, F.; Yang, Q.; Qin, T.; Li, G. Morphology-Controlled Growth of $NiCo_2O_4$ Ternary Oxides and Their Application in Dye-Sensitized Solar Cells as Counter Electrodes. *Sol. Energy* **2017**, *146*, 125–130. [CrossRef]

35. Wang, S.; Dong, W.; Fang, X.; Zhou, S.; Shao, J.; Deng, Z.; Tao, R.; Zhang, Q.; Hu, L.; Zhu, J. Enhanced Electrocatalytic Activity of Vacuum Thermal Evaporated Cu_xS Counter Electrode for Quantum Dot-Sensitized Solar Cells. *Electrochim. Acta* **2015**, *154*, 47–53. [CrossRef]

36. Dennyson Savariraj, A.; Viswanathan, K.K.; Prabakar, K. CuS Nano-flakes and Nano-platelets as Counter Electrode for Quantum Dots Sensitized Solar Cells. *Electrochim. Acta* **2014**, *149*, 364–369. [CrossRef]

37. Zakeeruddin, S.M.; Gratzel, M. Solvent-Free Ionic Liquid Electrolytes for Mesoscopic Dye Sensitized Solar Cells. *Adv. Funct. Mater.* **2009**, *19*, 2187–2202. [CrossRef]

38. Meng, K.; Surolia, P.K.; Byrne, O.; Thampi, K.R. Efficient CdS Quantum Dot Sensitized Solar Cells Made Using Novel Cu_2S Counter Electrode. *J. Power Sources* **2014**, *248*, 218–223. [CrossRef]

39. Li, D.-M.; Cheng, L.-Y.; Zhang, Y.-D.; Zhang, Q.-X.; Huang, X.-M.; Luo, Y.-H.; Meng, Q.-B. Development of Cu_2S/Carbon Composite Electrode for CdS/CdSe Quantum Dot Sensitized Solar Cell Modules. *Sol. Energy Mater. Sol. C* **2014**, *120*, 454–461. [CrossRef]

40. Liberatore, M.; Decker, F.; Burtone, L.; Zardetto, V.; Brown, T.M.; Reale, A.; Di Carlo, A. Using EIS for Diagnosis of Dye-Sensitized Solar Cells Performance. *J. Appl. Electrochem.* **2009**, *39*, 2291. [CrossRef]

41. Liu, F.; Zhu, J.; Li, Y.; Wei, J.; Lv, M.; Xu, Y.; Zhou, L.; Hu, L.; Dai, S. Earth-Abundant Cu_2SnSe_3 Thin Film Counter Electrode for High Efficiency Quantum Dot-Sensitized Solar Cells. *J. Power Sources* **2015**, *292*, 7–14. [CrossRef]

 *nanomaterials*

Article

Enhancing Output Power of Textured Silicon Solar Cells by Embedding Indium Plasmonic Nanoparticles in Layers within Antireflective Coating

Wen-Jeng Ho [1,*], Jheng-Jie Liu [1], Yun-Chieh Yang [1] and Chun-Hung Ho [2]

[1] Department of Electro-Optical Engineering, National Taipei University of Technology, Taipei 10608, Taiwan; jjliu@mail.ntut.edu.tw (J.-J.L.); t105658043@ntut.edu.tw (Y.-C.Y.)

[2] Realtek Semiconductor Corp., No. 2, Innovation Road II, Hsinchu Science Park, Hsinchu 300, Taiwan; cunhonho@gmail.com

* Correspondence: wjho@ntut.edu.tw or wjho@mail.ntut.edu.tw; Tel.: +886-2-2771-2171 (ext. 4639)

Received: 15 November 2018; Accepted: 30 November 2018; Published: 4 December 2018

Abstract: In this study, we sought to enhance the output power and conversion efficiency of textured silicon solar cells by layering two-dimensional indium nanoparticles (In NPs) within a double-layer (SiN_x/SiO_2) antireflective coating (ARC) to induce plasmonic forward scattering. The plasmonic effects were characterized using Raman scattering, absorbance spectra, optical reflectance, and external quantum efficiency. We compared the optical and electrical performance of cells with and without single layers and double layers of In NPs. The conversion efficiency of the cell with a double layer of In NPs (16.97%) was higher than that of the cell with a single layer of In NPs (16.61%) and greatly exceeded that of the cell without In NPs (16.16%). We also conducted a comprehensive study on the light-trapping performance of the textured silicon solar cells with and without layers of In NPs within the double layer of ARC at angles from 0° to 75°. The total electrical output power of cells under air mass (AM) 1.5 G illumination was calculated. The application of a double layer of In NPs enabled an impressive 53.42% improvement in electrical output power (compared to the cell without NPs) thanks to the effects of plasmonic forward scattering.

Keywords: indium nanoparticles (In NPs); textured silicon solar cells; antireflective coating (ARC); plasmonic forward scattering

---

1. Introduction

Most commercial solar cells are fabricated using a silicon-based wafer ranging in thickness from 150 to 200 micrometers. Light trapping in crystalline silicon solar cells is generally achieved using a pyramidal structure with an antireflective coating on the surface. This combination allows for the multiple reflection and scattering of incident light within the solar cell [1–8]. Metallic nanoparticles (NPs) [9–12] of silver (Ag NPs) [13–16], gold (Au NPs) [17–20], and aluminum (Al NPs) [21–24] have been applied to the front and/or rear-side surfaces of silicon solar cells to increase light trapping and enhance photovoltaic performance. Metallic NPs can be resonantly coupled with incident light, thereby allowing a portion of the light to be scattered into the absorber layer. Far-field forward light scattering from metallic NPs and a near-field enhanced localized field in the vicinity of the metallic NPs have been shown to boost the conversion efficiency of photovoltaic devices [9,25]. Researchers have investigated the use of various metallic materials in these devices. They have also controlled the dimensions, shapes, spacing, and surrounding dielectric environments of the NPs to enhance resonant plasmonic scattering [9,11]. In addition, a dielectric-based (TiO_2) photonic structure using colloidal-lithographed processing was also proposed for light trapping in thin film photovoltaics [26].

Indium nanoparticles (In NPs) exhibit plasmonic resonance in the ultraviolet range (below 280 nm) and broadband plasmonic light scattering from visible and near-infrared wavelengths [27–29], which makes them capable of boosting the conversion efficiency of photovoltaic devices. However, there has been relatively little research on the embedding of In NP sheets within a double-layer antireflective coating (DL-ARC) to enhance the photovoltaic performance of textured silicon solar cells [30–33].

In this study, we examined the plasmonic light scattering of In NPs of various dimensions, which were embedded in a coating of SiO_2 with a DL-ARC structure (SiN_x/SiO_2) for use in textured silicon solar cells. The plasmonic effects of samples with single and double layers of In NPs were characterized according to Raman scattering, absorbance, optical reflectance, and external quantum efficiency (EQE). We also measured the EQE and photovoltaic current-voltage (I-V) as a function of incident angle using cells with and without In NPs. We then calculated the total output electrical power of cells under AM 1.5 G illumination. The application of a double layer of In NPs enabled an impressive 53.42% improvement in output electrical power (compared to the cell without NPs) thanks to the effects of plasmonic forward scattering. The novelty of this study includes (a) the enhancement in output power and conversion efficiency of textured silicon solar cells by layering two-dimensional In NPs within a DL-ARC, and (b) a comprehensive study on the light-trapping performance of the textured silicon solar cells with and without layers of In NPs within the DL-ARC at angles from 0° to 75°, both of which are issues beyond the scope of previous studies [30–33].

2. Experiments

2.1. Characterization of Plasmonic Effects from Indium Nanoparticle Layers Embedded in SiO_2 Coating

Quartz substrates were used as a test template to characterize the plasmonic effects of indium nanoparticles in the UV-VIS band, due to their high transparency (low absorption) at UV-band wavelengths. Figure 1 presents a schematic diagram of (a) an SiO_2 coating (90 nm) deposited on a quartz substrate, (b) a single layer of In NPs embedded in a SiO_2 coating (90 nm), and (c) a double layer of In NPs embedded in a SiO_2 coating (90 nm). The SiO_2 layer and In NPs were deposited using electron-beam (e-beam) evaporation. The In NPs were formed by depositing indium films at thicknesses of 3.8, 5, and 7 nm directly on the quartz substrate or the SiO_2 coating and then applying rapid thermal annealing (RTA) at 200 °C under H_2. The average surface coverage and average diameter of the In NPs were as follows: 3.8 nm (36.75% and 20.13 nm), 5 nm (41.83% and 25.03 nm), and 7 nm (46.46% and 32.14 nm). These results were calculated using image-J software from corresponding SEM images. Sample (b) was fabricated by applying a layer of In NPs on the quartz substrate and then capping it with a 90-nm coating of SiO_2. Sample (c) was fabricated by applying an initial layer of In NPs on the quartz substrate and covering it with a 20-nm spacer layer of SiO_2 before applying a second layer of In NPs over the spacer layer and capping it with an additional 70-nm coating of SiO_2.

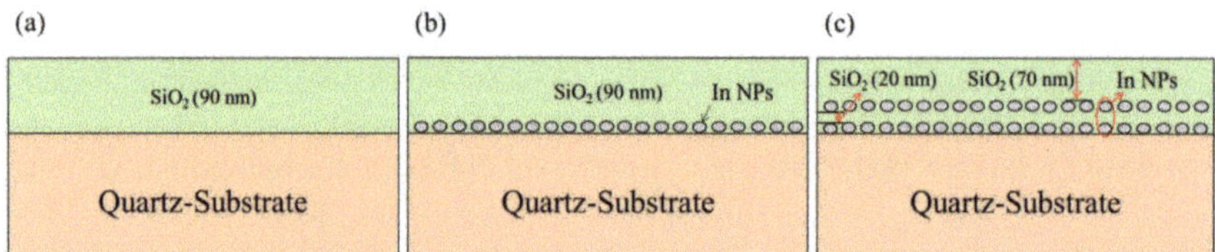

Figure 1. Schematic diagrams of proposed samples: (**a**) SiO_2 coating (90 nm) deposited on quartz substrate; (**b**) single layer of In NPs embedded in SiO_2 coating (90 nm) on quartz substrate; and (**c**) double layer of In NPs embedded in SiO_2 coating (90 nm).

The plasmonic effects of the In NPs (single and double layers) were examined using Raman scattering and absorbance measurements. Raman scattering spectra were collected using a Raman spectrometer (UniRAM, UniNanoTech Co., Yongin-si, Giheung-gu, Korea) with a 532-nm laser (approximately 3 mW), with the signal accumulated over a period of 30 s. The observed shifts in the Raman peaks and variations in Raman signal intensity revealed that the plasmonic effects depend on the number of NP layers and the dimensions of the In NPs they comprise. Absorbance spectra were collected using a miniature spectrometer (USB4000-VIS-NIR, Ocean Optics, Inc., Largo, FL, USA) with a deuterium tungsten light source (200–2000 nm) and reflective integrating sphere (diameter of 5 cm). The obtained absorbance spectrum revealed the intensity of surface plasmon resonance (SPR) and the SPR absorption band induced by In NPs. Both measurements were also obtained from the sample without In NPs (as a control) to confirm the plasmonic effects of In NPs layered within the SiO_2 coating.

2.2. Fabrication and Characterization of Plasmonic Textured Silicon Solar Cells

Boron-doped (P-type) crystalline silicon wafers with a (100) orientation and resistivity of 10 Ω-cm were cut to a thickness of 150 μm to use them as a base material for textured silicon solar cells. Following standard Radio Corporation of America (RCA) cleaning, the saw-damaged surface of the silicon wafers was removed by dipping them in a solution of H_2O/KOH (Potassium). The surface of the wafer was then etched by dipping it in a solution of H_2O/KOH/IPA (Isopropanol) at 80 °C for 20 min to create a surface texture in the form of randomly-arranged pyramidal structures. The textured wafers then underwent RCA cleaning prior to the application of an n^+-Si emitter layer with a sheet resistance of approximately 80 Ω/sq using a $POCl_3$ diffusion process in a tube diffusion chamber at 850 °C over a period of 3 min. The wafer was then cut into samples of 10 mm^2. The oxide layer that formed on the surface of the samples was removed using hydrogen fluoride (HF) solution prior to the deposition of an Al film with a depth of 2 μm on the rear surface using e-beam evaporation. The as-deposited wafer was then annealed at 450 °C for 5 min to form a back electrode with a good ohmic contact to the p-silicon. Plasma-enhanced chemical vapor deposition (PECVD) was used to deposit a 70-nm silicon nitride film on the front surface as an antireflective coating. Finally, top contact grid-electrodes were formed from a Ti film (20 nm) and Al film (10,000 nm) using photolithographic etching, e-beam evaporation, and lift-off processes. The resulting textured silicon solar cells (as shown in Figure 2a) underwent characterization in terms of optical and electrical performance for use as a reference by which to evaluate the performance of the plasmonic solar cells.

Figure 2 presents schematic diagrams showing the silicon solar cells tested in this study: 2(a) presents the bare reference silicon solar cell; 2(b) shows the solar cell with an SiO_2 coating (90 nm) without In NPs (DL-ARC; SiN_x/SiO_2); 2(c) shows the solar cell with an SiO_2 coating (90 nm) embedded with a single layer of In NPs; 2(d) shows the solar cell with an SiO_2 coating (90 nm) embedded with a double layer of In NPs (same dimensions in each layer) separated by a 20-nm SiO_2 spacer layer. Note that the total thickness of the SiO_2 coatings and embedded layers was maintained at 90 nm.

As described in Section 2.1, the layers of In NPs were fabricated by depositing indium film using e-beam evaporation with thicknesses of 3.8 nm, 5 nm, and 7 nm, followed by annealing in an RTA chamber at 200 °C under H_2 for 30 min. A scanning electron microscope (SEM; Hitachi S-4700, Hitachi High-Tech Fielding Corporation, Tokyo, Japan) was used to characterize the sample surfaces and cross-sections. Optical reflectance (Lambda 35, PerkinElmer, Inc., Waltham, MA, USA) and external quantum efficiency (EQE; Enli Technology Co., Ltd., Kaohsiung, Taiwan) measurements were used to assess the plasmonic effects of the In NPs layers embedded within the SiO_2 coating. The photovoltaic performance of the textured silicon solar cells (with and without In NPs layers) was assessed in terms of photovoltaic current-voltage (I-V) under AM 1.5 G illumination. The solar simulator (XES-151S, San-Ei Electric Co., Ltd., Osaka, Japan) was calibrated using a National Renewable Energy Laboratory (NREL)-certified crystalline silicon reference (PVM-894, PV Measurements Inc., Boulder, CO, USA) prior to measurement.

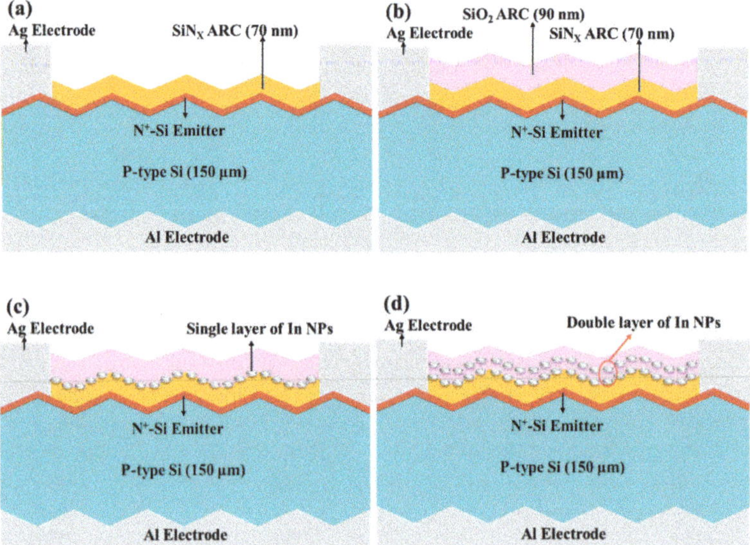

Figure 2. Schematic illustrations showing cells evaluated in this study: (**a**) reference textured silicon solar cell; (**b**) reference cell with SiO$_2$ coating (90 nm); (**c**) reference cell with SiO$_2$ coating (90 nm) embedded with single layer of In NPs; and (**d**) reference cell with SiO$_2$ coating (90 nm) embedded with double layer of In NPs.

2.3. EQE and Photovoltaic Performance of Plasmonic Textured Silicon Solar Cells under Incident Light of Various Angles

We evaluated the textured silicon solar cells with and without In NP layers in terms of EQE response and photovoltaic I-V curves under illumination by an incident light source of various angles (θ), ranging from 0° to 75°, as shown in Figure 3. The light source was fixed above a stage that could be rotated from 0° to 90°. The light source was calibrated using an NREL-certified crystalline silicon reference cell at 0° prior to measurement. The output power of the cells was calculated at each incident angle to compare the total output power under daylight illumination (from AM 0700 to PM 1700). The incident angles were meant to simulate illumination at various times, as follows: 0° (noon), 45° (AM 0900/PM 1500), and 75° (AM 0700/PM 1700).

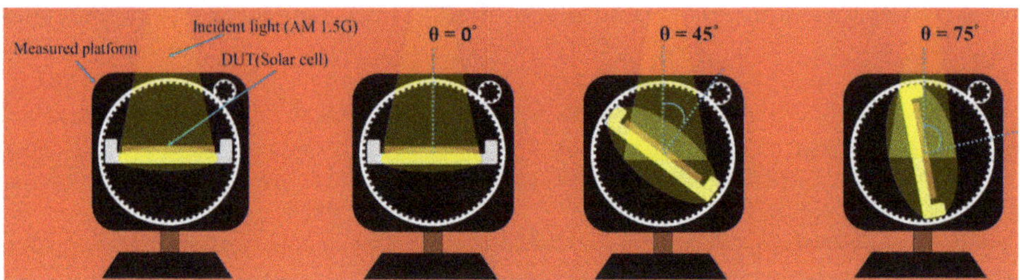

Figure 3. Varying incident light source angle (θ) while illuminating textured silicon solar.

3. Results and Discussion

Figure 4a presents the Raman spectra of samples with the following configurations: (1) quartz substrate/SiO$_2$ layer; (2) quartz substrate/SiO$_2$ coating embedded with single layers of In NPs of various sizes (3.8 nm, 5 nm, and 7 nm); and (3) quartz substrate/SiO$_2$ coating embedded with double layers of In NPs of various sizes (3.8 nm, 5 nm, and 7 nm). Compared to the sample with a quartz substrate/SiO$_2$ coating, the samples with In NPs presented shifts in the Raman peaks at 1181, 1362, and 1485 cm^{-1}. Generally, peaks in the Raman signal from metallic nanoparticles are an indication of SPR under a light source of specific wavelengths. We also observed an increase in the intensity of the Raman signals with an increase in the particle size and the number of NP layers. Thus, the most pronounced plasmonic effects were observed in the samples with a double layer of In NPs of 7 nm. Figure 4b presents the absorption spectra of samples with the following configurations: (1) quartz substrate/SiO$_2$ coating; (2) quartz substrate/SiO$_2$ coating embedded with single layers of In NPs of various sizes (3.8 nm, 5 nm, and 7 nm); and (3) quartz substrate/SiO$_2$ coating embedded with double layers of In NPs of various sizes (3.8 nm, 5 nm, and 7 nm). The fact that the peak absorption occurred at approximately 200 nm indicates that the principal absorption of incident light occurred in the quartz substrate due to the bandgap of the quartz substrate (approximately 6 eV). Compared to the quartz substrate/SiO$_2$ coating sample, the sample with a single layer of In NPs presented a higher absorption band between 220 and 300 nm, with peak absorption at approximately 260 nm. The samples with a double layer of In NPs presented far higher absorption values due to the higher density of the indium nanoparticles and the effects of light coupling between the two nanoparticle layers.

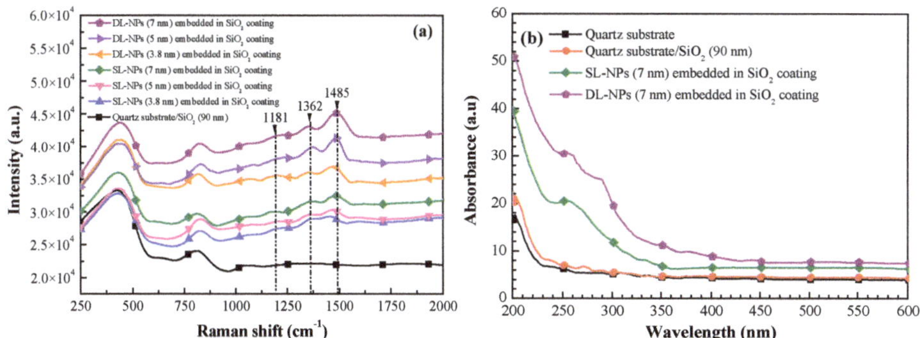

Figure 4. (**a**) Raman spectra; (**b**) absorption spectra of all tested samples.

Figure 5a,b present top-view SEM images of textured silicon solar cells without and with In NPs, respectively. These images show that the minimum and maximum spacing between pyramids on the textured surface was 4 μm and 8 μm, respectively. The minimum and maximum heights were 4 and 7 μm, respectively. Figure 5c presents a particle profile of In NPs (7 nm) within the textured surface. This profile was generated from the inset of Figure 5b. The size distribution and coverage were calculated using Image-J software (National Institutes of Health, Bethesda, MD, USA). Figure 5d presents a side-view SEM image of a sample with a double layer of In NPs (7 nm) embedded within the SiO$_2$ coating on a GaAs substrate. The GaAs substrate was used to examine the layer(s) of indium nanoparticles embedded in the SiO$_2$ coating due to the ease with which it can be cleaved to a strip-bar for side-view SEM examination. In this 2D profile, it is easy to differentiate the first and second layers of indium nanoparticles within the SiO$_2$ coating.

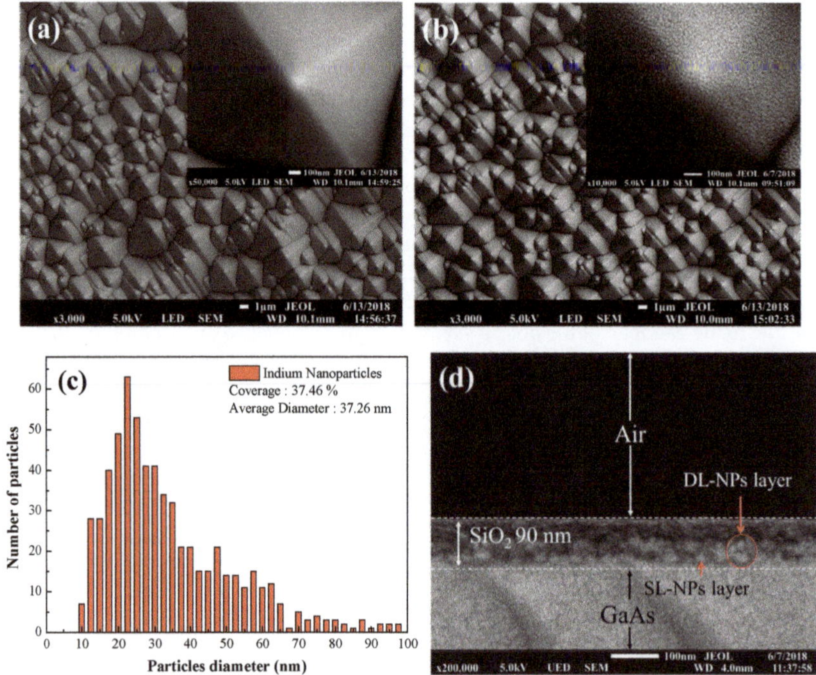

Figure 5. Top-view SEM images of textured silicon solar cells: (**a**) without and (**b**) with In NPs; (**c**) particle profile of In NPs (7 nm) within textured surface; (**d**) side-view SEM image of double layer of In NPs (7 nm) embedded within SiO$_2$ coating on GaAs substrate. The inset in Figure 5a,b is an enlarge graph of a pyramid structure on the cells without and with In NPs, respectively.

Figure 6a presents the optical reflectance of the reference textured silicon solar cell (Ref. Cell), the cell with an SiO$_2$ coating without In NPs (ARC Cell), and cells with an SiO$_2$ coating embedded with a single layer of indium nanoparticles of 3.8, 5 and 7 nm (SL-NPs Cell). The average weighted reference (R_W) was calculated from the wavelength range of 380–1000 nm, as listed in Table 1. For the sake of clarity, we calculated the R_W of the cells as follows:

$$R_W = \frac{\int_{380 \text{ nm}}^{1000 \text{ nm}} R(\lambda)\varphi_{\text{ph}}(\lambda)d\lambda}{\int_{380 \text{ nm}}^{1000 \text{ nm}} \varphi_{\text{ph}}(\lambda)d\lambda} \times 100\% \tag{1}$$

where $R(\lambda)$ is the optical reflectance at a given wavelength (λ) and φ_{ph} (λ) is the photon flux of AM 1.5 G at that wavelength (λ). The R_W of the cells with In NPs was lower than that of the reference cell due to SPR absorption in the wavelength range of 200–350 nm (Figure 4b) and plasmonic forward scattering beyond 600 nm, both of which were induced by the In NPs. The low R_W indicates that the NPs enabled more of the incident light to be trapped in the silicon. Samples with larger nanoparticles (7 nm) presented lower R_W values (3.34%) than the samples with 3.8-nm nanoparticles (3.78%). Again, we can see that larger In NPs were able to trap more of the incident light. We therefore fabricated samples with two layers of larger In NPs (7 nm) for further study and comparison. Figure 6b presents the optical reflectance of the reference cell, the cell with an SiO$_2$ coating (no NPs), and cells with single and double layers of In NPs of 7 nm (DL-NPs Cell). We calculated the R_W of all tested cells over a wavelength range of 380–1000 nm, the results of which are listed in Table 1. The lowest R_W value (2.32%) was obtained from the cell with the double layer of In NPs embedded within the SiO$_2$ coating.

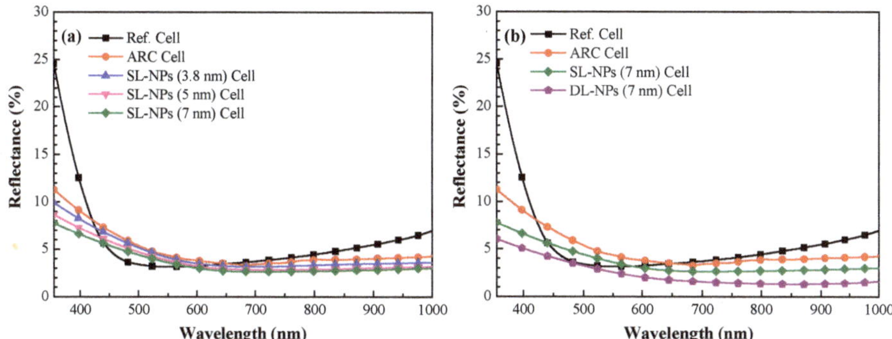

Figure 6. Optical reflectance: (**a**) reference cell, cell with SiO$_2$ coating (no In NPs), cell with single layer of In NPs (3.8 nm, 5 nm, and 7 nm) embedded in SiO$_2$ coating; (**b**) cell with double layer of In NPs (7 nm) embedded in SiO$_2$ coating.

Table 1. Average Weighted Reference (R_W) and Average Weighted External Quantum Efficiency (EQE_W) of Proposed Cells.

Silicon Solar Cell	R_W (%) @ 380–1000 nm	EQE_W (%) @ 380–1000 nm
Ref. Cell	4.58	88.82
DL-ARC	4.15	89.87
SL-In NPs (3.8 nm) Cell	3.78	90.13
SL-In NPs (5 nm) Cell	3.54	90.45
SL-In NPs (7 nm) Cell	3.34	91.35
DL-In NPs (7 nm) Cell	2.32	92.74

Figure 7a presents the EQE response of the reference solar cell, the cell with an SiO$_2$ coating (no In NPs), and cells with a single layer of In NPs of various sizes (3.8, 5, and 7 nm) embedded in an SiO$_2$ coating. The EQE values of cells with In NPs were higher than those without In NPs across the entire wavelength range, due to the effects of plasmonic forward scattering induced by the NPs. The EQE values of cells with larger NPs were slightly higher than those with smaller NPs. The EQE response values are in good agreement with the optical reflectance results. Figure 7b presents the EQE response of the reference cell, the cell with a SiO$_2$ coating (no In NPs), and cells with either a single layer of In NPs (7 nm) or a double layer of In NPs (7 nm). For the sake of clarity, we calculated the average weighted EQE (EQE_W) of the cells as follows:

$$EQE_W = \frac{\int_{380\,\text{nm}}^{1000\,\text{nm}} EQE(\lambda)\varphi_{\text{ph}}(\lambda)d\lambda}{\int_{380\,\text{nm}}^{1000\,\text{nm}} \varphi_{\text{ph}}(\lambda)d\lambda} \times 100\% \tag{2}$$

where $EQE(\lambda)$ is the EQE at a given wavelength (λ) and $\varphi_{\text{ph}}(\lambda)$ is the photon flux of AM 1.5 G at that wavelength (λ). The EQE_W values were as follows: double layer of 7-nm In NPs (92.74%), single layer of 7-nm In NPs (91.35%), SiO$_2$ coating without NPs (88.97%), and reference cell (88.82%). Table 1 summarizes the EQE_W of the cells calculated over a wavelength range of 380–1000 nm. The EQE values of cells with double layers of In NPs exceeded those of cells with a single layer due to the higher density of NPs and more pronounced plasmonic forward scattering.

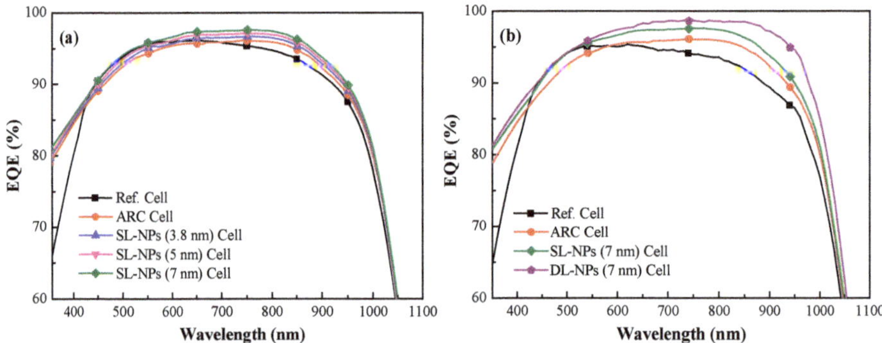

Figure 7. EQE response values: (**a**) reference cell, cell with SiO_2 coating (no NPs), and cell with single layer of In NPs (3.8 nm, 5 nm, and 7 nm); (**b**) reference cell, cell with SiO_2 coating (no NPs), cell with single layer of In NPs (7 nm), and cell with double layer of In NPs (7 nm).

Figure 8a presents the photovoltaic J-V curves obtained from the reference cell, the cell with an SiO_2 coating (no In NPs), and cells with a single layer of In NPs of various sizes (3.8, 5, and 7 nm) under normal incident illumination ($\theta = 0°$). The short-circuit current densities (J_{sc}) and conversion efficiency (η) values were as follows: single layer of 7-nm NPs (40.26 mA/cm^2 and 16.61%), single layer of 5-nm NPs (39.95 mA/cm^2 and 16.51%), single layer of 3.8-nm NPs (39.77 mA/cm^2 and 16.44%), SiO_2 coating without NPs (39.61 mA/cm^2 and 16.34%), and reference cell (39.19 mA/cm^2 and 16.16%). The J_{sc} values of cells with NPs were higher due to plasmonic forward scattering than those without NPs. The J_{sc} values of cells with larger NPs were slightly higher than those of cells with smaller NPs, due to stronger plasmonic forward scattering.

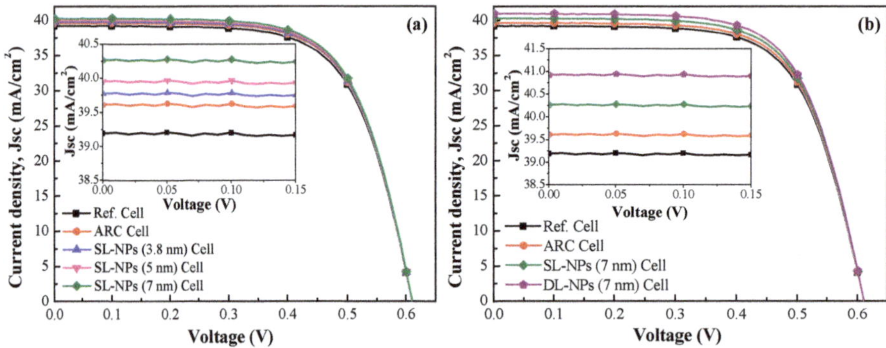

Figure 8. Photovoltaic I-V curves: (**a**) reference cell, cell with SiO_2 coating (no NPs), and cell with single layer of In NPs (3.8 nm, 5 nm, and 7 nm); (**b**) reference cell, cell with SiO_2 coating (no NPs), and cell with double layer of In NPs (7 nm). The inset in Figure 8a,b is an enlarge graph of J_{sc} of all evaluated cells at voltage 0−0.15 V.

Figure 8b presents the photovoltaic J-V curves of cells with a single layer of 7-nm In NPs and a double layer of 7-nm In NPs. The photovoltaic performance of the proposed cells is summarized in Table 2. Adding the second layer of In NPs increased the J_{sc} value from 40.26 to 40.92 mA/cm^2, and the η value from 16.61% to 16.97%, compared to a single layer of In NPs (7 nm). Adding two layers of In NPs (7 nm) increased the J_{sc} value from 39.61 to 40.92 mA/cm^2, and the η value from 16.34% to 16.97%, compared to DL-ARC without In NPs. These results demonstrate that using larger In NPs and including multiple layers of In NPs facilitates the trapping of incident light and enhances J_{sc} and η, due to stronger plasmonic forward scattering.

Table 2. Photovoltaic Performance of Proposed Cells Under AM 1.5 G Illumination at Normal Incidence.

Silicon Solar Cell	J_{sc} (mA/cm^2)	V_{oc} (mV)	Fill Factor (%)	η (%)	ΔJ_{sc} (%)	$\Delta\eta$ (%)
Ref. Cell	39.19	609.40	67.68	16.16	—	—
ARC Cell	39.61	609.50	67.70	16.34	1.07	1.11
SL-In NPs (3.8 nm) Cell	39.77	609.51	67.85	16.44	1.47	1.73
SL-In NPs (5 nm) Cell	39.95	609.52	67.84	16.51	1.93	2.16
SL-In NPs (7 nm) Cell	40.26	609.53	67.72	16.61	2.73	2.78
DL-In NPs (7 nm) Cell	40.92	609.50	68.05	16.97	4.41	5.01

Figure 9 presents the EQE_W and J_{sc} of the reference cell, the cell with an SiO$_2$ coating (no NPs), and cells with single and double layers of In NPs under incident angles of 0°–75°. Increasing the incident angle resulted in a gradual decrease in EQE_W and J_{sc} values in all tested cells. Compared to cells without NPs, we obtained higher EQE_W and J_{sc} values from cells with single and double layers of In NPs at all incident angles. At a high incident angle of 75°, the double layer cells presented a J_{sc} decrement of 25.2% (from 40.92 to 30.06 mA/cm^2), compared to the decrement of 76.3% (from 39.61 to 9.38 mA/cm^2) from cells without NPs. Overall, J_{sc} was proportional to EQE and η was proportional to J_{sc} in all of the photovoltaic devices. This means that a higher J_{sc} would no doubt result in a higher electrical output, as well as a higher conversion efficiency.

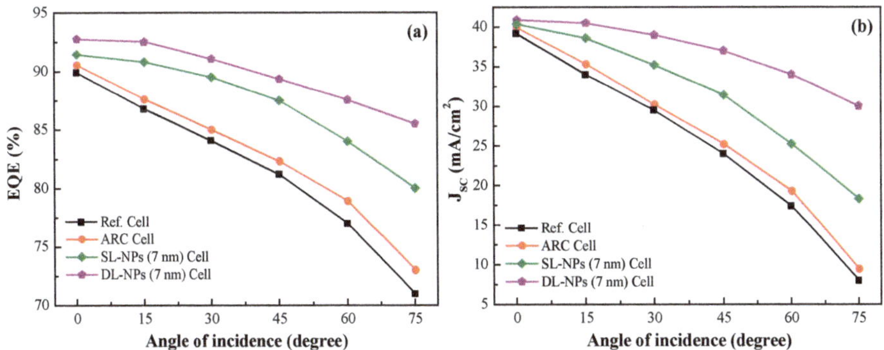

Figure 9. (a) EQE and (b) J_{sc} as a function of incident angle of reference cell, cell with SiO$_2$ coating (no NPs), cell with single layer of In NPs (7 nm), and cell with double layer of In NPs (7 nm).

Figure 10a presents the calculated electrical output power of all evaluated solar cells under illumination from −75° to 0° (sun rising; i.e., AM 0700 to noon) and then from 0° to 75° (sun descending; i.e., noon to PM 1700). At all illumination times/angles, the output power of cells with a double layer of In NPs exceeded that of cells with a single layer and cells without NPs. Figure 10b presents the daily output energy of all evaluated solar cells. For the sake of clarity, we calculated the electrical output power (P_E) and the daily output energy (E_{day}) of the cells as follows:

$$P_E = V_{oc} \times J_{sc} \times FF \tag{3}$$

where V_{oc} is the open-circuit voltage, J_{sc} is the short-circuit current density, and FF is the fill factor. The total P_E values of cells with an area of 10 mm^2 were as follows: double layer of 7-nm NPs (141.38 mW), single layer of 7-nm NPs (114.13 mW), SiO$_2$ coating (94.92 mW), and reference cell (92.15 mW).

$$E_{day} = \sum_{i=AM0700}^{i=PM1700} (P_E)_i \times 1 \text{ hour} \tag{4}$$

the E_{day} values of the cell with an SiO$_2$ coating and the reference cell were 94.92 and 92.15 mW·h, respectively. Using these as reference values, the inclusion of a single layer of In NPs (7 nm) increased E_{day} by 20.24% and 23.85%, respectively. The inclusion of a double layer of In NPs (7 nm) increased E_{day} by 48.95% and 53.42%, respectively.

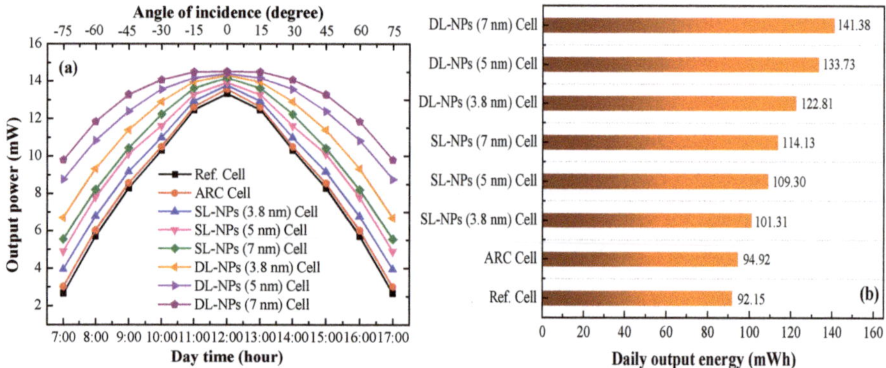

Figure 10. (**a**) Output electrical power; (**b**) output electrical energy of all tested solar cells under illumination at incidence angles from $-75°$ to $0°$ (sun rising) and from $0°$ to $75°$ (sun descending).

4. Conclusions

In this study, we examined the light trapping effects of In NPs according to optical reflectance and EQE measurements, with a particular focus on the dimensions of the NPs and the number of layers of NPs. A double layer of In NPs within the antireflective coating resulted in pronounced plasmonic forward scattering, which greatly enhanced the output power and conversion efficiency of the textured silicon solar cells. The inclusion of a double layer of In NPs increased the conversion efficiency from 16.16% to 16.97%, compared to the reference cell without In NPs. We also examined the light-trapping performance of cells with and without In NPs at incidence angles from $0°$ to $75°$. At all angles, the output power delivered from cells with a double layer of In NPs exceeded that of cells with a single layer and those without NPs. The cumulative output power (one day) delivered by the cell with an area of 10 mm^2 with a double layer of In NPs was 141.38 mW, which greatly exceeds the 94.92 mW of the cell without In NPs.

Author Contributions: All of the authors conceived the experiments; W.-J.H. designed, analyzed, and wrote the first draft of the paper; J.-J.L. and Y.-C.Y. performed the experiments; C.-H.H. conducted EQE measurements; all authors contributed to the discussion.

Funding: The Ministry of Science and Technology of the Republic of China under Grant MOST 106-2221-E-027-101-MY3.

Acknowledgments: The authors thank Professor Ching-Fuh Lin at National Taiwan University for his support in obtaining EQE measurements.

Conflicts of Interest: The authors declare no conflict of interest.

References

1. Blakers, A.; Zin, N.; McIntosh, K.R.; Fong, K. High efficiency silicon solar cells. *Energy Procedia* **2013**, *33*, 1–10. [CrossRef]
2. Cho, C.; Oh, J.; Lee, B.; Kim, B. Combined effects of pyramid-like structures and antireflection coating on Si solar cell efficiency. *J. Nanosci. Nanotechnol.* **2015**, *15*, 7624–7631. [CrossRef] [PubMed]
3. Young, K.; Wen, Q.; Hanany, S.; Imada, H.; Koch, J.; Matsumura, T.; Suttmann, O.; Schütz, V. Broadband millimeter-wave anti-reflection coatings on silicon using pyramidal sub-wavelength structures. *J. Appl. Phys.* **2017**, *121*, 213103-1–213103-10. [CrossRef]

4. Al-Husseini, A.M.; Lahlouh, B. Silicon pyramid structure as a reflectivity reduction mechanism. *J. Appl. Sci.* **2017**, *17*, 374–383. [CrossRef]

5. Baker-Finch, S.C.; McIntosh, K.R. Reflection of normally incident light from silicon solar cells with pyramidal texture. *Prog. Photovolt Res. Appl.* **2011**, *19*, 406–416. [CrossRef]

6. Chiao, S.C.; Zhou, J.L.; Macleod, H.A. Optimized design of an antireflection coating for textured silicon solar cells. *Appl. Opt.* **1993**, *32*, 5557–5560. [CrossRef]

7. Bouhafs, D.; Moussi, A.; Chikouche, A.; Ruiz, J.M. Design and simulation of antireflection coating systems for optoelectronic devices: Application to silicon solar cells. *Sol. Energy Mater. Sol. Cells* **1998**, *52*, 79–93. [CrossRef]

8. Ali, K.; Khan, S.A.; Mat Jafri, M.Z. Effect of double layer (SiO_2/TiO_2) anti-reflective coating on silicon solar cells. *Int. J. Electrochem. Sci.* **2014**, *9*, 7865–7874.

9. Sanz, J.M.; Ortiz, D.; Alcaraz de la Osa, R.; Saiz, J.M.; González, F.; Brown, A.S.; Losurdo, M.; Everitt, H.O.; Moreno, F. UV plasmonic behavior of various metal nanoparticles in the near- and far-field regimes: Geometry and substrate effects. *J. Phys. Chem. C* **2013**, *117*, 19606–19615. [CrossRef]

10. Deka, N.; Islam, M.; Sarswat, P.K.; Kumar, G. Enhancing solar cell efficiency with plasmonic behavior of double metal nanoparticle system. *Vacuum* **2018**, *152*, 285–290. [CrossRef]

11. Kelly, K.L.; Coronado, E.; Zhao, L.L.; Schatz, G.C. The optical properties of metal nanoparticles: The influence of size, shape, and dielectric environment. *J. Phys. Chem. B* **2003**, *107*, 668–677. [CrossRef]

12. Tsai, F.J.; Wang, J.Y.; Huang, J.J.; Kiang, Y.W.; Yang, C.C. Absorption enhancement of an amorphous Si solar cell through surface plasmon-induced scattering with metal nanoparticles. *Opt. Express* **2010**, *18*, A207–A220. [CrossRef] [PubMed]

13. Lee, K.C.; Lin, S.J.; Lin, C.H.; Tsai, C.S.; Lu, Y.J. Size effect of Ag nanoparticles on surface plasmon resonance. *Surf. Coat. Technol.* **2008**, *202*, 5339–5342. [CrossRef]

14. Tan, H.; Santbergen, R.; Smets, A.H.M.; Zeman, M. Plasmonic light trapping in thin-film silicon solar cells with improved self-assembled silver nanoparticles. *Nano Lett.* **2012**, *12*, 4070–4076. [CrossRef] [PubMed]

15. Pillai, S.; Catchpole, K.R.; Trupke, T.; Green, M.A. Surface plasmon enhanced silicon solar cells. *J. Appl. Phys.* **2007**, *101*, 093105. [CrossRef]

16. Winans, J.D.; Hungerford, C.; Shome, K.; Rothberg, L.J.; Fauchet, P.M. Plasmonic effects in ultrathin amorphous silicon solar cells: Performance improvements with Ag nanoparticles on the front, the back, and both. *Opt. Express* **2015**, *23*, A92–A105. [CrossRef] [PubMed]

17. Sardana, S.K.; Chava, V.S.N.; Thouti, E.; Chander, N.; Kumar, S.; Reddy, S.R.; Komarala, V.K. Influence of surface plasmon resonances of silver nanoparticles on optical and electrical properties of textured silicon solar cell. *Appl. Phys. Lett.* **2014**, *104*, 073903. [CrossRef]

18. Wang, P.H.; Millard, M.; Brolo, A.G. Optimizing plasmonic silicon photovoltaics with Ag and Au nanoparticle mixtures. *J. Phys. Chem. C* **2014**, *118*, 5889–5895. [CrossRef]

19. Yu, P.; Blondeau, J.P.; Andreazza, C.; Ntsoenzok, E.; Roussel, J.; Dutheil, P.; Thomann, A.L.; Caillard, A.; Mustapha, E.; Meot, J. Influence of gold nanoparticles (Au NPs) for performance improvement of a-Si:H photovoltaic cells. *Plasmonics* **2015**, *10*, 691–701. [CrossRef]

20. Lim, S.H.; Mar, W.; Matheu, P.; Derkacs, D.; Yu, E.T. Photocurrent spectroscopy of optical absorption enhancement in silicon photodiodes via scattering from surface plasmon polaritons in gold nanoparticles. *J. Appl. Phys.* **2007**, *101*, 104309. [CrossRef]

21. Temple, T.L.; Bagnall, D.M. Optical properties of gold and aluminium nanoparticles for silicon solar cell applications. *J. Appl. Phys.* **2011**, *109*, 084343-1–084343-13. [CrossRef]

22. Parashar, P.K.; Sharma, R.P.; Komarala, V.K. Plasmonic silicon solar cell comprised of aluminum nanoparticles: Effect of nanoparticles' self-limiting native oxide shell on optical and electrical properties. *J. Appl. Phys.* **2016**, *120*, 143104. [CrossRef]

23. Parashar, P.K.; Sharma, R.P.; Komarala, V.K. Mediating broad band light trapping in silicon solar cell by aluminum nanoparticles with native oxide shell. *Mater. Today Proc.* **2017**, *4*, 12708–12715. [CrossRef]

24. Zhang, Y.; Chen, X.; Ouyang, Z.; Lu, H.; Jia, B.; Shi, Z.; Gu, M. Improved multicrystalline Si solar cells by light trapping from Al nanoparticle enhanced antireflection coating. *Opt. Mater. Express* **2013**, *3*, 489–495. [CrossRef]

25. Atwater, H.A.; Polman, A. Plasmonics for improved photovoltaic devices. *Nat. Mater.* **2010**, *9*, 205–213. [CrossRef] [PubMed]

26. Olalla, S.S.; Manuel, J.M.; Sirazul, H.; Tiago, M.; Andreia, A.; Hugo, A.; Elvira, F.; Rodrigo, M. Colloidal-lithographed TiO$_2$ photonic nanostructures for solar cell light trapping. *J. Mater. Chem. C* **2017**, *5*, 6852–6861. [CrossRef]

27. Magnan, F.; Gagnon, J.; Fontaine, F.G.; Boudreau, D. Indium@silica core-shell nanoparticles as plasmonic enhancers of molecular luminescence in the UV region. *Chem. Commun.* **2013**, *49*, 9299–9301. [CrossRef]

28. Das, R.; Soni, R.K. Indium nanoparticles for ultraviolet surface-enhanced Raman spectroscopy. *AIP Conf. Proc.* **2018**, *1953*, 030123. [CrossRef]

29. Haynes, C.L.; Duyne, R.P.V. Plasmon-sampled surface-enhanced Raman excitation spectroscopy. *J. Phys. Chem. B* **2003**, *107*, 7426–7433. [CrossRef]

30. Lee, Y.Y.; Ho, W.J.; Liu, J.J.; Lin, C.H. Light-trapping performance of silicon thin-film plasmonics solar cells based on indium nanoparticles and various TiO$_2$ space layer thicknesses. *Jpn. J. Appl. Phys.* **2014**, *53*, 06JE11. [CrossRef]

31. Ho, W.J.; Lin, J.C.; Liu, J.J.; Yeh, C.W.; Syu, H.J.; Lin, C.F. Plasmonic light scattering in textured silicon solar cells with indium nanoparticles from normal to non-normal light incidence. *Materials* **2017**, *10*, 737. [CrossRef] [PubMed]

32. Ho, W.J.; Su, S.Y.; Lee, Y.Y.; Syu, H.J.; Lin, C.F. Performance-enhanced textured silicon solar cells based on plasmonic light scattering using silver and indium nanoparticles. *Materials* **2015**, *25*, 6668–6676. [CrossRef] [PubMed]

33. Ho, W.J.; Lee, Y.Y.; Lin, C.H.; Yeh, C.W. Performance enhancement of plasmonics silicon solar cells using Al$_2$O$_3$/In NPs/TiO$_2$ antireflective surface coating. *Appl. Sur. Sci.* **2015**, *354*, 100–105. [CrossRef]

 *nanomaterials*

Article

High Performance Ultrathin MoO$_3$/Ag Transparent Electrode and Its Application in Semitransparent Organic Solar Cells

Linlin Shi [1], Yanxia Cui [1,*], Yupeng Gao [1], Wenyan Wang [1], Ye Zhang [1], Furong Zhu [1,2] and Yuying Hao [1]

[1] Key Lab of Advanced Transducers and Intelligent Control System of Ministry of Education, College of Physics and Optoelectronics, Taiyuan University of Technology, Taiyuan 030024, China; shilinlinsll@sina.cn (L.S.); gaoyupeng@boe.com.cn (Y.G.); wangwenyan@tyut.edu.cn (W.W.); zhangye@tyut.edu.cn (Y.Z.); frzhu@hkbu.edu.hk (F.Z.); haoyuyinghyy@sina.com (Y.H.)

[2] Department of Physics and Institute of Advanced Materials, Hong Kong Baptist University, Kowloon 22100, Hong Kong, China

* Correspondence: yanxiacui@gmail.com; Tel.: +86-351-317-6639

Received: 10 June 2018; Accepted: 24 June 2018; Published: 27 June 2018

Abstract: In this paper, we demonstrate high performance ultrathin silver (Ag) transparent electrodes with a thin MoO$_3$ nucleation layer based on the thermal evaporation method. The MoO$_3$/Ag transparent electrodes fabricated at different deposition rates were compared systematically on aspects of the transmission spectrum, surface resistance, and surface morphology. Our study indicates that with the presence of the MoO$_3$ nucleation layer, an Ag film of only 7 nm thick can achieve percolation and the film is porous instead of forming isolated islands. In addition, the increase of the deposition rate can yield obvious improvement of the surface morphology of the Ag film. Specifically, with the help of a 1 nm thick MoO$_3$ nucleation layer, the Ag film of 9 nm thick realized under the deposition rate of 0.7 nm/s has a surface resistance of about 20 ohm/sq and an average transmittance in the visible light range reaching 74.22%. Such a high performance of transmittance is superior to the reported results in the literature, which inevitably suffer obvious drop in the long wavelength range. Next, we applied the ultrathin MoO$_3$/Ag transparent electrode in organic solar cells. The optimized semitransparent organic solar cell displays a power conversion efficiency of 2.76% and an average transmittance in the visible range of 38% when light is incident from the Ag electrode side.

Keywords: transparent electrode; Ag film; nucleation layer; organic solar cell

---

1. Introduction

Organic solar cells (OSCs), showing advantages in terms of rich material resources, light weight, good compatibility with roll-to-roll and large area fabrications, colorfulness, etc., have become a very hot research topic in the field of solar energy utilization [1]. In OSCs, Indium Tin Oxide (ITO) is the most common transparent electrode. However, the ITO transparent electrodes suffer a series of limitations due to their poor mechanical flexibility, high cost, and the rareness of indium [2]. To replace ITO, other transparent conductive materials with high transmittance, low surface resistance, good flexibility, and low cost are being developed. Different materials have been proposed to replace ITO as transparent electrodes, e.g., highly conductive polymers (e.g., Poly(3,4-ethylenedioxythiophene): poly(styrenesulfonate), shorted as PEDOT:PSS) [3], metal nanostructures [4–12], graphene [13,14], carbon nanotubes [15], etc. Among these conductive materials, noble metals including silver (Ag) and gold (Au), which have high conductivity and are inert from oxidation, have been frequently used to make transparent electrodes. Compared to Au, Ag has relatively lower resistivity [16]; besides, Ag is more inexpensive than Au. Therefore, Ag based transparent electrodes have attracted intensive attention in the field of optoelectronics [16–18].

Ag nanowires show high transmission and electrical conductivity [17], but they can produce interference and scattering effects in the range of visible light. Besides, it is not easy to attach them onto the photosensitive layer, resulting in poor performance of the device and other issues [18,19]. In contrast, a single layer of Ag film which also shows good conductivity does not generate any negative optical effects to affect the device performance. However, an Ag film with a preferable surface resistance has a large thickness due to its poor wettability on insulating substrates such as silicon dioxide, thereby exhibiting low transmittance [20,21]. Doping the Ag film with a small amount of calcium or aluminum is an effective method to greatly reduce the percolation thickness to form a continuous film, resulting in high transmittance at the same time with a preferable surface resistance [22,23]. However, the doping process in the experiment can hardly be controlled in obtaining a fixed ratio of the alloy film, especially when the thermal evaporation method is used.

Researchers have also attempted to realize thin Ag film by introducing a nucleation layer which can facilitate Ag percolation. Metals, like nickel [24], germanium [25,26], chromium [27], etc., have been frequently used as the nucleation layer for Ag growth, but these high glossy metals bring about unavoidable optical loss and they also alter the plasmonic properties of the Ag films. Metal oxides are another great choice of the nucleation layer for Ag growth. Some poor transparent conductive oxides, e.g., ZnO [28,29], TiO_2 [30,31], SnO_x [32,33], Nb_2O_5 [34], and so on, have been researched frequently for this purpose. However, their depositions require the magnetic sputtering method or electron/ion-beam assisted techniques, causing possible damage of the organic films. In this regard, transition metal oxides like MoO_3 [35–37], WO_3 [38–40], and V_2O_5 [41] produced by the thermal evaporation method, have been proposed as a nucleating layer for Ag growth, compatible with the process of organic devices.

In this paper, we select MoO_3 for Ag percolation using the thermal evaporation method because MoO_3 is also a good hole transport material which can act as a bi-functional layer in the designed OSCs. In some reports [42–44], the MoO_3/Ag (hereafter, called MA) transparent electrodes were capped with an additional MoO_3 layer (hereafter, the triple layer structure is called MAM) for further reducing the reflection of light at the visible range based on the principle of admittance match [45,46]. However, this is not always necessary, depending on the working wavelength range of real applications. It has been noticed that different research groups have reported quite different device performances of MA based electrodes. In 2015, Lee et al. fabricated an MAM electrode with each layer 30 nm/11 nm/24 nm thick which had a peak transmittance of 75% at 500 nm while the transmittance rapidly dropped to ~50% at 760 nm [47]. In 2016, Travkin et al. reported another MAM electrode with each layer 10 nm/10 nm/12 nm thick, showing a transmittance of 90% at 380 nm; but the transmittance slump at long wavelength range was much more severe, with the transmittance decreasing to only ~35% at 760 nm [16]. In contrast, Xu et al. reported an MA composite film with each layer 10 nm/10 nm thick which displayed relatively good transmittance at long wavelength range (~60% at 760 nm) [48]. The deviation between these results could be ascribed to their different deposition rates of Ag. In this paper, the MA electrodes fabricated at different deposition rates were compared systematically on the aspects of the transmission spectrum, surface resistance, and surface morphology. Our study indicates that with the presence of a 1 nm thick MoO_3 nucleation layer, the Ag film of only 7 nm thick can achieve percolation and the film is porous instead of forming isolated islands. In other words, the 10 nm thick MoO_3 layer utilized in most of the literature is not compulsory for Ag percolation. It is concluded that the increase of the deposition rate can yield obvious improvement of the surface morphology of the Ag film. The MA electrode with a 9 nm thick Ag layer shows a surface resistance of about 20 ohm/sq, and its average transmittance in the visible light range between 400 nm and 760 nm reaches 74.22%. Particularly, our optimal MA electrode displays a negligible transmittance drop at long wavelength range with the transmittance maintained at 70% at 760 nm, superior to all reported results in the literature. Furthermore, we apply the optimal MA transparent electrodes in semitransparent OSCs.

2. Materials and Methods

2.1. Fabrication of the Transparent Electrode

Prior to beginning the fabrication process, the glasses were cleaned in an ultrasonic cleaner successively with deionized water, acetone, and isopropyl alcohol for 10 min in each round, and then they were dried in a vacuum oven at 100 °C for 10 min. Both MoO_3 and Ag were deposited using a thermal evaporator under a base pressure below 8×10^{-4} Pa, and the thicknesses of the films were monitored during deposition. The deposition rate of MoO_3 was fixed at 0.01 nm/s and the thickness of MoO_3 as 1 nm. The deposition rates of the Ag films were varied from 0.1 to 0.9 nm/s and the thickness of Ag film was also tuned to yield the optimal optical and electrical performance at the same time.

2.2. Fabrication of Semitransparent Organic Solar Cell Devices

The fabricated semitransparent OSC has a configuration of $ITO/ZnO/PTB7:PC_{70}BM/MoO_3/Ag$. The donor PTB7 (Poly{4,8-bis[(2-ethylhexyl)oxy]benzo[1,2-b:4,5-b′]dithiophene-2,6-diyl-alt-3-fluoro-2-[(2-ethylhexyl)carbonyl]thieno[3,4-b]thiophene-4,6-diyl}) and the acceptor $PC_{70}BM$ (phenyl-C_{70}-butyric-acid-methyl-ester) were purchased from 1-Material. The active layer solution was prepared by dissolving PTB7 and $PC_{70}BM$ in chlorobenzene with 3% DIO at concentrations of 10 mg/mL and 15 mg/mL, respectively. PTB7 and $PC_{70}BM$ were completely dissolved after the solution mixture was stirred vigorously at 60 °C for 12 h but the stirring needs to be maintained until use. ITO glasses were scrubbed with detergent first, and then cleaned by deionized water, acetone, ethanol, and isopropyl alcohol, successively, in an ultrasonic cleaner. After the clean ITO substrate was transferred into the glove box, the electron transporting layer of ZnO and the active layer of $PTB7:PC_{70}BM$ were made consecutively by the spin-coating method both with a speed of 1000 rpm. The realized thicknesses of ZnO and $PTB7:PC_{70}BM$ were around 30 and 100 nm, respectively. Later, the samples were placed in the chamber of the thermal evaporator. After the base pressure reached 8×10^{-4} Pa, the depositions of the hole transporting layer MoO_3 and the ultrathin Ag electrode were carried out successively. The deposition rate of MoO_3 was 0.01 nm/s, and that of Ag was 0.7 nm/s. Here, for good hole transporting property, the thickness of the MoO_3 layer was tuned to 2 nm while the optimal thickness of the Ag film (7 nm) obtained in the step of making transparent electrodes was adopted in the OSC devices. In this paper, the active area of each device is 0.04 cm^2.

2.3. Measurement

Transmittance properties of the Ag based transparent electrodes and the semitransparent OSC were measured by a commercially available angle-resolved reflection/transmission spectroscopy system (R1, IdeaOptics Instruments, Shanghai, China). The surface resistance of the fabricated electrodes was characterized by a four point probe system (Loresta AX MCP-T370, Mitsubishi Chemical Analytech, Kanagawa, Japan). Scanning electron microscopy (SEM) images of the surface of ultrathin Ag films were recorded using a thermally assisted field emission SEM system (LEO 1530, Carl Zeiss NTS GmbH, Oberkochen, Germany). Current density-voltage curve characteristics and External Quantum Efficiency (EQE) spectra of the device were also analyzed. The current density-voltage characteristics of OSCs under dark and AM 1.5 G solar illumination provided by a solar simulator (Sun 3000, ABET, Milford, CT, USA, calibrated using a standard silicon cell) were measured using the Source Meter of Keithley 2400 (Keithley, Solon, OH, USA). The wavelength dependent external quantum efficiencies of the OSCs were recorded using a CSC1011 (ZOLIX instruments, Beijing, China) which was equipped with a short arc xenon lamp source (UXL-553, Ushio, Cypress, CA, USA).

3. Results and Discussion

Before the fabrication and characterization of Ag electrodes in the form of MA, studies related to the Ag electrodes without any nucleation layer were carried out first for comparison. Figure 1a,b show the SEM images of the Ag thin films of 7 nm which were directly prepared on clean glass substrates at a deposition rate of 0.1 and 0.9 nm/s, respectively. It is known that without the nucleation layer, the growth of the Ag films on glass substrate begins with island-like particles following the Volmer-Weber mode [49,50], which can be clearly reflected from both SEM images in Figure 1a,b. No matter that the deposition rate of Ag is low or high, one can see from the SEM images that island-shaped particles are formed by directly depositing Ag on glass. By comparison, one also sees that the average grain size of Ag islands obtained under a high deposition rate of 0.9 nm/s is larger than that obtained with a low deposition rate of 0.1 nm/s, indicating that the high deposition rate is able to suppress the island growth and thereby facilitate percolation. In other words, the Ag growth forms a percolating layer more easily with a high deposition rate. This can be further verified through the characterization of transmittance properties of Ag films with varied thicknesses. Here, Figure 1c,d show the transmittance spectra of Ag films of 7 nm, 8 nm, 9 nm, 10 nm, 11 nm, and 15 nm thick with the deposition rate of 0.1 and 0.9 nm/s, respectively. It can be seen that in Figure 1c, for the low deposition rate case, the transmittance dip due to surface plasmon resonance of Ag nanoparticles always takes place when the Ag thickness is equal or below 11 nm, and the surface plasmon induced dip bears a red shift with the increase of the film thickness, ascribed to the increase of the average grain size. When the Ag thickness increases to 15 nm, the transmittance dip disappears, and a smooth spectrum is produced, reflecting that a continuous film which can conduct electrons (with a measured surface resistance of 34.02 ohm/sq) is formed. While in Figure 1d, for the high deposition rate case, the transmittance dip due to surface plasmon resonance only takes place when the Ag thickness is below 11 nm. The Ag films with thickness of both 11 and 15 nm do not show any transmittance dip in the visible range, indicating that the conductive film starts being formed at a thickness of 11 nm (with a measured surface resistance of 71.39 ohm/sq) at a deposition rate of 0.9 nm/s, which is 4 nm smaller than that at the deposition rate of 0.1 nm/s. The measured surface resistance of the 15 nm thick Ag films under 0.9 nm/s is also greatly reduced (10.26 ohm/sq) compared with the case of 0.1 nm/s deposition rate. Moreover, due to better percolation, the transmittance efficiencies for the continuous films in Figure 1d are higher than that in Figure 1c. However, even under a high deposition rate, the transmittances at the long wavelength range of the 11 nm thick Ag film are still quite low (~42% at 760 nm), similar to the reported results [16].

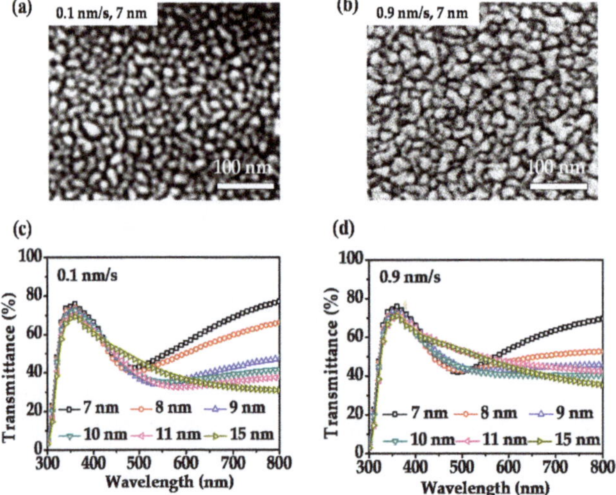

Figure 1. Scanning electron microscope (SEM) images of 7 nm thick Ag thin films directly prepared on glass substrate at the deposition rates of 0.1 nm/s (**a**) and 0.9 nm/s (**b**). Transmission spectra of Ag films with varied thicknesses prepared at the deposition rates of 0.1 nm/s (**c**) and 0.9 nm/s (**d**).

Next, the study on ultrathin Ag film growth was carried out with the MoO_3 nucleation layer. Here, different from most of the literature which used a MoO_3 layer with a thickness of 10 nm or so, we found that 1 nm thick MoO_3 is sufficiently good for functioning as the nucleation layer. Figure 2a,b show the SEM images of the Ag thin films of 7 nm which were prepared on a 1 nm thick MoO_3 coated glass substrates (i.e., the MA films) at a deposition rate of 0.1 and 0.9 nm/s, respectively. It is observed that under the deposition rate of either 0.1 or 0.9 nm/s, that the SEM images of 7 nm thick Ag samples with a nucleation layer as shown in Figure 2a,b do not own clear island boundaries as those observed in Figure 1a,b. This means that the growth of Ag no longer follows the Volmer-Weber mode. Instead, most parts of the Ag surface are connected with some isolated black domains (corresponding to pores of the film) witnessed in Figure 2a,b. This is because the transition metal oxide MoO_3 more easily provides oxygen bonds to Ag atoms than the insulator medium SiO_2 [51,52], resulting in better adhesion of the Ag atoms on the MoO_3 layer. Besides, the SEM images displayed in Figure 2a,b also deviate from each other. It was found that a 0.1 nm/s deposition rate tends to produce a continuous Ag film with some tiny bumps as reflected from the change of color in Figure 2a. In comparison, the deposition rate of 0.9 nm/s yields a quasi-flat surface; as seen there is less obvious color change taking place in Figure 2b than in Figure 2a. Such observation tells us that the process of high deposition rate lowers the selectivity of the Ag atoms toward the target substrate during growth.

We further characterized the transmittance spectra of MA films when the Ag layer is fixed at 7 nm and its deposition rate is tuned from 0.1 to 0.9 nm/s, as shown in Figure 2c. Later, the MA film with the Ag layer of X nm is called X nm MA film. It can be seen that under a deposition rate of 0.1 nm/s, there is a very tiny dip occurring in the transmittance spectrum, and at higher rates the transmittance spectra are all smooth, displaying no dips over the investigated wavelength range. The tiny dip for the 0.1 nm/s case could be ascribed to the surface plasmon resonance produced by the bump-like geometries on the Ag surface as observed from the SEM image in Figure 2a. The transmittance efficiencies over the whole range are more or less the same for the cases of 0.5 nm/s, 0.7 nm/s, and 0.9 nm/s, which are a bit higher than those obtained under the lower deposition rate of 0.3 nm/s. Compared with the results of the thinnest continuous Ag film obtained under 0.9 nm/s without any nucleation layer (that is of 11 nm thick) as shown in Figure 1d, it was found that the participation of the 1 nm thick MoO_3 layer enables a thinner Ag layer to become continuous (7 nm) and at the same time lifts its transmission in the visible range much higher (~68% at 760 nm). The high transmission of the MA film is because the Ag layer is porous, allowing more incoming light to transmit through compared with a perfect continuous film without any pores.

The surface resistance of the above five different MA films were also characterized by the four point probe system as shown in Figure 2d. It was found that the conductivity property of the 7 nm MA film was improved significantly with the increase of the deposition rate. The improvement of the surface resistance of 7 nm MA film from 0.1 to 0.3 nm/s achieves around 1000 ohm/sq and the improvement from 0.3 to 0.5 nm/s is also quite high (~260 ohm/sq). The 7 nm MA films prepared at a deposition rate no smaller than 0.3 nm/s show surface resistance below 100 ohm/sq, and the higher the deposition rate, the lower the surface resistance. It was found that the lowest surface resistance of a 7 nm MA film is 40 ohm/sq, realized at the Ag deposition rate of 0.9 nm/s.

A series of studies was then carried out for MA films with the thickness of the Ag layer varying from 7 to 15 nm. In Figure 2d, the surface resistances of all different samples are displayed. Similar to the situation taking place for 7 nm MA films, the increase of the deposition rate causes a decrease of the surface resistance for all other MA films except the 15 nm thick cases. Moreover, the decrease of surface resistance with the increase of deposition rate becomes less apparent for thicker MA films. In detail, surface resistances of 9 nm MA films do not show obvious differences between the cases of 0.7 and 0.9 nm/s and both surface resistances are around 20 ohm/sq. The 10 nm (or 11 nm) MA films perform almost the same on the aspect of conductivity when the deposition rate is no smaller than 0.3 nm/s and their surface resistances are about 17 ohm/sq (or 12 ohm/sq). It was also seen that the surface resistances between the 15 nm MA films fabricated at different rates are almost identical and their surface resistances are extremely small (around 10 ohm/sq). From Figure 2d, one can easily draw a conclusion that depositing the Ag layer with a rate above 0.5 nm/s is favorable for improving the electrical property of the transparent electrodes. Combined with the conclusion drawn from Figure 2c, we know that such a choice yields improvement of both the optical and electrical properties of the MA transparent electrodes.

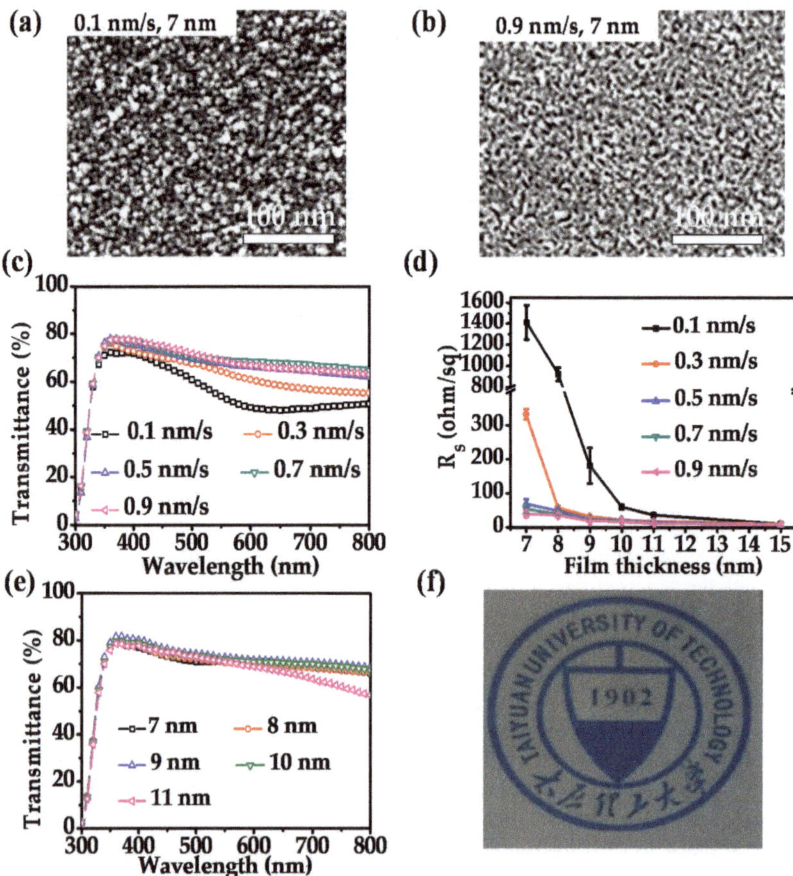

Figure 2. SEM images of MoO$_3$/Ag (MA) films prepared at the deposition rates of 0.1 nm/s (**a**) and 0.9 nm/s (**b**); (**c**) Transmission spectra of 7 nm Ag thin films deposited on MoO$_3$ nucleation layer at the rates of 0.1, 0.3, 0.5, 0.7, and 0.9 nm/s, respectively; (**d**) Surface resistances of MA films with varied Ag thicknesses which were deposited at different rates; (**e**) Transmission spectra of MA films with varied Ag thicknesses at the deposition rate of 0.7 nm/s; (**f**) Photograph of the 9 nm MA transparent electrode placed over an image.

We also characterized the transmittance spectra of all the MA electrodes studied in Figure 2d (not shown). For the thickness dependent transmittance spectra of the MA electrodes, we show the situation of 0.7 nm/s rate as an example in Figure 2e because it is seen from Figure 2c that this rate produces 7 nm MA films with more or less the same spectra with respect to the rates of 0.5 and 0.9 nm/s. It was found that the performance of an 11 nm MA film is inferior to the other four cases. Table 1 shows the calculated T_a and the measured R_s of the MA films with different thicknesses fabricated with 0.7 nm/s.

It is clearly seen that the 9 nm MA film exhibits the highest average transmittance of 74.22%; but its R_s is 19.68 ohm/sq, a bit higher than the 11 and 15 nm MA films. Figure 2f is the photograph of our 9 nm MA transparent electrode placed over a paper with the logo of our university, which exhibits quite good transparency. Though the 11 nm MA film has the lowest transmittance among the five samples, its surface resistance is the best (only 12.18 ohm/sq). In applications of optoelectronic devices, the exact thickness of the Ag layer of the MA electrode should be determined case by case because the device configuration, the working wavelength range, and the semiconductor materials are varied. The ultimate decision is made based on the device performances.

Table 1. The average transmittance T_a calculated from measured transmittance spectra in Figure 2e and the measured R_s of the MoO_3/Ag (MA) films with varied Ag thicknesses fabricated with 0.7 nm/s deposition rate.

Ag Thickness (nm)	7	8	9	10	11
T_a (%)	71.00	71.72	74.22	72.28	69.39
R_s (ohm/sq)	53.74 ± 9.16	38.71 ± 6.59	19.68 ± 1.77	17.48 ± 2.27	12.18 ± 0.74

Finally, we applied the MA transparent electrodes in semitransparent OSCs. The device configuration is $ITO/ZnO/PTB7:PC_{70}BM/MoO_3$/Ag with the thickness of the top Ag layer varying from 7 to 11 nm. The device structure is shown in Figure 3a, with X (nm) denoting the thickness of the thin Ag layer. An opaque OSC which has the same configuration as the semitransparent OSCs but with the top Ag layer of 100 nm thick was first demonstrated as the control. The fabricated control opaque OSC, with the current density versus voltage curve shown in Figure 3b, exhibits a performance with PCE of 7.32% and the corresponding open circuit voltage (V_{oc}), short circuit current (J_{sc}), fill factor (FF), series resistance (R_s), and shunt resistance (R_{sh}) are 0.73 V, 15.31 mA/cm^2, 66.00%, 3.22 $\Omega \cdot$cm^2, and 735.94 $\Omega \cdot$cm^2, respectively. Figure 3c displays the current density versus voltage curves for the MA electrode based semi-transparent OSCs with their device characteristics summarized in Table 2 as well. It was found that increasing the thickness of the Ag electrode does not cause apparent change of the open circuit voltage. However, the short circuit current varies greatly with the increase of the thickness of the Ag electrode. The device with a 9 nm thick silver layer has the highest J_{sc} of 7.47 mA/cm^2, that is about half of that of the opaque control. The 9 nm device which performs the best is mainly because this electrode has the highest average transmittance as shown in Table 1, which can also be seen from the transmittance spectra of different MA based OSCs as shown in Figure 3d. As can be seen, the semitransparent OSC with a 9 nm thick silver layer indeed shows the highest transmittances over the investigated wavelength range and its average transmittance T_a over the visible range was calculated to be 38%. Ascribed mainly to such a high transmittance and corresponding insufficient photon absorption, the semitransparent OSCs have a much lower PCE (2.76% with a 9 nm thick Ag electrode) with respect to the opaque OSC which bears zero transmission, in accordance with the measurement of the external quantum efficiency (EQE) spectra as shown in Figure 3e. It can also be clearly observed from Figure 3d that, among these five semitransparent devices, the one with an 11 nm thick Ag electrode performs the poorest on the aspect of transmittance, in accordance with the results in Table 1.

In addition, we found that compared with the control opaque OSC, the semitransparent OSCs do not show any obvious difference on shunt resistance. Figure 3f shows the comparison between curves of the current density versus voltage for one 9 nm MA electrode semitransparent OSC and the control. It was found that the reverse saturation current for the semitransparent OSC is a bit higher that of the control. Such a difference is acceptable as the standard deviation of shunt resistances are around 100 $\Omega \cdot$cm^2. The almost unchanged shunt resistance between control and semitransparent OSCs indicates that the MA electrode does not induce extra surface defects and thus yields more or less the same surface recombination as the 100 nm thick electrode. We also found that the series resistances of the semitransparent OSCs are much higher than that of the control. The increase of the series resistance leads to a relatively poor fill factor of the semitransparent OSC, which would definitely cause insufficient charge extraction efficiency and further reduce the PCE of semitransparent OSCs on the premise of insufficient photon absorption. The poor R_s of MA based OSCs is produced because the ultrathin silver electrode has a higher surface resistance compared with that of the 100 nm thick silver electrode.

Table 3 summarizes the performance parameters of MA (or MAM) based semitransparent OSCs reported in the literature and this work. By comparison, we concluded that the MA based semitransparent OSC demonstrated in this work is much superior to the two reported devices [42,43] which were also illuminated from the Ag side. Compared with the reported semitransparent OSC illuminated from the ITO side [44], the PCE of our semitransparent OSC is lower but our average transmittance is higher than twice that in the reference.

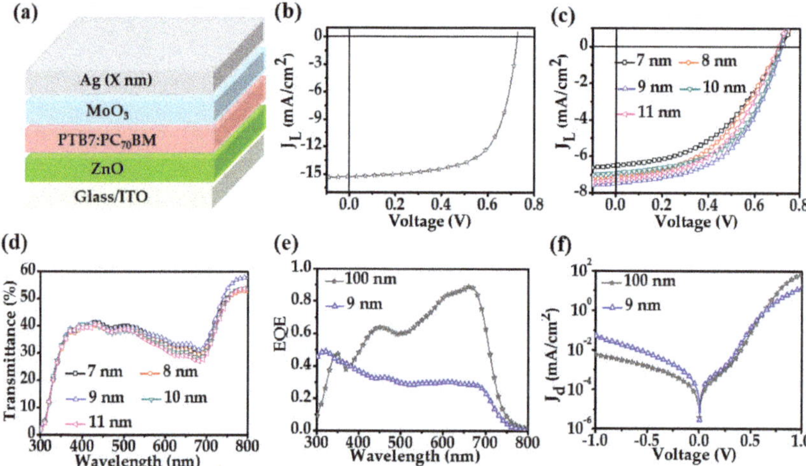

Figure 3. (**a**) Structure of the semitransparent organic solar cells (OSCs); (**b**) Curve of current density versus voltage for the control opaque OSC; (**c,d**) Curves of current density versus voltage and transmittance spectra for the MA electrode based semitransparent OSCs devices with varied Ag thicknesses when light is illuminated from the Ag electrode side; (**e,f**) External quantum efficiency (EQE) spectra and curves of current density versus voltage under dark for the one 9 nm MA electrode based semitransparent OSC and the control.

Table 2. Performance parameters of the MA electrode based semitransparent OSCs when the Ag thickness is varied under AM 1.5 G illumination at 100 mW/cm^2 from the Ag electrode side.

Device (X/nm)		J_{sc} (mA/cm^2)	V_{oc} (V)	FF (%)	PCE (%)	R_s ($\Omega \cdot$cm^2)	R_{sh} ($\Omega \cdot$cm^2)
7	Average	6.48 ± 0.08	0.71 ± 0.008	47.75 ± 1.89	2.19 ± 0.07	33.93 ± 5.77	925.87 ± 121.89
	Best	(6.53)	(0.71)	(48.00)	(2.23)	(32.59)	(835.11)
8	Average	6.96 ± 0.14	0.71 ± 0.006	44.33 ± 0.58	2.18 ± 0.07	39.54 ± 2.09	669.53 ± 68.20
	Best	(7.12)	(0.71)	(45.00)	(2.26)	(40.29)	(675.17)
9	Average	6.93 ± 0.40	0.72 ± 0.005	54.00 ± 2.45	2.71 ± 0.05	22.56 ± 3.07	999.04 ± 131.60
	Best	(7.47)	(0.72)	(51.00)	(2.76)	(23.42)	(873.40)
10	Average	6.71 ± 0.16	0.71 ± 0.010	53.17 ± 0.75	2.54 ± 0.09	22.40 ± 2.21	859.18 ± 102.82
	Best	(6.85)	(0.71)	(54.00)	(2.63)	(21.68)	(883.63)
11	Average	7.17 ± 0.17	0.71 ± 0.010	47.50 ± 2.38	2.40 ± 0.12	37.01 ± 6.54	722.22 ± 137.24
	Best	(7.11)	(0.72)	(49.00)	(2.51)	(42.8)	(883.82)

Table 3. Performance parameters of MA (or MAM) based semitransparent OSCs reported in the literature and this work. * means the illumination is from the ITO side in that work.

MA(M) Data		Performance of Semitransparent OSCs					Reference
MA(M) Thickness (nm)	*T* at 760 nm (%)	T_a (%)	J_{sc} (mA/cm^2)	V_{oc} (V)	FF (%)	PCE (%)	
1/10/20	65.0	NA	2.72	0.57	61.6	0.96	[42]
6/10/40	~40.0	NA	4.64	0.54	55.0	1.39	[43]
6/7/40	NA	18.3	14.7	0.68	46.3	4.7	[44] *
2/9	70.0	38.0	7.47	0.72	51.0	2.76	This work

4. Conclusions

In this work, we systematically studied the effect of the deposition rate of Ag on the performances of ultrathin MoO_3/Ag transparent electrodes based on the thermal evaporation method. It was found that the Ag film fabricated on top of the glass substrate shows minor improvement on film growth with the increase of deposition rate. At a deposition rate of 0.9 nm/s, the Ag film on glass substrates still have a percolation thickness of ~11 nm. In contrast, with the help of a 1 nm thick MoO_3 nucleation layer, under a deposition rate of 0.7 nm/s, the MA film with Ag layer of 7 nm can already conduct electrons, corresponding to a surface resistance of ~54 ohm/sq. It was demonstrated that the MA electrode with the Ag layer of 9 nm fabricated at 0.7 nm/s deposition rate owns the highest average transmittance of light over the visible range (~74.22%) compared to MA electrodes with a Ag layer of other thicknesses. The transmittance performance of our optimized MA electrode is superior to all reported results in the literature, which may be ascribed to the better control of the deposition rate of the Ag film realized through the tantalum boat in our experiments. This transmittance optimized sample has a surface resistance of around 20 ohm/sq. For MA electrodes with a silver layer thicker than 9 nm, the transmittance becomes slightly lower but the surface resistance can be improved. In applications of optoelectronic devices, the exact thickness of the Ag layer of the MA electrode should be determined case by case and the ultimate decision should be made based on the device performance. For example, we applied the fabricated ultrathin MoO_3/Ag transparent electrodes in semitransparent organic solar devices. In our device, MoO_3 acts as not only the hole transport layer but also the nucleation layer beneficial to the formation of the Ag thin film. Through optimization, we obtained a semitransparent OSC with a PCE of 2.76% and an average transmission in the visible range of 38% when light is incident from the Ag electrode side. The power conversion efficiency of the semitransparent OSCs can be further optimized by introducing a MoO_3 capping layer on top of the Ag electrode (in other words, using MAM electrodes) by tuning the transmittance spectrum of the MA electrode matching well with the absorption of the active layer. However, this manipulation would inevitably bring about a reduction in the transmittance of light through the device, therefore compromises between the optical and electrical performances should be made if MAM is utilized.

Author Contributions: Conceptualization and Supervision, Y.C.; Methodology, L.S., Y.G., W.W. and Y.Z.; Writing—Original Draft Preparation, L.S. and Y.G.; Writing—Review and Editing, Y.H. and F.Z.; and Funding Acquisition, Y.C. and Y.H.

Funding: This research was funded by National Natural Science Foundation of China (61775156, 61475109 and 61605136), Young Talents Program of Shanxi Province, Young Sanjin Scholars Program, the Excellent Young Scientists Fund of the Natural Science Foundation of Shanxi Province (201701D211002), Henry Fok Education Foundation Young Teachers fund, Key Research and Development (International Cooperation) Program of Shanxi Province (201603D421042), and Platform and Base Special Project of Shanxi Province (201605D131038).

Conflicts of Interest: The authors declare no conflict of interest.

References

1. Sangar, A.; Merlen, A.; Torchio, P.; Vedraine, S.; Flory, F.; Escoubas, L.; Patrone, L.; Delafosse, G.; Chevallier, V.; Moyen, E.; et al. Fabrication and characterization of large metallic nanodots arrays for organic thin film solar cells using anodic aluminum oxide templates. *Sol. Energy Mater. Sol. Cells* **2013**, *117*, 657–662. [CrossRef]
2. Luo, G.; Cheng, X.; He, Z.; Wu, H.; Cao, Y. High-performance inverted polymer solar cells based on thin copper film. *J. Photonics Energy* **2014**, *5*, 057206. [CrossRef]
3. Xia, Y.J.; Sun, K.; Ouyang, J.Y. Solution-processed metallic conducting polymer films as transparent electrode of optoelectronic devices. *Adv. Mater.* **2012**, *24*, 2436–2440. [CrossRef] [PubMed]
4. Sergeant, N.P.; Hadipour, A.; Niesen, B.; Cheyns, D.; Heremans, P.; Peumans, P.; Rand, B.P. Design of Transparent Anodes for Resonant Cavity Enhanced Light Harvesting in Organic Solar Cells. *Adv. Mater.* **2012**, *24*, 728–732. [CrossRef] [PubMed]
5. Salinas, J.F.; Yip, H.L.; Chueh, C.C.; Li, C.Z.; Maldonado, J.L.; Jen, A.K. Optical Design of Transparent Thin Metal Electrodes to Enhance In-Coupling and Trapping of Light in Flexible Polymer Solar Cells. *Adv. Mater.* **2012**, *24*, 6362–6367. [CrossRef] [PubMed]
6. Ghosh, D.S.; Chen, T.L.; Formica, N.; Hwang, J.; Bruder, I.; Pruneri, V. High figure-of-merit Ag/Al:ZnO nano-thick transparent electrodes for indium-free flexible photovoltaics. *Sol. Energy Mater. Sol. Cells* **2012**, *107*, 338–343. [CrossRef]
7. Cao, W.; Zheng, Y.; Li, Z.; Wrzesniewski, E.; Hammond, W.T.; Xue, J. Flexible organic solar cells using an oxide/metal/oxide trilayer as transparent electrode. *Org. Electron.* **2012**, *13*, 2221–2228. [CrossRef]
8. Yang, K.; Rahman, M.A.; Jeong, K.; Nam, H.; Kim, J.; Lee, J. Semitransparent, thin metal grid-based hybrid electrodes for polymer solar cells. *Mater. Sci. Semicond. Process.* **2014**, *23*, 104–109. [CrossRef]
9. Morgenstern, F.S.F.; Kabra, D.; Massip, S.; Brenner, T.J.K.; Lyons, P.E.; Coleman, J.N.; Friend, R.H. Ag-nanowire films coated with ZnO nanoparticles as a transparent electrode for solar cells. *Appl. Phys. Lett.* **2011**, *4*, 183307. [CrossRef]
10. Kim, A.; Won, Y.; Woo, K.; Kim, C.; Moon, J. Highly Transparent Low Resistance ZnO/Ag Nanowire/ZnO Composite Electrode for Thin Film Solar Cells. *ACS Nano* **2013**, *7*, 1081–1091. [CrossRef] [PubMed]
11. Krantz, J.; Stubhan, T.; Richter, M.; Spallek, S.; Litzov, I.; Matt, G.J.; Spiecker, E.; Brabec, C.J. Spray-Coated Ag Nanowires as Top Electrode Layer in Semitransparent P3HT:PCBM-Based Organic Solar Cell Devices. *Adv. Funct. Mater.* **2013**, *23*, 1711–1717. [CrossRef]
12. Lee, J.Y.; Connor, S.T.; Cui, Y.; Peumans, P. Solution-processed metal nanowire mesh transparent electrodes. *Nano Lett.* **2008**, *8*, 689–692. [CrossRef] [PubMed]
13. Zhu, Y.; Sun, Z.Z.; Yan, Z.; Jin, Z.; Tour, J.M. Rational design of hybrid graphene films for high-performance transparent electrodes. *ACS Nano* **2011**, *5*, 6472–6479. [CrossRef] [PubMed]
14. De Arco, L.G.; Zhang, Y.; Schlenker, C.W.; Ryu, K.; Thompson, M.E.; Zhou, C.W. Continuous, Highly Flexible, and Transparent Graphene Films by Chemical Vapor Deposition for Organic Photovoltaics. *ACS Nano* **2010**, *4*, 2865–2873. [CrossRef] [PubMed]
15. Xiao, G.; Tao, Y.; Lu, J.; Zhang, Z. Highly conductive and transparent carbon nanotube composite thin films deposited on polyethylene terephthalate solution dipping. *Thin Solid Films* **2010**, *518*, 2822–2824. [CrossRef]
16. Travkin, V.V.; Luk'yanov, A.Y.; Drozdov, M.N.; Vopilkin, E.A.; Yunin, P.A.; Pakhomov, G.L. Ultrathin metallic interlayers in vacuum deposited MoO_x/metal/MoO_x electrodes for organic solar cells. *Appl. Surf. Sci.* **2016**, *390*, 703–709. [CrossRef]
17. Cattin, L.; Morsli, M.; Dahou, F.; Yapi Abe, S.; Khelilc, A.; Bernède, J.C. Investigation of low resistance transparent MoO_3/Ag/MoO_3 multilayer and application as anode in organic solar cells. *Thin Solid Films* **2010**, *518*, 4560–4563. [CrossRef]
18. Leem, D.; Eawards, A.; Faist, M.; Nelson, J.; Bradley, D.D.C.; de Mello, J.C. Efficient Organic Solar Cells with Solution-Processed Ag Nanowire Electrodes. *Adv. Mater.* **2011**, *23*, 4371–4375. [CrossRef] [PubMed]
19. Espinosa, N.; Søndergaard, R.; Jørgensen, M.; Krebs, F.C. Flow Synthesis of Ag Nanowires for Semitransparent Solar Cell Electrodes: A Life Cycle Perspective. *ChemSusChem* **2016**, *9*, 893–899. [CrossRef] [PubMed]
20. Pastorelli, F.; Romero-Gomez, P.; Betancur, R.; Martinez-Otero, A.; Mantilla-Perez, P.; Bonod, N.; Martorell, J. Enhanced Light Harvesting in Semitransparent Organic Solar Cells using an Optical Metal Cavity Configuration. *Adv. Energy Mater.* **2015**, *5*, 1400614. [CrossRef]

21. Hösel, M.; Angmo, D.; Søndergaard, R.R.; dos Reis Benatto, G.A.; Carlé, J.E.; Jørgensen, M.; Krebs, F.C. High-Volume Processed, ITO-Free Superstrates and Substrates for Roll-to-Roll Development of Organic Electronics. *Adv. Sci.* **2014**, *1*, 50–55. [CrossRef] [PubMed]

22. Xu, L.H.; Ou, Q.D.; Li, Y.Q.; Zhang, Y.B.; Zhao, X.D.; Xiang, H.Y.; Chen, J.D.; Zhou, L.; Lee, S.T.; Tang, J.X. Microcavity-Free Broadband Light Outcoupling Enhancement in Flexible Organic Light-Emitting Diodes with Nanostructured Transparent Metal-Dielectric Composite Electrodes. *ACS Nano* **2016**, *10*, 1625–1632. [CrossRef] [PubMed]

23. Gu, D.; Zhang, C.; Wu, Y.K.; Guo, L.J. Ultrasmooth and Thermally Stable Ag-Based Thin Films with Subnanometer Roughness by Aluminum Doping. *ACS Nano* **2014**, *8*, 10343–10351. [CrossRef] [PubMed]

24. Liu, H.; Wang, B.; Leong, E.S.P.; Yang, P.; Zong, Y.; Si, G.; Teng, J.; Maier, S.A. Enhanced Surface Plasmon Resonance on a Smooth Ag Film with a Seed Growth Layer. *ACS Nano* **2010**, *4*, 3139–3146. [CrossRef] [PubMed]

25. Logeeswaran, V.J.; Kobayashi, N.P.; Islam, M.S.; Wu, W.; Chaturvedi, P.; Fang, N.X.; Wang, S.Y.; Williams, R.S. Ultrasmooth Ag Thin Films Deposited with a Germanium Nucleation Layer. *Nano Lett.* **2009**, *9*, 178–182. [CrossRef] [PubMed]

26. Zhang, J.; Fryauf, D.M.; Garrett, M.; Logeeswaran, V.J.; Sawabe, A.; Islam, M.S.; Kobayashi, N.P. Phenomenological Model of the Growth of Ultrasmooth Ag Thin Films Deposited with a Germanium Nucleation Layer. *Langmuir* **2015**, *31*, 7852–7859. [CrossRef] [PubMed]

27. Cioarec, C.; Melpignano, P.; Gherardi, N.; Clergereaux, R.; Villeneuve, C. Ultrasmooth Ag thin film electrodes with high polar liquid wettability for OLED microcavity application. *Langmuir* **2011**, *27*, 3611–3617. [CrossRef] [PubMed]

28. Sahu, D.R.; Lin, S.-Y.; Huang, J.-L. ZnO/Ag/ZnO multilayer films for the application of a very low resistance transparent electrode. *Appl. Surf. Sci.* **2006**, *252*, 7509–7514. [CrossRef]

29. Zhao, G.Q.; Kim, S.M.; Lee, S.-G.; Bae, T.-S.; Mun, C.W.; Lee, S.H.; Yu, H.S.; Lee, G.-H.; Lee, H.-S.; Song, M.; Yun, J. Bendable Solar Cells from Stable, Flexible, and Transparent Conducting Electrodes Fabricated Using a Nitrogen-Doped Ultrathin Copper Film. *Adv. Funct. Mater.* **2016**, *26*, 4180–4191. [CrossRef]

30. Ito, S.; Takeuchi, T.; Katayama, T.; Sugiyama, M.; Matsuda, M.; Kitamura, T.; Wada, Y.; Yanagida, S. Conductive and Transparent Multilayer Films for Low-Temperature-Sintered Mesoporous TiO_2 Electrodes of Dye-Sensitized Solar Cells. *Chem. Mater.* **2003**, *15*, 2824–2828. [CrossRef]

31. Bauch, M.; Dimopoulos, T. Design of ultrathin metal-based transparent electrodes including the impact of interface roughness. *Mater. Des.* **2016**, *104*, 37–42. [CrossRef]

32. Behrendt, A.; Friedenberger, C.; Gahlmann, T.; Trost, S.; Becker, T.; Zilberberg, K.; Polywka, A.; Gorrn, P.; Riedl, T. Highly robust transparent and conductive gas diffusion barriers based on tin oxide. *Adv. Mater.* **2015**, *27*, 5961–5967. [CrossRef] [PubMed]

33. Yu, S.H.; Jia, C.H.; Zheng, H.W.; Ding, L.H.; Zhang, W.F. High quality transparent conductive $SnO_2/Ag/SnO_2$ tri-layer films deposited at room temperature by magnetron sputtering. *Mater. Lett.* **2012**, *85*, 68–70. [CrossRef]

34. Fahland, M.; Vogt, T.; Schoenberger, W.; Schiller, N. Optical properties of metal based transparent electrodes on polymer films. *Thin Solid Films* **2008**, *516*, 5777–5780. [CrossRef]

35. Della, G.E.; Peng, Y.; Hou, Q.; Spiccia, L.; Bach, U.; Jasieniak, J.J.; Cheng, Y.-B. Ultra-thin high efficiency semitransparent perovskite solar cells. *Nano Energy* **2015**, *13*, 249–257. [CrossRef]

36. Yang, Y.; Chen, Q.; Hsieh, Y.-T.; Song, T.-B.; Marco, N.D.; Zhou, H.; Yang, Y. Multilayer Transparent Top Electrode for Solution Processed Perovskite/Cu(InGa)(SeS)$_2$ Four Terminal Tandem Solar Cells. *ACS Nano* **2015**, *9*, 7714–7721. [CrossRef] [PubMed]

37. Jin, H.; Tao, C.; Velusamy, M.; Aljada, M.; Zhang, Y.; Hambsch, M.; Burn, P.L.; Meredith, P. Efficient, Large Area ITO-and-PEDOT-free Organic Solar Cell Sub-modules. *Adv. Mater.* **2012**, *24*, 2572–2577. [CrossRef] [PubMed]

38. Jeon, K.; Youn, H.; Kim, S.; Shin, S.; Yang, M. Fabrication and characterization of $WO_3/Ag/WO_3$ multilayer transparent anode with solution-processed WO_3 for polymer light-emitting diodes. *Nanoscale Res. Lett.* **2012**, *7*, 253. [CrossRef] [PubMed]

39. Li, H.; Lv, Y.; Zhang, X.; Wang, X.; Liu, X. High-performance ITO-free electrochromic films based on bi-functional stacked $WO_3/Ag/WO_3$ structures. *Sol. Energy Mater. Sol. Cells* **2015**, *136*, 86–91. [CrossRef]

40. Hong, K.; Kim, K.; Kim, S.; Lee, I.; Cho, H.; Yoo, S.; Choi, H.W.; Lee, N.-Y.; Tak, Y.-H.; Lee, J.-L. Optical Properties of WO_3/Ag/WO_3 Multilayer As Transparent Cathode in Top-Emitting Organic Light Emitting Diodes. *J. Phys Chem. C* **2011**, *115*, 3453–3459. [CrossRef]

41. Lee, K.-T.; Lee, J.Y.; Seo, S.; Guo, L.J. Colored ultrathin hybrid photovoltaics with high quantum efficiency. *Light-Sci. Appl.* **2014**, *3*, e215. [CrossRef]

42. Tao, C.; Xie, G.; Liu, C.; Zhang, X.; Dong, W.; Meng, F.; Kong, X.; Shen, L.; Ruan, S.; Chen, W. Semitransparent inverted polymer solar cells with MoO_3/Ag/MoO_3 as transparent electrode. *Appl. Phys. Lett.* **2009**, *95*, 53303. [CrossRef]

43. Cho, J.M.; Lee, S.K.; Moon, S.-J.; Jo, J.; Shin, W.S. MoO_3/Ag/MoO_3 top anode structure for semitransparent inverted organic solar cells. *Curr. Appl. Phys.* **2014**, *14*, 1144–1148. [CrossRef]

44. Upama, M.B.; Wright, M.; Elumalai, N.K.; Mahmud, M.A.; Wang, D.; Chan, K.H.; Xu, C.; Haque, F.; Uddin, A. High performance semitransparent organic solar cells with 5% PCE using non-patterned MoO_3/Ag/MoO_3 anode. *Curr. Appl. Phys.* **2017**, *17*, 298–305. [CrossRef]

45. Köstlin, H.; Frank, G. Optimization of transparent heat mirrors based on a thin silver film between antireflection films. *Thin Solid Films* **1982**, *89*, 287–293. [CrossRef]

46. Lee, C.-C.; Chen, S.-H.; Jaing, C.-C. Optical monitoring of silver-based transparent heat mirrors. *Appl. Opt.* **1996**, *35*, 5698–5703. [CrossRef] [PubMed]

47. Lee, S.; Kang, T.E.; Han, D.; Kim, H.; Kim, B.J.; Lee, J.; Yoo, S. Polymer/small-molecule parallel tandem organic solar cells based on MoO_x–Ag–MoO_x intermediate electrodes. *Sol. Energy Mater. Sol. Cells* **2015**, *137*, 34–43. [CrossRef]

48. Xu, W.-F.; Chin, C.-C.; Hung, D.-W.; Wei, P.-K. Transparent electrode for organic solar cells using multilayer structures with nanoporous Ag film. *Sol. Energy Mater. Sol. Cells* **2013**, *118*, 81–89. [CrossRef]

49. Barrows, A.T.; Masters, R.; Pearson, A.J.; Rodenburg, C.; Lidzey, D.G. Indium-free multilayer semi-transparent electrodes for polymer solar cells. *Sol. Energy Mater. Sol.* **2016**, *144*, 600–607. [CrossRef]

50. Lee, I.; Lee, J.-L. Transparent electrode of nanoscale metal film for optoelectronic devices. *J. Photonics Energy* **2015**, *5*, 057609. [CrossRef]

51. Galhenage, R.P.; Yan, H.; Tenney, S.A.; Park, N.; Henkelman, G.; Albrecht, P.; Mullins, D.R.; Chen, D.A. Understanding the Nucleation and Growth of Metals on TiO_2: Co Compared to Au, Ni, and Pt. *J. Phys. Chem. C* **2013**, *117*, 7191–7201. [CrossRef]

52. Diebold, U.; Pan, J.-M.; Madey, T.E. Ultrathin metal film growth on TiO_2(110): An overview. *Surf. Sci.* **1995**, *331–333*, 845–854. [CrossRef]

 *nanomaterials*

Article

Significant Carrier Extraction Enhancement at the Interface of an InN/p-GaN Heterojunction under Reverse Bias Voltage

Vladimir Svrcek [1,*], Marek Kolenda [2], Arunas Kadys [2], Ignas Reklaitis [2], Darius Dobrovolskas [2], Tadas Malinauskas [2], Mickael Lozach [1], Davide Mariotti [3], Martin Strassburg [4] and Roland Tomašiūnas [2]

[1] Research Center for Photovoltaics, National Institute of Advanced Industrial Science and Technology (AIST), Tsukuba 305-8568, Japan; mickael.lozach@aist.go.jp
[2] Institute of Photonics and Nanotechnology, Faculty of Physics, Vilnius University, Sauletekio 3, 10257 Vilnius, Lithuania; mkolendaus@gmail.com (M.K.); arunas.kadys@ff.vu.lt (A.K.); ignas.reklaitis@gmail.com (I.R.); darius.dobrovolskas@gmail.com (D.D.); tadas.malinauskas@ff.vu.lt (T.M.); rolandas.tomasiunas@ff.vu.lt (R.T.)
[3] Nanotechnology & Integrated Bio-Engineering Centre (NIBEC), Ulster University, Shore Road, Newtownabbey BT37 0QB, UK; d.mariotti@ulster.ac.uk
[4] OSRAM Opto Semiconductors GmbH, Leibnizstr. 4, D-93055 Regensburg, Germany; Martin.Strassburg@osram-os.com
* Correspondence: vladimir.svrcek@aist.go.jp; Tel.: +81-29-861-5429

Received: 23 October 2018; Accepted: 10 December 2018; Published: 12 December 2018

Abstract: In this paper, a superior-quality InN/p-GaN interface grown using pulsed metalorganic vapor-phase epitaxy (MOVPE) is demonstrated. The InN/p-GaN heterojunction interface based on high-quality InN (electron concentration 5.19×10^{18} cm^{-3} and mobility 980 cm^2/(V s)) showed good rectifying behavior. The heterojunction depletion region width was estimated to be 22.8 nm and showed the ability for charge carrier extraction without external electrical field (unbiased). Under reverse bias, the external quantum efficiency (EQE) in the blue spectral region (300–550 nm) can be enhanced significantly and exceeds unity. Avalanche and carrier multiplication phenomena were used to interpret the exclusive photoelectric features of the InN/p-GaN heterojunction behavior.

Keywords: InN/p-GaN heterojunction; interface; photovoltaics

---

1. Introduction

Single-nitride solar cells possess key and important features that can facilitate their large-scale application, whereby the nontoxicity of the primary elements offers opportunities towards future "green" technologies. Of the nitride family, an attractive candidate is indium nitride (InN), which presents a narrow-energy band gap (E_g) of 0.7 eV at room temperature, useful for the near-infrared region (NIR). Indium is also a key element in indium gallium nitride (InGaN) alloys, providing a bandgap for a broad spectral response from 0.7 eV to 3.4 eV without deteriorating its high carrier mobility [1,2]. The constituents of InN present a large difference in mass. Namely, the light nitrogen anions beside the heavy indium cations result in a phonon dispersion between high-lying optical phonon energies and low-lying acoustic phonon energies. Such a large gap [3] can effectively block Klemens decay [4,5]. Consequently, InN has been recognized as one of the most suitable materials with high potential for the realization of so-called hot carrier absorbers related to third-generation solar cell concept [6]. In principle, hot carrier absorbers are expected to reduce significant thermal losses by directly extracting the hot carriers [6,7]. Theoretically, their population is maintained inside the absorber by inhibiting the inherently ultrafast cooling process [6].

However, the realization of high-quality epitaxial InN layers remains a challenge. High-quality growth means very low dislocation density, and reduction of the unintended n-type doping due to indium or nitrogen vacancies and oxygen-related defects [8]. To date, the best results were obtained through epitaxial growth of InN on p-GaN by molecular beam epitaxy (MBE) [9]. From the point of view of optoelectronic device production, metalorganic vapor-phase epitaxy (MOVPE) is more suitable [10–12]. Nevertheless, MOVPE is restricted by the low decomposition rate of NH_3, resulting in InN island growth on GaN, which deteriorates the quality of the InN films [13]. These nucleation islands also strongly affect the junction interface morphology by producing dislocation defects and causing inefficient photo-response of the devices (e.g., detectors, solar cells) [12].

In addition, the growth of high-quality GaN is favorable for fabrication of GaN avalanche photodiodes for optical detection in the ultraviolet spectral region, due to a low operation voltage and the possibility of Geiger-mode operation [14,15]. The combination of GaN with InN can expand the detection spectral range and enhance the avalanche properties. Recently, illuminated and biased GaN p-i-n avalanche photodiodes have been widely investigated to achieve low multiplication noise and high gain [16–19]. Under reverse bias, the extended depletion region generates a strong electric field, which effectively separates the electron–hole pairs. Although GaN avalanche photodiode properties have been widely reported [14–19], to our knowledge an InN/p-GaN junction has never been investigated in an avalanche regime. It should be stressed that only electron–hole pairs that are generated at the interface or very close to a high-quality InN/p-GaN junction can contribute to the external current.

The large lattice mismatch between GaN and InN hinders the growth of high-quality interfaces and leads to the formation of a diffusive interface with poor photoluminescence (PL)/electroluminescence (EL) emission and carrier photogeneration efficiency [20–23]. A diffusive interface is often attributed to the migration of indium and gallium atoms through the spinodal decomposition process resulting in the formation of clusters and localization of charge carriers. [24,25]. The intrinsic phonon properties of high-quality InN are thus of relevance to self-heating and phonon engineering. The phonon decay into two or more phonons is subject to energy and wave vector conservation, while basically two phonon decay phenomena are referred to as Klemens and Ridley phonon decay mechanisms [4,26]. The most probable decay channels are those for which the vibrations created have a high density of states, whereby the prevention of the Klemens optical phonon decay mechanism occurs at the mini-Brillouin-zone boundaries of nanostructured interfaces. The important feature in this context is a poor transmission of phonons across the interface. The development of crystal growth technology, resulting in persistent improvements in the crystal and interface quality, thus permits the examination of intrinsic properties and refinement of the important phonon decay mechanisms. In this context, the InN/GaN interface quality is a key factor for the development of the new generation of InN-based optoelectronic devices.

In this contribution, we demonstrate high-quality InN films with low defect concentration and limited-roughness interface with magnesium-doped p-GaN, both grown by pulsed MOVPE. The formed InN/p-GaN heterojunction shows good rectifying behavior with InN characteristic electron concentration, among the best values obtained by pulsed MOVPE. An efficient photoelectric InN/p-GaN heterojunction is demonstrated by measuring carrier extraction under unbiased and reverse biased conditions.

2. Materials and Methods

The InN/p-GaN heterojunction investigated in this study was grown on a 5-µm-thick undoped GaN buffer layer and sapphire substrate by MOVPE (AIXTRON 3 × 2 CCS Flip Top reactor, AIXTRON SE, Herzogenrath, Germany). Trimethylgallium (TMGa), trimethylindium (TMIn) and ammonia (NH_3) have been used as precursors for gallium (Ga), indium (In), and nitrogen (N), respectively.

Bis(cyclopentadienyl) magnesium (Cp$_2$Mg) was used as a precursor for Mg p-type doping. The GaN templates on sapphire were grown using a standard low-temperature GaN buffer layer followed by high-temperature GaN growth (i.e., two-step growth), which were the necessary processes to achieve a high-quality GaN film with low dislocation density and high carrier mobility (600 cm^2/(V s) at 300 K) [27]. Then, a 600 nm p-GaN layer was grown on the un-doped 5-μm-thick GaN layer followed by 20 min annealing at 850 °C under a nitrogen (N$_2$) ambient atmosphere to activate Mg acceptor states in the p-GaN. Finally, a 300-nm-thick InN layer was grown on the p-GaN under 400 mbar pressure using TMIn precursor using a multiple flow-interruption technique, that is, pulsed. One growth cycle duration was 27 s, where TMIn and NH$_3$ were supplied into the reactor for the first 7 s and then only NH$_3$ was supplied for the next 20 s. The overall number of growth cycles was 600. The growth process of the InN layers was split into two steps. InN began to grow at 570 °C for the first few dozen growth cycles; then the growth temperature was increased at a constant rate until 610 °C (first step). For the remaining growth cycles, the temperature was kept constant at 610 °C (second step). The V/III ratio was stabilized at 99/68 during the entire InN growth process.

Scanning electron microscopy (SEM; Apollo 300, CamScan, Cambridge, UK (now successor Applied Beams, LLC, Beaverton, OR, USA)) was used for evaluating the heterojunction cross-section. Electron beam induced current (EBIC; Digiscan II priedu, model 778, Gatan, Inc., Pleasanton, CA, USA) was used to determine the electrically active areas of the heterojunction cross-section.

In order to determine the crystallinity of the structure, X-ray diffraction (XRD) measurements were carried out using a Rigaku SmartLab X-ray diffractometer (Rigaku, Tokyo, Japan).

The optical properties were investigated by photoluminescence (PL) at room temperature using a continuous wave (cw) He-Ne laser (633 nm) as an excitation source. The PL emission was collected into a 0.3 m spectrometer (Andor, Shamrock 303, Oxford Instruments, Abingdon, UK) and detected by an InGaAs detector array (Andor, iDus DU491A-2.2, Oxford Instruments, Abingdon, UK. The Hall measurements were performed on the Van der Pauw Ecopia HMS-3000 Hall Measurement System (Bridge Technology, Chandler Heights, AZ, USA).

AM 1.5G standard solar spectra were simulated using Wacom Electric Co. solar simulator (JIS, IEC standard conforming, CLASS AAA, Tokyo, Japan) calibrated to give 100 mW/cm^2 using two reference solar cells: a-Si and c-Si. The electrical data were recorded using a Keithley 2400 source meter (Tektronics, IL USA). The external quantum efficiency (EQE) characteristics were measured by CEP-25BXS (Bunkoh-Keiki Co., Ltd., Tokyo, Japan) in an extended spectral region of 300–2000 nm.

3. Results and Discussion

The X-ray diffraction (XRD) 2-theta scan measurements of the InN/p-GaN structure (Figure 1a) revealed a pronounced peak at 31.46° corresponding to the diffraction of hexagonal InN (0002); compared to the relaxed InN (0002) [27–29] (vertical line at 31.33° in Figure 1a), our measurements show our InN layer slightly strained. The peak at 34.5° corresponds to hexagonal GaN (0002). Improved crystallinity is due to the pulsed nature of the growth process, which enabled higher temperature growth. This was confirmed by the rocking curve for the InN (0002) reflection (Figure 1b) with a full width at half maximum (FWHM) of ~0.284° (1022 arcs). This value is by far better than those reported in the literature for MOVPE growth: 1800 arcs (at 550 °C) [30], 5601 arcs (at 550 °C) [14], 1300 arcs [31] and 0.27° (at 550 °C) obtained for c-oriented prismatic InN nanowalls grown on c-GaN/sapphire [32]. The dislocation density estimated from the rocking curve FWHM [33] showed a total (screw and edge) value of ~6 × 10^{10} cm^{-3}.

Figure 2a shows the structure of the investigated p-GaN/InN heterojunction and Figure 2b reports the SEM image of the junction edge. An electron-beam-induced current (EBIC) image is used to show the electrically active areas of the junction, that is, the depletion region with a built-in electric field. The non-equilibrium charge carriers (electrons, holes) in the depletion region, due to the built-in electric field, are immediately separated providing electrical current in the external circuit, expressed via bright spots on the image. Areas without an electric field remain dark (no electrical current), indicating fast recombination of the electron–hole pair as a consequence. In Figure 2c, we superimposed the EBIC on the SEM image to underline the precise location of the electrically active area, in our case, the junction between InN and p-GaN. The depletion region is quite shallow (<100 nm) and it follows the surface topology.

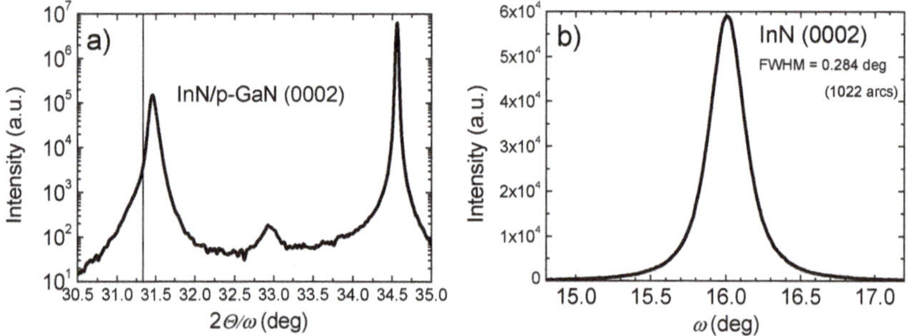

Figure 1. XRD measurement results of the InN/p-GaN structure at (0002) reflection: the 2-theta scan (**a**) and the rocking curve (**b**).

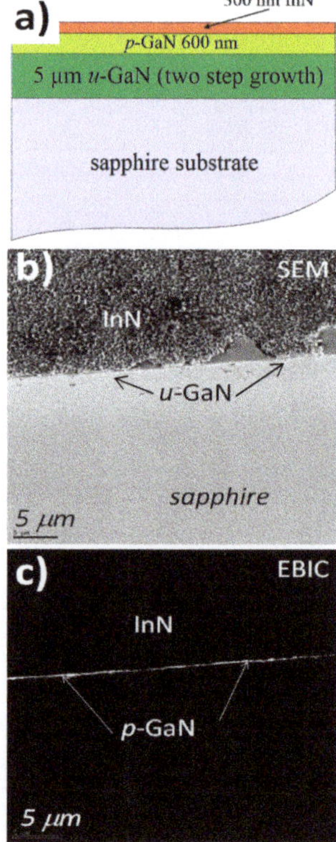

Figure 2. (**a**) Structure of the Mg doped p-GaN/InN junction; (**b**) SEM image (bird's-eye view) of a cleaved p-GaN/InN heterojunction. (**c**) Electron-beam-induced current (EBIC) image superimposed on the SEM image. The bright line indicates the depleted region of the junction. Both SEM and EBIC images were obtained at 5 kV.

Furthermore, the extent of the depletion region at the InN/p-GaN interface has been investigated theoretically. The built-in electrical field across the p–n interface is enhanced due to the gradient of spontaneous polarization. Figure 3a shows hole and electron concentration at the InN/p-GaN interface, and Figure 3b reports the built-in electrical field as a function of depth at the InN/p-GaN interface. The initial conditions for the simulations and doping concentration were evaluated from the equilibrium charge carrier concentrations obtained by Hall measurements. From our simulations, we estimate the thickness of the depletion region to be about 22.8 nm. The distribution of carrier concentration in the depletion region is broader towards p-GaN (18.3 nm) than InN (4.5 nm). We must underline that still the good quality of the InN film presented an unintended n-doping; to our knowledge, the measured electron concentration $n_e = 5.19 \times 10^{18}$ cm^{-3} and mobility $\mu_e = 980$ cm^2 V^{-1} s^{-1} are among the best values obtained by MOVPE growth ($n_e = (3 \div 5) \times 10^{18}$ cm^{-3}, $\mu_e = (542 \div 980)$ cm^2 V^{-1} s^{-1} [34–36]).

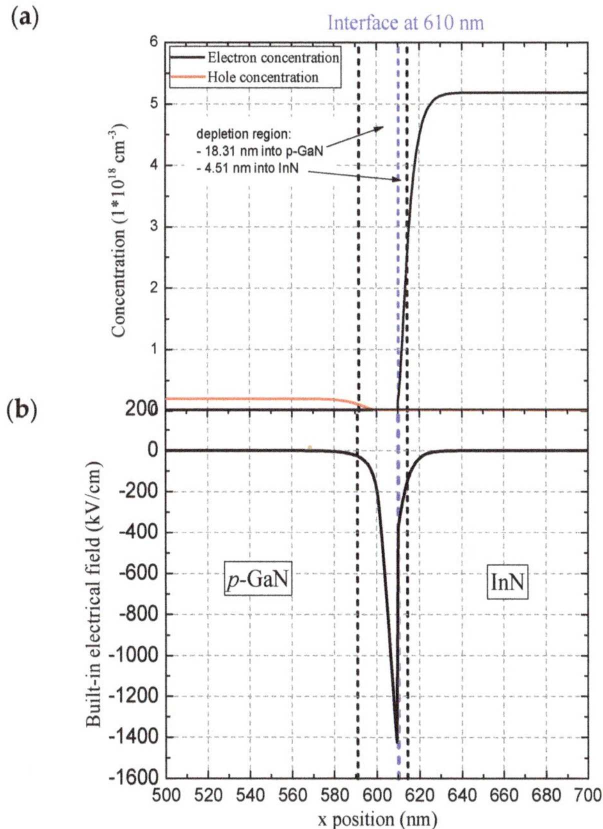

Figure 3. (**a**) Hole/electron concentration and (**b**) built-in electrical field as a function of depth at the InN/p-GaN interface.

The pulsed MOVPE growth method is quite commonly used to grow group III-nitride heterostructures [37–39], but in this case it allowed us to grow good crystalline-quality InN layers using a higher growth temperature (610 °C) (see Materials and Methods paragraph). The growth temperature in the MOVPE growth of InN is the most critical parameter to control the film quality. Because of the low InN dissociation temperature and high equilibrium nitrogen (N$_2$) vapor pressure over the InN film, the preparation of InN requires a low growth temperature. Due to the low (400–500 °C) growth temperatures, the growth of InN is restricted by a low decomposition rate of NH_3 and reduced migration of adatoms on the surface, which leads to metallic In formation on the surface [40,41]. In our sample, no metallic In droplets were observed on the surface of the InN layer. Due to higher growth temperature, the NH_3 dissociation rate was enhanced and there was no shortage in active nitrogen atoms (N). The object input of high V/III ratio is to provide a sufficient amount of reactive nitrogen. The TMIn pulse length controls the thickness of the deposited ultrathin InN layer, while the pause length controls the time allowed for surface migration of In adatoms on the surface and the amount of additional reactive nitrogen. The ratio between the pause and pulse determines the effective V/III ratio, which can be expressed as $\frac{V}{III} = \left(\frac{V}{III}\right)_{nom}\left(1 + \frac{t_p}{t_{TMIn}}\right)$ [41]. Here, $(V/III)_{nom}$ is the nominal V/III ratio, in our case ~38,000 during the TMIn pulse of 7 s, t_{TMIn} and t_P are the TMIn pulse and pause lengths, respectively. A high V/III ratio and a high growth temperature for InN allowed us to achieve good electrical properties, such as low electron concentration and high mobility. This could be explained by the reduction in the native defect concentration, such as N vacancies [42]. In addition, the gradual increase of growth temperature (i.e., the temperature ramp) results in the repeated deposition of ultrathin InN layers, where each successive layer is deposited at a slightly higher temperature until the temperature reaches 610 °C. Such temperature ramping facilitates the formation of larger islands with better alignment due to the increased diffusion of In adatoms, determined by the ultrathin InN layers deposited at the lowest growth temperatures [41]. The temperature ramp approach is similar to using two-step growth with a low-temperature InN buffer [43] or a graded composition InGaN buffer [9], which are reported to improve the structural quality in the InN epilayers.

Furthermore, photoluminescence (PL) measurements of the structure at room temperature were performed. The PL spectra consist of two parts corresponding to the near-infrared (InN) (Figure 4a) and to the UV (p-GaN) (Figure 4b) spectral regions. The InN PL spectrum is centered at 1580 nm (0.75 eV) which is consistent with the established band-to-band PL of InN [44]. The PL spectrum of the p-GaN is a typical GaN room-temperature PL spectrum with near band-edge emission around 3.4 eV (360 nm) [45].

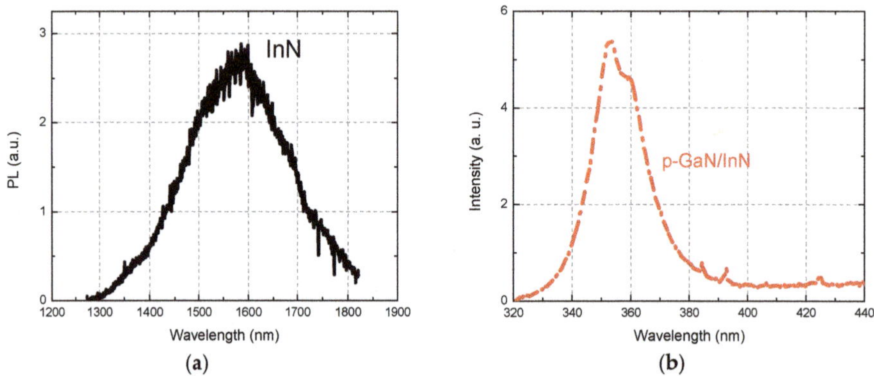

Figure 4. Room-temperature photoluminescence (PL) spectra of InN/p-GaN heterojunction grown on sapphire substrates: (**a**) near-infrared from 1200 nm to 1900 nm, (**b**) UV range from 320 nm to 440 nm.

In order to study the capability of the heterojunction to generate a photo-current, two different types of measurements were performed: (i) photovoltaic response and (ii) EQE measurements. Figure 5a reveals the current density–voltage (J–V) characteristic of the InN/p-GaN heterojunction under dark and AM1.5G illumination conditions. The current density was measured for the applied bias voltage from −1 V to +1 V. The characteristics are asymmetrical with clear diode-like rectification behavior. Within the region of the turn-on voltage and when the voltage is above ~0.2 V under dark condition and above 0.06 V under illumination, the J–V characteristics follow a power law $V \sim I^2$;

consequently, for single-type carrier injection, the current conduction is expected to be space-charge limited. At larger applied voltages >0.5 V, the J–V does not deviate considerably from linearity, indicating low series resistance and retaining a low density of interface states. Figure 5b shows corresponding EQE spectra of the unbiased InN/p-GaN heterojunction. The carrier collection peaked in the wavelength region of 400–500 nm.

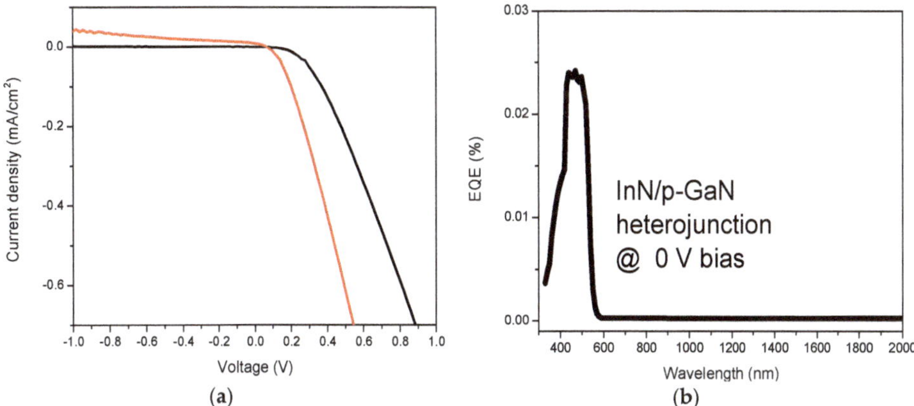

Figure 5. (**a**) Current voltage (J–V) characteristics of an InN/p-GaN heterojunction under dark (black) and AM1.5G illumination (red). (**b**) Corresponding external quantum efficiency (EQE) spectra of the unbiased InN/p-GaN heterojunction.

Figure 6a shows the EQE spectra of the InN/p-GaN junction at different reverse biases: 0 V, −10 V and −20 V. Photo-response in the region 300–600 nm originates from the absorption in both p-GaN and InN, as explained below. Local EQE maxima are peaked at wavelengths of 430 nm, 450 nm and 511 nm. Increased reverse bias results in a general rising trend of the full spectrum where the peak at 511 nm (2.4 eV) appears to be the strongest. The sub-maximum at around 360 nm originates from the bandgap transitions and is peaked at the same wavelength of the PL emission in Figure 4b; this implies that conditions are favorable for both charge carriers generated in the p-GaN to reach the contacts, that is, electrons down the barrier towards the n-contact and holes to the p-contact (see Figure 7a). The same applies to the broad band at 375–550 nm (3.3–2.2 eV), which correlates with known multiple transition mechanisms: (i) in the UV range peaked at 3.20–3.26 eV, the Mg doping introduces shallow acceptors with an ionization energy of 200 meV; (ii) in the blue spectra range peaked at 2.7–2.9 eV, the compensating deep donors formed at high p-doping levels are responsible for the donor-to-band or donor-acceptor-pair transitions [45]. The dissociation of electron–hole pairs enhanced under reverse bias is likely responsible for that. Surprisingly, the EQE in the 300–600 nm range is enhanced significantly above 100%. At −20 V reverse bias we could observe also a contribution to the EQE for the longer wavelengths (>1500 nm) (Figure 6b). The band peak position for carrier extraction found at 1780 nm (~0.7 eV) corresponded to the InN energy bandgap [44]. Figure 6c highlights the nonlinear increase of the photocurrent as a function of reverse bias. This suggests that the p–n junction is normally operated and the generated photocurrent increases exponentially with the applied bias attaining a 2kT/q slope in the semi-log (I–V) plot due to recombination effects [46].

In order to analyse the junction existing at the interface of InN/p-GaN, we have constructed a band diagram based on our measurement results and calculations (Figure 7a). The schematic energy band diagram of the InN/p-GaN heterojunction under irradiation is shown in Figure 7a, which reflects the type-I straddling configuration [47]. Nitride semiconductors are pyroelectrics, that is, they present a strong polarization due to fixed charges in the crystal structure [48,49]. The band diagram in Figure 7a shows the band bending behavior at the InN and p-GaN interface that underlines the heterojunction formation. The absorption of photons leads to the formation of excess electrons in the n-side and excess holes in the p-side of the device, generating a voltage drop Voc across the junction. The broader depletion region in the p-GaN maintains space charge neutrality, which results in efficient collection of the carriers generated by UV and blue–visible light (Figure 6). For electrons there is no offset, while for holes, the potential barrier of 0.3 eV is obtained (Figure 7a). Nevertheless, with 1600–1800 nm wavelengths, by 1 sun irradiation, electrons and holes are generated in the InN, and when a sufficient reverse bias (−20 V) is applied, holes can overcome the interface barrier and generate photocurrent for the near-infrared (NIR) region (Figure 7b). In the reverse bias, the presence of recombination centers affects the overall I–V characteristics by the generation of additional electron–hole pairs within the depletion region, which greatly exceeds the unbiased state. The generation rate of carriers in the depletion region will reach a maximum when $E_g/2$. Since the intrinsic concentration is a function of the energy gap of the material, InN with a small energy gap will therefore exhibit high generation rates [50].

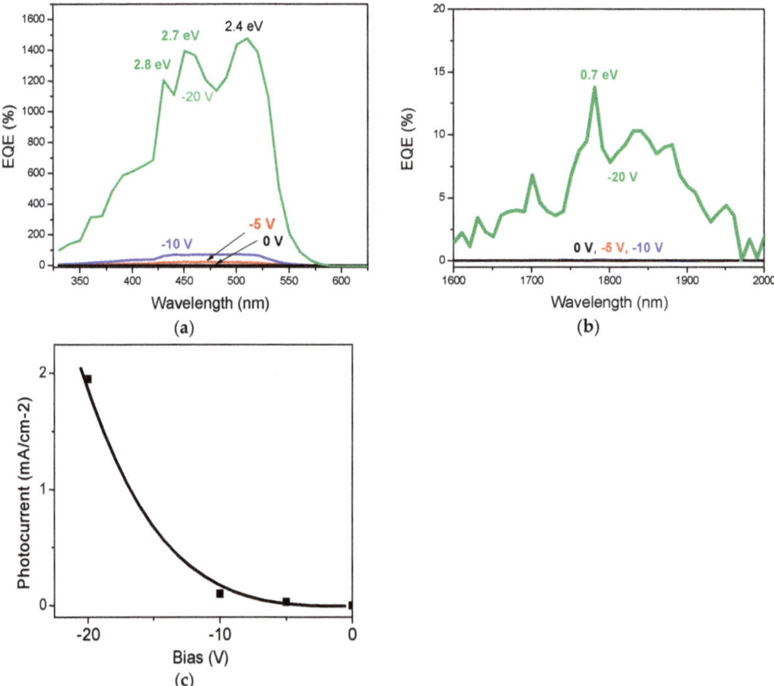

Figure 6. (a) EQE of the InN/p-GaN junction at the reverse bias of 0, −5, −10 and −20 V, (b) EQE spectra of the InN/p-GaN heterojunction at −20 V reverse bias in the NIR spectral region, (c) corresponding photocurrent density as a function of the reverse bias whereby the line is a guide to the eye.

Although enhancement of the EQE over unity under different excitation conditions has been widely reported in the literature [51,52], the origin of the increase up to 1400% in EQE in InN/p-GaN heterojunction under −20 V reverse bias is not yet clear and may be attributed to various factors. The enhancement might originate from an enlargement of the depletion region contributing to the enhanced absorption and charge collection efficiency at the junction. Bias voltage actually increases the ionized donor concentration of GaN and decreases its work function while harvesting photo carriers in a globally larger volume for the light in the spectral region 300–600 nm. Likewise for carrier multiplication (CM), a larger depletion region ensures that any photogenerated carriers are efficiently collected [6,7,53].

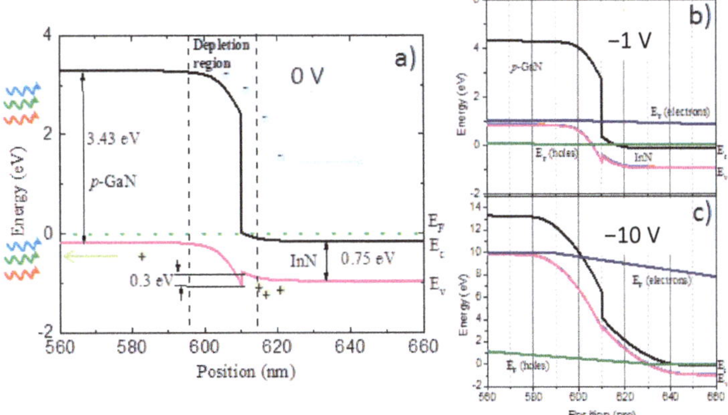

Figure 7. (**a**) Schematic of the unbiased (0 V) interface band alignment at room temperature for InN/p-GaN heterojunction. The irradiation of the junction from the GaN side is indicated. (**b**,**c**) Biased band alignments for the InN/p-GaN heterojunction at −1 V and −10 V reverse bias, respectively. The E_C stands for conduction band, E_V for valence band and E_F for Fermi levels of electrons and holes.

The key aspect is to extract multiple excitons and hot carriers before they thermalize to the band edge and rapidly recombine. In principle, that could be possible in our case when the relative high-voltage reverse bias is applied. For the hot carrier contribution, however, InN should be most likely in a cubic structure featuring a very large gap between high-lying optical phonon and low-lying acoustic phonon energies. If the gap is wide enough, the Klemens decay of optical phonons is prevented [4,54]. However, a pronounced XRD peak at 31.3° for the InN film reveals the hexagonal lattice type, likewise for the GaN (0002) peak at 34.5° (Figure 1), weakening our expectation for the hot carrier contribution.

Our results show that the photocurrent is increasing nonlinearly with the reverse bias (Figure 6c) exhibiting a peak at 2.4 eV (see Figure 6b) which corresponds roughly to three times the InN bandgap. This observation might support further the contribution from CM, which was demonstrated in bulk InN to occur at photon energies roughly up to three times its bandgap [53,55]. When the reverse bias approaches the breakdown level, amplification factors at the electrical level can take place [56]. In order to get a better picture about these processes, we have simulated InN/p-GaN heterojunction at different reverse biases (Figure 7b,c). Because the depletion layer is relatively thin (in the range of nanometers), electrons can directly tunnel across the depletion region from p-GaN valence band (E_V) into the n-InN conduction band. Due to the increased reverse bias (>−10 V), photocurrent generation proceeds at the breakdown level leading to the separation of multiple excitons and efficient electron–hole generation, therefore contributing significantly to the EQE of the heterojunction. Because of the large conduction band offset of the nitride materials, the high level of Mg doping improves the tunneling and therefore overall photocurrent generation at the InN/p-GaN heterojunction [57].

4. Conclusions

In summary, efficient electron–hole pair generation and extraction in a superior-quality InN/p-GaN heterojunction interface grown by pulsed MOVPE was achieved. The considerably improved crystallinity was obtained thanks to pulsed nature and higher-temperature growth process. The InN/p-GaN heterojunction depletion region was simulated, demonstrating 22.8 nm width extended by four times more into p-GaN than InN. Experimentally, good rectifying behavior and photo-carrier extraction without external electrical field was observed. Under increased reverse bias voltage at the enlarged depletion region width the EQE exceeded unity in the blue spectral region (300–550 nm) due to carrier multiplication and photoconductivity. The optical generation of additional electron–hole pairs within the depletion region still makes a significant impact onto the unbiased state.

Author Contributions: Conceptualization, V.S. and R.T.; methodology, M.S., V.S., D.M. and R.T; formal analysis, M.K., A.K., D.D., T.M., M.L. and I.R.; investigation, M.K., A.K., D.D., T.M., M.L. and I.R.; data curation, M.K., A.K., D.D., V.S., T.M., M.L. and I.R.; writing—original draft preparation, V.S.; writing—review and editing, V.S., D.M. and R.T.; supervision, V.S. and R.T.; funding acquisition, D.M. and R.T.

Funding: This research was funded by a grant (project FLINGO No. M-ERA.NET-2/2016) from the Research Council of Lithuania. DM would like to thanks the support of EPSRC (EP/M024938/1).

Conflicts of Interest: The authors declare no conflict of interest.

References

1. Wu, J. Unusual properties of the fundamental band gap of InN. *Appl. Phys. Lett.* **2002**, *80*, 3967–3969. [CrossRef]

2. Millot, M. Determination of effective mass in InN by high-field oscillatory magnetoabsorption spectroscopy. *Phys. Rev. B* **2011**, *83*, 125204. [CrossRef]

3. Thomas, E.L.; Groishnyy, T.; Maldovan, M. Phononics: Colloidal crystals go hypersonic. *Nat. Mater.* **2006**, *5*, 773–774. [CrossRef] [PubMed]

4. Klemens, P. Anharmonic Decay of Optical Phonons. *Phys. Rev.* **1966**, *148*, 845. [CrossRef]

5. Pomeroy, J.W.; Kuball, M.; Lu, H.; Schaff, W.J.; Wang, X.; Yoshikawa, A. Phonon lifetimes and phonon decay in InN. *Appl. Phys. Lett.* **2005**, *86*, 223501. [CrossRef]

6. Conibeer, G.; Shrestha, S.; Huang, S.; Patterson, R.; Xia, H.; Feng, Y.; Zhang, P.; Gupta, N.; Tayebjee, M.; Smyth, S.; et al. Hot carrier solar cell absorber prerequisites and candidate material systems. *Sol. Energy Mater. Sol. Cells* **2015**, *135*, 124–129. [CrossRef]

7. Shrestha, S.; Aliberti, P.; Conibeer, G.J. Energy selective contacts for hot carrier solar cells. *Sol. Energy Mater. Sol. Cells* **2010**, *94*, 1546–1550. [CrossRef]

8. Lozac'h, M.; Nakano, Y.; Sang, L.; Sakoda, K.; Sumiya, M. Study of Defect Levels in the Band Gap for a Thick InGaN Film. *Jpn. J. Appl. Phys.* **2012**, *51*, 121001. [CrossRef]

9. Islam, S.M.; Protasenko, V.; Rouvimov, S.; Xing, H.; Jena, D. High-quality InN films on GaN using graded InGaN buffers by MBE. *Jpn. J. Appl. Phys.* **2016**, *55*, 05FD12. [CrossRef]

10. Wu, G.G.; Li, W. Ch.; Shen, Ch. S.; Gao, F.B.; Liang, H.W.; Wang, H.; Song, L.J.; Du, G.T. Near infrared electroluminescence from n-InN/p-GaN light-emitting diodes. *Appl. Phys. Lett.* **2012**, *10*, 103504. [CrossRef]

11. Rajpalke, M.K.; Bhat, T.N.; Roul, B.; Kumar, M.; Krupanidhi, S.B. Current transport in nonpolar a-plane InN/GaN heterostructures Schottky junction. *J. Appl. Phys.* **2012**, *112*, 023706. [CrossRef]

12. Hsu, L.H.; Kuo, C.T.; Huang, J.K.; Hsu, S.C.; Lee, H.Y.; Kuo, H.C.; Lee, P.T.; Tsai, Y.L.; Hwang, Y.C.; Su, C.F.; et al. InN-based heterojunction photodetector with extended infrared response. *Opt. Express* **2015**, *23*, 31150–31162. [CrossRef] [PubMed]

13. Dimakis, A. Surface structure and surface kinetics of InN grown by plasma-assisted atomic layer epitaxy. *J. Appl. Phys.* **2005**, *97*, 113520. [CrossRef]

14. Kim, J.; Ji, M.H.; Detchprohm, T.; Ryou, J.H.; Dupuis, R.D.; Sood, A.K.; Dhar, N.K. $Al_xGa_{1-x}N$ Ultraviolet Avalanche Photodiodes with Avalanche Gain Greater Than105. *IEEE Photonics Technol. Lett.* **2015**, *27*, 642. [CrossRef]

15. Dupuis, R.D.; Ryou, J.H.; Shen, S.C.; Yoder, P.D.; Zhang, Y.; Kim, H.J.; Choi, S.; Lochner, Z. Growth and fabrication of high-performance GaN-based ultraviolet avalanche photodiodes. *J. Cryst. Growth* **2008**, *310*, 5217–5222. [CrossRef]

16. McClintock, R.; Pau, J.L.; Minder, K.; Bayram, C.; Kung, P.; Razeghi, M. Hole-initiated multiplication in back-illuminated GaN avalanche photodiodes. *Appl. Phys. Lett.* **2007**, *90*, 141112. [CrossRef]

17. Ji, M.H.; Kim, J.; Detchprohm, T.; Zhu, Y.Z.; Shen, S.C.; Dupuis, R.D. p-i-p-i-n Separate Absorption and Multiplication Ultraviolet Avalanche Photodiodes. *IEEE Photonics Technol. Lett.* **2018**, *30*, 181–184. [CrossRef]

18. Padmanabhan, P.; Hancock, B.; Nikzad, S.; Bell, L.D.; Kroep, K.; Charbon, E. A Hybrid Readout Solution for GaN-Based Detectors Using CMOS Technology. *Sensors* **2018**, *18*, 449. [CrossRef]

19. Cai, C.; Ge, M.; Xue, J.J.; Hu, L.Q.; Chen, D.J.; Lu, H.; Zhang, R.; Zheng, Y.D. An Improved Design for Solar-Blind AlGaN Avalanche Photodiodes. *IEEE Photonics J.* **2017**, *9*, 6803507. [CrossRef]

20. Watanabe, K.; Nakanishi, N.; Yamazaki, T.; Yang, J.R.; Huang, S.Y.; Inoke, K.; Hsu, J.T.; Tu, R.C.; Shiojiri, M. Atomic-scale strain field and In atom distribution in multiple quantum wells InGaN/GaN. *Appl. Phys. Lett.* **2003**, *82*, 715.

21. Cho, H.K.; Lee, J.Y.; Sharma, N.; Humphreys, C.J.; Yang, G.M.; Kim, C.S.; Song, J.H.; Yu, P.W. Response to "Comment on 'Effect of growth interruptions on the light emission and indium clustering of InGaN/GaN multiple quantum wells'" [Appl. Phys. Lett. 81, 3100 (2002)]. *Appl. Phys. Lett.* **2002**, *81*, 3102. [CrossRef]

22. Feng, S.-W.; Cheng, Y.-C.; Chung, Y.-Y.; Yang, C.C.; Mao, M.-H.; Lin, Y.-S.; Ma, K.-J.; Chyi, J.-I. Multiple-component photoluminescence decay caused by carrier transport in InGaN/GaN multiple quantum wells with indium aggregation structures. *Appl. Phys. Lett.* **2002**, *80*, 4375. [CrossRef]

23. Lin, Y.-S.; Ma, K.-J.; Hsu, C.; Feng, S.-W.; Cheng, Y.-C.; Liao, C.-C.; Yang, C.C.; Chou, C.-C.; Lee, C.-M.; Chyi, J.-I. Dependence of composition fluctuation on indium content in InGaN/GaN multiple quantum wells. *Appl. Phys. Lett.* **2000**, *77*, 2988. [CrossRef]

24. Cheng, Y.C.; Tseng, C.H.; Hsu, C.; Ma, K.J.; Feng, S.W.; Lin, E.C.; Yang, C.C.; Chyi, J.I. Mechanisms for photon-emission enhancement with silicon doping in InGaN/GaN quantum-well structures. *J. Electron. Mater.* **2003**, *32*, 375. [CrossRef]

25. Kisailus, D.; Choi, J.H.; Lange, F.F. GaN nanocrystals from oxygen and nitrogen-based precursors. *J. Cryst. Growth* **2003**, *252*, 106–120. [CrossRef]

26. Ridley, B.K. The LO phonon lifetime in GaN. *J. Phys. Condens. Matter* **1996**, *8*, L511. [CrossRef]

27. Nakamura, S. GaN Growth Using GaN Buffer Layer. *Jpn. J. Appl. Phys.* **1991**, *30*, L1705–L1707. [CrossRef]

28. Cimalla, V.; Förster, C.; Kittler, G.; Cimalla, I.; Kosiba, R.; Ecke, G.; Ambacher, O.; Goldhahn, R.; Shokhovets, S.; Georgakilas, A.; et al. Status of high efficiency and high power Thin GaN-LED development. *Phys. Status Solidi C* **2003**, *6*, 2818–2821. [CrossRef]

29. Davydov, V.Y.; Klochikhin, A.A.; Seisyan, R.P.; Emtsev, V.V.; Ivanov, S.V.; Bechstedt, F.; Furthmüller, J.; Harima, H.; Mudryi, A.V.; Aderhold, J.; et al. Absorption and Emission of Hexagonal InN. Evidence of Narrow Fundamental Band Gap. *Phys. Status Solidi B* **2002**, *229*, R1–R3. [CrossRef]

30. Tuna, Ö.; Behmenburg, H.; Giesen, C.; Kalisch, H.; Jansen, R.H.; Yablonskii, G.P.; Heuken, M. Dependence of InN properties on MOCVD growth parameters. *Phys. Status Solidi C* **2011**, *8*, 2044–2046. [CrossRef]

31. Kuo, C.T.; Hsua, L.H.; Lai, Y.Y.; Cheng, S.Y.; Kuo, H.C.; Lin, C.C.; Cheng, Y.J. Dominant near infrared light-emitting diodes based on p-NiO/n-InN heterostructure on SiC substrate. *Appl. Surf. Sci.* **2017**, *405*, 449–454. [CrossRef]

32. Barick, B.K.; Saroj, K.R.; Prasad, N.; Sutar, D.S.; Dhar, S. Identifying threading dislocation types in ammonothermally grown bulk α-GaN by confocal Raman 3-D imaging of volumetric stress distribution. *J. Cryst. Growth* **2018**, *490*, 104. [CrossRef]

33. Bushuykin, P.A.; Andreev, B.A.; Yu, V.; Davydov, K.; Lobanov, D.N.; Kuritsyn, D.I.; Yablonskiy, A.N.; Averkiev, N.S.; Savchenko, G.M.; Krasilnik, Z.F. New photoelectrical properties of InN: Interband spectra and fast kinetics of positive and negative photoconductivity of InN. *J. Appl. Phys.* **2018**, *123*, 195701. [CrossRef]

34. Bhuiyan, A.G.; Hashimoto, A.; Yamamoto, A. Indium nitride (InN): A review on growth, characterization, and properties. *J. Appl. Phys.* **2003**, *94*, 2779. [CrossRef]

35. Lund, C.; Catalano, M.; Wang, L.; Wurm, C.; Mates, T.; Kim, M.; Nakamura, S.; DenBaars, S.P.; Mishra, U.K.; Keller, S. Metal-organic chemical vapor deposition of N-polar InN quantum dots and thin films on vicinal GaN. *J. Appl. Phys.* **2018**, *123*, 055702. [CrossRef]

36. Jamil, M.; Zhao, H.; Higgins, J.B.; Tansu, N. MOVPE and photoluminescence of narrow band gap (0.77 eV) InN on GaN/sapphire by pulsed growth mode. *Phys. Status Solidi A* **2008**, *205*, 2886–2891. [CrossRef]

37. Bayram, C.; Fain, B.; Péré-laperne, N.; Mc Clintock, R.; Razeghi, M. Pulsed metalorganic chemical vapor deposition of high quality AlN/GaN superlattices for intersubband transitions. *Proc. SPIE* **2009**, *7222*, 722212.

38. Johnson, M.C.; Konsek, S.L.; Zettl, A.; Bourret-Courchesne, E.D. Nucleation and growth of InN thin films using conventional and pulsed MOVPE. *J. Cryst. Growth* **2004**, *272*, 400–406. [CrossRef]

39. Zhang, Y.; Zhou, X.; Xu, S.; Wang, Z.; Chen, Z.; Zhang, J.; Zhang, J.; Hao, Y. Effects of growth temperature on the properties of InGaN channel heterostructures grown by pulsed metal organic chemical vapor deposition. *AIP Adv.* **2015**, *5*, 127102. [CrossRef]

40. Bhuiyan, A.G.; Sugita, K.; Hashimoto, A.; Yamamoto, A. InGaN solar cells: Present state of the art and important challenges. *IEEE J. Photovolt.* **2012**, *2*, 276–293. [CrossRef]

41. Mickevičius, J.; Dobrovolskas, D.; Steponavičius, T.; Malinauskas, T.; Kolenda, M.; Kadys, A.; Tamulaitis, G. Engineering of InN epilayers by repeated deposition of ultrathin layers in pulsed MOCVD growth. *Appl. Surf. Sci.* **2018**, *427*, 1027–1032. [CrossRef]

42. Wang, H.; Jiang, D.S.; Zhu, J.J.; Zhao, D.G.; Liu, Z.S.; Wang, Y.T.; Zhang, S.M.; Yang, H. The influence of growth temperature and input V/III ratio on the initial nucleation and material properties of InN on GaN by MOCVD. *Semicond. Sci. Technol.* **2009**, *24*, 055001. [CrossRef]

43. Ruffenach, S.; Moret, M.; Briot, O.; Gil, B. Recent advances in the MOVPE growth of indium nitride. *Phys. Status Solidi A* **2010**, *207*, 9–18. [CrossRef]

44. Wu, J.; Walukiewicz, W.; Shan, W.; Yu, K.M.; Ager, J.W.; Li, S.X.; Haller, E.E.; Lu, H.; Schaff, W.J. Temperature dependence of the fundamental band gap of InN. *J. Appl. Phys.* **2003**, *94*, 4457. [CrossRef]

45. Reshchikov, M.A. Luminescence properties of defects in GaN. *J. Appl. Phys.* **2005**, *97*, 061301. [CrossRef]

46. Dervos, C.T.; Skafidas, P.D.; Mergos, J.A.; Vassiliou, P. p-n Junction Photocurrent Modelling Evaluation under Optical and Electrical Excitation. *Sensors* **2004**, *4*, 58–70. [CrossRef]

47. King, P.D.C.; Veal, T.D.; Kendrick, C.E.; Bailey, L.R.; Durbin, S.M.; McConvill, C.F. InN/GaN valence band offset: High-resolution x-ray photoemission spectroscopy measurements. *Phys. Rev. B* **2008**, *78*, 033308. [CrossRef]

48. Ambacher, O.; Majewski, J.; Miskys, C.; Link, A.; Hermann, M.; Eickhoff, M.; Stutzmann, M.; Bernardini, F.; Fiorentini, V.; Tilak, V.; et al. Pyroelectric properties of Al(In)GaN/GaN hetero- and quantum well structures. *J. Phys. Condens. Matter* **2002**, *14*, 3399. [CrossRef]

49. Perlin, P.; Leszczyński, M.; Prystawko, P.; Wisniewski, P.; Czernetzki, R.; Skierbiszewski, C.; Nowak, G.; Purgal, W.; Weyher, J.L.; Kamler, G.; et al. Low dislocation density, high power InGaN laser diodes. *MRS Internet J. Nitride Semicond. Res.* **2004**, *9*. [CrossRef]

50. Sze, S.M. *Physics of Semiconductor Devices*, 2nd ed.; Wiley: New York, NY, USA, 1981.

51. Yang, Y.; Peng, X.; Hyatt, S.; Yu, D. Broadband Quantum Efficiency Enhancement in High Index Nanowire Resonators. *Nano Lett.* **2015**, *15*, 3541–3546. [CrossRef]

52. Liu, F.; Yan, C.; Sun, K.; Zhou, F.; Hao, X.; Green, M.A. Light-Bias-Dependent External Quantum Efficiency of Kesterite Cu_2ZnSnS_4 Solar Cells. *ACS Photonics* **2017**, *4*, 1684–1690. [CrossRef]

53. Aliberti, P.; Feng, Y.; Shrestha, S.K.; Green, M.A.; Conibeer, G.; Tu, L.W.; Tseng, P.H.; Clady, R. Effects of non-ideal energy selective contacts and experimental carrier cooling rate on the performance of an indium nitride based hot carrier solar cell. *Appl. Phys. Lett.* **2011**, *99*, 223507. [CrossRef]

54. Chen, F.; Cartwright, A.N.; Lu, H.; Schaff, J. Time-resolved spectroscopy of recombination and relaxation dynamics in InN. *Appl. Phys. Lett.* **2003**, *83*, 4984. [CrossRef]

55. Jensen, S.A.; Versluis, J.; Canovas, E.; Pijpers, J.J.H.; Sellers, I.R.; Bonn, M. Carrier Multiplication in Bulk Indium Nitride. *Appl. Phys. Lett.* **2012**, *101*, 222113. [CrossRef]

56. Feng, S.; Chen, Y.; Lai, C.; Tu, L.; Han, J. Anisotropic strain relaxation and the resulting degree of polarization by one-and two-step growth in nonpolar *a*-plane GaN grown on *r*-sapphire substrate. *J. Appl. Phys.* **2013**, *114*, 233103–233109. [CrossRef]

57. Krishnamoorthy, S.; Akyol, F.; Park, P.S.; Rajan, S. Low resistance GaN/InGaN/GaN tunnel junctions. *Appl. Phys. Lett.* **2013**, *102*, 11350. [CrossRef]

 *nanomaterials*

Article

Emphasizing the Operational Role of a Novel Graphene-Based Ink into High Performance Ternary Organic Solar Cells

Minas M. Stylianakis [1,*,†], Dimitrios M. Kosmidis [1,†], Katerina Anagnostou [1], Christos Polyzoidis [1], Miron Krassas [1,2], George Kenanakis [3], George Viskadouros [1,4], Nikolaos Kornilios [1], Konstantinos Petridis [1,5] and Emmanuel Kymakis [1,*]

[1] Department of Electrical & Computer Engineering, Hellenic Mediterranean University (HMU), Estavromenos, 71410 Heraklion, Greece; kosdimitris@hmu.gr (D.M.K.); katerinanag@hmu.gr (K.A.); polyzoidis@hmu.gr (C.P.); kmiron@hmu.gr (M.K.); viskadouros@hmu.gr (G.V.); kornil@hmu.gr (N.K.); c.petridischania@gmail.com (K.P.)

[2] Department of Materials Science and Technology, University of Crete, 71003 Heraklion, Greece

[3] Institute of Electronic Structure and Laser, Foundation for Research and Technology-Hellas, N. Plastira 100, 70013 Heraklion, Greece; gkenanak@iesl.forth.gr

[4] Department of Mineral Resources Engineering, Technical University of Crete, 73100 Chania, Greece

[5] Department of Electronic Engineering, Hellenic Mediterranean University (HMU), 73132 Chania, Greece

[*] Correspondence: stylianakis@hmu.gr (M.M.S.); kymakis@hmu.gr (E.K.); Tel.: +30-2810-379775 (M.M.S.)

[†] These authors contributed equally to this work.

Received: 27 November 2019; Accepted: 28 December 2019; Published: 2 January 2020

Abstract: A novel solution-processed, graphene-based material was synthesized by treating graphene oxide (GO) with 2,5,7-trinitro-9-oxo-fluorenone-4-carboxylic acid (TNF-COOH) moieties, via simple synthetic routes. The yielded molecule N-[(carbamoyl-GO)ethyl]-N′-[(carbamoyl)-(2,5,7-trinitro-9-oxo-fluorene)] (GO-TNF) was thoroughly characterized and it was shown that it presents favorable highest occupied molecular orbital (HOMO) and lowest unoccupied molecular orbital (LUMO) energy levels to function as a bridge component between the polymeric donor poly(({4,8-bis[(2-ethylhexyl)oxy]benzo[1,2-b:4, 5-b′]dithiophene-2,6-diyl}{3-fluoro-2-[(2-ethylhexyl)carbonyl] thieno[3,4-b]thiophenediyl})) (PTB7) and the fullerene derivative acceptor [6,6]-phenyl-C_{71}-butyric-acid-methylester ($PC_{71}BM$). In this context, a GO-TNF based ink was prepared and directly incorporated within the binary photoactive layer, in different volume ratios (1%–3% ratio to the blend) for the effective realization of inverted ternary organic solar cells (OSCs) of the structure ITO/PFN/PTB7:GO-TNF:$PC_{71}BM$/MoO_3/Al. The addition of 2% v/v GO-TNF ink led to a champion power conversion efficiency (PCE) of 8.71% that was enhanced by ~13% as compared to the reference cell.

Keywords: ternary organic solar cells; graphene ink; functionalization; air-processed; cascade effect; charge transfer

---

1. Introduction

Due to the highly increased global demand for low-cost energy generation over the last three decades, significant research efforts have been made towards the development and progress of organic solar cells (OSCs), in order to boost their competitiveness over silicon technology [1,2]. Owing to several attractive properties, including light weight, flexibility, low manufacturing costs, and compatibility with large-area processes, OSCs is considered as one of the most prominent photovoltaic technologies for sustainable energy production [3].

In this context, several polymeric donor:fullerene-based acceptor combinations have been flourished providing a rapid increase in the OSC devices efficiency over 9% [4–7]. On top of that, very recently, alternative optimized architectures such as tandem structures, novel donors and non-fullerene acceptor design and synthesis, as well as ternary systems have escalated the performance of OSCs over 14% [8–11].

Unlike to the typical binary OSC configuration, that is based on a donor–acceptor bulk heterojunction (BHJ) blend, the ternary one contains a third component which can function as: (i) second donor, (ii) second acceptor, and (iii) non-volatile additive [12]. The operation of a ternary OSC device relies on one of the four existing dominant mechanisms including: (1) charge transfer, (2) Forster resonance energy transfer, (3) parallel-linkage, and (4) alloyed donor structure mechanism [12,13]. Thus, according to the above mechanisms, small molecules [14–18], polymers [19–21], dye molecules [22,23], graphene-based materials [24,25] or 2D materials [12,26,27] could be chosen and incorporated as additives within the binary active layer.

This study discusses for the first time the design and synthesis of a novel graphene-based material (GO-TNF) through simple chemical processes as well as its direct incorporation in ink form within the binary active layer (PTB7:PC$_{71}$BM) for the realization of inverted ternary OSC devices. GO-TNF consists of graphene oxide (GO) as core and TNF side groups linked with ethylenediamine (EDA) aliphatic spacers. Since the energy levels of the synthesized graphene-based molecule and these of PTB7 and PC$_{71}$BM perfectly match, GO-TNF ink was incorporated in different ratios ranging from 1% to 3%. Upon the incorporation of GO-TNF, charge transfer operational mechanism dominated (cascade effect), while the photovoltaic performance was boosted in all ternary devices compared to the reference cell. The champion device, containing 2% *v/v* GO-TNF ink, exhibited a significant enhancement by ~13%, leading to a power conversion efficiency (PCE) of 8.71%. This increase is mainly due to the improvement of the nanomorphology between the donor:acceptor blend's interfaces, as well as to electron mobility enhancement.

2. Materials and Methods

2.1. Materials

Initially, 9-oxo-fluorene-4-carboxylic acid 97%, 1,2-ethylenediamine (EDA) puriss. p.a., absolute, ≥99.5% (GC), graphite synthetic, H$_2$SO$_4$ 95–97%, fuming HNO$_3$ 70%, SOCl$_2$ ReagentPlus >99%, were purchased from Sigma Aldrich (Taufkirchen, Germany). PTB7 was purchased from Solaris Chem (Vaudreuil-Dorion, QC, Canada), while PC$_{71}$BM and PFN were both purchased from Solenne BV (Groningen, The Netherlands). Finally, MoO$_3$ and Al were bought from Kurt J. Lesker (East Sussex, UK), while the glass-ITO substrates were purchased from Naranjo Substrates (Groningen, The Netherlands).

2.2. Materials' Synthetic Procedures

The preparation of GO-TNF took place into several steps, as it is depicted in Figure 1 and analyzed in Appendix A. First, 9-oxo-fluorene-4-carboxylic acid was nitrated using a mixture of concentrated sulfuric acid (H$_2$SO$_4$, 95–97%) and fuming nitric acid (HNO$_3$, 70%), yielding 2,5,7-trinitro-9-oxo-fluorene-4-carboxylic acid (TNF-COOH). Afterwards, the carboxyl group of TNF was chlorinated using thionyl-chloride (SOCl$_2$), to get 2,5,7-trinitro-9-oxo-fluorene-4-acyl-chloride (TNF-COCl). The linkage of TNF-COCl with 1,4-ethylenediamine (EDA) was held via a typical nucleophilic substitution reaction to obtain TNF-EDA. In the second parallel step, GO was prepared via a modified Hummers' method [28] and was subsequently acylated, using SOCl$_2$ to get GO-COCl [25]. The final GO-TNF was extracted upon the coupling between GO-COCl and TNF-EDA through a nucleophilic substitution reaction. Finally, GO-TNF ink was prepared as described in the SI (see Appendix A).

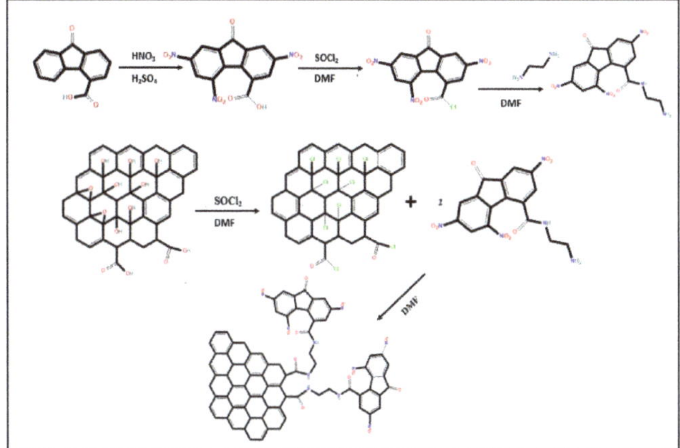

Figure 1. Schematic representation of the chemical synthetic procedure.

2.3. OSC Device Fabrication

All OSC devices were fabricated in a typical sandwich inverted geometry consisting of a bottom indium tin oxide (ITO) coated glass substrates electrode, poly [(9,9-bis(3'-(N,N-dimethy -lamino)propyl)-2,7-fluorene)-alt-2,7-(9,9–dioctylfluorene)] (PFN) as the ETL, a PTB7:PC$_{71}$BM BHJ thin film as the active layer, a MoO$_3$ as the HTL and a top metal (Al) electrode. GO-TNF ink was directly incorporated within the binary photoactive layer, in ratios ranging from 1 to 3% ratio to the polymer for the fabrication of the ternary devices. The devices' fabrication in detail is reported in the SI. The schematic representation of the device and the respective energy level diagram are depicted in Figure 2.

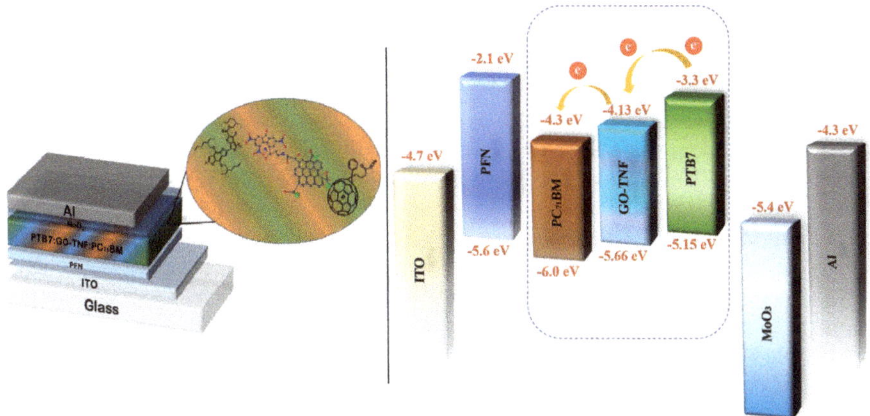

Figure 2. Schematic representation of the ternary OSC device (**left**) and energy levels diagram (**right**).

2.4. Characterization Techniques

ATR FT-IR (transmittance) experiments were carried out with a Bruker Vertex 70v FT-IR vacuum spectrometer equipped with a A225/Q Platinum ATR unit with single reflection diamond crystal which allows the infrared analysis of unevenly shaped solid samples and liquids through total reflection measurements, in a spectral range of 4000–700 cm^{-1}. UV-vis absorption spectra were taken using a Shimadzu UV-2401 PC spectrophotometer, over the wavelength range of 270–800 nm. The photoluminescence (PL) measurements of the devices' active layers were carried out at room temperature and resolved by using a UV grating and a sensitive, calibrated and liquid N_2-cooled CCD camera in the wavelength range from 600 to 950 nm. The excitation source employed was a He-Cd CW laser at 325 nm with a full power of $P_0 = 35$ mW. Raman measurements were performed at room temperature using a Horiba LabRAM HR Evolution confocal micro-spectrometer, in backscattering geometry (180°), equipped with an air-cooled solid-state laser operating at 532 nm with 100 mW output power. The laser beam was focused on the samples using a 10× Olympus microscope objective (numerical aperture of 0.25), providing a ~55 mW power on each sample. XRD patterns were collected on a Panalytical Expert Pro X-ray diffractometer, using Cu K_α radiation ($\lambda = 1.5406$ Å). Thermogravimetric analysis (TGA) was performed on 5–10 mg samples over the temperature range from 40 to 800 °C at a heating rate of 10 °C/min utilizing a Perkin-Elmer Diamond Pyris model under N_2 atmosphere. The morphology of the surfaces was examined with an Atomic Force Microscope (Park Systems XE7, Park Systems Corporate Headquarters, Suwon, Korea). SEM images were taken through a JEOL JSM-7000F field emission scanning electron microscope. Cyclic voltammetry measurements were conducted on an Autolab PGSTAT302N. The photovoltaic performance of the devices was evaluated at room temperature within glove box (MBRAUN) conditions (($O_2 < 0.1$ ppm), moisture-free ($H_2O < 0.1$ ppm)), and under standard illumination conditions with an Air Mass 1.5 Global (A.M. 1.5 G) solar simulator at an intensity of 1000 Wm^{-2} using an Agilent B1500A Semiconductor Device Analyzer, calibrated through a reference monocrystalline silicon solar cell supplied by Newport Corporation. The external quantum efficiency (EQE) measurements were conducted immediately after device fabrication, using an integrated system (Enlitech, Taiwan) and a lock-in amplifier with a current preamplifier, under short-circuit conditions. To enhance the credibility of our measurements, the solar simulator (a Xenon lamp) spectrum was calibrated using a monocrystalline photodetector of known spectral response. OSC devices were measured using a Xe lamp and an optical chopper at low frequencies (~200 Hz) in order to maximize the sound/noise (S/N) ratio. At least ten identical devices with six photovoltaics cells each were fabricated so that the reproducibility of the J-V characteristics is ensured.

3. Results and Discussions

3.1. ATR FT-IR Spectroscopy

ATR FT-IR spectra of GO and GO-TNF, in powder form, are presented in Figure 3. Pristine GO (black line) shows a broad and strong peak at 3390 cm^{-1}, which is attributed to O-H stretching vibration of the OH- moieties. Furthermore, stretching vibration of C=O moieties is appeared at 1706 cm^{-1}, while the remaining graphitic domains (C=C) stretching vibration are shown at 1568 cm^{-1}. In addition, C-O-H bending vibration due to COOH groups are presented at 1394 cm^{-1}. The peaks at 1143 cm^{-1} and 1027 cm^{-1} represent C-OH stretching vibration of the hydroxide domains and the stretching vibration of C-O-C groups, respectively. On the other hand, GO-TNF (red line), exhibits a broad peak of low intensity at 3331 cm^{-1}, indicating a N-H stretching vibration. Next, a peak at 1697 cm^{-1} is attributed to C=O stretching vibration deriving from the carbonyl moiety of trinitrofluorenone. Moreover, two peaks occurred at 1652 cm^{-1} and 1575 cm^{-1} are due to amidic C=O stretching vibration. NO_2 asymmetric and symmetric stretch vibrations are shown at 1525 cm^{-1} and 1334 cm^{-1}, respectively, as well as the peak at 1446 cm^{-1} corresponds to the aliphatic a-CH_2 bending vibration of ethylenediamine moiety. Finally, C-N stretching vibration of the ethylene diamine moieties appears at 1105 cm^{-1}. ATR FT-IR spectra of the intermediate TNF and EDA-TNF are reported in the SI.

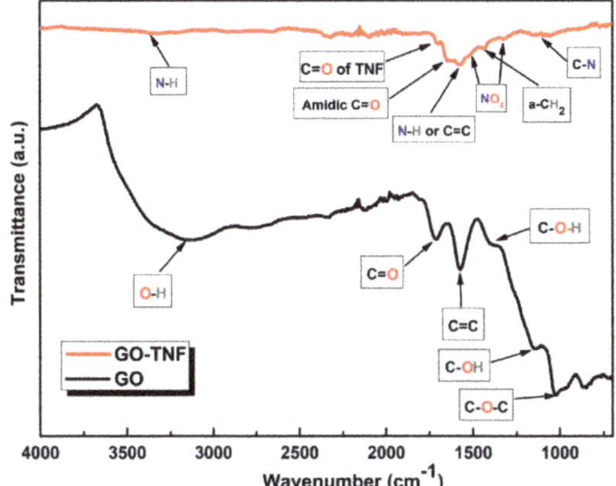

Figure 3. ATR FT-IR spectra of graphene oxide (GO) (black line) and GO-TNF (red line) in transmission mode.

3.2. UV-Visible Measurements

In Figure 4, UV-vis spectra of GO and GO-TNF in solid state are presented. Due to the strong attachment of TNF moieties to the edges of the lattice of GO, the absorption spectrum of GO-TNF is broader than that of the pristine GO, exhibiting a shoulder at ~365 nm. This fact indicates that there is a strong interaction between the GO lattice and TNF moieties, which is mainly attributed to the enhanced electron delocalization caused by TNF [24].

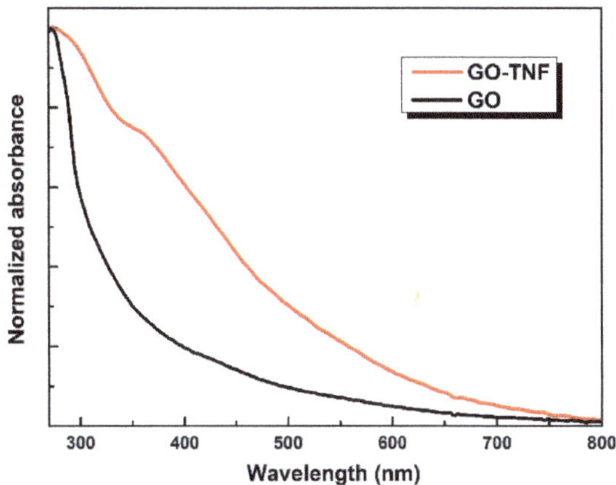

Figure 4. UV-vis spectra of GO (black line) and GO-TNF (red line).

3.3. Raman Spectroscopy

Raman spectra of GO and GO-TNF are shown in Figure 5. No shift is observed for both D and G peaks of GO compared to the respective ones of GO-TNF; D bands occurred at ~1340 cm⁻¹, while G bands at ~1580 cm⁻¹. However, a difference in the relative intensity ratio (I_D/I_G) was observed from 0.92 for GO to 1.04 for GO-TNF, indicating that the linking between GO and TNF increased disorder and defects in the graphitic lattice [25].

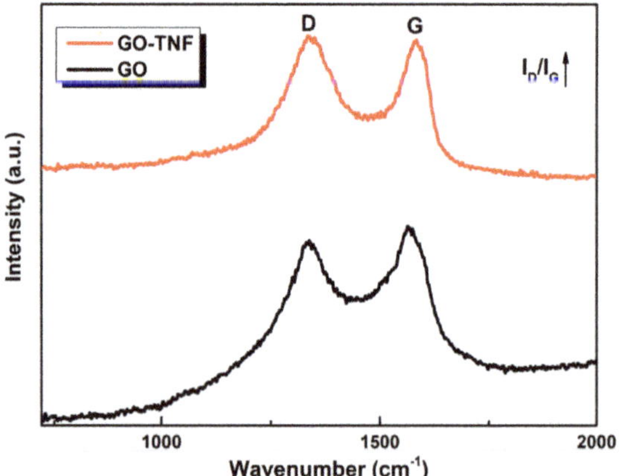

Figure 5. Raman spectra of GO and GO-TNF.

3.4. Photoluminescence (PL) Spectroscopy

Photoluminescence (PL) measurements were conducted to evaluate charge transfer mechanism upon the incorporation of GO-TNF within the active layer, and the respective PL spectra are depicted in Figure 6. In this context, PTB7 and PTB7:GO-TNF thin films were excited at 471 nm presenting an emission band around 760 nm corresponding to radiative decay of photogenerated excitons from the excited state to ground state [29]. When 2% *v/v* GO-TNF ink was added, PL intensity quenching is significant owing to the better energy offset between the LUMO levels of PTB7 and GO-TNF that enhances the charge transfer mechanism. In our case, the incorporation of GO-TNF ink with an optimum concentration of 2% *v/v*, facilitates exciton dissociation at the PTB7:PC$_{71}$BM interface thus leading to a higher number of electrons that can be collected by the cathode, which is in agreement with the champion current density value achieved so far (17.65 mA cm^{-2}) [30].

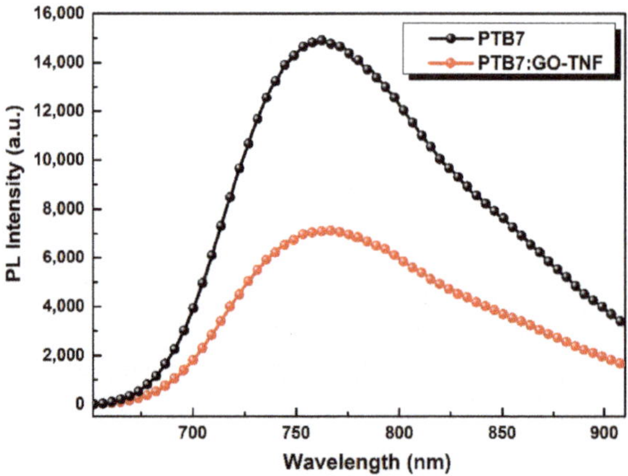

Figure 6. PL spectra of PTB7 (black) and PTB7:GO-TNF (2%) (red).

3.5. XRD Measurements

The crystallinity of pristine GO and GO-TNF was investigated by X-ray diffraction (XRD) in a 2θ range from 5° to 60° (Figure 7). GO displays a narrow peak at 9.51° which is attributed to the main reflection (002) of its stacks with an interlayer d-spacing of ~8.2 Å, while a second weak peak appearing at 42.69° is due to the turbostratic band of disordered carbon materials [31]. On the other hand, GO-TNF exhibits a broad peak at 24.99° referring to (002) reflection with a slightly increased d-spacing of 9.1 Å which is attributed to GO covalent bonding with TNF moieties.

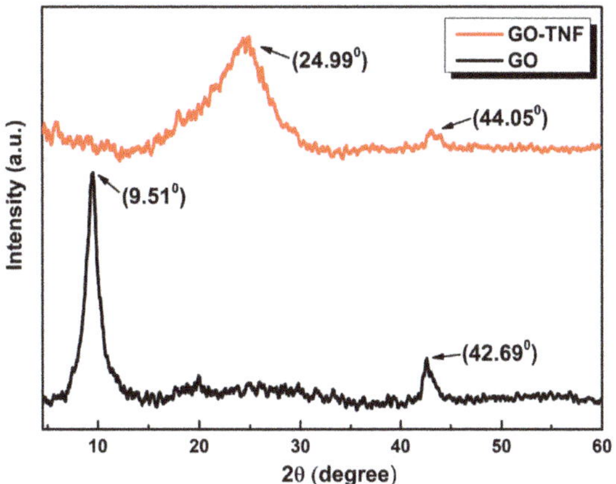

Figure 7. X-ray diffraction patterns of GO (black) and GO-TNF (red).

3.6. Thermogravimetric (TGA) Analysis

Figure 8 displays the TGA curves of GO and GO-TNF obtained under inert atmosphere with a heating rate of 10 °C/min, while the maximum temperature limit was set at 800 °C. First, GO exhibited a moderate weight loss of 5% at a 210 °C, which was followed by a steep weight loss of 37% at about 270 °C due to oxygen functional groups' pyrolysis. Its total mass loss was 40% at 800 °C. On the other hand, GO-TNF presented an improved thermal stability when compared to GO, since the total loss did not exceed 22% of its initial weight. The improved thermal stability of GO-TNF was attributed to the successful amide bond formation between GO and TNF that enhances thermal stability [32].

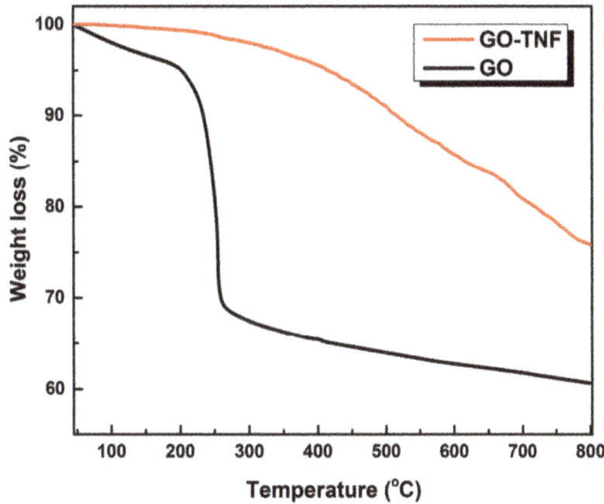

Figure 8. TGA curves of GO (black) and GO-TNF (red) taken under N_2 atmosphere and 10 °C/min heating rate.

3.7. Cyclic Voltammetry Measurements

To determine the energy levels of GO-TNF, cyclic voltammetry measurements were carried out using an electrolytic solution of TBAPF6 in CH_3CN 0.1M, with a scan rate of 10 mVs^{-1}, between the potential sweep window of −2 V to +2 V, as demonstrated in Figure 9. The energy HOMO and LUMO levels of GO-TNF were calculated using the empirical relations below [33]:

$$E_{HOMO} = -(E_{(onset,ox\ vs\ Fc+)/Fc]} + 5.1)(eV) \tag{1}$$

$$E_{LUMO} = -(E_{(onset,red\ vs\ Fc+)/Fc]} + 5.1)(eV) \tag{2}$$

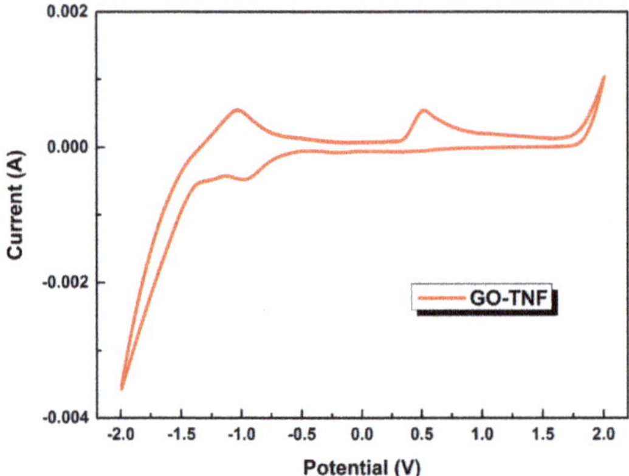

Figure 9. Cyclic voltammogram of GO-TNF.

The HOMO level was approximately −5.66 eV as calculated by the oxidation peak onset 0.53 V, while its LUMO level was extracted from the onset of the reduction peak (0.85 V) and was −4.13 eV.

3.8. Microscopic Characterization

The morphology of GO-TNF was examined using field emission scanning electron microscopy (FE-SEM). Representative SEM images of its flakes coated on silicon substrates are demonstrated in Figure S11. The size of the wrinkled GO-TNF flakes varies ranging from 100 nm to 1 μm, while it should be reported that no charging was observed during SEM imaging, thus indicating that the formed network was electrically conductive.

3.9. Photovoltaic Performance Evaluation

Several OSC devices were fabricated via the incorporation of GO-TNF ink within the binary PTB7:PC$_{71}$BM photoactive layer (Figure 1) in various ratios (1%, 2%, and 3%) and *J-V* characteristic curves were exported to evaluate its operational role into device's photovoltaic (PV) performance (Figure 10). It is obvious that the presence of GO-TNF was beneficial, as stated below in Table 1.

Table 1. Photovoltaic characteristics summary of the OSC devices based on PTB7:GO-TNF:PC$_{71}$BM ternary blends [a].

GO-TNF Content (%)	J_{sc} (mA/cm^2)	Calc. J_{sc} (mA/cm^2)	V_{oc} (V)	FF (%)	PCE (%)
0	16.20 ± 0.45	15.72	0.760 ± 0.010	61.8 ± 0.7	7.61 ± 0.11
1	16.54 ± 0.54	16.21	0.760 ± 0.005	63.0 ± 0.4	7.92 ± 0.26
2	17.21 ± 0.44	16.78	0.760 ± 0.011	64.0 ± 0.1	8.37 ± 0.34
3	16.53 ± 0.35	16.21	0.760 ± 0.009	62.4 ± 0.6	7.84 ± 0.17

[a] The data is averaged from 10 identical devices with 6 cells each.

All ternary devices showed an improved performance, especially the device containing 2% *v/v* GO-TNF ink. In particular, the champion device exhibited a current density (J_{sc}) of 17.65 mA/cm^2 and a power conversion efficiency (PCE) of 8.71% that show an improvement of ~10% and ~13% respectively when compared to the reference device.

To further confirm the experimental J_{sc} improvement due to the incorporation of GO-TNF, external quantum efficiency measurements were conducted to determine the calculated J_{sc}. Figure 10 depicts the external quantum efficiency (EQE) curves of the reference, as well as the champion ternary OSC device incorporating 2% v/v GO-TNF ink.

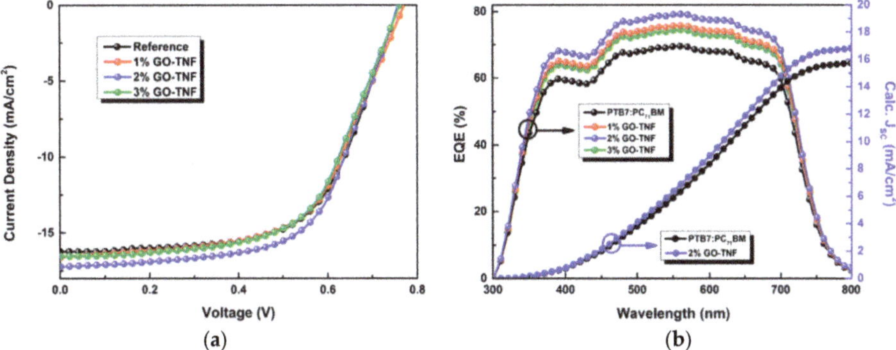

Figure 10. *J-V* characteristics (**a**) and EQE curves (**b**) of the reference (PTB7:PC$_{71}$BM) and the devices incorporating different GO-TNF ink content. The calculated J_{sc} curves (inset in EQE) correspond to the reference and the champion device with 2% GO-TNF ink content.

It can be seen that the EQE enhancement is in full accordance to J_{sc} increase, as calculated from the respective *J-V* curves depicted in Figure 10a. In addition, the absence of any new peak is in agreement with the operational role of GO-TNF, which does not contribute in exciton generation but only in electron transfer, hence confirming the charge transfer mechanism (cascade effect). The accuracy of the PV measurements was checked, by calculating the J_{sc} values of the OSCs from the integration of the EQE spectra. The calculated J_{sc} was found to be -15.72 mA cm^{-2} and -16.78 mA cm^{-2} for the binary and the 2% GO-TNF ink content ternary devices, respectively, which are within the standard deviation from the J_{sc} obtained from the *J-V* curves. It should be also noted that any concentration of GO-TNF ink higher than 3% wt. resulted in a short circuit, probably due to the occurrence of local shunts. This undesired effect could be linked with the concentration of GO-TNF ink in the blend that becomes enough to allow a direct bridging with the ITO electrode.

To get a more accurate insight into the influence of GO-TNF blend into the charge transfer process in the ternary approach, hole-only and electron-only cells were fabricated to calculate the hole and electron mobility, respectively. Measurements were based on space charge limited current method. Hole-only cells and electron-only cells were fabricates using the architecture indium tin oxide ITO/PEDOT:PSS/PTB7:GO-TNF:PC$_{71}$BM/MoO$_3$/Al for holes and ITO/PFN/PCDTBT:GO-TNF:PC$_{71}$BM/Ca/Al for electrons, respectively. The evaluation of the charge carrier mobilities was based on the Mott–Gurney equation [34]:

$$J_{SCLC} = \frac{9}{8}\varepsilon_r\varepsilon_0\mu\frac{(V - V_{bi})^2}{d^3}$$

where ε_r is the relative dielectric constant, ε_0 is the permittivity of free space, μ is the charge carrier mobility, V is the applied voltage, V_{bi} is the built-in potential, and d is the active layer thickness.

Supplementary Figure S12 illustrates *J-V²* characteristics under dark conditions for (a) electron-only and (b) hole-only devices, respectively, where the black line refers to the control device (PTB7:PC$_{71}$BM), while the red line corresponds to the champion ternary one PTB7:GO-TNF (2%): PC$_{71}$BM. According to *J-V²* characteristics, although hole mobility did not present any significant change upon the addition of GO-TNF, the respective electron mobility has shown a significant improvement, passing from 7.80×10^{-5} cm^2 V^{-1} s^{-1} to 9.93×10^{-5} cm^2 V^{-1} s^{-1} (Table 2). This improvement in electron mobility obviously originates from the presence of GO-TNF which has favourable energy levels, located between the energy levels of the donor and acceptor materials, respectively. Hence, the observed electrons' mobility enhancement is directly associated with the cascade effect facilitating electrons' transition from PTB7 to the ITO electrode [35].

In Appendix A, supplementary Figure S13 represents a simple one diode equivalent circuit model corresponding to the ternary OSC device incorporating 2% GO-TNF ink. The diode "D" corresponds to the electrical equivalent of the optical losses at the surface of bulk heterojunction. "R_{sh}" represents the leakage and recombination losses whereas "R_S" represents the sum of the internal resistance, including the resistance of the active layer and ohmic contact.

Due to photoluminescence, the generated excitons are diffused to the nearest D: A interface and are dissociated to form polaron-pairs. Polaron pairs can be either recombined or dissociated into free carriers and are subsequently extracted to the electrodes through a cascade-diffusion process. GO-TNF plays a key role in this cascade process, facilitating charge transfer from the polymer to the fullerene reducing the possibility of free carriers' recombination. This can be proven, by calculating the overall resistance at the maximum power point which represents the decrease of the overall resistance that derives from the higher recombination rate, resulting in larger R_{sh} value (Supplementary Figure S14).

Table 2. Hole and electron mobilities of PTB7:PC$_{71}$BM and ternary blend PTB7:GO-TNF:PC$_{71}$BM *.

Active Layer	μ_h (cm^2 V^{-1} s^{-1})	μ_e (cm^2 V^{-1} s^{-1})	Ratio (μ_h/μ_e)
PTB7:PC$_{71}$BM (reference)	1.28×10^{-4}	7.80×10^{-5}	1.64
1% GO-TNF	1.31×10^{-4}	8.71×10^{-5}	1.50
2% GO-TNF	1.39×10^{-4}	9.93×10^{-5}	1.39
3% GO-TNF	1.34×10^{-4}	8.03×10^{-5}	1.67

* The data were averaged from 10 identical devices with 6 cells each.

3.10. Morphology Characterization of the Active Layer

The morphology of the reference PTB7:PC$_{71}$BM as well as the champion ternary blend (2%) surfaces were both examined by atomic force microscopy (AFM), with a scan size of 1 mm by 1 mm, as displayed in Figure 11. The said ternary active layer exhibited a slightly smoother surface than the binary one, giving a root-mean-square (RMS) roughness of 1.15 nm and 1.22 nm, respectively. The fact that the morphology was improved upon the incorporation of 2% GO-TNF is in full accordance to J_{sc} increase, indicating that the possibility of energetic disorders formation is smaller in case of the ternary device. On the other hand, V_{oc} slightly changed upon the addition of GO-TNF, which is normal, since GO-TNF was incorporated in very low concentrations that cannot significantly affect V_{oc} values.

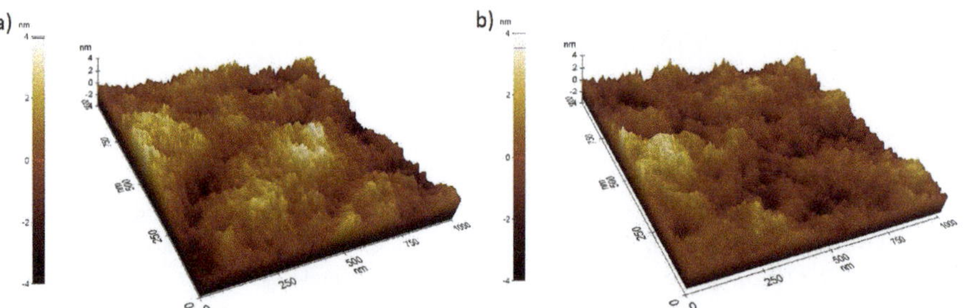

Figure 11. AFM images of (**a**) the reference PTB7:PC$_{71}$BM active layer and (**b**) the champion ternary device containing 2% of GO-TNF ink.

4. Conclusions

Herein, we reported the synthesis and characterization of a new graphene-trinitrofluorenone derivative, named GO-TNF, as well as its operational role as the cascade material between the LUMO energy levels of PTB7 and PC$_{71}$BM, in efficient ternary OSCs. Due to its ideal energy levels, it was directly incorporated, as the third component, within the binary photoactive layer providing a significant improvement in current density of the champion device (2% *v/v* GO-TNF ink) by ~10%. Respectively, the PCE value of the same device was higher by ~13%, leading to a champion efficiency of 8.71%. Our efforts proved that the rational design of graphene-based molecules could provide the opportunity for novel materials synthesis with tunable properties to be incorporated as additives, even as interlayers into organic, as well as hybrid solar cells, thus contributing to the realization of new generation high performance PV systems.

Supplementary Materials: The following are available online at http://www.mdpi.com/2079-4991/10/1/89/s1, Figure S1: The reaction mechanism of the nitration of 9-oxo-fluorene-4-carboxylic acid, Figure S2: The reaction mechanism of acyl chloride synthesis, Figure S3: The reaction mechanism of the amide bond formation, Figure S4: Schematic of 9-oxo-4-carboxyl-fluorenone nitration, Figure S5: ATR FT-IR spectrum of TNF-COOH, Figure S6: Schematic representation of TNF-COCl synthesis, Figure S7: The reaction representation of TNF-EDA synthesis, Figure S8: ATR FT-IR spectrum of TNF-EDA, Figure S9: The reaction representation of GO-COCl preparation, Figure S10: (i) Schematic of the final reaction of GO-TNF, (ii) GO-COCl and TNF-EDA in powder form (a), setup of the reaction (b), GO-TNF in powder form (c), Figure S11: FE-SEM images of GO-TNF, Figure S12: *J-V*^{2} characteristics of the fabricated (a) electron-only and (b) hole-only devices, Figure S13: The one diode equivalent circuit model corresponding to the ternary OSC device incorporating 2% GO-TNF ink, Figure S14: The effect of shunt resistance.

Author Contributions: Conceptualization, M.M.S. and D.M.K.; Methodology, M.M.S., D.M.K. and M.K.; Formal Analysis, M.M.S., C.P., G.K., G.V., N.K. and K.P.; Investigation, M.M.S. and D.M.K., Data Curation, M.M.S., K.A., M.K., G.K., G.V., N.K. and K.P.; Writing—Original Draft Preparation, M.M.S., D.M.K., K.A. and C.P.; Writing—Review and Editing, M.M.S., M.K., G.V. and K.P.; Supervision, M.M.S. and E.K.; Project Administration, M.M.S. Funding Acquisition, M.M.S. All authors have read and agreed to the published version of the manuscript.

Funding: This research is co-financed by Greece and the European Union (European Social Fund-ESF) through the Operational Programme «Human Resources Development, Education and Lifelong Learning» in the context of the project "Reinforcement of Postdoctoral Researchers" (MIS-5001552), implemented by the State Scholarships Foundation (IKY), Grant No. 13992.

Acknowledgments: The authors would like to thank Aleka Manousaki for the FE-SEM images.

Conflicts of Interest: The authors declare no conflict of interest.

Appendix A

The synthetic procedure of GO-TNF and its intermediate stages, as well as explanations of experimental details to understanding and reproducing the research are available.

References

1. Inganäs, O. Organic Photovoltaics over Three Decades. *Adv. Mater.* **2018**, *30*, 1800388. [CrossRef] [PubMed]
2. Lee, C.; Lee, S.; Kim, G.U.; Lee, W.; Kim, B.J. Recent Advances, Design Guidelines, and Prospects of All-Polymer Solar Cells. *Chem. Rev.* **2019**, *119*, 8028–8086. [CrossRef] [PubMed]
3. Zhang, J.; Tan, H.S.; Guo, X.; Facchetti, A.; Yan, H. Material Insights and Challenges for Non-Fullerene Organic Solar Cells Based on Small Molecular Acceptors. *Nat. Energy* **2018**, *3*, 720–731. [CrossRef]
4. He, Z.; Zhong, C.; Huang, X.; Wong, W.Y.; Wu, H.; Chen, L.; Su, S.; Cao, Y. Simultaneous Enhancement of Open-Circuit Voltage, Short-Circuit Current Density, and Fill Factor in Polymer Solar Cells. *Adv. Mater.* **2011**, *23*, 4636–4643. [CrossRef]
5. He, Z.; Zhong, C.; Su, S.; Xu, M.; Wu, H.; Cao, Y. Enhanced power-conversion efficiency in polymer solar cells using an inverted device structure. *Nat. Photon.* **2012**, *6*, 591–595. [CrossRef]
6. Green, M.A.; Emery, K.; Hishikawa, Y.; Warta, W.; Dunlop, E.D. Solar cell efficiency tables (Version 45). *Prog. Photovolt. Res. Appl.* **2013**, *21*, 1–11. [CrossRef]
7. Li, G.; Zhu, R.; Yang, Y. Polymer solar cells. *Nat. Photon.* **2012**, *6*, 153–161. [CrossRef]

8. Meng, L.; Zhang, Y.; Wan, X.; Li, C.; Zhang, X.; Wang, Y.; Ke, X.; Xiao, Z.; Ding, L.; Xia, R.; et al. Organic and Solution-Processed Tandem Solar Cells with 17.3% Efficiency. *Science* **2018**, *361*, 1094–1098. [CrossRef]

9. Yuan, J.; Zhang, Y.; Zhou, L.; Zhang, G.; Yip, H.L.; Lau, T.K.; Lu, X.; Zhu, C.; Peng, H.; Johnson, P.A.; et al. Single-Junction Organic Solar Cell with over 15% Efficiency Using Fused-Ring Acceptor with Electron-Deficient Core. *Joule* **2019**, *3*, 1140–1151. [CrossRef]

10. Liu, T.; Luo, Z.; Chen, Y.; Yang, T.; Xiao, Y.; Zhang, G.; Ma, R.; Lu, X.; Zhan, C.; Zhang, M.; et al. A Nonfullerene Acceptor with a 1000 Nm Absorption Edge Enables Ternary Organic Solar Cells with Improved Optical and Morphological Properties and Efficiencies over 15%. *Energy Environ. Sci.* **2019**, *12*, 2529–2536. [CrossRef]

11. Liu, X.; Yan, Y.; Yao, Y.; Liang, Z. Ternary Blend Strategy for Achieving High-Efficiency Organic Solar Cells with Nonfullerene Acceptors Involved. *Adv. Funct. Mater.* **2018**, *28*, 1–20. [CrossRef]

12. Stylianakis, M.M.; Konios, D.; Petridis, C.; Kakavelakis, G.; Stratakis, E.; Kymakis, E. Ternary Solution-Processed Organic Solar Cells Incorporating 2D Materials. *2D Mater.* **2017**, *4*, 042005. [CrossRef]

13. Gasparini, N.; Salleo, A.; McCulloch, I.; Baran, D. The Role of the Third Component in Ternary Organic Solar Cells. *Nat. Rev. Mater.* **2019**, *4*, 229–242. [CrossRef]

14. Song, X.; Gasparini, N.; Nahid, M.M.; Paleti, S.H.K.; Wang, J.L.; Ade, H.; Baran, D. Dual Sensitizer and Processing-Aid Behavior of Donor Enables Efficient Ternary Organic Solar Cells. *Joule* **2019**, *3*, 846–857. [CrossRef]

15. Seco, C.R.; Vidal-Ferran, A.; Misra, R.; Sharma, G.D.; Palomares, E. Efficient Non-Polymeric Heterojunctions in Ternary Organic Solar Cells. *ACS Appl. Energy Mater.* **2018**, *1*, 4203–4210. [CrossRef]

16. Qin, R.; Guo, D.; Li, M.; Li, G.; Bo, Z.; Wu, J. Perylene Monoimide Dimers Enhance Ternary Organic Solar Cells Efficiency by Induced D-A Crystallinity. *ACS Appl. Energy Mater.* **2019**, *2*, 305–311. [CrossRef]

17. Fu, H.; Li, C.; Bi, P.; Hao, X.; Liu, F.; Li, Y.; Wang, Z.; Sun, Y. Efficient Ternary Organic Solar Cells Enabled by the Integration of Nonfullerene and Fullerene Acceptors with a Broad Composition Tolerance. *Adv. Funct. Mater.* **2019**, *29*, 1–8. [CrossRef]

18. Cheng, P.; Li, Y.; Zhan, X. Efficient Ternary Blend Polymer Solar Cells with Indene-C60 Bisadduct as an Electron-Cascade Acceptor. *Energy Environ. Sci.* **2014**, *7*, 2005–2011. [CrossRef]

19. Lu, L.; Xu, T.; Chen, W.; Landry, E.S.; Lu, L. Ternary blend polymer solar cells with enhanced power conversion efficiency. *Nat. Photon.* **2014**, *8*, 716–722. [CrossRef]

20. Lu, L.; Chen, W.; Xu, T.; Yu, L. High-Performance Ternary Blend Polymer Solar Cells Involving Both Energy Transfer and Hole Relay Processes. *Nat. Commun.* **2015**, *6*, 1–7. [CrossRef]

21. Gasparini, N.; Lucera, L.; Salvador, M.; Prosa, M.; Spyropoulos, G.D.; Kubis, P.; Egelhaaf, H.J.; Brabec, C.J.; Ameri, T. High-Performance Ternary Organic Solar Cells with Thick Active Layer Exceeding 11% Efficiency. *Energy Environ. Sci.* **2017**, *10*, 885–892. [CrossRef]

22. Ke, L.; Gasparini, N.; Min, J.; Zhang, H.; Adam, M.; Rechberger, S.; Forberich, K.; Zhang, C.; Spiecker, E.; Tykwinski, R.R.; et al. Panchromatic Ternary/Quaternary Polymer/Fullerene BHJ Solar Cells Based on Novel Silicon Naphthalocyanine and Silicon Phthalocyanine Dye Sensitizers. *J. Mater. Chem. A* **2017**, *5*, 2550–2562. [CrossRef]

23. Stylianakis, M.M.; Konios, D.; Viskadouros, G.; Vernardou, D.; Katsarakis, N.; Koudoumas, E.; Anastasiadis, S.H.; Stratakis, E.; Kymakis, E. Ternary Organic Solar Cells Incorporating Zinc Phthalocyanine with Improved Performance Exceeding 8.5%. *Dyes Pigm.* **2017**, *146*, 408–413. [CrossRef]

24. Bonaccorso, F.; Balis, N.; Stylianakis, M.M.; Savarese, M.; Adamo, C.; Gemmi, M.; Pellegrini, V.; Stratakis, E.; Kymakis, E. Functionalized Graphene as an Electron-Cascade Acceptor for Air-Processed Organic Ternary Solar Cells. *Adv. Funct. Mater.* **2015**, *25*, 3870–3880. [CrossRef]

25. Stylianakis, M.M.; Konios, D.; Kakavelakis, G.; Charalambidis, G.; Stratakis, E.; Coutsolelos, A.G.; Kymakis, E.; Anastasiadis, S.H. Efficient Ternary Organic Photovoltaics Incorporating a Graphene-Based Porphyrin Molecule as a Universal Electron Cascade Material. *Nanoscale* **2015**, *7*, 17827–17835. [CrossRef] [PubMed]

26. Sygletou, M.; Tzourmpakis, P.; Petridis, C.; Konios, D.; Fotakis, C.; Kymakis, E.; Stratakis, E. Laser Induced Nucleation of Plasmonic Nanoparticles on Two-Dimensional Nanosheets for Organic Photovoltaics. *J. Mater. Chem. A* **2016**, *4*, 1020–1027. [CrossRef]

27. Kakavelakis, G.; Del Rio Castillo, A.E.; Pellegrini, V.; Ansaldo, A.; Tzourmpakis, P.; Brescia, R.; Prato, M.; Stratakis, E.; Kymakis, E.; Bonaccorso, F. Size-Tuning of WSe2 Flakes for High Efficiency Inverted Organic Solar Cells. *ACS Nano* **2017**, *11*, 3517–3531. [CrossRef]

28. Stylianakis, M.M.; Viskadouros, G.; Polyzoidis, C.; Veisakis, G.; Kenanakis, G.; Kornilios, N.; Petridis, K.; Kymakis, E. Updating the Role of Reduced Graphene Oxide Ink on Field Emission Devices in Synergy with Charge Transfer Materials. *Nanomaterials* **2019**, *9*, 137. [CrossRef]

29. Nagarjuna, P.; Bagui, A.; Gupta, V.; Singh, S.P. A Highly Efficient PTB7-Th Polymer Donor Bulk Hetero-Junction Solar Cell with Increased Open Circuit Voltage Using Fullerene Acceptor CN-PC$_{70}$BM. *Org. Electron.* **2017**, *43*, 262–267. [CrossRef]

30. Fan, R.; Huai, Z.; Sun, Y.; Li, X.; Fu, G.; Huang, S.; Wang, L.; Yang, S. Enhanced Performance of Polymer Solar Cells Based on PTB7-Th:PC 71 BM by Doping with 1-Bromo-4-Nitrobenzene. *J. Mater. Chem. C* **2017**, *5*, 10985–10990. [CrossRef]

31. Zhu, Y.; Murali, S.; Cai, W.; Li, X.; Suk, J.W.; Potts, J.R.; Ruoff, R.S. Graphene and Graphene Oxide: Synthesis, Properties, and Applications. *Adv. Mater.* **2010**, *22*, 3906–3924. [CrossRef] [PubMed]

32. Tang, X.Z.; Li, W.; Yu, Z.Z.; Rafiee, M.A.; Rafiee, J.; Yavari, F.; Koratkar, N. Enhanced thermal stability in graphene oxide covalently functionalized with 2-amino-4, 6-didodecylamino-1, 3, 5-triazine. *Carbon* **2011**, *49*, 1258–1265. [CrossRef]

33. Cardona, C.M.; Li, W.; Kaifer, A.E.; Stockdale, D.; Bazan, G.C. Electrochemical considerations for determining absolute frontier orbital energy levels of conjugated polymers for solar cell applications. *Adv. Mater.* **2011**, *23*, 2367–2371. [CrossRef] [PubMed]

34. Azimi, H.; Senes, A.; Scharber, M.C.; Hingerl, K.; Brabec, C.J. Charge Transport and Recombination in Low-Bandgap Bulk Heterojunction Solar Cell using Bis-adduct Fullerene. *Adv. Energy Mater.* **2011**, *1*, 1162–1168. [CrossRef]

35. Lu, L.; Kelly, M.; You, W.; Yu, L. Status and prospects for ternary organic photovoltaics. *Nat. Photon.* **2015**, *9*, 491–500. [CrossRef]

 *nanomaterials*

Article

A Compact and Smooth CH₃NH₃PbI₃ Film: Investigation of Solvent Sorts and Concentrations of CH₃NH₃I towards Highly Efficient Perovskite Solar Cells

Liang Chen [1,2,*], Hao Zhang [1], Jiyuan Zhang [3] and Yong Zhou [2,3]

[1] Hunan Key Laboratory of Applied Environmental Photocatalysis,
    Hunan Collaborative Innovation Center of Environmental and Energy Photocatalysis,
    Changsha University, 98 Hongshan Road, Changsha 410022, China; zhanghao0122@126.com
[2] Jiangsu Key Laboratory for Nano Technology, National Laboratory of Solid State Microstructures,
    College of Engineering and Applied Sciences, and Collaborative Innovation Center of Advanced
    Microstructures, Nanjing University, 22 Hankou Road, Nanjing 210093, China; zhouyong1999@nju.edu.cn
[3] Kunshan Innovation Institute, Nanjing University, Kunshan, Jiangsu 215347, China; jyzxni@163.com
* Correspondence: z20161153@ccsu.edu.cn

Received: 10 October 2018; Accepted: 29 October 2018; Published: 1 November 2018

Abstract: Four solvents (isopropanol (IPA), *n*-butyl alcohol (NBA), *n*-amyl alcohol (NAA), and *n*-hexyl alcohol (NHA)) were investigated to prepare CH₃NH₃I (methylammonium iodide, MAI) solutions to transform PbI₂ film into CH₃NH₃PbI₃ (MAPbI₃) film. It was found that the morphology of the perovskite MAPbI₃ film was not only affected by the chain of the solvent molecule, but also by the concentration of MAI. The use of solvents with a long alkyl chain (NAA and NHA) allowed the MAPbI₃ to grow via an *in situ* transformation step, which easily made the perovskite films compact, but with a high surface roughness due to the growth of unexpected nanorods/nanoplates. The solvent with a short alkyl chain (IPA) led to the dissolution−crystallization growth mechanism, resulting in rapid generation of perovskite films with a number of pinholes. A high-quality (compact, smooth, pinhole-free) perovskite film was obtained with NBA and an optimized MAI concentration of 8 mg/mL. The corresponding perovskite solar cells achieved a maximum power conversion efficiency (PCE) of 16.66% and average PCE of 14.76% (for 40 cells).

Keywords: solvent; compact; smooth; perovskite solar cells

---

1. Introduction

Organic-inorganic lead halide perovskite solar cells have attracted intensive attention due to the remarkable progress in power conversion efficiency (PCE), increasing rapidly to over 22% in a few years [1–18]. The PCEs of perovskite solar cells with various architectures (mesoporous or planar heterojunction structures) are strongly dependent on the morphology of the perovskite film [19–26]. The growth mechanism relating to the chemical reaction kinetics significantly influences the perovskite polymorph [27]. There are several fabrication methods of perovskite films, including two-step deposition techniques [6,7], one-step solution spin-coating [8,28,29], and vacuum-evaporation deposition [9,30]. Among them, the two-step solution-process deposition technique is an easy method to obtain cells with excellent photovoltaic performance in a reproducible way [6,7].

In a typical two-step sequential solution process, PbI₂ film is first deposited on a TiO₂-coated substrate by spin-coating with a dimethylformamide (DMF) solution of PbI₂, and is subsequently transformed into a perovskite CH₃NH₃PbI₃ (methylammonium lead iodide, MAPbI₃) film through exposure to a CH_3NH_3I (methylammonium iodide, MAI) solution in isopropanol (IPA) [6,7]. However, this two-step solution method often results in poor grain structure, incomplete surface coverage, and high surface roughness [19–23,25], which limits the further improvement of photovoltaic performance of perovskite solar cells. In a previous investigation, it was proposed that the perovskite grain morphology was governed by its growth mechanism. The growth mechanism is in line with a competing relation between an *in situ* transformation and dissolution-crystallization mechanism, which is determined by the rate and amount of PbI_2 dissolving in the MAI solution [27]. It was confirmed that the MAI concentration significantly affects the solubility of PbI_2 in the MAI solution and the morphology of perovskite grains. When the $MAPbI_3$ crystals grow based on the *in situ* transformation mechanism, the formed $MAPbI_3$ grains will maintain parallel geometries to PbI_2. On the other hand, in the dissolution–crystallization mechanism, the large amount of dissolved PbI_2 results in the rapid generation rate of perovskite grains, the collapse of lead backbones, and the occurrence of tetragonal-shaped $MAPbI_3$ [27]. Hence, perovskite films forming in different solutions might display diverse morphologies. Unfortunately, the role of the solvent on the $MAPbI_3$ film formation has been neglected. Thus, it is essential to systematically probe the relationship between the chemical properties of the solvent and the perovskite crystal growth mechanism.

Herein, besides commonly used IPA, *n*-butyl alcohol (NBA), *n*-amyl alcohol (NAA), and *n*-hexyl alcohol (NHA) (the corresponding molecular formula are shown in Figure S1 in Supporting Information) were also selected as solvents to prepare MAI solutions to transform PbI_2 into $MAPbI_3$. From IPA to NHA, the polarity of the solvents decreases with the increasing length of the molecule chain. Considering that PbI_2 is a polar molecule, its solubility is oppositely correlated with the length of the alkyl chain of the solvent molecule in the MAI solution. Therefore, the effect of the length of the alkyl chain of the solvents on the perovskite film morphology could be investigated systemically.

In this work, it was found that the morphology of the perovskite $MAPbI_3$ film was significantly affected by the alkyl chain of the solvent (i.e., solvent polarity) and the MAI concentration. On one hand, the coverage and compactness of the perovskite film was improved through the use of a solvent with a longer alkyl chain of MAI. This may be attributed to the slow chemical reaction rate of PbI_2 and MAI, resulting from the low polarity of the solvent with a longer alkyl chain. On the other hand, the long-alkyl-chain solvent may also generate unexpected nanorods/nanoplates on the film surface that increase the roughness of the perovskite film, which is negatively related to the photovoltaic performance of the perovskite solar cell. In a word, both the high concentration of MAI and long reaction time facilitated PbI_2 to form PbI_3^- or PbI_4^{2-}, which also led to large crystal $MAPbI_3$ nanorods/nanoplates growing on the surface with long loading time [27,31]. The best-quality (highly compact, pinhole-free, and smooth) perovskite films were fabricated with an optimized MAI concentration of 8 mg/mL and using NBA as solvent. The resulting perovskite solar cell based on this film achieved a maximum PCE of 16.66% and average PCE of 14.76% (for 40 cells).

2. Experimental Section

2.1. Materials

Methylamine (33 wt% in absolute ethanol), hydroiodic acid (57 wt% in water, 99.99%), and Li-bis(trifluoromethanesulfonyl) imide (Li-TFSI) were purchased from Sigma-Aldrich (St. Louis, MO, USA). 4-tert-butylpyridine (TBP) was purchased from Aladdin (Shanghai, China). Acetonitrile (99.8%), chlorobenzene (99.9%), dimethylformamide (DMF) (99.9%), PbI_2 (99.999%), IPA, NBA, NAA, and NHA were purchased from Alfa Aesar (Ward Hill, MA, USA). 2,2′,7,7′-tetrakis-(*N*,*N*-di-p-methoxyphenylamine)-9,9′-spirobifluorene (spiro-OMeTAD) ($\geq$99.0%) was purchased from Shenzhen Feiming Science and Technology Co., Ltd. (Shenzhen, China).

2.2. MAI Preparation

A solution was prepared by reacting methylamine (27.8 mL, 33 wt% in absolute ethanol) with hydroiodic acid (30 mL, 57 wt% in water) at 0 °C for 2 h. Subsequently, a precipitate was obtained through rotary evaporation of the solution at 60–70 °C. The as-prepared product was washed with diethyl ether and ethanol, and then dried in a vacuum oven at 60 °C overnight.

2.3. Devices Fabrication

A TiO_2 compact layer was prepared on fluorine-doped tin oxide glass (FTO) by spin-coating a sol-gel solution at 5000 rpm for 30 s and sintering at 500 °C for 30 min. The sol-gel solution consisted of 18 ml ethanol, 1.8 mL tetrabutyl titanate, and 0.38 mL diethanolamine. Subsequently, the TiO_2 microstructure film was fabricated by spin-coating with dispersion (Deysol HR TiO_2/ethanol 1:5 wt) and heated at 500 °C for 30 min.

Next, 60 μL of 1.0 M PbI_2 solution in DMF was dropped on the TiO_2-coated FTO substrate, and the substrate was then spun at 4000 rpm for 25 s. The as-prepared PbI_2 film was heated at 70 °C for 10 min, and then naturally cooled to room temperature. Subsequently, 150 μL MAI solution was used to fully covered the PbI_2 film. After a certain loading time (1–3 min), PbI_2 was transformed into $MAPbI_3$, and the solution was then dried by spinning. The $MAPbI_3$ film was heated at 100 °C for 30 min. When the $MAPbI_3$ film cooled down to room temperature, 30 μL of hole-transport materials (HTM) solution was spun on the film at 3000–4000 rpm for 45 s. The HTM solution was prepared by dissolving 68 mM spiro-OMeTAD, 26 mM Li-TFSI, and 55 mM TBP in acetonitrile and chlorobenzene (v/v = 1:10). The above processes were operated in an Ar glove box. Finally, a back-contact electrode (120 nm silver film) was evaporated on HTM layer. In this work, the active area of a cell was 0.09 cm^2. A series of solar cells were prepared by tuning the solvents and concentration of MAI in different MAI solution. The solvent sorts are IPA, NBA, NAA and NHA; the concentrations of MAI in the four-selected solvents range from 6 to 12 mg/mL.

2.4. Characterization

The crystallographic phases of the as-prepared products were determined by powder X-ray diffraction (XRD) (Rigaku Ultima III, Tokyo, Japan) using Cu-Ka radiation (λ = 0.154178 nm) with a scan rate of 10 °/min at 40 kV and 40 mA. The morphologies of the films were examined with a field emission scanning electron microscope (FE-SEM) (JEOL, Tokyo, Japan, JSM-6700F with an accelerating voltage of 5 kV). Atomic force microscope (AFM) analysis was performed on a MFP3D microscope (Asylum Research, MFP-3D-SA, Santa Barbara, CA, USA). The absorption properties of films were investigated by ultraviolet-visible (UV-vis) spectrometer (Shimadzu UV-2550, Kyoto, Japan) and all the data were corrected by deducting the baseline value of the TiO_2-coated FTO substrate. Photocurrent density-voltage (I-V) measurements were carried out with a Keithley (Johnston, OH, USA) 236 source measurement unit under AM 1.5 illumination cast with an Oriel 92251A-1000 sunlight simulator (Irvine, CA, USA) calibrated with the standard reference of a Newport 91150 silicon solar cell. The external quantum efficiency (EQE) spectra were measured under monochromatic irradiation with a xenon lamp and a monochromator. Photoluminescence (PL) spectrum was obtained with a 532 nm pulsed laser as excition source at a frequency of 9.743 MHz.

3. Results and Discussion

Figure 1a$_1$–a$_4$ shows the morphologies of various perovskite films prepared using different solvents with the same MAI concentration of 6 mg/mL. As the solvent changes from IPA to NHA accompanying with the alkyl chain increasing, the perovskite films become pinhole-free and compact. The presence of the pinholes may result in subsequently coated HTM solutions to deleteriously infiltrate into the perovskite absorber layer. It increased the recombination of electron and hole due to the large contact area between HTM and perovskite [13]. Furthermore, for NAA and NHA solvents, some nanorods/nanoplates were observed to grow on the film surface, which will be detrimental to the performance of the perovskite solar cell. The presence of the nanorods/nanoplates requires thick HTM layers to cover the perovskite in order to eliminate the short-circuiting [22]. Thus, the thick HTM layer may result in lower short-circuit photocurrent density (J_{sc}) and fill factor (FF) due to the low conductivity of the HTM.

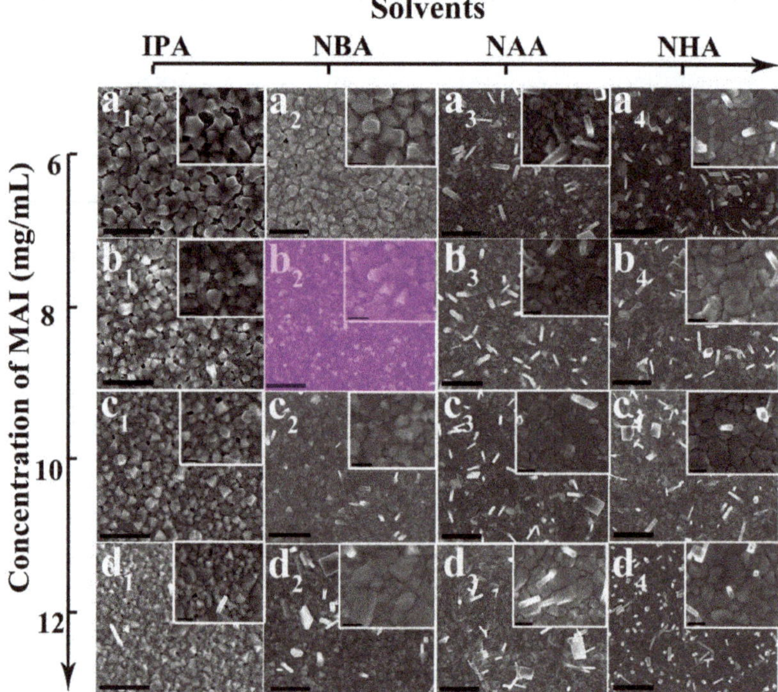

Figure 1. Field emission scanning electron microscope (FE-SEM) images of $CH_3NH_3PbI_3$ (MAPbI$_3$) films obtained from 6–12 mg/mL CH_3NH_3I (MAI) in (a_1–d_1) isopropanol (IPA), (a_2–d_2) *n*-butyl alcohol (NBA), (a_3–d_3) *n*-amyl alcohol (NAA), and (a_4–d_4) *n*-hexyl alcohol (NHA). The scale bars of a_1–d_4 are 1 µm, and those of inserts are 200 nm.

Through careful observation of the film (inserts of Figure 1a_1–a_4), the perovskite grains were found to gradually transit from nanocuboids to an unspecific shape without any corners. The morphology change trend indicated different growth mechanisms of perovskite grains between the four used solvents. For the *in situ* transformation mechanism, the resulting perovskite grains should keep parallel geometries to PbI$_2$ due to the preservation of the inorganic lead framework. For the dissolution-crystallization mechanism, grains would be reconstructed into a tetragonal shape because of the tetragonal phase MAPbI$_3$ [27]. In order to clearly reveal the effect of the solvent molecule, we specifically explored the growth mechanisms of the perovskite grains in IPA and NHA solutions. Two series of films were fabricated through exposing PbI$_2$ films to the 6 mg/mL MAI in IPA or NHA with a loading time of 0–3 min. Note that the 0 min loading time signifies that the film was quickly dried through spinning as soon as the PbI$_2$ film was covered with the MAI solution. X-ray diffraction (XRD) patterns (Figure S2) and ultraviolet-visible (UV-vis) absorption spectra (Figure S3) were performed to trace the reaction process of the PbI$_2$ film with MAI in IPA and NHA. In the case of IPA, only the film prepared with a 0-min loading time shows a PbI$_2$ diffraction peak, whereas no PbI$_2$ peak was observed in the other films with longer loading times (Figure S2a). This demonstrates that the reaction of PbI$_2$

and MAI can be finished in 1 min in IPA. The smaller differences between the UV-vis absorption spectra of those films with loading times of 1–3 min for IPA also demonstrate that the PbI_2 film was totally transformed into the $MAPbI_3$ film in 1 min (Figure S3a). The dissolution-crystallization mechanism is responsible for the fast reaction [18]. The PbI_2 peak intensity of the film in NHA was gradually decreased with increasing loading time, and totally disappeared at 3 min (Figure S2b). This means that the formation of the PbI_2 residue-free perovskite film took 3 min of loading time. The UV-vis absorption spectra of those films showed that the absorption obviously increased at 2 min, and differences decreased between 2 and 3 min in NHA (Figure S3b), which indicated that the formation of $MAPbI_3$ film was completed in 2–3 min. The slower rate of perovskite film generation provides evidence for the *in situ* transformation mechanism in NHA [27]. In addition, long reaction times (2–3 min) would make a certain amount of PbI_2 dissolve in NHA, resulting in the formation of perovskite nanorods/nanoplates.

The corresponding FE-SEM images of the resulting films are presented in Figure S4. For IPA, the number of perovskite nanocuboids increased with loading time from 0 min (Figure S4a) to 1 min (Figure S4b), and the interconnection between nanocuboids became closer. The morphologies of films with 2 min (Figure S4c) and 3 min loading (Figure S4d) are not obviously different from that of 1 min loading, and the films still have incomplete coverage. The final tetragon-shaped morphology (Figure S4a–d) implies the collapse of the lead backbone occurred in the IPA solution with the present 6 mg/mL MAI. This indicates that the growing $MAPbI_3$ crystals followed a dissolution-crystallization step. Notably, the randomly stacked nanocuboids might be responsible for difficulties in obtaining pinhole-free and full-coverage perovskite films from IPA solutions.

For NHA, with 0 min loading time, the film was found to be already full-covered but the grains did not interconnect closely (Figure S4e). The interconnection became increasingly closer with the extension of loading time, and the perovskite films finally became very compact (Figure S4f–h). The applied PbI_2 films consisted of unspecific-shaped particles (Figure S5), and the subsequent perovskite grains of the full-covered films were still unspecific-shaped rather than tetragonal. This indicated the dominant growth mechanism was *in situ* transformation in the NHA solution with the present 6 mg/mL MAI [27]. With the *in situ* transformation, the $MAPbI_3$ grains grow up *in situ* to fill the void between the pristine PbI_2 particles (Figure S5), resulting in the film easily becoming compact and complete. However, perovskite nanorods grow upon the surface with loading times of 2 min or longer. The reason is that PbI_2 is partly soluble in the MAI solution by forming a Pb complex (PbI_3^- or PbI_4^{2-}), and the dissolving Pb complex increases with longer loading times so that large and tetragon-shaped $MAPbI_3$ crystals grow on the surfaces of underlying film via the recrystallization step [27,31], i.e.,

$$PbI_2 + xI^- \rightleftharpoons PbI_{2+x}^{x-} (x = 1, 2)$$

$$CH_3NH_3^- + PbI_{2+x}^{x-} \rightleftharpoons CH_3NH_3PbI_3 + (x - 1)I^- (x = 1, 2)$$

Therefore, the *in situ* transformation mechanism occurred in NHA solution, whereas the dissolution-crystallization mechanism appeared in the IPA solution. It could be concluded that the molecules with a longer alkyl chain gave rise to slower chemical reaction rates. The results might be ascribed to the polarity of NHA being lower than that of IPA due to the different alkyl-chain lengths. Given that PbI_2 is a polar molecule, it easily dissolves in the IPA solution, and then forms nanocuboids of $MAPbI_3$. For NHA, the *in situ* transformation mechanism might bear responsibility for the formation of full-covered film, as the grain gradually grows *in situ* and maintains the inorganic lead framework (unspecific-shaped PbI_2 particles). In other words, the coverage and compactness of the perovskite film could be easily achieved by using a longer-alkyl-chain solvent (i.e., low polar solvent) for MAI.

The morphology of the perovskite $MAPbI_3$ film does not only depend on solvent polarity but also on the MAI concentration. Therefore, the relationship between the concentration of MAI and quality (e.g., compactness and smoothness) of the $MAPbI_3$ film was also studied. The surface morphologies of the various $MAPbI_3$ films obtained from different concentrations of MAI are described in Figure 1 and summarized in Table 1. The obtained perovskite films could be classified into three basic types: (1) incomplete-covered film, (2) compact film roughed with nanorods/nanoplates, and (3) compact and smooth film. All the perovskite films were proved to be PbI_2-free by XRD patterns (Figure S6). For IPA, the resulting $MAPbI_3$ films were incomplete-covered (Figure $1a_1$–d_1). With increasing MAI concentrations, the coverage of the perovskite film increased, but the size of the perovskite grains decreased. These findings are in agreement with the literature [1,27]. No perovskite nanorods grew on the surface of films except sporadically on the film obtained from the 12 mg/mL MAI in IPA. For NBA, the perovskite grains growing in 6 mg/mL MAI were observed to still be nanocuboids (Figure $1a_2$). Although the perovskite film contained a number of pinholes, its coverage was better than that with the IPA solution. The film prepared with 8 mg/mL MAI in NBA became very compact, and no obvious perovskite nanorods were observed (Figure $1b_2$). Thus, this pinhole-free film was quite smooth. Since the high-quality film consisted of perovskite grains with tetragonal and unspecific morphologies, both dissolution–crystallization and *in situ* transformation mechanisms worked together with 8 mg/mL MAI in NBA. Perovskite nanorods or nanoplates began to form on the surfaces of two films obtained from 10 and 12 mg/mL MAI in NBA (Figure $1c_2$,d_2). Figure $1a_3$–d_4 shows that the surfaces of all compact films prepared with 6–12 mg/mL in NAA or NHA also contained nanorods or nanoplates. These nanorods/nanoplates grew on the compact film surface through $CH_3NH_3^-$ reacting with the Pb complex (PbI_3^- or PbI_4^-) in solutions of MAI, increasing the surface roughness. Figure S7 shows the corresponding atomic force microscope (AFM) images and root mean squares (RMS) of three perovskite films respectively obtained from 6, 8, and 10 mg/mL MAI in NBA, which represented incomplete-covered film, smooth and compact film, and compact film roughed with nanorods/nanoplates, respectively. With the film changing from incomplete to compact, the RMS decreased due to the disappearance of pinholes. Figure S7c indicates that the increase of the RMS originates from the height of nanorods/nanoplates over the compact film surface.

Table 1. Morphologies of $CH_3NH_3PbI_3$ ($MAPbI_3$) films obtained from 6–12 mg/mL CH_3NH_3I (MAI) in four-selected solvents.

Conentration \ Solvents	Isopropanol (IPA)	*n*-butyl Alcohol (NBA)	*n*-amyl Alcohol (NAA)	*n*-hexyl Alcohol (NHA)
6 (mg/mL)	Incomplete-covered	Incomplete-covered	compact with nanorods	compact with nanorods
8 (mg/mL)	Incomplete-covered	compact	compact with nanorods	compact with nanorods
10 (mg/mL)	Incomplete-covered	compact with nanorods	compact with nanorods and nanoplates	compact with nanorods and nanoplates
12 (mg/mL)	Incomplete-covered	compact with nanorods and nanoplates	compact with nanorods and nanoplates	compact with nanorods and nanoplates

A series of perovskite solar cells were assembled to study the photovoltaic performance of those PbI_2 residue-free perovskite films. Figure 2 performs the photovoltaic parameters of the corresponding devices as functions of the MAI concentrations and solvent sorts. Those photovoltaic parameters were obtained from photocurrent density-voltage (I-V) measurements under reverse scan. On one hand, the perovskite solar cells obtained from 6–8 mg/mL MAI in IPA suffered lower J_{sc} (16–18 mA/cm^2) and open-circuit voltage (V_{oc}, 0.93–0.95 V) due to serious recombination. The reason for this is that HTM infiltrated into the incomplete-covered films through the pinholes (Figure $1a_1$,b_1), which increased the contact area between HTM and perovskite. The higher J_{sc} (20–21 mA/cm^2) and V_{oc} (0.97–1.01 V) of the films from 10–12 mg/mL MAI in IPA resulted from the improvement of perovskite film coverage (Figure $1c_1$,d_1). On the other hand, unsatisfactory and slightly different PCEs (range 10–12%) were achieved by those cells obtained from NAA and NHA solutions, resulting from lower J_{sc} (17–19 mA/cm^2) and fill factor (FF, 54–64%). This might be explained by the fact that thick HTM layers were required to completely cover the compact perovskite films with dense nanorods/nanoplates (Figure $1c_2$–d_4 and Figure S7c), which increased the series resistance of devices. A solar cell obtained from an NBA solution of 8 mg/mL MAI (Figure $1b_2$) achieved promising J_{sc} (22.79 mA/cm^2), V_{oc} (1.06 V), FF (69%), and PCE (16.66%), which is ascribed to the compact and smooth perovskite film without pinholes or nanorods/nanoplates. In addition, it is well known that perovskite solar cells often show an unusual hysteresis in the I-V curves. All perovskite solar cells were measured with both reverse and forward scan directions to estimate their photovoltaic performance. The corresponding detailed parameters and I-V cures are presented in Figure 3a and Supporting Information (Figures S8–S11). Figure 3a shows the I-V curves and detailed parameters of the best-performing solar cell. The reverse scan curve shows the highest PCE (16.66%) and the forward scan curve indicates that the same cell achieved a PCE of 15.96%. The small difference between the two PCEs implies less hysteresis in the device prepared with 8 mg/mL MAI in NBA. In order to further study the performance of the cell under work conditions, the photocurrent density was measured at 0.833 V for 110 s. The stabilized photocurrent density was 18.38 mA/cm^2, and a stable high PCE of 15.31% was obtained (Figure 3b). This stable power output efficiency suggests the possibility to prepare efficient and stable perovskite solar cells with 8 mg/mL MAI in NBA.

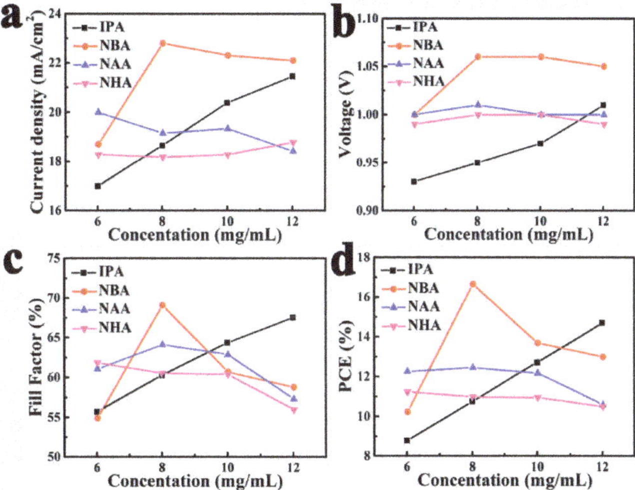

Figure 2. Detail parameters and power conversion efficiencies (PCEs) of the cells prepared by using MAPbI$_3$ films obtained from 6–12 mg/mL MAI in IPA, NBA, NAA, or NHA.

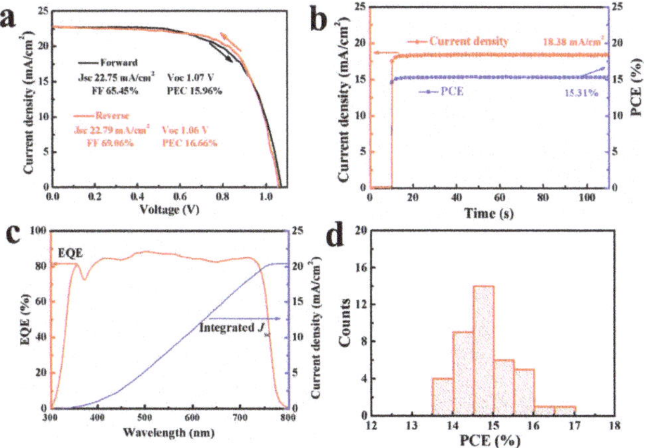

Figure 3. (**a**) Density-voltage (I–V) cures of the best-performing cell, the voltage scan rate is 50 mV/s (**b**) Photocurrent density and PCE as a function of time for the same cell under 0.833 V, (**c**) External quantum efficiency (EQE) spectrum of the best-performing cell, (**d**) Histograms of PCEs measured for 40 cells prepared by using 8 mg/mL MAI in NBA.

To check the J_{sc} of the best-performing cell, the external quantum efficiency (EQE) was measured (Figure 3c). The J_{sc} can be predicted by integrating the EQE spectrum with the incident photo flux density distribution using Equation (1),

$$\text{Integrated } J_{SC} = \int qF(\lambda F(\lambda)EQ)d\lambda \tag{1}$$

where q is the electron charge and $F(\lambda)$ is the incident photon flux density (AM 1.5, ASTM G173) at wavelength λ. The integrated J_{sc} of the cell was 20.46 mA/cm^2, which is close to the value obtained from the I–V measurement (Figure 3a). To investigate the reproducibility of the performance of cells fabricated using 8 mg/mL MAI in NBA, 40 individual devices were measured to obtain the corresponding PCEs. Statistical analysis of these PCEs showed an average value of 14.76% with a small standard deviation of 0.67% for all 40 cells (Figure 3d).

4. Conclusions

It was demonstrated that the solvent significantly affects the growth mechanism of perovskite grains. The morphology of perovskite films was controlled by manipulating the solvent polarity and concentration of MAI solutions. The use of solvents with long alkyl-chain molecule (i.e., low polarity solvent) allowed the MAPbI$_3$ to grow via an *in situ* transformation step, which easily made the perovskite films compact, but increased the surface roughness due to the growth of unexpected nanorods/nanoplates. Using the solvent with a short alkyl chain (i.e., high polarity solvent) led to the dissolution-crystallization growth mechanism, resulting in rapidly generation of perovskite films with smooth surface, but with a number of pinholes. The best quality (highly compact, pinhole-free, and smooth) perovskite films were fabricated with an optimized MAI concentration of 8 mg/mL and NBA as the solvent, where both growth mechanisms worked together. The corresponding perovskite solar cell achieved a promising PCE of 16.66% and average PCE of 14.76% (for 40 cells). The present work may give some clues to other groups to further investigate the mechanisms of perovskite grain growth and prepare high-quality perovskite film through simple methods.

Supplementary Materials: Supplementary materials are available online at http://www.mdpi.com/2079-4991/8/11/897/s1.

Author Contributions: L.C. performed the experiments and wrote the paper. H.Z. carried out UV-vis spectrometer measurements. J.Z. and Y.Z. contributed the idea of the paper and analyzed the experimental results. All authors read and agreed to the final version for publication.

Funding: This research was funded by National Natural Science Foundation of China (No. 51802029), Scientific Research Fund of Hunan Provincial Education Department (No. 17B028), Natural Science Foundation of Hunan Province, China (No. 2017JJ3343) and the Initial Founding of Scientific Research for The Introduction of Talents of Changsha University (No. SF1606). Project of Changsha bureau of science and technology (No. k1705062). NSF of Jiangsu Province (Nos. BK20160412 and BK20130425).

Conflicts of Interest: The authors declare no conflict of interest.

References

1. Green, M.A.; Ho-Baillie, A.; Snaith, H.J. The emergence of perovskite solar cells. *Nat. Photonics* **2014**, *8*, 506. [CrossRef]
2. Zhou, H.; Chen, Q.; Li, G.; Luo, S.; Song, T.-B.; Duan, H.-S.; Hong, Z.; You, J.; Liu, Y.; Yang, Y. Interface engineering of highly efficient perovskite solar cells. *Science* **2014**, *345*, 542. [CrossRef] [PubMed]
3. Yang, W.S.; Noh, J.H.; Jeon, N.J.; Kim, Y.C.; Ryu, S.; Seo, J.; Seok, S.I. High-performance photovoltaic perovskite layers fabricated through intramolecular exchange. *Science* **2015**, *348*, 1234. [CrossRef] [PubMed]
4. Mei, A.; Li, X.; Liu, L.; Ku, Z.; Liu, T.; Rong, Y.; Xu, M.; Hu, M.; Chen, J.; Yang, Y.; et al. A hole-conductor–free, fully printable mesoscopic perovskite solar cell with high stability. *Science* **2014**, *345*, 295. [CrossRef] [PubMed]
5. Jeon, N.J.; Noh, J.H.; Yang, W.S.; Kim, Y.C.; Ryu, S.; Seo, J.; Seok, S.I. Compositional engineering of perovskite materials for high-performance solar cells. *Nature* **2015**, *517*, 476. [CrossRef] [PubMed]

6.   Im, J.-H.; Jang, I.-H.; Pellet, N.; Grätzel, M.; Park, N.-G. Growth of $CH_3NH_3PbI_3$ cuboids with controlled size for high-efficiency perovskite solar cells. *Nat. Nanotechnol.* **2014**, *9*, 927. [CrossRef] [PubMed]

7.   Burschka, J.; Pellet, N.; Moon, S.-J.; Humphry-Baker, R.; Gao, P.; Nazeeruddin, M.K.; Grätzel, M. Sequential deposition as a route to high-performance perovskite-sensitized solar cells. *Nature* **2013**, *499*, 316. [CrossRef] [PubMed]

8.   Saliba, M.; Matsui, T.; Seo, J.-Y.; Domanski, K.; Correa-Baena, J.-P.; Nazeeruddin, M.K.; Zakeeruddin, S.M.; Tress, W.; Abate, A.; Hagfeldt, A.; et al. Cesium-containing triple cation perovskite solar cells: improved stability, reproducibility and high efficiency. *Energy Environ. Sci.* **2016**, *9*, 1989–1997. [CrossRef] [PubMed]

9.   Yang, W.S.; Park, B.-W.; Jung, E.H.; Jeon, N.J.; Kim, Y.C.; Lee, D.U.; Shin, S.S.; Seo, J.; Kim, E.K.; Noh, J.H.; et al. Iodide management in formamidinium-lead-halide-based perovskite layers for efficient solar cells. *Science* **2017**, *356*, 1376. [CrossRef] [PubMed]

10.   Kim, Y.Y.; Park, E.Y.; Yang, T.-Y.; Noh, J.H.; Shin, T.J.; Jeon, N.J.; Seo, J. Fast two-step deposition of perovskite via mediator extraction treatment for large-area, high-performance perovskite solar cells. *J. Mater. Chem. A* **2018**, *6*, 12447–12454. [CrossRef]

11.   Shabdan, E.; Hanford, B.; Ilyassov, B.; Dikhanbayev, K.; Nuraje, N. Perovskite Solar Cell. In *Multifunctional Nanocomposites for Energy and Environmental Applications*; John Wiley & Sons, Inc.: Hoboken, NJ, USA, 2018; Volume 1, pp. 91–111. [CrossRef]

12.   Tsai, H.; Asadpour, R.; Blancon, J.-C.; Stoumpos, C.C.; Durand, O.; Strzalka, J.W.; Chen, B.; Verduzco, R.; Ajayan, P.M.; Tretiak, S.; et al. Light-induced lattice expansion leads to high-efficiency perovskite solar cells. *Science* **2018**, *360*, 67. [CrossRef] [PubMed]

13.   Zhang, Y.; Kim, S.-G.; Lee, D.-K.; Park, N.-G. $CH_3NH_3PbI_3$ and $HC(NH_2)_2PbI_3$ Powders Synthesized from Low-Grade PbI2: Single Precursor for High-Efficiency Perovskite Solar Cells. *ChemSusChem* **2018**, *11*, 1813–1823. [CrossRef] [PubMed]

14.   Bai, Y.; Meng, X.; Yang, S. Interface Engineering for Highly Efficient and Stable Planar p-i-n Perovskite Solar Cells. *Adv. Energy Mater.* **2017**, *8*, 1701883. [CrossRef]

15.   Luo, D.; Yang, W.; Wang, Z.; Sadhanala, A.; Hu, Q.; Su, R.; Shivanna, R.; Trindade, G.F.; Watts, J.F.; Xu, Z.; et al. Enhanced photovoltage for inverted planar heterojunction perovskite solar cells. *Science* **2018**, *360*, 1442. [CrossRef] [PubMed]

16.   Rajagopal, A.; Yao, K.; Jen, A.K.Y. Toward Perovskite Solar Cell Commercialization: A Perspective and Research Roadmap Based on Interfacial Engineering. *Adv. Mater.* **2018**, *30*, 1800455. [CrossRef] [PubMed]

17.   Singh, T.; Miyasaka, T. Stabilizing the Efficiency Beyond 20% with a Mixed Cation Perovskite Solar Cell Fabricated in Ambient Air under Controlled Humidity. *Adv. Energy Mater.* **2017**, *8*, 1700677. [CrossRef]

18.   Turren-Cruz, S.-H.; Saliba, M.; Mayer, M.T.; Juárez-Santiesteban, H.; Mathew, X.; Nienhaus, L.; Tress, W.; Erodici, M.P.; Sher, M.-J.; Bawendi, M.G.; et al. Enhanced charge carrier mobility and lifetime suppress hysteresis and improve efficiency in planar perovskite solar cells. *Energy Environ. Sci.* **2018**, *11*, 78–86. [CrossRef]

19.   Zhang, T.; Yang, M.; Zhao, Y.; Zhu, K. Controllable Sequential Deposition of Planar $CH_3NH_3PbI_3$ Perovskite Films via Adjustable Volume Expansion. *Nano Lett.* **2015**, *15*, 3959–3963. [CrossRef] [PubMed]

20.   Wu, C.-G.; Chiang, C.-H.; Tseng, Z.-L.; Nazeeruddin, M.K.; Hagfeldt, A.; Grätzel, M. High efficiency stable inverted perovskite solar cells without current hysteresis. *Energy Environ Sci.* **2015**, *8*, 2725–2733. [CrossRef]

21.   Li, W.; Fan, J.; Li, J.; Mai, Y.; Wang, L. Controllable Grain Morphology of Perovskite Absorber Film by Molecular Self-Assembly toward Efficient Solar Cell Exceeding 17%. *J. Am. Chem. Soc.* **2015**, *137*, 10399–10405. [CrossRef] [PubMed]

22.   Li, G.; Ching, K.L.; Ho, J.Y.L.; Wong, M.; Kwok, H.-S. Identifying the Optimum Morphology in High-Performance Perovskite Solar Cells. *Adv. Energy Mater.* **2015**, *5*, 1401775. [CrossRef]

23.   Zhang, H.; Mao, J.; He, H.; Zhang, D.; Zhu, H.L.; Xie, F.; Wong, K.S.; Grätzel, M.; Choy, W.C.H. A Smooth $CH_3NH_3PbI_3$ Film via a New Approach for Forming the PbI2 Nanostructure Together with Strategically High CH_3NH_3I Concentration for High Efficient Planar-Heterojunction Solar Cells. *Adv. Energy Mater.* **2015**, *5*, 1501354. [CrossRef]

24.   Eperon, G.E.; Burlakov, V.M.; Docampo, P.; Goriely, A.; Snaith, H.J. Morphological Control for High Performance, Solution-Processed Planar Heterojunction Perovskite Solar Cells. *Adv. Funct. Mater.* **2013**, *24*, 151–157. [CrossRef]

25. Wu, Y.; Islam, A.; Yang, X.; Qin, C.; Liu, J.; Zhang, K.; Peng, W.; Han, L. Retarding the crystallization of PbI2 for highly reproducible planar-structured perovskite solar cells via sequential deposition. *Energy Environ. Sci.* **2014**, *7*, 2934–2938. [CrossRef]

26. Zhao, Y.; Zhu, K. Solution Chemistry Engineering toward High-Efficiency Perovskite Solar Cells. *J. Phys. Chem. Lett.* **2014**, *5*, 4175–4186. [CrossRef] [PubMed]

27. Yang, S.; Zheng, Y.C.; Hou, Y.; Chen, X.; Chen, Y.; Wang, Y.; Zhao, H.; Yang, H.G. Formation Mechanism of Freestanding $CH_3NH_3PbI_3$ Functional Crystals: In Situ Transformation vs. Dissolution-Crystallization. *Chem. Mater.* **2014**, *26*, 6705–6710. [CrossRef]

28. Xiao, M.; Huang, F.; Huang, W.; Dkhissi, Y.; Zhu, Y.; Etheridge, J.; Gray-Weale, A.; Bach, U.; Cheng, Y.-B.; Spiccia, L. A Fast Deposition-Crystallization Procedure for Highly Efficient Lead Iodide Perovskite Thin-Film Solar Cells. *Angew. Chem. Int. Ed.* **2014**, *53*, 9898–9903. [CrossRef] [PubMed]

29. Ahn, N.; Son, D.-Y.; Jang, I.-H.; Kang, S.M.; Choi, M.; Park, N.-G. Highly Reproducible Perovskite Solar Cells with Average Efficiency of 18.3% and Best Efficiency of 19.7% Fabricated via Lewis Base Adduct of Lead(II) Iodide. *J. Am. Chem. Soc.* **2015**, *137*, 8696–8699. [CrossRef] [PubMed]

30. Chen, C.-W.; Kang, H.-W.; Hsiao, S.-Y.; Yang, P.-F.; Chiang, K.-M.; Lin, H.-W. Efficient and Uniform Planar-Type Perovskite Solar Cells by Simple Sequential Vacuum Deposition. *Adv. Mater.* **2014**, *26*, 6647–6652. [CrossRef] [PubMed]

31. Zheng, E.; Wang, X.-F.; Song, J.; Yan, L.; Tian, W.; Miyasaka, T. PbI_2-Based Dipping-Controlled Material Conversion for Compact Layer Free Perovskite Solar Cells. *ACS Appl. Mater. Int.* **2015**, *7*, 18156–18162. [CrossRef] [PubMed]

 *nanomaterials*

Article

Study on the Coupling Mechanism of the Orthogonal Dipoles with Surface Plasmon in Green LED by Cathodoluminescence

Yulong Feng [1], Zhizhong Chen [1,2,*], Shuang Jiang [1], Chengcheng Li [1], Yifan Chen [1], Jinglin Zhan [1], Yiyong Chen [1], Jingxin Nie [1], Fei Jiao [1,3], Xiangning Kang [1], Shunfeng Li [2], Tongjun Yu [1], Guoyi Zhang [1,2] and Bo Shen [1]

[1] State Key Laboratory for Artificial Microstructure and Mesoscopic Physics, School of Physics, Peking University, Beijing 100871, China; fengyulong@pku.edu.cn (Y.F.); jiangshuang@pku.edu.cn (S.J.); 1501110124@pku.edu.cn (C.L.); 1501110129@pku.edu.cn (Y.C.); 1601110180@pku.edu.cn (J.Z.); chenyiyong@pku.edu.cn (Y.C.); niejingxin@pku.edu.cn (J.N.); fjiao@pku.edu.cn (F.J.); xnkang@pku.edu.cn (X.K.); tongjun@pku.edu.cn (T.Y.); gyzhang@pku.edu.cn (G.Z.); bshen@pku.edu.cn (B.S.)

[2] Dongguan Institute of Optoelectronics, Peking University, Guangdong, Dongguan 523808, China; shunfengli@gmail.com

[3] State Key Laboratory of Nuclear Physics and Technology, School of Physics, Peking University, Beijing 100871, China

* Correspondence: zzchen@pku.edu.cn; Tel.: +86-10-6275-2169

Received: 23 March 2018; Accepted: 12 April 2018; Published: 16 April 2018

Abstract: We analyzed the coupling behavior between the localized surface plasmon (LSP) and quantum wells (QWs) using cathodoluminescence (CL) in a green light-emitting diodes (LED) with Ag nanoparticles (NPs) filled in photonic crystal (PhC) holes. Photoluminescence (PL) suppression and CL enhancement were obtained for the same green LED sample with the Ag NP array. Time-resolved PL (TRPL) results indicate strong coupling between the LSP and the QWs. Three-dimensional (3D) finite difference time domain (FDTD) simulation was performed using a three-body model consisting of two orthogonal dipoles and a single Ag NP. The LSP–QWs coupling effect was separated from the electron-beam (e-beam)–LSP–QW system by linear approximation. The energy dissipation was significantly reduced by the z-dipole introduction under the e-beam excitation. In this paper, the coupling mechanism is discussed and a novel emission structure is proposed.

Keywords: localized surface plasmon; green LED; cathodoluminescence; FDTD

1. Introduction

Despite GaN-based blue light-emitting diodes (LEDs) achieving rather high external quantum efficiency (EQE), the green gap is still a key issue for high-quality illumination [1,2]. Because of the imperfect structure and high polarization field in high-indium content InGaN quantum wells (QWs), the EQE for a green LED is quite low compared with that for a blue LED. Various approaches have been applied to solve this problem, such as the use of nonpolar/semipolar substrates, the use of a Si substrate, band engineering, and the introduction of a surface plasmon (SP) [3–6]. The polarization reduction and incorporation of indium into the InGaN alloy have been improved by these methods except for the last one, the SP method. Because of the high density of states (DOS) of the SP, the spontaneous emission rate (SER) in QWs can be very high, leading to a higher internal quantum efficiency (IQE) [6]. Two main perspectives on light emission enhancement via SP–QW coupling have emerged. One perspective is that the SP modes in the metal are excited by spontaneous emission (SE) in QWs and radiate to air in dipole mode [7,8], thus the energy transferred to SP is divided into "SP radiation" and "metal dissipation". The other perspective is that the spontaneous emission rate in QWs is greatly enhanced by the strong near-field strength of the SP excited by SE in QWs [9–11]. Both of these perspectives can explain the light emission enhancement by SP–QW coupling. As to SP radiation, less than 50% of the coupled energy can be radiated into air, and the emission enhancement is effective for emitters with low original EQE [8]. The latter perspective introduces the possibility of emission enhancement for a green LED [10,11]. However, the energy distribution dynamics in the SP–QW coupling system remain unclear.

Although evidence of the SER enhancement has been reported for several decades, the problem of energy dissipation in the metal is not yet well resolved. The energy dissipation in the metal corresponds to the Ohmic loss and to the electron-hole pair creation [12]. Many techniques have been reported to reduce the energy dissipation effects, including using a localized surface plasmon (LSP), placing the emitters into a metal gap, adjusting the dipole orientation, and coupling the QWs between each other [8,10,11,13,14]. An enhancement of the SER exceeding 1000 times with a quantum efficiency above 50% indicates that the energy dissipation is greatly reduced by the gap modes [10]. However, the complicated fabrication procedure for the gap structure prevents its further application. LSP–QW coupling between the radial or orbital dipoles and Ag NPs is quite different in the simulation [13]. The enhancements of radiated power of a radial and an orbital dipole are induced through coupling with the lower-order (dipole) and higher-order LSP resonance, respectively. Because the conventional InGaN QWs are thin and flat, the dipoles are mainly orbitally oriented [9,13], that is, some energy dissipation always occurs for planar QW structure.

Recently, several reports of electron beam (e-beam)-excited SP in metal nanoparticles (NPs) by cathodoluminescence (CL) have been reported [15–18]. Because CL measurements combine the ultrahigh spatial resolution of an electron microscope with broadband optical sensitivity, they can be used to study the optical process in metal NPs. There is a direct link between CL and radiative modes ("bright modes") or the radiative electromagnetic local density of states (LDOS) [16]. Some simulation results for CL measurements have also been performed by regarding the e-beam as a dipole source along the incident direction [15,18]. Although Ag NP–QW structures excited by an e-beam have been reported [19–21], these works did not consider the SPs induced by the e-beam. In fact, the e-beam, approximated by a vertical dipole induces the high near-field strength near the Ag NPs [15–18], will greatly influence the coupling between the LSP and the QWs. Moreover, the dipoles representing the QW and the e-beam are in-plane and out-plane, respectively. The configuration of a single Ag NP and two orthogonal dipoles can be used to study the energy transfer process of the LSP–QW coupled system.

In this work, we fabricated LSP–QW coupled samples of an array of Ag NPs embedded in photonic crystal (PhC) holes in the p-GaN layer of a green LED. Photoluminescence (PL) and time-resolved photoluminescence (TRPL) and CL measurements were carried out. A novel three-dimensional (3D) finite difference time domain (FDTD) numerical simulation model for CL and PL measurements was also put forward using Lumerical software (FDTD Solutions v8.17, Vancouver, BC, Canada) to illustrate the difference in the LSP–QW coupling mechanism under light and e-beam excitations [22]. The LSP-two orthogonal dipoles coupling mechanism is discussed and an effective way to reduce the energy dissipation in Ag NPs is proposed.

2. Experimental

GaN-based green LED structures with a peak wavelength of 545 nm were grown by metal organic chemical vapor deposition (MOCVD) on the double-polished c-plane sapphire substrate. The structure consisted of 10 pairs of InGaN/GaN (2.5 nm/12.5 nm) multiple quantum wells (MQWs), on which is 160 nm thick p-GaN. As shown in Figure 1, LSP–QW coupled samples (Ag–PhC) were fabricated based on the PhC structure in the p-GaN layer. The scanning electron microscope (SEM) images of the Ag arrayed LED were recorded using an FEI NanoSEM 430 (FEI, Hillsboro, OR, USA). First, a 120 nm thick SiO_2 mask was deposited onto the surface of p-GaN. The PhC patterns were obtained via nano-imprinting using an Obducat Eitre 3 instrument (Obducat, Lund, Sweden) with a standard PhC stamp whose period was 545 nm. Then, the hexagonal nanohole array was subsequently obtained using an induced coupled plasma (ICP) etcher to etch the pores to a depth of 150 nm, i.e., the bottom of each pore was 10 nm away from QWs. Then, a 30 nm thick Ag film was deposited onto the patterned surface. After thermal annealing at 600 °C for 10 min under a N_2 atmosphere, Ag NPs were formed in the PhC holes, followed by the lift-off process to remove the Ag NPs on the SiO_2 mask. The Ag NPs were spherical cap-shaped. Statistically, the diameter and height were 160 ± 10 and ~80 nm, respectively. For comparison, reference samples (denoted PhC) with the same pattern but without Ag NPs were also prepared.

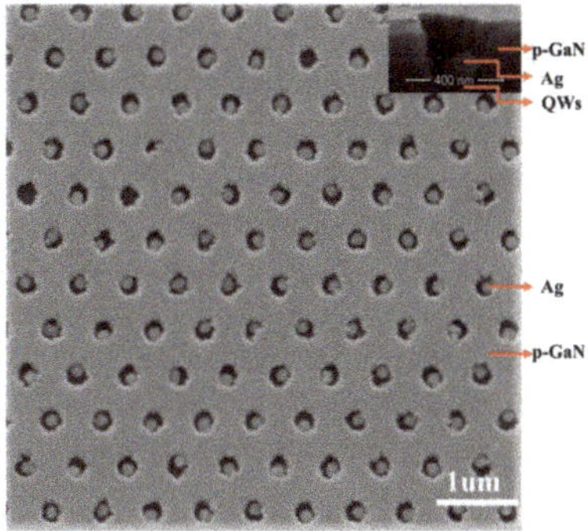

Figure 1. SEM image of a Ag–photonic crystal (PhC) sample. The period of the PhC is 545 nm. The diameter and height of the Ag NPs are 160 ± 10 and ~80 nm, respectively. The inset shows the cross-sectional image of a single Ag NP in the hole.

PL measurements using a 405 nm laser diode with a power of 150 mW and a spot diameter of ~1 mm and CL measurements using a Gatan Mono-CL2 system (Gatan, Pleasanton, CA, USA) were performed at room temperature on the Ag–PhC and PhC samples with a configuration of top excitation and top detection. The electron acceleration voltage was set to 15 kV with a beam current of 158 pA. The TRPL measurements were conducted using a LifeSpec-Red picosecond lifetime spectrometer with a pulsed 372 nm laser (Edinburgh Instruments, Livingston, UK). The pulse duration was 69 ps. This instrument was made use of a time-correlated single photon counting (TCSPC) technique with a time resolution of ~30 ps within a range of 10 ns.

3. Results and Discussion

TRPL measurement is an efficient method to confirm LSP–QW coupling. Figure 2A shows the TRPL results at the peak wavelength 545 nm of the Ag–PhC sample and the PhC sample. The decay curves were fitted by a double exponential function as reported in the previous work [14]. The fast decay time, which corresponds to the rapid carrier recombination in InGaN QWs, were obtained as 0.23 and 0.76 ns for the Ag–PhC and PhC samples, respectively. The 3.3-fold reduction of the decay time indicates that LSP coupling with QWs substantially enhances the SER [14]. Contrarily, the PL intensity of the Ag–PhC decreased by 1.7 times compared with that of PhC sample, as shown in Figure 1B. This decrease is attributed to the large energy dissipation in Ag NPs as a result of their small aspect parameter (α), which is defined as the ratio between the height of the Ag NP and its radius [14,23]. When the α is greater than 1.5, the emission enhancement will be realized with Ag NPs with diameters ranging from 90 to 200 nm [23]. Because the LSP–QW coupled energy was either radiated into the air or dissipated in the Ag NPs, the ratio between radiated energy and the dissipated energy would determine whether the final PL intensity was enhanced or suppressed.

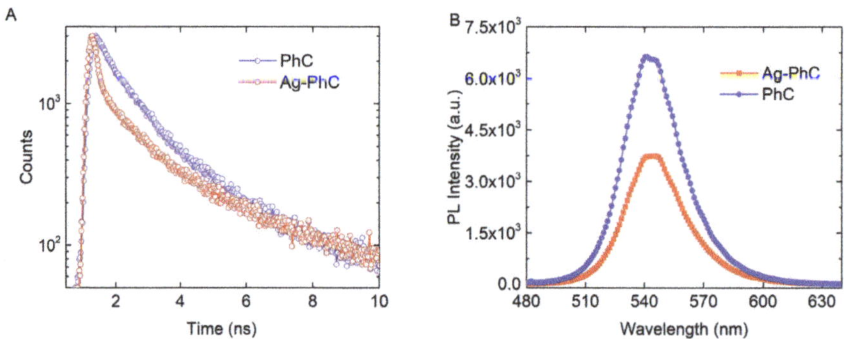

Figure 2. (**A**) Time-resolved photoluminescence (TRPL) and (**B**) PL spectra for Ag–PhC and PhC samples.

A schematic setup for the CL measurement is shown in Figure 3A. An e-beam was highly focused and directed onto the surface of samples. CL measurement was performed using a 15 kV acceleration voltage and a beam current of 158 pA. The electron penetrating depth into the Ag NPs was greater than 20 nm [19,20]. The light emitted from the samples was collected by a retractable parabolic mirror and collimated to an optical monochromator, after which the signal was detected by a charge-coupled device (CCD). To ensure the maximum light collection efficiency, the samples were placed at the focal plane of the parabolic mirror, approximately ~1 mm away from the mirror. Figure 3C,D show the panchromatic CL (PanCL) image of an Ag–PhC sample and a PhC sample, respectively. The PanCL image, where all of the emitted light from the sample was collected by the e-beam scanning point by point, can clearly depict the LSP-induced luminescence around the Ag NPs. The holes are much brighter than the platform of the Ag–PhC sample. The dark spots in the holes are related to the Ag NPs. However, the spot size is much smaller than the actual size of Ag NPs in Figure 1. This discrepancy arises because of the penetration ability of high-energy electrons [19] and the strong coupling between the LSPs induced by the e-beam [16–18] and the QWs. The darker cloudy areas in Figure 3C,D are mainly attributed to the InGaN phase separation in QWs. CL spectra were recorded by scanning the electron beam over the entire surface of interest under the same conditions, as shown in Figure 3B. To confirm that the emission originates from QWs rather than from the Ag NPs, samples without QWs were fabricated. It was found that CL spectrum of the sample without QWs mainly originated from the emission of GaN and its intensity was approximately two orders of magnitude smaller than that of the sample with QWs. The CL intensity of the Ag–PhC sample is 2.91 times greater than that of the PhC sample, whereas the PL intensity of PhC sample is 1.7 times that of the Ag–PhC sample. Compared with laser excitation, e-beam excitation enhances the emission intensity of the Ag–PhC sample by a factor of 4.95. Considering the penetration electron energy loss in Ag NPs, even if all the electrons can penetrate through the Ag NP to excite the QW directly, it is impossible for the enhancement factor to be as high as 4.95. The high near-field strength of the LSP in the Ag NPs induced by continuously injected e-beam should be considered in the LSP–QW coupling process [16–18]. Moreover, how the energy dissipation is reduced by the e-beam excitation is interesting.

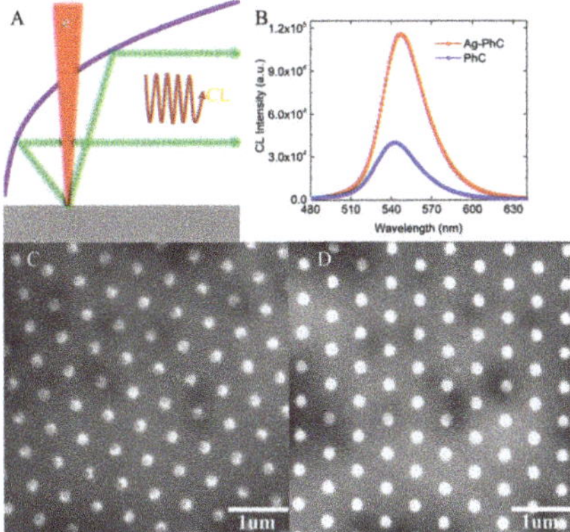

Figure 3. (**A**) Schematic setup for the cathodoluminescence (CL) measurement; (**B**) CL spectra for Ag–PhC and PhC samples. Panchromatic CL images for (**C**) the Ag–PhC sample and (**D**) the PhC sample. CL intensity for the Ag–PhC sample is enhanced 2.91 times compared with that for PhC sample.

To distinguish the different LSP–QW coupling mechanisms by e-beam excitation and laser excitation, 3D-FDTD numerical simulations were carried out [22]. In FDTD, Maxwell's equations are solved in discretized space and time. Figure 4 shows the schematic structure of Ag–PhC sample used in the 3D-FDTD simulation. Since the separation between Ag NPs is greater than 200 nm, the coupling between Ag NPs can be ignored [24]. Therefore, only one Ag NP needs to be considered. The dispersion relation of the Ag NP adopts "Ag (silver)-Palik (0–2 um)" data provided by the FDTD material database [22]. The Ag NP was placed in the hole center of PhC on p-GaN ($n = 2.55$) to simplify the simulation. The sizes of Ag NP and PhC are consistent with the SEM result as shown in Figure 1. The space between Ag NP and the first QW is 10 nm. The perfectly matched layer (PML) absorbing boundaries were adopted on all sides. To improve the simulation accuracy, an override region was applied over the Ag NPs and x-dipole with the mesh size of 2 nm. Outside that region, automatic graded meshing was used. Moreover, the simulation span was 6 um*6 um (x–y plane), which is large enough for the light to propagate.

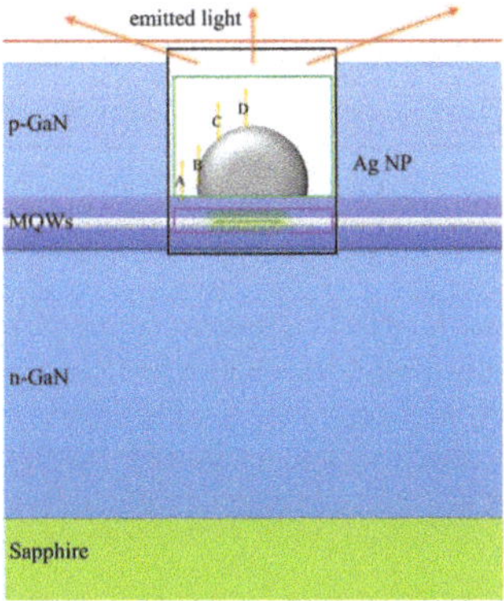

Figure 4. The schematic structure of the Ag–PhC sample in 3D finite difference time domain (FDTD) simulation. The purple, green, and black boxes were used to collect the total power radiated by the dipole, the dissipation power in the Ag NP, and the scatted energy, respectively. The red line (plane) was used to record the radiated power from top surface.

Given the symmetry in the QW plane, where the radiating dipoles lie, only one dipole polarized along x direction (x-dipole) was placed below the Ag NP to represent the QW. The e-beam, however, was represented by a dipole polarized along its trajectory (z dipole) in order to simplify the model [18]. Both the x-dipole and z-dipole adopted the built-in broadband source model, with the wavelength range from 480 to 630 nm. In FDTD, broadband sources can be used to perform simulations in which wideband frequency data is required. Since the e-beam impinging on different positions of the Ag NP may lead to different excitation, the z-dipole was successively placed with an interval of 30 nm at the positions of A, B, C, and D, as shown by the yellow arrows in Figure 4. To calculate the power transition, four monitors were used in the simulations. The purple, green, and black boxes were used to collect the power of x-dipole (QW), dissipated power, and scattering power, respectively, while the upper red plane monitor was used to record the energy successfully extracted into the air.

The actual power emitted by a dipole can vary dramatically depending on what dielectric envelopments are nearby. The field induced by a dipole or reflected from the surrounding structures can feed back on itself, causing it to radiate more or less power than expected in a homogenous material [10,11,13]. As a result, the calculated power should be normalized to the power that a dipole would emit in a homogenous material (such as vacuum), rather than the actually emitted power. In FDTD, the quantum mechanical decay rate can be related to the power collected by a monitor box in the same environment. Therefore, the decay rate can be normalized by [22],

$$\frac{\gamma}{\gamma_0} = \frac{P}{P_0} \tag{1}$$

where γ and P are the decay rate and radiated power by a dipole in an inhomogeneous environment, and γ_0 and P_0 are the corresponding parameters in a homogeneous environment (here GaN). In particular, when a dipole is surrounded by a monitor box (i.e., the purple box in Figure 4), γ would become the well-known Purcell factor (F_p), which is defined as the ratio of the radiation power of a dipole near Ag NPs to that in a bulk dielectric material (GaN). Thus, the EQE—namely the ratio of power measured in the far field to the total power injected into the x-dipole (QW)—as well as the IQE and light extraction efficiency (LEE) can be defined as [11,22],

$$\eta_{EQE} = \eta_{IQE}\eta_{LEE} = \left(\frac{F_p\gamma_{rad}}{F_p\gamma_{rad} + \gamma_{non}} \frac{\gamma_{scat}}{\gamma_{diss} + \gamma_{scat}} \right) \frac{\gamma_{out}}{\gamma_{scat}} \tag{2}$$

where γ_{scat} and γ_{diss} are the scattering and dissipation rate, γ_{rad} and γ_{non} are the radiative and nonradiative decay rate, and γ_{out} is light extraction rate. According to our temperature-dependent PL measurements, the internal quantum efficiency is calculated as 26%. A rough estimation of the ratio of the nonradiative decay rate (γ_{non}) to the radiative decay rate (γ_{rad}) for the PhC sample is 3:1. Based on the data drawn from the aforementioned monitors, EQE, IQE, LEE, and F_p can be calculated, respectively.

In the PL simulation, because the Ag NPs are opaque to light excitation [19] or the laser wavelength cannot match the SP resonant wavelength [14], z-dipole is not included, similar to our previous work [11,23]. Figure 5 shows the simulated PL spectra for the Ag–PhC and PhC samples. The IQE and LEE of the Ag–PhC sample as a function of wavelength are normalized to those of the PhC sample, as also shown in Figure 5. According to the broadband simulation results, the final emitted light spectrum can be obtained through post processing by multiplying the EQE with the actual normalized QW spectrum that is obtained without any Ag NPs nearby [11]. The simulated PL spectra show that the intensity of Ag–PhC sample decreases by 2.5 times, which agrees with the experimental result. However, the enhancement is not completely consistent with that of the experimental result, which is attributed to the approximations, including the dipole-QW approximation [9], the single Ag NP approximation [8], and the simplification of a single dipole for the QWs [11]. Moreover, the transmitted diffraction across the Ag NPs may also affect the consistence. The Purcell factor at 545 nm is calculated as 18.7, which indicates that LSP–QW coupling is strong and that the SER is greatly enhanced. However, due to the dissipation by Ag NP [11], the IQE of the Ag–PhC sample is only 54% of that of PhC sample at 545 nm. Furthermore, the LEE of Ag–PhC is also decreased by a factor of 1.8, as shown in Figure 5B. Obviously, the decreases of both IQE and LEE lead to the suppression of PL intensity.

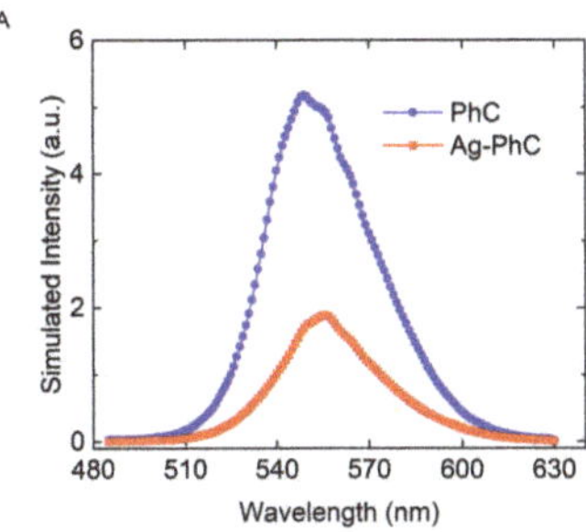
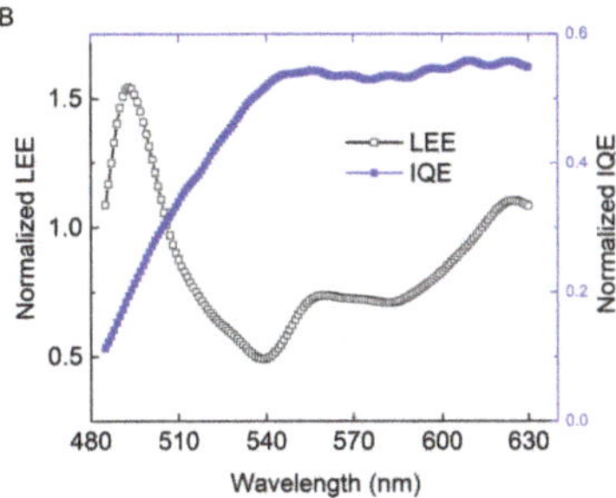

Figure 5. (**A**) Simulated PL spectra for Ag–PhC and PhC samples, (**B**) internal quantum efficiency (IQE) and light extraction efficiency (LEE) of Ag–PhC sample normalized to those of the PhC sample.

To simulate the CL measurement for the Ag–PhC sample, the z-dipole is added to the system as shown in Figure 4. Since the e-beam excitation is dependent on its impinging position as discussed above, the simulations are carried out at e-beam impinging points A, B, C, and D individually. Considering that the electron beam itself does not radiate light, after a z-dipole is added to simulate the electron beam impinging at specific impinging points, the energy radiated by the z-dipole must be subtracted. With the presence of both the z-dipole and x-dipole, the net powers flowing through the green box, purple box, black box, and the upper red plane monitor (P_{gBox}, P_{pBox}, P_{bBox} and P_{rPlane} respectively) have the relationship of

$$P_{rPlane} = P_{up-xDipole} + P_{up-zDipole} \tag{3}$$

$$P_{pBox} = P_{xDipole} \tag{4}$$

$$P_{gBox} = P_{zDipole} - \left(P_{Ag-xDipole} + P_{Ag-zDipole} \right) \tag{5}$$

$$P_{bBox} = P_{xDipole} + P_{zDipole} - \left(P_{Ag-xDipole} + P_{Ag-zDipole} \right) \tag{6}$$

where $P_{xDipole}$ and $P_{zDipole}$ are the power radiated by the x-dipole and z-dipole, which can be directly recorded in the simulations. P_{up} is the power extracted upward into the air, and P_{Ag} is the dissipated power by Ag NP. These two quantities can also be recorded directly; however, they are the sum of the x-dipole and z-dipole components. To distinguish the efficiencies of the x-dipole (QW) from the two orthogonal dipoles system, simulations at each point without x-dipole have also been performed. Similarly, Equations (3)–(6) without x-dipole component is rewritten as

$$P'_{rPlane} = P'_{up-xDipole} + P'_{up-zDipole} \tag{7}$$

$$P'_{pBox} = P'_{xDipole} = 0 \tag{8}$$

$$P'_{gBox} = P'_{zDipole} - P'_{Ag-zDipole} \tag{9}$$

$$P'_{bBox} = P'_{xDipole} + P'_{zDipole} - P'_{Ag-zDipole} \tag{10}$$

where the prime ($'$) indicates all the powers recorded in the case without the x-dipole.

As mentioned above, for the laser beam with a power of 150 mW and a spot diameter of ~1 mm, the power density of the laser beam is on the order of ~100 mW/mm^2. Additionally, for the e-beam with a beam current of 158 pA at 15 kV impinging on the 160 nm diameter Ag NP, the power density of the e-beam is on the order of ~100 W/mm^2. Considering the dipole-QW and dipole-e-beam approximations and that the energy of e beam cannot be fully converted to the QW radiation, the ratio of the amplitude of z-dipole to that of x-dipole was roughly set to be only 10 since the power of a dipole is proportional to the square of its amplitude. By default, the power recorded by different detectors in Equations (3)–(6) and (7)–(10) are normalized to the sum of power from all sources (P_{source}). For consistency, all power in Equations (3)–(10) were renormalized by multiplying a correction factor of P_{source}/P_0, where P_0 is defined in Equation (1). Figure 6A clearly shows that $P_{zDipole}$ and $P'_{zDipole}$ vs. wavelength almost coincide in the wavelength range from 480 nm to 630 nm at point B. We set $P_{zDipole} = (1 + \beta)P'_{zDipole}$, where β is a modification coefficient for the x-diploe. According to the simulation results, β is a small quantity. On the contrary, the power of x-dipole ($P_{xDipole}$) changes greatly with the presence of the z-dipole, as shown in Figure 6B. Figure 6C shows the Purcell factor for the x-dipole and z-dipole without the Ag NP. It is noted that the direct interaction between x-dipole and z-dipole is far weaker than that shown in Figure 6B. Therefore, the great change in Figure 6B is attributed to the SP strongly excited by z-dipole rather than z-dipole itself. Besides, the peak on the left of 545 nm in Figure 6B is enhanced with the presence of the z-dipole, indicating that high-order modes in LSP are excited and radiated [13]. According to the simulation, the electric field mapping under the Ag NP also exhibits a quadrupole characteristic at shorter wavelength peak, which again confirms the excitation of the high-order modes.

Table 1 lists the powers of z-dipole and x-dipole at different positions at 545 nm for Ag–PhC sample. The simulation results at all the points show that the presence of x-dipole can hardly affect z-dipole whereas the z-dipole strongly influences on the x-dipole through the excitation of the LSP. Thus, it is reasonable to assume that the dissipated power and extracted power of z-dipole linearly changes with $P_{zDipole}$, that is

$$P_{Ag-zDipole} = (1 + \beta)P'_{Ag-zDipole} \tag{11}$$

$$P_{up-zDipole} = (1 + \beta)P'_{up-zDipole} \tag{12}$$

By subtracting Equations (4)–(10) from Equations (3)–(6), the effects of x-dipole can be separated from the e-beam–LSP–QW coupling system as follows

$$P_{xDipole} = P_{pBox} \tag{13}$$

$$P_{Ag-xDipole} = P_{gBox} - (1 + \beta)P'_{gBox} \tag{14}$$

$$P_{up-xDipole} = P_{rPlane} - (1 + \beta)P'_{rPlane} \tag{15}$$

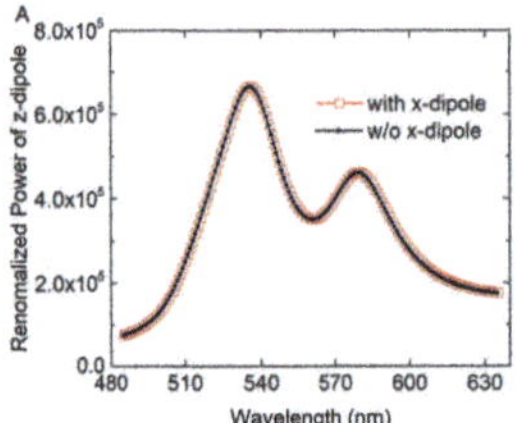
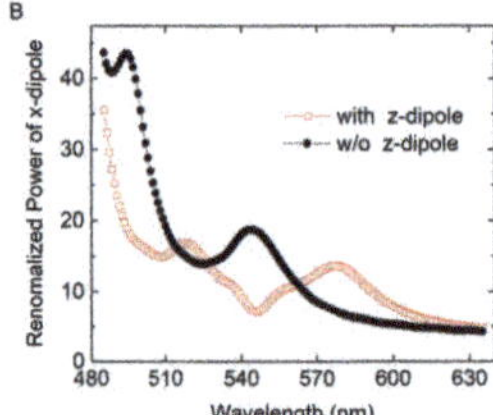
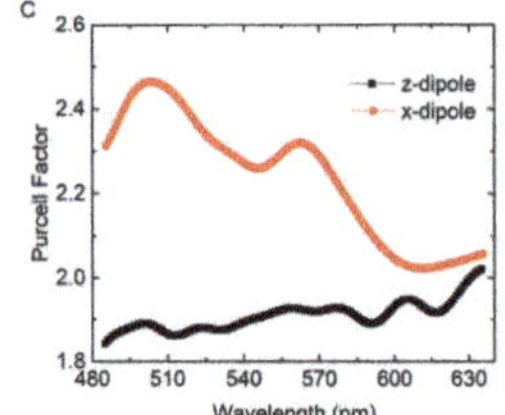

Figure 6. Renormalized powers vs. wavelengths of (**A**) z-dipole with and without x-dipole and (**B**) x-dipole (i.e., Fp in this case) with and without z-dipole at impinging point B; (**C**) Purcell factor for the x-dipole and z-dipole without the Ag NP.

Table 1. Powers of z-dipole and x-dipole and efficiencies of x-dipole at different positions at 545 nm for Ag–PhC sample.

Position	Power of z-Dipole		Power (*Fp*) of x-Dipole		Efficiency of x-Dipole		
	With x-Dipole	w/o x-Dipole	With z-Dipole	w/o z-Dipole	LEE	IQE	EQE
A	189.04	196.55	11.38	18.7	2.69	1.15	3.02
B	530115	530081	7.11	18.7	2.41	1.43	3.36
C	1346570	1346580	15.32	18.7	1.86	1.52	2.75
D	44715.1	44427.7	17.12	18.7	0.64	0.43	0.27

Based on Equations (1), (2), and (13)–(15), the EQE for the x-dipole (QW) in the e-beam–LSP–QW system can be obtained. The typical impinging positions A, B, C and D in the Ag–PhC sample are calculated and their weights are considered. The efficiencies of the x-dipole at 545 nm are also listed in the Table 1. Figure 7A shows the averaged CL spectra of Ag–PhC sample. The CL intensity of the Ag–PhC is enhanced by 2.4 times compared with that of the PhC sample, which agrees with the experimental result. The incomplete consistency is attributed to the approximations, as mentioned above. It is found that the EQE of the x-dipole in most areas around the Ag NP are enhanced by more than 2 times, except at its central area. The EQE at position D is only 0.27 times that of the PhC sample, which agrees with the panchromatic CL images in Figure 3C. In order to better understand the CL enhancement, the EQE at point B was divided to IQE and LEE, as shown in Figure 7B. Both the LEE and the IQE of the x-dipole are enhanced in the emission range compared with those of the PhC sample, which is quite different from that of the PL case. The LEE and IQE is enhanced by 2.41 and 1.43 times at 545 nm, respectively. It is notable that the F_p of x-dipole at point B is the smallest in the four points in the CL simulation, as shown in Table 1. When the F_p is enhanced to a certain value, it is no longer important for IQE enhancement. Since the term $\frac{F_p\gamma_{rad}}{F_p\gamma_{rad}+\gamma_{non}}$ in Equation (2) approaches 1, IQE is dominated by the scattering ratio (i.e., $\frac{\gamma_{scat}}{\gamma_{diss}+\gamma_{scat}}$). Compared with the simulated PL results, the scattering ratio of the x-dipole is enhanced 2.65 times. That is, when the z-dipole is added, less dissipation in the Ag NP can be obtained even when the F_p decreases. According to References [8,13], with the presence of the z-dipole, the blue-shifted resonant peak in Figure 6B and the enhanced IQE in Figure 7B indicates that the original higher-order nonradiating modes and lower-order radiating modes are suppressed and enhanced, respectively. A high F_p may lead to severe luminescence quenching effect via the high-order LSP modes [8].

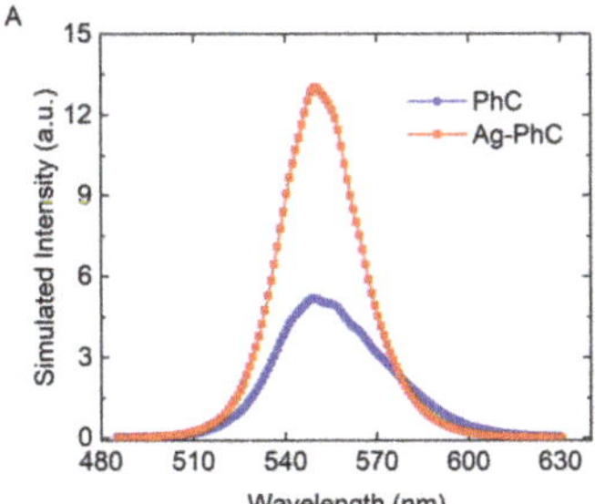
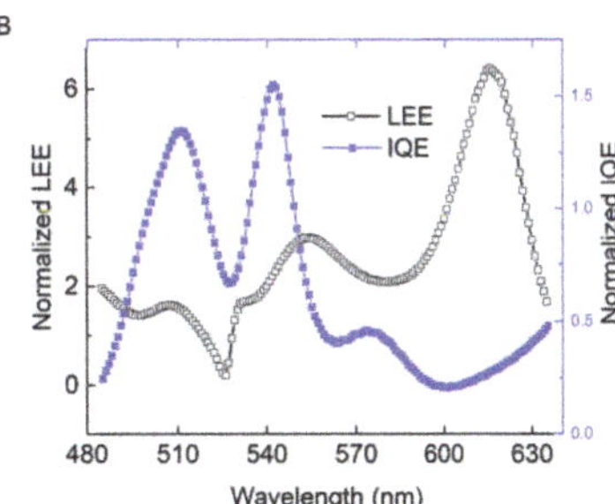

Figure 7. (**A**) Simulated CL spectra for Ag–PhC and PhC samples; (**B**) IQE and LEE of x-dipole normalized to those of the PhC sample at impinging point B for the Ag–PhC sample.

In addition, the LEE in CL simulation is also surprisingly enhanced by 4.3 times at 545 nm at point B compared with the PL results in Figure 5B. The LEE enhancement of 68% of the Ag–PhC sample compared with the PhC sample has been obtained in our previous work, which was explained by the resultant modification of the waveguide mode, which combines the effects of the LSP and the PhC [11].

In this work, the enhancement may occur because the excessive LSP modes excited by the z-dipole also modify the waveguide mode in the GaN LEDs. The penetration ability of high-energy electrons is stronger than that of the laser, which will enhance the CL intensity as well.

As described above, the z-dipole introduction enhances both the IQE and LEE of the QWs in the LSP–QWs coupling process. The dissipated energy is reduced and the waveguide modes are extracted effectively. It is reasonable to make an orthogonal emission dipole in LSP–QW coupling systems to enhance the light output of the devices. The InGaN QWs epitaxy on pyramids or V-pit structures seems to provide the possibility of two orthogonal dipoles coupling to LSP [25], where dipole radiators within the quantum wells on the facets are similar with the z-dipole in this work compared with the dipole radiators within normal quantum wells.

4. Conclusions

In summary, PL and CL measurements were performed on a green LED with Ag NPs filled in photonic crystal holes. The suppression of PL and the enhancement of CL were observed compared with the PhC sample. In the FDTD simulations, the two orthogonal dipoles were used to couple with the LSP. The x-dipole (QW) effect was separated in the e-beam–LSP–QW system by linear approximation. The simulation showed that both the IQE and the LEE were enhanced by the z-dipole added to the LSP–QW system. The enhancements were attributed to the LSP excited by z-dipole coupled to some LSP modes excited by the x-dipole. A novel LED device was proposed with orthogonal emission dipoles based on the simulation and experimental results.

Acknowledgments: This work was supported by National Key Research and Development Program under Grant No. 2016YFB0400100, National Natural Science Foundation of China under Grant No. 61674005, National Natural Science Foundation of China under Grant No. 61334009, Beijing Municipal Science & Technology Commission under Grant No. Z161100001616010 and Science and Technology Major Project of Guangdong Province under Grant No. 2016B010111001.

Author Contributions: Y.F. designed the experiment, prepared the samples, performed the measurements and the simulation and wrote the manuscript. Z.C. designed the experiment and wrote the manuscript. S.J. contributed to the conception of the study and the CL measurement. C.L. contributed to the PL measurement. C.L. contributed to the PL measurement. Y.C. contributed to preparing the sample. J.Z. contributed to the PL measurement and data analysis. Y.C. contributed to data analysis. J.N. contributed to preparing the sample. F.J. contributed to preparing the sample and the PL measurement. X.K. contributed to the data analysis. S.L. contributed to preparing the samples. T.Y. contributed to the data analysis. B.S. and G.Z. supervised the study and reviewed the manuscript.

References

1. Krames, M.R.; Shchekin, O.B.; Mueller-Mach, R.; Mueller, G.O.; Zhou, L.; Harbers, G.; Craford, M.G. Status and future of high-power light-emitting diodes for solid-state lighting. *J. Disp. Technol.* **2007**, *3*, 160–175. [CrossRef]
2. Saito, S.; Hashimoto, R.; Hwang, J.; Nunoue, S. InGaN light-emitting diodes on c-face sapphire substrates in green gap spectral range. *Appl. Phys. Express* **2013**, *6*, 111004. [CrossRef]
3. Yamamoto, S.; Zhao, Y.; Pan, C.C.; Chung, R.B.; Fujito, K.; Sonoda, J.; DenBaars, S.P.; Nakamura, S. High-efficiency single-quantum-well green and yellow-green light-emitting diodes on semipolar (20–21) GaN substrates. *Appl. Phys. Express* **2010**, *3*, 122102. [CrossRef]
4. Hwang, J.; Hashimoto, R.; Saito, S.; Nunoue, S. Effects of local structure on optical properties in green-yellow InGaN/GaN quantum wells. *Proc. SPIE* **2013**, *8625*, 455–461.
5. Zhang, J.L.; Xiong, C.B.; Liu, J.L.; Quan, Z.J.; Wang, L.; Jiang, F.Y. High brightness InGaN-based yellow light-emitting diodeswith strain modulation layers grown on Si substrate. *Appl. Phys. A* **2014**, *114*, 1049–1053. [CrossRef]
6. Okamoto, K.; Kawakami, Y. High-efficiency InGaN/GaN light emitters based on nanophotonics and plasmonics. *IEEE J. Sel. Top. Quantum Electron.* **2009**, *15*, 1199–1209. [CrossRef]

7. Huang, C.W.; Tseng, H.Y.; Chen, C.Y.; Liao, C.H.; Hsieh, C.; Chen, K.Y.; Lin, H.Y.; Chen, H.S.; Jung, Y.L.; Kiang, Y.W.; et al. Fabrication of surface metal nanoparticlesand their induced surface plasmon coupling with subsurface InGaN/GaNquantum wells. *Nanotechnology* **2011**, *22*, 475201. [CrossRef] [PubMed]

8. Sun, G.; Khurgin, J.B. Plasmon enhancement of luminescence by metal nanoparticles. *IEEE J. Sel. Top. Quantum Electron.* **2011**, *17*, 110–118. [CrossRef]

9. Gontijo, I.; Boroditsky, M.; Yablonovitch, E.; Keller, S.; Mishra, U.K.; DenBaars, S.P.; Krames, M. Coupling of InGaN quantum-well photoluminescence to silver surface plasmons. *Phys. Rev. B* **1999**, *60*, 11564–11567. [CrossRef]

10. Akselrod, G.M.; Argyropoulos, C.; Hoang, T.B.; Ciracì, C.; Fang, C.; Huang, J.N.; Smith, D.R.; Mikkelsen, M.H. Probing the mechanisms of large Purcell enhancement in plasmonic nanoantennas. *Nat. Photonics* **2014**, *8*, 835–840. [CrossRef]

11. Jiang, S.; Chen, Z.Z.; Feng, Y.L.; Jiao, Q.Q.; Fu, X.X.; Ma, J.; Li, J.Z.; Jiang, S.X.; Yu, T.J.; Zhang, G.Y. The coupling behavior of multiple dipoles and localized surface plasmon in Ag nanoparticles array. *Plasmonics* **2016**, *11*, 125–130. [CrossRef]

12. Maier, S.A. *Plasmonics: Fundamentals and Applications*; Springer: New York, NY, USA, 2007.

13. Kuo, Y.; Chang, W.Y.; Chen, H.S.; Kiang, Y.W.; Yang, C.C. Surface plasmons coupling with a radiating dipole near a Ag nanoparticle embedded in GaN. *Appl. Phys. Lett.* **2013**, *102*, 161103. [CrossRef]

14. Jiang, S.; Hu, Z.; Chen, Z.Z.; Fu, X.X.; Jiang, X.Z.; Jiao, Q.Q.; Yu, T.J.; Zhang, G.Y. Resonant absorption and scattering suppression of localized surface plasmons in Ag particles on green LED. *Opt. Express* **2013**, *21*, 12100–12110. [CrossRef] [PubMed]

15. Chaturvedi, P.; Hsu, K.H.; Kumar, A.; Fung, K.H.; Mabon, J.C.; Fang, N.X. Imaging of plasmonic modes of silver nanoparticles using high-resolution cathodoluminescence spectroscopy. *ACS Nano* **2009**, *3*, 2965–2974. [CrossRef] [PubMed]

16. De Abajo, F.J.G. Optical excitations in electron microscopy. *Rev. Mod. Phys.* **2010**, *82*, 209–275. [CrossRef]

17. Coenen, T.; Schoen, D.T.; Brenny, B.J.M.; Polman, A.; Brongersma, M.L. Combined electron energy-loss and cathodoluminescence spectroscopy on individual and composite plasmonic nanostructures. *Phys. Rev. B* **2016**, *93*, 195429. [CrossRef]

18. Yang, C.; Manjavacas, A.; Large, N.; Nordlander, P. Electron energy-loss spectroscopy calculation in finite-difference time-domain package. *ACS Photonics* **2015**, *2*, 369–375.

19. Estrin, Y.; Rich, D.H.; Keller, S.; DenBaars, S.P. Temperature dependence of exciton-surface plasmon polariton coupling in Ag, Au, and Al films on $In_xGa_{1-x}N$/GaN quantum wells studied with time-resolved cathodoluminescence. *J. Appl. Phys.* **2015**, *117*, 043105. [CrossRef]

20. Estrin, Y.; Rich, D.H.; Keller, S.; DenBaars, S.P. Observations of exciton–surface plasmon polariton coupling and exciton–phonon coupling in InGaN/GaN quantum wells covered with Au, Ag, and Al films. *J. Phys. Condens. Matter* **2015**, *27*, 265802. [CrossRef] [PubMed]

21. Gao, N.; Huang, K.; Li, J.C.; Li, S.P.; Yang, X.; Kang, J.Y. Surface-plasmon-enhanced deep-UV light emitting diodes based on AlGaN multi-quantum wells. *Sci. Rep.* **2012**, *2*, 816. [CrossRef] [PubMed]

22. Lumerical Illuminating the Way. 2003. Available online: http://www.lumerical.com (accessed on 8 November 2017).

23. Jiang, S.; Chen, Z.Z.; Fu, X.X.; Jiao, Q.Q.; Feng, Y.L.; Yang, W.; Ma, J.; Li, J.Z.; Jiang, S.X.; Yu, T.J.; et al. Fabrication and effects of Ag nanoparticles hexagonal arrays in green LEDs by nanoimprint. *IEEE Photonics Technol. Lett.* **2015**, *27*, 1363–1366. [CrossRef]

24. Sun, G.; Khurgin, J.B. Comparative study of field enhancement between isolated and coupledmetal nanoparticles: An analytical approach. *Appl. Phys. Lett.* **2010**, *97*, 263110. [CrossRef]

25. Ajia, I.A.; Edwards, P.R.; Pak, Y.; Belekov, E.; Roldan, M.A.; Wei, N.N.; Liu, Z.Q.; Martin, R.W.; Roqan, I.S. Generated Carrier Dynamics in V-Pit-Enhanced InGaN/GaN Light-Emitting Diode. *ACS Photonics* **2018**, *5*, 820–826. [CrossRef]

Article

Effect of Excitation Wavelength on Optical Performances of Quantum-Dot-Converted Light-Emitting Diode

Caiman Yan [1], Xuewei Du [1], Jiasheng Li [1], Xinrui Ding [1], Zongtao Li [1,2,*] and Yong Tang [1]

[1] Key Laboratory of Surface Functional Structure Manufacturing of Guangdong High Education Institutes, South China University of Technology, Guangzhou 510641, China
[2] Foshan Nationstar Optoelectronics Company, Ltd., Foshan 528000, China
* Correspondence: meztli@scut.edu.cn; Tel.: +86-138-2446-0886

Received: 10 July 2019; Accepted: 25 July 2019; Published: 1 August 2019

Abstract: Light-emitting diode (LED) combined with quantum dots (QDs) is an important candidate for next-generation high-quality semiconductor devices. However, the effect of the excitation wavelength on their optical performance has not been fully explored. In this study, green and red QDs are applied to LEDs of different excitation wavelengths from 365 to 455 nm. The blue light is recommended for exciting QDs from the perspective of energy utilization. However, QD LEDs excited at 365 nm have unique advantages in eliminating the original peaks from the LED chip. Moreover, the green or red light excited by ultraviolet light has an advantage in colorimetry. Even for the 455 nm LED with the highest QD concentration at 7.0 wt%, the color quality could not compete with the 365 nm LED with the lowest QD concentration at 0.2 wt%. A 117.5% (NTSC1953) color gamut could be obtained by the 365 nm-excited RGB system, which is 32.6% higher than by the 455 nm-excited solution, and this can help expand the color gamut of LED devices. Consequently, this study provides an understanding of the properties of QD-converted LEDs under different wavelength excitations, and offers a general guide to selecting a pumping source for QDs.

Keywords: colorimetry; excitation wavelength; light-emitting diode; quantum dots

---

1. Introduction

Light-emitting diodes (LEDs) are fundamental devices for backlights and displays [1,2]. Nowadays, the common base pixel is composed of red, green, and blue (RGB) tricolor in both RGB backlights and direct display technology. Three different-color epitaxial chips are mainly applied in an RGB pixel: Blue, green, and red LED chips [3]. However, this solution cannot avoid the following problems: First, the blue, green, and red LEDs have different turn-on voltage characteristics, resulting in voltage balance problems in the circuit design [4]; second, owing to process limitations, the optical efficiency and production yield of red LEDs cannot compare with blue LEDs [5], resulting in high technology thresholds and increased costs; third, it is difficult to further improve the color quality in the current common RGB pixel because of the epitaxial material property limitation.

In order to solve the above difficulties, there are two advanced technologies under discussion. The first advanced technology is using only one LED chip, such as a blue or ultraviolet chip, to solve the voltage mismatch problem [6,7]. For the blue chip solution, the blue light is from a blue LED chip; the green and red light are obtained from the blue light to excite the green and red fluorescent materials. For the ultraviolet chip solution, the blue, green, and red light are all obtained from the ultraviolet-light-excitation fluorescent material [8]. Another technique is to maintain the mature commercial blue chip as blue light and introduce an ultraviolet chip to excite green and red fluorescent materials in order to achieve green and red light. Among these strategies, the core point is that using a short-wavelength light to excite fluorescent materials attains the primary color.

Fluorescent materials are key to improving display performance such as the color gamut. In the past decade, fluorescent materials have evolved from a single YAG:Ce^{3+} phosphor to green and red fluorescent materials such as β-sialon:Eu and CaAlSiN$_3$:Eu phosphors [9–13]. However, these fluorescent materials with a half-width wider than 50 nm are not ideal. Further improving the color gamut of LEDs is a hot spot for future commercial competition [14,15]. As a new type of luminescent material, quantum dots (QDs) are quite promising owing to their narrow half-peak width down to 30 nm, wide excitation wavelength, high color purity, and tunable emission [16,17].

LED combined with QDs is an important candidate for next-generation high-quality semiconductor devices [18,19]. By taking the advantages of the QD, a green or red LED could restore the natural color to the greatest extent [20]. Shin-Tson Wu et al. applied CdSe QDs for backlight applications, and a 115% color gamut in CIE 1931 was attained [21]. Similar high-quality colors from QD were reported by Shinae Jun [22], Jian Chen [23], and Huang-Yu Lin [24] et al. In order to obtain green and red LEDs, QDs are typically excited by short wavelengths such as blue or ultraviolet (UV) light [24,25]. However, the effect of the excitation wavelength on the optical performance of QD devices has not been systematically studied.

Herein, using CdSe QDs, this study compared the performance of green and red QD LEDs at different excitation wavelengths. The photoluminescence (PL) pattern of QD films was explored. Then, the optical performances of green and red LEDs were compared and analyzed at different excitation wavelengths. It turns out the excitation wavelength of the LED chip has an important impact on the overall device performance.

2. Methods

2.1. Materials and Equipments

The green and red QDs were purchased from Beijing Beida Jubang Science & Technology Co., Ltd. The PDMS package material was purchased from Dow Corning Corporation, USA. TEM images were taken by a transmission electron microscope (TEM, JEM-2100F, JEOL, Akishima, Japan) operated at an accelerating voltage of 200 kV while dropping the QD solution on the copper network. The PL spectrum was measured by a fluorescence spectrophotometer (RF-6000, Shimadzu, Kyoto, Japan) under 365 nm of excitation. The absorption and transmission spectra were collected by a UV-vis spectrometer with an integrating sphere accessory (TU-1901, Persee, Beijing, China). The electroluminescence (EL) spectra and optical parameters of the LED devices were measured with a calibrated integrating sphere system, a spectrometer (USB2000+, Ocean Optics, Largo, FL, USA), and a power supply (Keithley 2425, Keithley Instruments & Products, Cleveland, OH, USA).

2.2. Characterization of Quantum Dot

To determine the characteristics of the QDs used in this study, their PL and absorbance spectrum after dispersal in *n*-hexane were measured and are shown in Figure 1. The emission peaks of green QDs (GQDs) and red QDs (RQDs) are located at 525 and 626 nm, while their absorption peaks are at 515 and 610 nm, respectively. This difference between the emission and absorption peaks is attributed to the Stokes shift effect. The half width of the emission peak is 30 nm for GQDs and 28 nm for RQDs. This indicates that the QDs have a narrower half-width than most common fluorescent materials such as YAG:Ce^{3+} phosphor [26,27] and that the color purity can be guaranteed. The inset picture also indicates their excellent color purity. QDs can absorb light of a shorter wavelength than their PL emission peaks, and QDs are more capable of absorbing ultraviolet light than in absorbing blue light such as that at 455 nm.

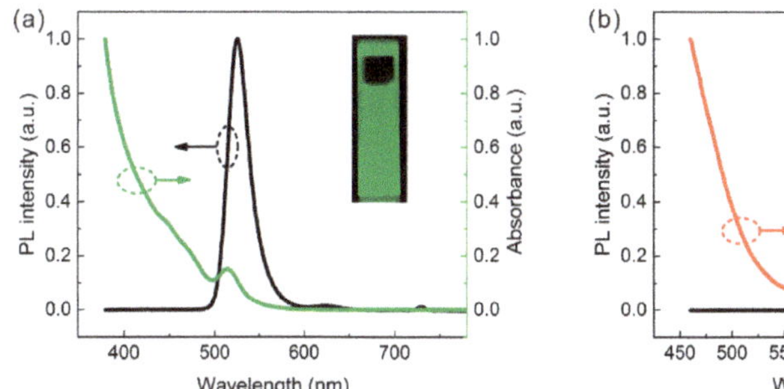

Figure 1. Photoluminescence (PL) and absorbance spectrum of (**a**) GQDs and (**b**) RQDs.

From the transmission electron microscope (TEM) graph, all of the QDs show good dispersibility in *n*-hexane. The QDs exhibit a distinct spherical shape, which is formed by the core-shell structure. For these samples, the CdSe acts as a core, and the ZnS acts as a shell. Further, by using ImageJ software to calculate the diameter of 100 QDs in the TEM graph, the particle size distribution is shown in Figure 2. Both particle size distributions show an approximately normal distribution. The average diameter of the GQDs and RQDs is 7.63 ± 1.16 nm and 9.55 ± 1.10 nm, respectively. The particle size of RQDs is larger than that of the GQDs, which is consistent with the reported discovery: the emission peak of CdSe-based QD is related to its particle size, and a larger diameter can result in a longer emission wavelength [28,29].

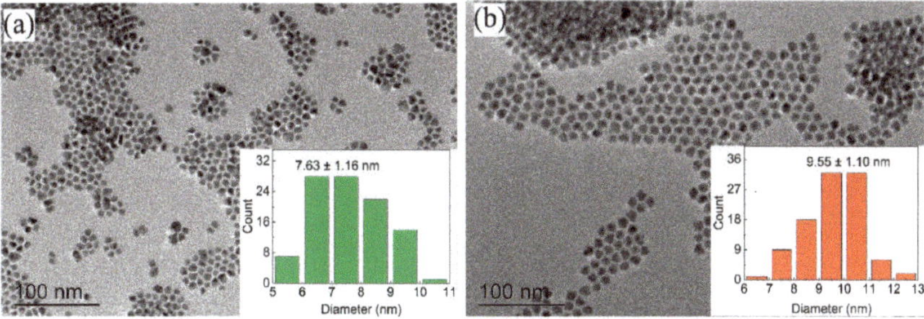

Figure 2. TEM graph of (**a**) GQDs and (**b**) RQDs; inset picture shows particle size distribution of QDs.

2.3. Fabrication of QD-Based LED with Different Excitation Wavelengths

To obtain a QD-based LED with different excitation wavelengths for this study, first, commercial LED chips of the same size (45 × 45 mil) with 365, 385, 405, and 455 nm of emission wavelengths were mounted on a copper substrate. The copper substrate serves as a circuit board and a heat sink. Then, the golden wire was connected by ultrasonic welding. For QD applications, the remote excitation to attain green and red light is the basic LED package structure in this study, as shown in Figure 3. Polydimethylsiloxane (PDMS) was selected as the package material. PDMS gel was injected into the cavity and then cured at 100 °C for 1 h.

To prepare QD remote films for LEDs, green CdSe/ZnS and red CdSe/ZnS/ZnS QDs were selected. After dispersal in *n*-hexane, the QDs were mixed with the PDMS at mass concentrations of 0, 0.2, 0.5, 1.2, 3.0, and 7.0 wt%. Then, the film was prepared using a designed mold. The thickness of the film was fixed at 500 μm by a gasket. After 1 h of heating at 100 °C, the mold could be released to obtain QD films. Finally, the QD remote film was assembled to form LED samples with different excitation wavelengths. Green and red QD-converted LEDs (GQD LED and RQD LED) were obtained.

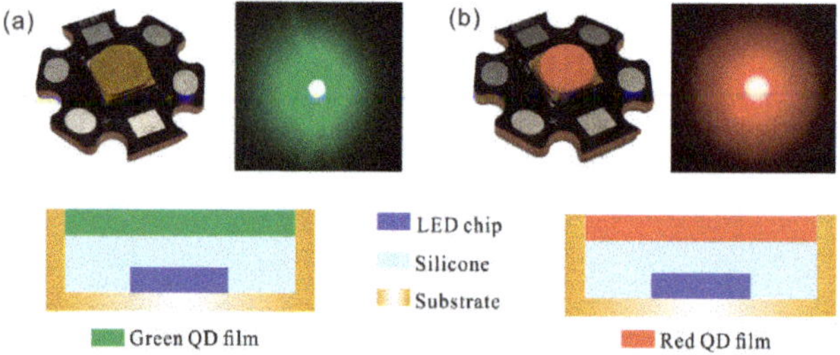

Figure 3. Physical picture and schematic diagram of (**a**) green quantum-dot LED and (**b**) red quantum-dot LED.

3. Results and Discussion

3.1. Characterization of QD Film

After the QDs were prepared as QD film, the properties of the QD films were explored. As shown in Figure 4a,b, the absorbance peaks of the GQD and RQD films were located at 512 nm and 610 nm, respectively. These are almost identical to the absorption peak of the QD solution, which means that the film maintains the original luminescence properties of the QDs. The absorbance of both GQD- and RQD-based film increases as the concentration of QDs increases. After adding the QDs, the absorbance below the absorption wavelength increases significantly, and the absorbance of the ultraviolet region is more pronounced than in the blue region. This regular pattern is also reflected in the transmission spectrum, as shown in Figure S1. Thus, with more QDs, more short-wavelength light could be absorbed.

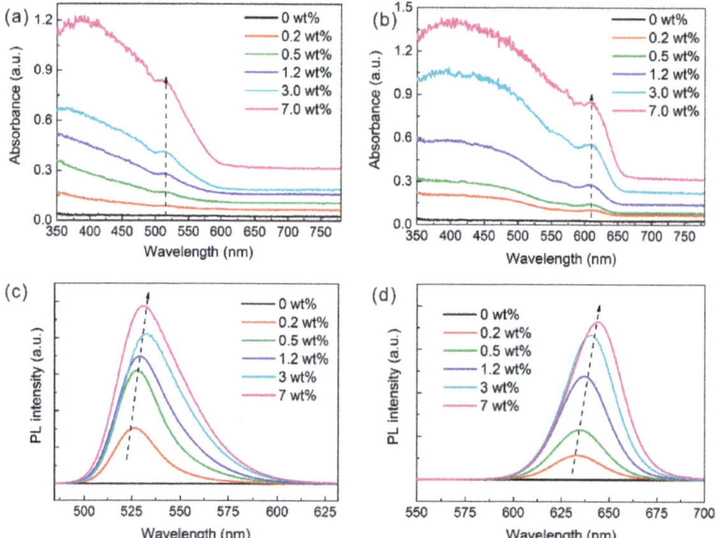

Figure 4. Absorption spectrum of (**a**) GQD and (**b**) RQD film; photoluminescence (PL) spectrum of (**c**) GQDs and (**d**) RQD film.

The PL spectra of different QD films excited at 365 nm are shown in Figure 4c,d. The higher the QD concentration, the greater the PL intensity that can be attained. However, a red-shift in the emission peak occurs at a higher QD concentration. For example, at a low concentration of 0.2 wt%, the emission wavelength of GQD film is 526 nm, but at a high concentration of 7 wt%, it is red-shifted to 531 nm, as shown in Figure S2. The reason for the red shift could be the agglomeration and reabsorption of QDs. Agglomeration leads to an increase in the average particle size of the QDs, and the QDs have special properties of reabsorption and reemission [30]. Since the red QD absorption is much stronger than the green QD absorption at 365 nm of excitation, the reabsorption effect is much more significant. Thus, the RQD suffers from a more severe red-shifted effect at high concentrations such as 7.0 wt%.

3.2. Light Conversion Optical Model of QD LED

With the combination of the LED chip and QD film, the entire device forms a green or red LED, which is the fundamental base of a display pixel. As shown in Figure 5, LED light transformation can be summarized into three stages:

(1) Absorption. The short-wavelength light from the LED chip is absorbed by the QDs. Thus, the QD absorption rate is defined as the ratio of the absorbed short-wavelength light intensity to the initial short-wavelength light intensity. The QD absorption rate can be used to measure the initial peak elimination ability of QDs for short-wavelength light;

(2) Conversion. Then, the QDs can turn the short-wavelength light into their own emission light. For example, the green QD transforms the blue light into green emissions. Thus, the QD conversion efficiency is defined as the ratio of the QD emission intensity to the absorbed short-wavelength light intensity. The QD conversion efficiency reflects the ability to convert short-wavelength light;

(3) Output. After mixing the residual short-wavelength light intensity and the QD emission peak intensity, the LED devices produce new light output. In other words, the final optical output is composed of residual short-wavelength light and QD emissions. Thus, the energy conversion efficiency is defined as the ratio of the final total light intensity to the initial light intensity. This can reveal the energy utilization of the total LED device.

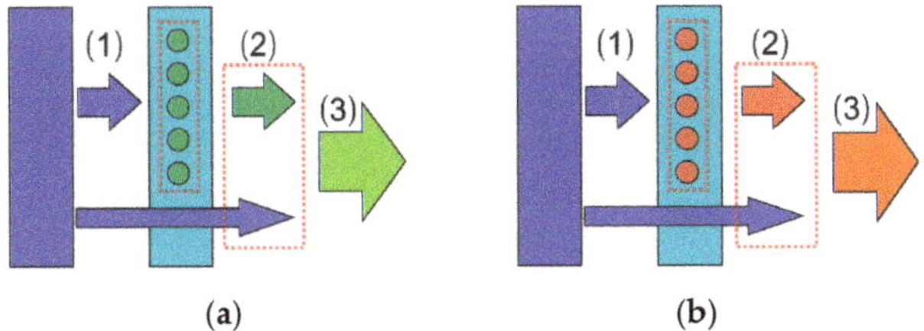

Figure 5. Light conversion optical models of (**a**) GQD LED and (**b**) RQD LED.

3.3. Performance of GQD LED with Different Excitation Wavelengths

The EL spectra of different GQD LEDs are explored and shown in Figure 6. At different excitation wavelengths, the following discoveries about QD concentrations are consistent: First, as the concentration of GQD increases, the peak of short-wavelength light decreases while the green peak first rises and then falls. Almost all of the LED samples at different excitation wavelengths show the top green peak intensity at 1.2 wt% of GQD concentration. Second, the green peak emission wavelength shows a red shift with increasing GQD concentration, which is consistent with the PL test of the QD film. Third, the overall radiant flux of the LED device decreased. This confirms the significant energy loss during QD conversion and this situation becomes more serious with high concentration QDs. Therefore, varieties in QD concentration at different excitation wavelengths show similar patterns in the LED spectrum. This means that even if the initial excitation wavelength of the device is changed, the basic shape of the EL spectrum stays the same.

To further analyze the effects of the excitation wavelength, the optical model described above is introduced here. The first stage is absorption. For the GQD absorption rate, all LEDs with different excitation wavelengths indicate that the higher the QD concentration, the more the short-wavelength light is absorbed, as shown in Figure 7a. However, the GQD absorption rate excited by UV light is much higher than that by blue light with the same QD concentration. Only a GQD concentration of 0.2 wt% can absorb nearly half of the 365 nm original light intensity (47.7%). Further, the initial incident radiant flux of LED chips and the QD concentration (1.2 wt%) were fixed to ensure a single variable of the excitation wavelength. Three initial incident radiant fluxes of the LED chips were studied: 20, 60, and 100 mW. Under a 20 mW incident radiant flux, the GQDs could absorb 69.5% of the 365 nm violet light, while only 53.3% of the 455 nm blue light could be absorbed, as shown in Figure 7d. The ability to absorb short wavelengths of GQD increased by 16.1% under 365 nm of excitation. Thus, a solution that combines the UV light and QDs has advantages in eliminating raw chip peaks.

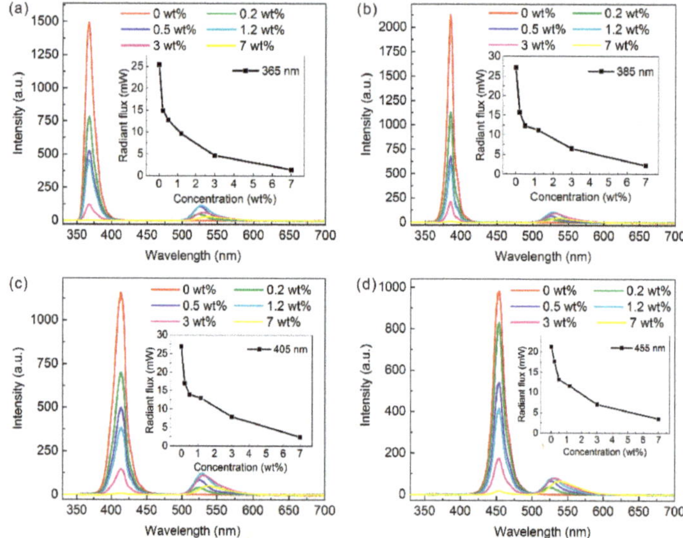

Figure 6. Electroluminescence (EL) spectrum of GQD LED with (**a**) 365 nm, (**b**) 385 nm, (**c**) 405 nm, and (**d**) 455 nm of LED-chip excitation wavelength under 20 mA. Inset graph shows radiation flux vs. GQD concentration.

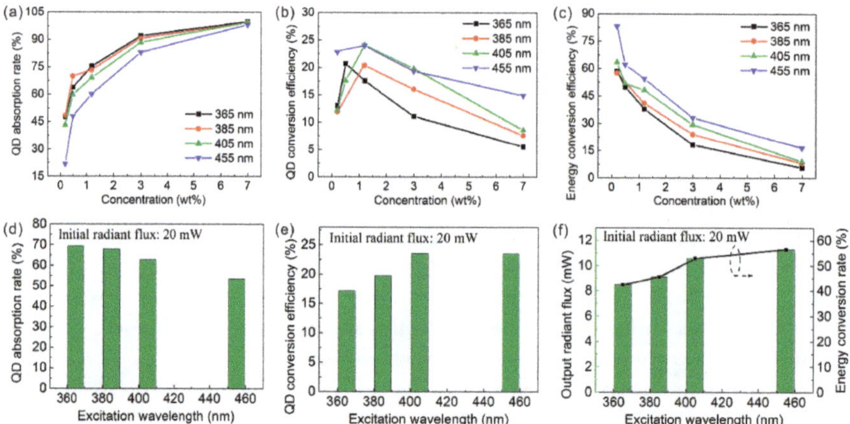

Figure 7. (**a**,**d**) QD absorption rate, (**b**,**e**) QD conversion efficiency, and (**c**,**f**) energy conversion efficiency of GQD LED with different excitation wavelengths.

Then, the absorbed short-wavelength light is converted to green light by GQDs. Among the excitation wavelengths, the QD conversion efficiency of GQD rises first and then drops as the QD concentration increases. Thus, a moderate QD concentration such as 1.2 wt% can maximize the QD conversion efficiency. The QD conversion efficiency tends to increase at longer excitation wavelengths. In other words, the QD conversion efficiency excited by ultraviolet light is lower than that by blue light. Taking the 20 mW incident radiant flux as an example, the QD conversion efficiency is only 17.1% under 365 nm of excitation but increases to 23.5% with 455 nm of excitation, a 6.4% increase. It should be pointed out that the QD conversion efficiency is not ideal; this could be owing to the severe reabsorption characteristics of QDs [31]. In short, typical blue light has an advantage in QD conversion efficiency.

The last part of the optical model is the output. At different excitation wavelengths, the energy conversion efficiency of all LEDs declines as the QD concentration increases. This may be owing to the agglomeration and reabsorption effect of the high QD concentration, which is more likely to produce significant heat [32,33]. Therefore, from the perspective of energy utilization, a low QD concentration is better for LED applications. However, at the same QD concentration, the energy conversion efficiency gradually increases as the excitation wavelength increases. The typical blue 455 nm excited LED has a maximum energy conversion efficiency of 83.1% with a GQD concentration of 0.2 wt%. As shown in Figure 7f, when the incident radiant flux is 20 mW, the output radiant flux of the LED with excitation wavelengths of 365, 385, 405, and 455 nm is 8.5, 9.1, 10.6, and 11.3 mW, respectively. Therefore, the energy conversion efficiency at 455 nm is 14.1% higher than that at 365 nm. Repeated experiments with different incident radiant fluxes such as 60 and 100 mW demonstrate the reliability of the above conclusions, as shown in Figure S3.

To determine the color quality of these different LEDs, their color coordinates are summarized and drawn in the chromaticity diagram in Figure 8. The colorimetric behavior of the GQD LED with different excitations can be divided into two categories: (A) 365 and 385 nm of excitation and (B) 405 and 455 nm of excitation. Under 365 and 385 nm of excitation, the x color coordinate keeps increasing, while the y color coordinate first increases and then decreases as the GQD concentration increases. This indicates that the green light gradually plays a leading role. The color coordinates only shifted within the green-light area. Interestingly, the LED excited at 365 nm at a low QD concentration of 0.2 wt% emits a relatively high-purity green light. However, 405 and 455 nm-based LEDs show distinct color qualities. Both the x and y color coordinates keep increasing as the QD concentration increases, which leads to a gradual light shift from the blue area to the green area. Even the green color quality of the 455 nm LED with the highest QD concentration at 7.0 wt% could not compete with the 365 nm LED with the lowest QD concentration at 0.2 wt%. If the LEDs with different excitation wavelengths aim to obtain the same green light color quality, the GQD concentration at 455 nm of excitation should be much higher than that at 365 nm of excitation. Thanks to the excellent color characteristics of the QD itself, the color coordinate of the GQD LED with 365 nm excitation and high GQD concentration at 7.0 wt% is (0.322, 0.663). This green light almost reaches the edge of the chromaticity diagram, which means that ultrawide color gamut performance is possible if this GQD is applied to displays.

Even with the same incident light power and the same GQD concentration, the LEDs clearly exhibit different color properties owing to different excitation wavelengths, as shown in Figure 9. The color coordinates of the green LED excited at 365, 385, 405, and 455 nm are (0.231, 0.706), (0.230, 0.674), (0.202, 0.318), and (0.162, 0.123) under a 20 mW incident radiant flux, respectively. Thus, both the x and y coordinates show a downward trend. In other words, under the excitation of a 365 nm LED chip, the entire GQD device emits high-purity green light. However, under the excitation of the 455-nm LED chip, the device emits blue light. This superiority in color purity excited at 365 nm is mainly owing to two reasons: the GQD LEDs excited at 365 nm have unique advantages in eliminating initial short peaks owing to the stronger absorption ability, and the naked eye is not sensitive to residual 365 nm ultraviolet light. But green light from QD emission is visible and the human eye is quite sensitive to green light. Thus, in terms of color purity, the excitation of QDs by ultraviolet light such as

365 nm can maximize the advantages of the quantum-dot color quality. This UV excitation solution is recommended for display applications.

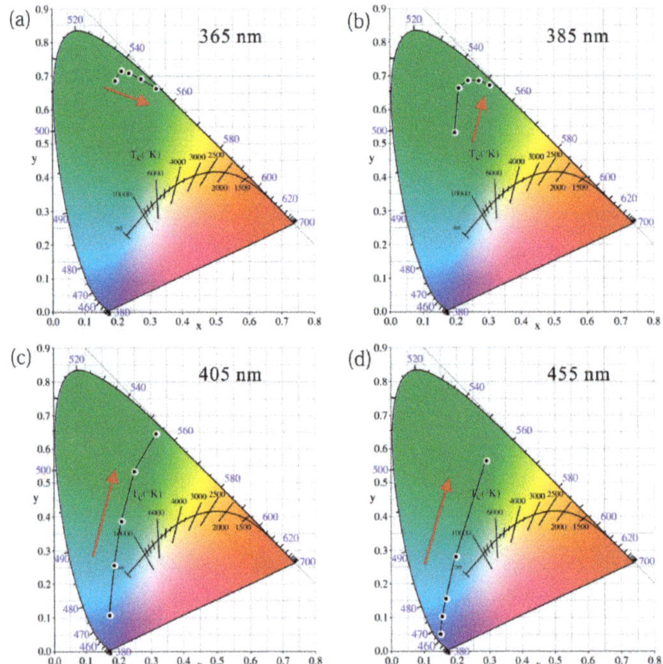

Figure 8. Chromaticity diagram of GQD LED with different QD concentrations under (**a**) 365 nm, (**b**) 385 nm, (**c**) 405 nm, and (**d**) 455 nm of excitation.

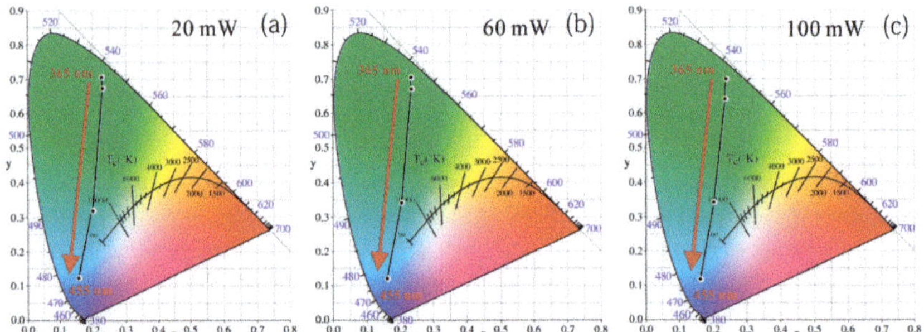

Figure 9. Chromaticity diagram of GQD LED with different excitation wavelengths under equal incident radiant fluxes of (**a**) 20 mW, (**b**) 60 mW, and (**c**) 100 mW.

3.4. Performance of RQD LEDs with Different Excitation Wavelengths

As for the RQDs, the EL spectrum is also explored and shown in Figure S4. The EL spectra reveal that LEDs based on RQDs exhibit laws similar to those of GQDs. Then, as with the GQD LED, the QD absorption rate, QD conversion efficiency, and energy conversion efficiency of the RQD LED were calculated and are shown in Figure 10. In general, the red LED has a pattern similar to that of the green LED. Unlike green QDs, almost all of the short-wavelength light can be absorbed with only 3.0 wt% concentration RQDs. This reveals that RQDs have a stronger advantage than GQDs in eliminating short peaks.

Under 20 mW of incident optical power, the RQD absorption rate excited at 365 nm is 5.0% higher than that at 455 nm, but the RQD conversion efficiency and energy conversion efficiency excited at 455 nm are 14.3% and 15.5% higher, respectively, than that at 365 nm. Like the GQD, the blue chip is recommended from the perspective of energy utilization because the maximum energy conversion efficiency is 69.3 wt% from the 455 nm LED. At different incident power such as 60 and 100 mW, the same law is exhibited in Figure S5.

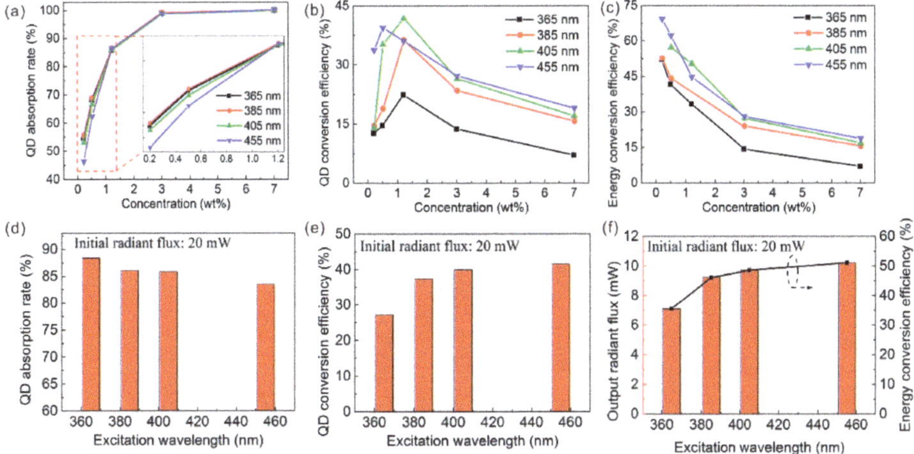

Figure 10. (**a,d**) QD absorption rate, (**b,e**) QD conversion efficiency, and (**c,f**) energy conversion efficiency of GQD LED with different excitation wavelengths.

A chromaticity diagram of the RQD LED under different wavelength excitations is also summarized. There is a significant difference in behavior in the color performance between a typical UV 365 nm LED and a blue 455 nm LED. Under 365 nm of excitation, because the original peak from the LED chip is more easily eliminated owing to the stronger absorption ability and the human eye is not sensitive to residual 365 nm ultraviolet light. The main color perceived by the human eye is visible red light, which leads to excellent red-light quality even at a low QD concentration of 0.2 wt%. In addition, the red color quality of the 385 nm excitation is similar to that of 365 nm LED but is slightly closer to the orange color at low concentrations. However, under 455 nm of excitation, the color shifts from blue and purple to red. The LED exhibits significant blue light at low concentrations such as 0.2 wt%. The 405 nm LED shows patterns similar to those of the blue excitation.

Even with the same incident light power and the same RQD concentration (1.2 wt%), the color performance of these LED samples is totally different, as shown in Figure 11. From 365 nm to 455 nm of excitation, the color shifts from red to purple regions, as shown in Figure S6. Taking advantage of the QD, the color coordinates of the 365 nm excitation basically reached the boundaries of the gamut, which means that the red color is highly pure. These experimental results reveal that the combination of UV light with RQDs has unparalleled advantages in colorimetry, which may shine in future display fields. Therefore, the red LED prepared by ultraviolet light to excite RQDs is superior to the blue-light excitation in attaining high-quality red light. This is consistent with the conclusion regarding GQDs.

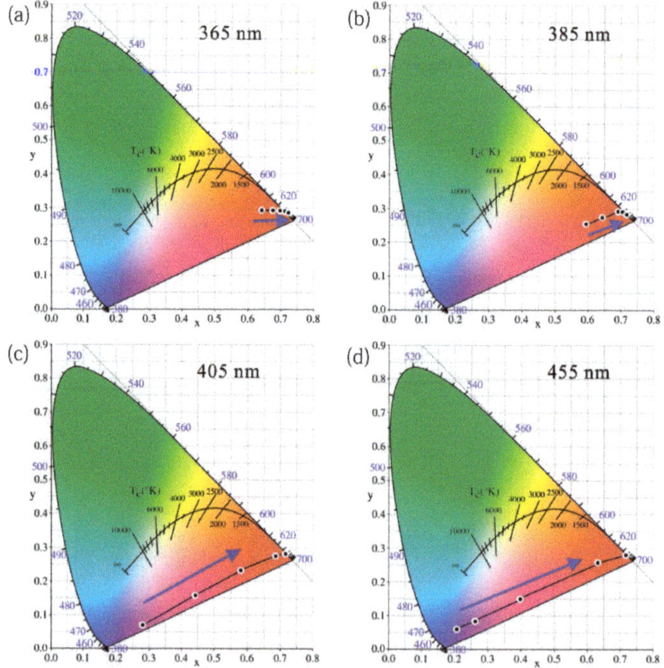

Figure 11. Chromaticity diagram of RQD LED with different QD concentrations under (**a**) 365 nm, (**b**) 385 nm, (**c**) 405 nm, and (**d**) 455 nm of excitation.

3.5. Comparison of Color Gamut between Blue (455 nm) Excitation and Ultraviolet (365 nm) Excitation Solutions

The color gamut is an important parameter when evaluating RGB display systems. The wider the gamut, the higher the fidelity of the colors of nature that can be attained. A comparison of the color gamut for different excitation wavelength solutions is analyzed in Table 1. The blue-color coordinates of all solutions are derived from the blue LED chip, and the green and red color coordinates are obtained from the QD-converted LEDs. With the highest QD concentration at 7.0%, the RGB color gamut excited at 455 nm is only 84.9% of the NTSC1953 standard. Interestingly, the color gamut of 365 nm-excited solution could cover 99.7% of the NTSC1953 standard area even at a low QD concentration (0.2 wt%). Moreover, as shown in Figure 12, the color gamut of a 365-nm-excited RGB system could be high as 117.5% by adjusting the QD concentration properly. This is a 32.6% improvement over the blue-light-excited solution. Thus, the UV-excited quantum-dot solution can achieve a wider color gamut and has great potential for development in quantum-dot applications, especially in the display field.

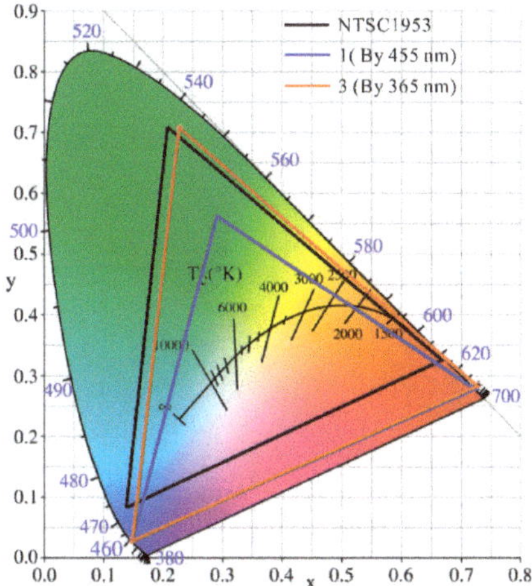

Figure 12. Comparison of color gamut between blue (455-nm) excitation and ultraviolet (365-nm) excitation solutions.

Table 1. Comparison of color gamut for wavelength solutions with different excitations.

Group	Excitation Wavelength (nm)	GQD (%)	RQD (%)	R(x, y)	G(x, y)	B(x, y)	NTSC1953 Standard (%)
1	455	7.0	7.0	(0.7210, 0.2794)	(0.2935, 0.5624)	(0.1492, 0.0298)	84.9
2	365	0.2	0.2	(0.6485, 0.2931)	(0.1947, 0.6855)	(0.1492, 0.0298)	99.7
3	365	1.2	7.0	(0.7289, 0.2834)	(0.2300, 0.7064)	(0.1492, 0.0299)	117.5

4. Conclusions

In this study, we investigated the effect of the excitation wavelength on the optical performances of QD-converted light-emitting diodes. Sample LEDs with excitation light sources of 365, 385, 405, and 455 nm were prepared, and their optical performances were compared. Under different excitation wavelengths, green and red QD-converted LEDs showed similar patterns. The QD conversion efficiency and energy conversion efficiency increase at longer excitation wavelengths. At an incident radiant flux of 20 mW, the QD conversion efficiency and energy conversion efficiency of the 455 nm-excited GQD LED were 6.4% and 14.1% higher than those of the 365 nm LED. Therefore, blue light is recommended to excite QDs with regard to energy utilization. However, the UV-excited QD LED solution has more advantages than blue-light excitation for color purity. Even for the 455 nm LED with the highest QD concentration at 7.0 wt%, the green color quality could not compete with a 365 nm LED with the lowest QD concentration of 0.2 wt%. This is because QD LEDs excited at 365 nm have unique advantages in eliminating short peaks because of higher absorption in the ultraviolet range of QD. Moreover, the naked eye is not sensitive to the residual 365 nm ultraviolet light. As a result, green and red light excited at 365 nm with a 7.0 wt% QD concentration can approach the boundary of the chromaticity diagram. Taking advantage of the 365 nm-excited solution, a 117.5% color gamut of the RGB system could be obtained compared with the NTSC1953 standard, which is an improvement of 32.6% over the blue-light-excited solution. This research helps understand the variation of quantum-dot-converted LEDs at different excitation wavelengths, and provides basic guidance for the selection of a pumping source for QDs.

Supplementary Materials: The following are available online at http://www.mdpi.com/2079-4991/9/8/1100/s1, Figure S1: Transmission spectrum of (a) GQD and (b) RQD film; Figure S2: PL shift phenomenon of (a) GQD film and (c) RQD film; Figure S3: (a,d) GQD absorption rate; (b,e) GQD conversion efficiency; (c,f) output radiant flux and energy conversion efficiency under 60- and 100-mW incident radiant flux; Figure S4: Electroluminescence (EL) spectrum of RQD LED with (a) 365 nm, (b) 385 nm, (c) 405 nm, and (d) 455-nm LED chip excitation wavelength under 20 mA; Figure S5: (a,d) RQD absorption rate; (b,e) RQD conversion efficiency; (c,f) output radiant flux and energy conversion efficiency under 60- and 100-mW incident radiant flux; Figure S6: Chromaticity diagram of RQD LED with different excitation wavelengths under equal incident radiant fluxes such as (a) 20 mW, (b) 60 mW, and (c) 100 mW.

Author Contributions: Formal analysis, C.Y. and X.D.; funding acquisition, Y.T. and Z.L.; investigation, X.D. and J.L.; methodology, C.Y. and X.D.; resources, J.L. and Z.L.; writing (original draft), C.Y.; writing (review & editing), Z.L. and Y.T.

Funding: This study was supported by the Science & Technology Program of Guangdong Province (2019B010130001), the National Natural Science Foundation of China (No. 51775199, No. 51735004) and the Natural Science Foundation of Guangdong Province (2018B030306008).

Conflicts of Interest: The authors declare no conflict of interest.

References

1. Pimputkar, S.; Speck, J.S.; Denbaars, S.P.; Nakamura, S. Prospects for LED lighting. *Nat. Photonics* **2009**, *3*, 180–182. [CrossRef]
2. Nann, T.; Skinner, W.M. Quantum dots for electro-optic devices. *ACS Nano* **2011**, *5*, 5291–5295. [CrossRef] [PubMed]
3. Hasan, J.; Ang, S.S. A RGB-driver for LED display panels. In Proceedings of the 2010 Twenty-Fifth Annual IEEE Applied Power Electronics Conference and Exposition (APEC), Palm Springs, CA, USA, 21–25 February 2010; pp. 750–754.
4. Nguyen, F. Challenges in the design of a RGB LED display for indoor applications. *Synth. Met.* **2001**, *122*, 215–219. [CrossRef]
5. Li, Z.; Cao, K.; Li, J.; Tang, Y.; Xu, L.; Ding, X.; Yu, B. Investigation of Light-Extraction Mechanisms of Multiscale Patterned Arrays With Rough Morphology for GaN-Based Thin-Film LEDs. *IEEE Access* **2019**, *7*, 73890–73898. [CrossRef]
6. Kurtin, J.; Puetz, N.; Theobald, B.; Stott, N.; Osinski, J. 12.5L: Late-News Paper: Quantum Dots for High Color Gamut LCD Displays using an On-Chip LED Solution. *SID Symp. Dig. Tech. Pap.* **2015**, *45*, 146–148. [CrossRef]
7. Chen, C.J.; Lien, J.Y.; Wang, S.L.; Chiang, R.K. P-91: Highly-Efficient LEDs with On-Chip Quantum-Dot Package for Wide Color Gamut LCD Display. In Proceedings of the SID Symposium Digest of Technical Papers, San Francisco, CA, USA, 22–27 May 2016; pp. 1465–1468.
8. Peng, Y.; Wang, S.; Li, R.; Li, H.; Cheng, H.; Chen, M.; Liu, S. Luminous efficacy enhancement of ultraviolet-excited white light-emitting diodes through multilayered phosphor-in-glass. *Appl. Opt.* **2016**, *55*, 4933–4938. [CrossRef] [PubMed]
9. Anandan, M. Progress of LED backlights for LCDs. *J. Soc. Inf. Disp.* **2008**, *16*, 287–310. [CrossRef]
10. Xie, R.J.; Hirosaki, N.; Takeda, T. Wide color gamut backlight for liquid crystal displays using three-band phosphor-converted white light-emitting diodes. *Appl. Phys. Express* **2009**, *2*, 022401. [CrossRef]
11. Wang, L.; Wang, X.; Kohsei, T.; Yoshimura, K.I.; Izumi, M.; Hirosaki, N.; Xie, R.J. Highly efficient narrow-band green and red phosphors enabling wider color-gamut LED backlight for more brilliant displays. *Opt. Express* **2015**, *23*, 28707–28717. [CrossRef] [PubMed]
12. Nitti, A.; Bianchi, G.; Po, R.; Swager, T.M.; Pasini, D. Domino Direct Arylation and Cross-Aldol for Rapid Construction of Extended Polycyclic π-Scaffolds. *J. Am. Chem. Soc.* **2017**, *139*, 8788–8791. [CrossRef] [PubMed]
13. Pasini, D.; Takeuchi, D. Cyclopolymerizations: Synthetic Tools for the Precision Synthesis of Macromolecular Architectures. *Chem. Rev.* **2018**, *118*, 8983–9057. [CrossRef] [PubMed]
14. Chen, H.W.; Zhu, R.D.; He, J.; Duan, W.; Hu, W.; Lu, Y.Q.; Li, M.C.; Lee, S.L.; Dong, Y.J.; Wu, S.T. Going beyond the limit of an LCD's color gamut. *Light Sci. Appl.* **2017**, *6*, e17043. [CrossRef] [PubMed]
15. Chen, H.; He, J.; Wu, S.T. Recent Advances on Quantum-Dot-Enhanced Liquid-Crystal Displays. *IEEE J. Sel. Top. Quantum Electron.* **2017**, *23*, 1–11. [CrossRef]
16. Song, W.S.; Yang, H. Efficient White-Light-Emitting Diodes Fabricated from Highly Fluorescent Copper Indium Sulfide Core/Shell Quantum Dots. *Chem. Mater.* **2012**, *24*, 1961–1967. [CrossRef]

17. Rao, L.; Tang, Y.; Yan, C.; Li, J.; Zhong, G.; Tang, K.; Yu, B.; Li, Z.; Zhang, J.Z. Tuning the emission spectrum of highly stable cesium lead halide perovskite nanocrystals through poly (lactic acid)-assisted anion-exchange reactions. *J. Mater. Chem. C* **2018**, *6*, 5375–5383. [CrossRef]

18. Li, J.; Tang, Y.; Li, Z.; Ding, X.; Yuan, D.; Yu, B. Study on Scattering and Absorption Properties of Quantum-Dot-Converted Elements for Light-Emitting Diodes Using Finite-Difference Time-Domain Method. *Materials* **2017**, *10*, 1264. [CrossRef] [PubMed]

19. Tang, Y.; Lu, H.; Li, J.; Li, Z.; Du, X.; Ding, X.; Yu, B. Improvement of Optical and Thermal Properties for Quantum Dots WLEDs by Controlling Layer Location. *IEEE Access* **2019**, *7*, 77642–77648. [CrossRef]

20. Coe-Sullivan, S.; Liu, W.; Allen, P.; Steckel, J.S. Quantum dots for LED downconversion in display applications. *ECS J. Solid State Sci. Technol.* **2013**, *2*, R3026–R3030. [CrossRef]

21. Luo, Z.; Chen, Y.; Wu, S.T. Wide color gamut LCD with a quantum dot backlight. *Opt. Express* **2013**, *21*, 26269–26284. [CrossRef]

22. Jang, E.; Jun, S.; Jang, H.; Lim, J.; Kim, B.; Kim, Y. White-light-emitting diodes with quantum dot color converters for display backlights. *Adv. Mater.* **2010**, *22*, 3076–3080. [CrossRef]

23. Chen, J.; Hartlove, J.; Hardev, V.; Yurek, J.; Lee, E.; Gensler, S. P-119: High efficiency LCDs using quantum dot enhancement films. In Proceedings of the SID Symposium Digest of Technical Papers, San Diego, CA, USA, 1–6 June 2014; pp. 1428–1430.

24. Lin, H.Y.; Sher, C.W.; Hsieh, D.H.; Chen, X.Y.; Chen, H.M.P.; Chen, T.M.; Lau, K.M.; Chen, C.H.; Lin, C.C.; Kuo, H.C. Optical cross-talk reduction in a quantum-dot-based full-color micro-light-emitting-diode display by a lithographic-fabricated photoresist mold. *Photonics Res.* **2017**, *5*, 411–416. [CrossRef]

25. Han, H.V.; Lin, H.Y.; Lin, C.C.; Chong, W.C.; Li, J.R.; Chen, K.J.; Yu, P.; Chen, T.M.; Chen, H.M.; Lau, K.M. Resonant-enhanced full-color emission of quantum-dot-based micro LED display technology. *Opt. Express* **2015**, *23*, 32504–32515. [CrossRef]

26. Loo, K.H.; Lai, Y.M.; Tan, S.C.; Chi, K.T. On the Color Stability of Phosphor-Converted White LEDs Under DC, PWM, and Bilevel Drive. *IEEE Trans. Power Electron.* **2012**, *27*, 974–984. [CrossRef]

27. Ding, X.; Li, M.; Li, Z.; Tang, Y.; Xie, Y.; Tang, X.; Fu, T. Thermal and optical Investigations of a laser-driven phosphor converter coated on a heat pipe. *Appl. Therm. Eng.* **2019**, *148*, 1099–1106. [CrossRef]

28. Dabbousi, B.O.; Rodriguez-Viejo, J.; Mikulec, F.V.; Heine, J.R.; Mattoussi, H.; Ober, R.; Jensen, K.F.; Bawendi, M.G. (CdSe) ZnS core−shell quantum dots: Synthesis and characterization of a size series of highly luminescent nanocrystallites. *J. Phys. Chem. B* **1997**, *101*, 9463–9475. [CrossRef]

29. Zhang, K.; Chang, H.; Fu, A.; Alivisatos, A.P.; Yang, H. Continuous distribution of emission states from single CdSe/ZnS quantum dots. *Nano Lett.* **2006**, *6*, 843–847. [CrossRef]

30. Li, J.S.; Tang, Y.; Li, Z.T.; Cao, K.; Yan, C.M.; Ding, X.R. Full spectral optical modeling of quantum-dot-converted elements for light-emitting diodes considering reabsorption and reemission effect. *Nanotechnology* **2018**, *29*, 295707. [CrossRef]

31. Li, J.S.; Tang, Y.; Li, Z.T.; Ding, X.R.; Rao, L.S.; Yu, B.H. Effect of quantum dot scattering and absorption on the optical performance of white light-emitting diodes. *IEEE Trans. Electron Devices* **2018**, *65*, 2877–2884. [CrossRef]

32. Woo, J.Y.; Kim, K.N.; Jeong, S.; Han, C.S. Thermal behavior of a quantum dot nanocomposite as a color converting material and its application to white LED. *Nanotechnology* **2010**, *21*, 495704. [CrossRef]

33. Li, Z.; Song, C.; Qiu, Z.; Li, J.; Cao, K.; Ding, X.; Tang, Y. Study on the Thermal and Optical Performance of Quantum Dot White Light-Emitting Diodes Using Metal-Based Inverted Packaging Structure. *IEEE Trans. Electron Devices* **2019**, *66*, 3020–3027. [CrossRef]

 *nanomaterials*

Article

An InGaN/GaN Superlattice to Enhance the Performance of Green LEDs: Exploring the Role of V-Pits

Mengling Liu [1,2], Jie Zhao [1,2], Shengjun Zhou [1,2,*], Yilin Gao [1,2], Jinfeng Hu [1,2], Xingtong Liu [1,2] and Xinghuo Ding [1,2]

[1] Key Laboratory of Hydraulic Machinery Transients (Wuhan University), Ministry of Education, Wuhan 430072, China; lml0305@whu.edu.cn (M.L.); zjie1994@whu.edu.cn (J.Z.); gyl1016@whu.edu.cn (Y.G.); 2017282080101@whu.edu.cn (J.H.); 2016202080010@whu.edu.cn (X.L.); xinghuo@whu.edu.cn (X.D.)

[2] Center for Photonic and Semiconductor, School of Power and Mechanical Engineering, Wuhan University, Wuhan 430072, China

* Correspondence: zhousj@whu.edu.cn; Tel.: +86-027-5085-3293

Received: 15 May 2018; Accepted: 18 June 2018; Published: 21 June 2018

Abstract: Despite the fact that an InGaN/GaN superlattice (SL) is useful for enhancing the performance of a GaN-based light-emitting diode (LED), its role in improving the efficiency of green LEDs remains an open question. Here, we investigate the influence of a V-pits-embedded InGaN/GaN SL on optical and electrical properties of GaN-based green LEDs. We recorded a sequence of light emission properties of InGaN/GaN multiple quantum wells (MQWs) grown on a 0- and 24-pair InGaN/GaN SL by using scanning electron microscopy (SEM) in combination with a room temperature cathodoluminescence (CL) measurement, which demonstrated the presence of a potential barrier formed by the V-pits around threading dislocations (TDs). We find that an increase in V-pit diameter would lead to the increase of V-pit potential barrier height. Our experimental data suggest that a V-pits-embedded, 24-pair InGaN/GaN SL can effectively suppress the lateral diffusion of carriers into non-recombination centers. As a result, the external quantum efficiency (EQE) of green LEDs is improved by 29.6% at an injection current of 20 mA after implementing the V-pits-embedded InGaN/GaN SL layer. In addition, a lower reverse leakage current was achieved with larger V-pits.

Keywords: green LEDs; InGaN/GaN superlattice; V-pits; external quantum efficiency

---

1. Introduction

GaN-based light-emitting diodes (LEDs) have been applied in many commercial areas, such as backlights for liquid crystal displays (LCDs), solid-state lighting, visible light communications, head-up displays, and optogenetics [1–7]. Generally, the white light emission is obtained by using a combination of phosphors with a blue LED. However, in this scheme, the light emission efficiency of a phosphor-converted white LED is limited due to the energy loss when higher energy photons are converted to lower energy photons. To solve this problem, the direct color mixing of red, green, and blue LEDs are generally considered as a sufficiently efficient method to eliminate optical conversion using phosphors. Nevertheless, it is difficult to obtain high-efficiency LEDs with an emission wavelength ranging from 500 nm to 600 nm by using GaN-based or AlGaInP-based materials, which is known as the "green gap" [8–10].

It has been reported that the "green gap" is caused by high-density threading dislocations (TDs), which result from a large mismatch in the lattice parameter between the GaN and the high-In-content InGaN. The TDs can severely reduce the quantum efficiency of green LEDs [11–14]. Additionally, highly strained InGaN/GaN green multiple quantum wells (MQWs), due to a large lattice mismatch between the high-In-content InGaN and the GaN, exhibit strong piezoelectric-field-induced quantum-confined stark effects (QCSE), resulting in a spatial separation of the wave functions of electrons and holes in the quantum well [15,16]. In the past few decades, much scientific effort has focused on approaches to improve the optical and electrical properties of green LEDs [17–19]. It was found that hexagonal V-pits having inverted pyramids with (10-11) faceted sidewalls were generally formed in InGaN/GaN MQWs [20–23]. Previous studies have demonstrated that the V-pits could impact the performance of LEDs, such as the leakage current, electrostatic discharge capabilities, and the radiative recombination efficiency [24–27]. An underlying InGaN/GaN superlattice (SL) embedded between an n-GaN and an MQW has been used to relax misfit strain [28–31]. However, the increasing SL pairs would accumulate strain energy and the partial strain relaxation would trigger the formation of V-pits [32]. The influence of V-pits and their energy barrier originating from (10-11) facets on optical and electrical properties of LEDs have been investigated [33,34]. It was observed that the presence of V-pits could efficiently suppress the lateral diffusion of excited carriers into non-radiative recombination centers at TDs [35]. Despite the fact that the insertion of an underlying InGaN/GaN SL before the growth of InGaN/GaN MQWs is known to enhance the device performance of LEDs, its role in improving the efficiency of green LEDs remains an open question.

In this work, we investigated the effect of a V-pits-embedded, 24-pair InGaN/GaN SL on the optical and electrical properties of green LEDs. Atomic force microscopy (AFM), scanning electron microscopy (SEM), and cathodoluminescence (CL) measurements were used to characterize the surface morphology and optical property of a green InGaN/GaN MQW. A detailed analysis of a V-pit's structure was performed by using high-angle dark-field (HAADF) scanning transmission electron microscopy (STEM). Additionally, the potential barrier height around the V-pit was also investigated by CL. We found that V-pits play an important role in improving the external quantum efficiency (EQE) and reducing the reverse leakage current of green LEDs.

2. Materials and Methods

2.1. Growth and Device Fabrication

The green MQW and LED samples used in this study were grown on a *c*-plane-patterned sapphire substrate by metal organic chemical vapor deposition (MOCVD). Trimethylindium (TMIn), trimethylgallium (TMGa), and ammonia (NH_3) were used as gallium (Ga), indium (In), and nitrogen (N) sources, respectively. Silane (SiH_4) and biscyclopentadienylmagnesium (CP_2Mg) were used as the n-dopant and p-dopant source. Hydrogen was used as the carrier gas for the growth of the GaN layer while nitrogen was used as the carrier gas for the growth of the InGaN layer. The epitaxial structure of the green LEDs is composed of a 25-nm-thick sputtered AlN nucleation layer, a 2-μm-thick undoped GaN buffer layer, a 2.5-μm-thick Si-doped n-GaN layer, a 24-pair $In_{0.04}Ga_{0.96}N$ (3 nm)/GaN (3 nm) SL at 800 °C, a 12-pair $In_{0.25}Ga_{0.75}N$ (3 nm)/GaN (12 nm) MQW, a 60-nm-thick low temperature p-GaN layer, a 60-nm-thick p-AlGaN/GaN SL electron blocking layer, and a 285-nm-thick p-GaN capping layer. Figure 1 shows schematic illustration of the green LED epitaxial structure. To investigate the effect of the InGaN/GaN SL on optical and electrical properties of green LEDs, the 24-pair $In_{0.04}Ga_{0.96}N$ (3 nm)/GaN (3 nm) was sandwiched between the $In_{0.25}Ga_{0.75}N$/GaN MQW active layer and the n-GaN layer. A green LED without an InGaN/GaN SL was also grown for reference. Additionally, the green MQWs without and with an InGaN/GaN SL were prepared for investigation of the surface morphology.

To form the mesa structure of the green LEDs and expose the n-GaN layer for the deposition of an n-type electrode, a standard photolithography process and BCl_3-based inductively coupled plasma etching were performed. Next, an indium tin oxide (ITO) transparent conductive layer was deposited onto the p-GaN layer for current spreading followed by thermal annealing at 500 °C for 30 min in an N_2 atmosphere. Finally, Cr/Au (30/300 nm) layers were evaporated onto the ITO layers and n-GaN layer for formation of the n-type and the p-type electrodes using an e-beam evaporator. Finally, the epitaxial green LED wafers were diced into chips with a size of 305×330 μm^2.

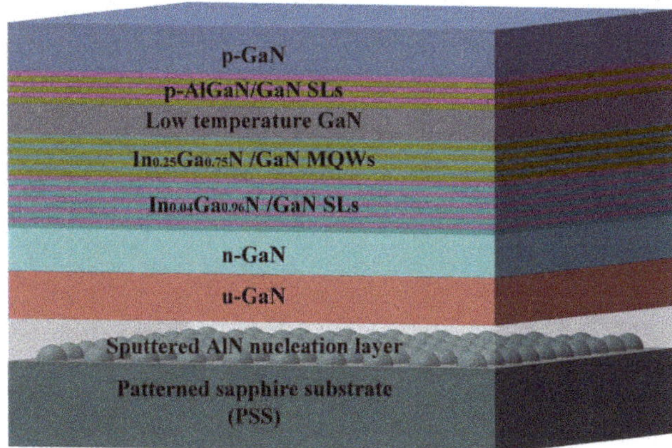

Figure 1. Schematic illustration of green light-emitting diode (LED) epitaxial structure. MQW, multiple quantum well; SL, superlattice.

2.2. Material Characterization and Measurement

AFM (Bruker, Karlsruhe, Germany), performed on a Bruker Multimode 8 in tapping mode, was used to determine the V-pit densities of green MQW samples. The surface morphology of green MQW samples was characterized by using an FEI Nova 3D (FEI, Hillsboro, OR, USA). CL measurements performed on an FEI Quanta 200F field emission SEM (FEI, Hillsboro, OR, USA) fitted with a Gatan Mono CL3+ (Gatan, Pleasanton, CA, USA) system under a vacuum of 10^{-6} Torr were used to further analyze the properties of V-pits. Cross-sectional and plan-view TEM samples were prepared by focus ion beam (FIB) milling using Ga ions at 30 kV, and then TEM images were taken with an FEI Talos F200X system at 200 kV (FEI, Hillsboro, OR, USA). X-ray diffraction (XRD) performed on a BEDE D1 (BEDE, Durham, UK) was used to characterize the crystalline quality of the green LEDs. Temperature-dependent photoluminescence (PL) experiments were conducted at temperatures from 5 K to 300 K using a He–Cd laser ($\lambda = 325$ nm) as the excitation source. The detection part of the PL system consisted of a charge-coupled device detector (Princeton Instruments PIX IS256, Trenton, NJ, USA) connected to a spectrometer (Princeton Instruments SP2500i, Trenton, NJ, USA). The light output power versus current and the current versus voltage characteristics of the green LEDs were measured by using a probe station system (NationStar, Foshan, China) [36].

3. Results and Discussion

Figure 2 shows AFM and morphological SEM images of green MQW samples without and with an InGaN/GaN SL. The V-pit densities of the sample without and with an InGaN/GaN SL were 1.54×10^8 and 1.75×10^8 cm^{-2}, respectively. A previous study revealed that stacking mismatch boundaries induced by stacking faults would trigger the formation of V-pits due to the strain relaxation [21]. As a result, a higher V-pits density was observed in the green MQW grown on an InGaN/GaN SL because of the accumulated strain energy. Additionally, the V-pit diameter of the MQW sample without and with an InGaN/GaN SL was determined to be about 99 and 280 nm, respectively, as shown in Figure 2c,d. Accordingly, the corresponding ratios of the V-pit area were 0.02 and 0.12 with a V-pit diameter of 99 and 280 nm, respectively.

To further analyze the optical properties of V-pits, the CL spectra of the green MQWs were measured around the V-pits and at a point away from the V-pits at an acceleration voltage of 2 kV at room temperature. As shown in Figure 3a,b, the CL spectra exhibited two emission components, which comprised a main energy peak identical to that taken from the point away from the V-pits and a higher-energy component corresponding to the emission energy of the sidewall MQW of the V-pits. In the case of the CL spectra for the green MQWs without an InGaN/GaN SL, the main energy is about 2.30 eV, and the higher-energy component is approximately 2.42 eV. The energy gap between the main energy and the higher-energy component was 120 meV, which indicated that a potential barrier was formed by the V-pit. However, in the case of the InGaN/GaN SL, the energy gap between the main energy and the higher-energy component was about 233 meV as shown in Figure 3b. The potential barrier height of the green MQW with an InGaN/GaN SL was 113 meV higher than that of the green MQW without an InGaN/GaN SL. The extended potential barrier height was attributed to the larger V-pit diameter. A CL intensity measurement with various acceleration voltages was also performed to obtain depth-resolved information on the optical property of the InGaN layer. Figure 3c shows the peak shift of the InGaN emission toward higher energy with increasing acceleration voltage from the outer to the inner region of the InGaN layer. The penetration depth of an electron is determined by the acceleration voltage of the electron beam during the CL measurement. It was previously reported that strain hinders the incorporation of In atoms in the InGaN lattice and is the driving force for the compositional pulling effect in InGaN films [37,38]. As the layer thickness increases, the InGaN film partially relaxes allowing more In atoms to be incorporated in the lattice. For electron energy from 1 kV to 5 kV, the region of maximum energy deposition progressively moves from the near-surface region to the deeper InGaN/GaN QW. The observed blue-shifts of the CL peak position with the increasing acceleration voltage from 1 kV to 5 kV can be explained by a decrease of In content over the penetration depth due to a compositional pulling effect. At an acceleration voltage of 5 kV, the penetration depth of an electron is estimated to be 125 nm. The thickness of the $In_{0.25}Ga_{0.75}N$/GaN green MQW (180 nm) is much larger than the depth of the electron penetration depth at 5 kV. Depth-resolved CL spectra of a green InGaN/GaN MQW measured at various acceleration voltages can exclude the possibility of part of the CL arising from the InGaN/GaN SL, which might result in a blue shift not related to the potential barrier around the dislocation.

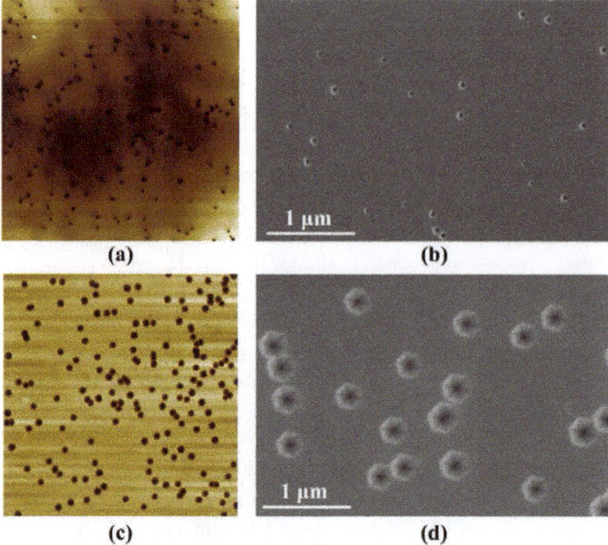

Figure 2. The 10 µm × 10 µm AFM and morphological SEM images of samples. (**a,b**) without an InGaN/GaN SL; (**c,d**) with an InGaN/GaN SL.

As shown in Figure 3d,e, the CL intensity distribution images of the green LEDs without and with an InGaN/GaN SL along with SEM images were taken from the same area of the samples. It was obviously found that the V-pits, directly observable on the SEM images, coincide with dark spots in the CL images corresponding to the non-radiative recombination areas where the TD is located [39]. Generally, the excited charge carriers in the InGaN quantum wells are prone to diffuse laterally until they recombine radiatively or non-radiatively. The excited charge carriers can excite over the small V-pits with a relatively low potential barrier height and non-radiatively recombine at TDs. Therefore, it was observed in Figure 3d,e that the lateral size of the dark spot around small V-pits was larger than its physical size, suggesting that the non-radiative recombination regions are extended outside the V-pits. However, the presence of larger V-pits with a higher potential barrier could effectively suppress the lateral diffusion of the carriers into TDs, minimizing the size of the dark spot.

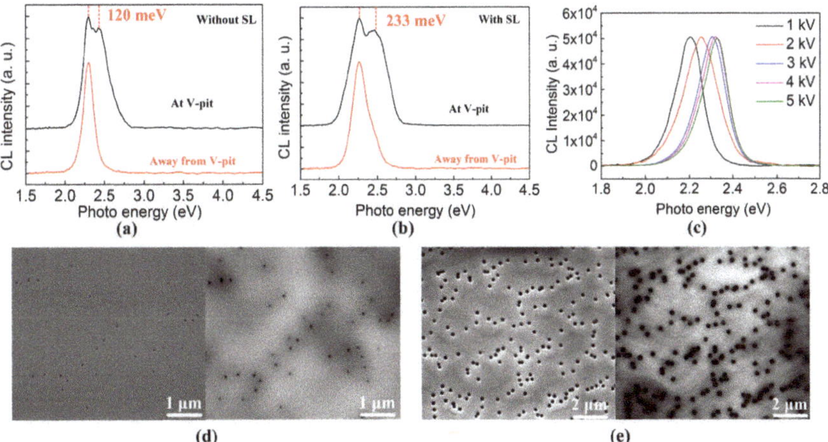

Figure 3. CL spectra measured from a green InGaN/GaN MQW (**a**) without an InGaN/GaN SL; and (**b**) with an InGaN/GaN SL at an acceleration voltage of 2 kV; (**c**) Depth-resolved CL spectra of the green InGaN/GaN MQW measured at various acceleration voltages. The SEM and CL images taken from the same area of the samples (**d**) without an InGaN/GaN SL; (**e**) with an InGaN/GaN SL.

Figure 4a,b show the cross-sectional HAADF-STEM images of a green $In_{0.25}Ga_{0.75}N$/GaN MQW grown on an $In_{0.04}Ga_{0.96}N$/GaN SL, where the V-pits connecting with TDs extend from the underlying SL and pass through the MQW to the low temperature p-GaN and p-AlGaN electron blocking layer. We found that large V-pits start to form at the last few SL pairs while small V-pits form at the first few quantum well pairs. Besides this, it was interesting to observe in Figure 4b that two adjacent V-pits merged and formed a wider W-defect section. A plan-view TEM analysis was performed to confirm the distribution of V-pits on the epilayer by bright-field imaging in the vicinity of the (0001) zone axis. Figure 4c shows the plan-view TEM images of a portion of a green LED epitaxial layer, including the low temperature p-GaN layer, the InGaN/GaN MQW, and a portion of the n-GaN layer along the (0001)-growth direction. The plan-view TEM images associated with selected area diffraction (SAD) patterns confirm that the V-pits were projected along (0001), and each side of the hexagon corresponds to intersections of the (0001) and (10-11) planes.

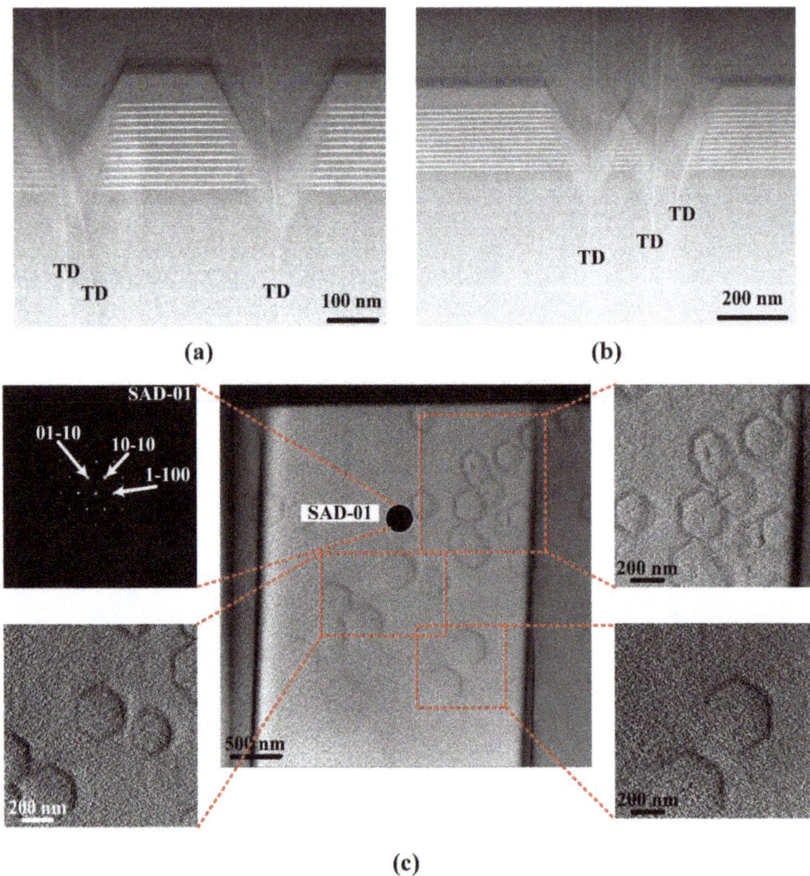

Figure 4. (**a,b**) The HAADF-STEM images showing a representative cross-section of V-pits; (**c**) Plan-view TEM images of a green LED specimen including layers of low temperature p-GaN, InGaN/GaN MQW, and a portion of n-GaN with the selected area diffraction (SAD) pattern of SAD-01.

To elucidate the impact of the underlying InGaN/GaN SL on the crystalline quality of the green LEDs, we performed an X-ray diffraction (XRD) measurement on the green LEDs. Figure 5 shows symmetric (002) and asymmetric (102) ω-scan rocking curves of green LEDs without and with an InGaN/GaN SL. The full widths at half maximum (FWHMs) of the symmetric (002) rocking curves of green LEDs without and with an InGaN/GaN SL were about 265 and 165.5 arcsec, respectively. The FWHMs of the symmetric (102) rocking curve of green LEDs without and with an InGaN/GaN SL were about 194.2 and 181.5 arcsec, respectively. The FWHMs of symmetric (102) and (002) rocking curves of green LEDs with an InGaN/GaN SL are much smaller than those of green LEDs without an InGaN/GaN SL. We can reasonably conclude that the green LED with an InGaN/GaN SL has superior crystallite quality as compared with the green LED without an InGaN/GaN SL.

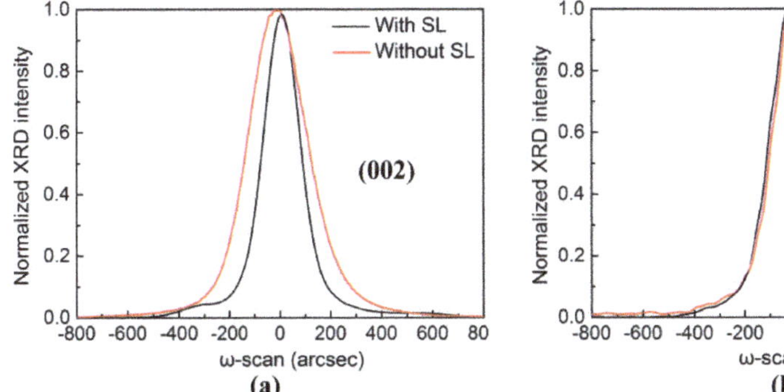

Figure 5. (**a**) Symmetric (002) and (**b**) asymmetric (102) XRD ω-scan rocking curves of green LEDs without and with an InGaN/GaN SL.

Figure 6a,b show the PL spectra of green LEDs without and with an InGaN/GaN SL. A S-shaped (decrease-increase-decrease) temperature dependence of the peak energy with increasing temperature from 5 K to 300 K was observed in Figure 6a,b, which was attributed to the inhomogeneity and carrier localization in the InGaN/GaN MQW [40]. The peak wavelength shifts of the green LED without and with an InGaN/GaN SL were 3.6 and 2.5 nm, respectively, when the temperature was increased from 5 K to 300 K. The smaller peak wavelength shift of the green LED with an InGaN/GaN SL was attributed to the reduced strain in the MQWs due to the insertion of an InGaN/GaN SL. The temperature dependence of the normalized integrated PL intensity, as shown in Figure 6c, can be described by the empirical Arrhenius equation [41]

$$I(T) = \frac{I_0}{1 + Cexp(-\frac{E_a}{k_B T})},$$ (1)

where I_0 is the PL intensity at low temperature, C is a constant proportional to the density of non-radiative recombination centers, E_a is the activation energy of non-radiative recombination centers, T is the temperature, and k_B is the Boltzmann constant. The fitting curves give the activation energy E_a of 15.5 meV for the green MQW without an InGaN/GaN SL and 16.4 meV for the green MQW with an InGaN/GaN SL. The higher activation energy E_a of non-radiative recombination centers is indicative of the superior light emission efficiency of the green MQW with an InGaN/GaN SL. The constant C is 2.4 and 0.8 for the green MQWs without and with an InGaN/GaN SL, respectively, indicating that there was a lower density of non-radiative recombination centers in the green MQW with an InGaN/GaN SL.

By assuming that the non-radiative channels are suppressed at a low temperature, the ratio of the spectrally integrated PL intensity at room temperature (300 K) to that at a low temperature can be used to estimate the internal quantum efficiency (IQE) of green LEDs [42,43]. The IQE of green LEDs is given by the following equation

$$IQE = \frac{I_{300K}}{I_{0K}}.$$ (2)

In a realistic measurement, I_{0K} could be replaced by I_{5K}. Here, the *IQE* of green LEDs without and with an InGaN/GaN SL were estimated to be 20.07% and 29.35%, respectively. The improvement in *IQE* of the green LED with an InGaN/GaN SL was attributed to the large V-pits having a higher potential barrier height, which could effectively suppress the lateral diffusion of the carriers into TDs and thus result in a reduction in the non-radiative combination rate.

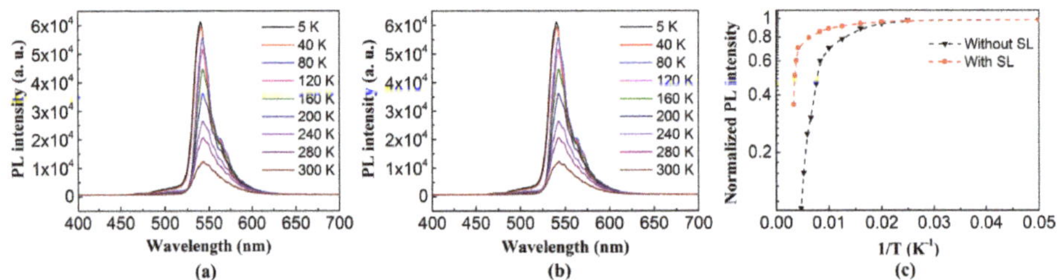

Figure 6. PL spectra of green LEDs (**a**) without and (**b**) with an InGaN/GaN SL; (**c**) Arrhenius plot of the normalized integrated PL intensity of the green MQWs without an SL and with an SL.

Figure 7a shows the light output power as a function of injection current for green LEDs. At 20 mA, the light output power of the green LEDs without and with an InGaN/GaN SL was 8.2 and 10.6 mW, respectively. The external quantum efficiency (EQE) of green LEDs is described by

$$EQE = \frac{P/h\upsilon}{I/e} = \frac{P\lambda}{I} \times \frac{e}{hc} = \frac{P\lambda}{1240 \times I} \tag{3}$$

where P is the light output power of green LEDs, I is the injection current, and λ is the light emission wavelength. The light emission wavelength of the grown green LEDs was 542 nm. The corresponding EQEs were estimated to be 17.9% and 23.2% for the 542 nm green LEDs without and with an InGaN/GaN SL, respectively, at 20 mA. The *EQE* of the green LED with an InGaN/GaN SL was 29.6% higher than that of the green LED without an InGaN/GaN SL. The result revealed that larger V-pits having a higher potential barrier height can more effectively suppress the non-radiative recombination of the carrier at TDs, thereby leading to an improvement in the *EQE* of a green LED. Figure 7b shows the current-voltage characteristic curves of green LEDs. At 20 mA, the forward voltages of the green LEDs without and with an InGaN/GaN SL were 3.93 and 2.71 V, respectively. Owing to the lower polarization charge densities at the InGaN/GaN interfaces from the V-pit sidewalls, the effective barrier height for holes injected from the V-pit sidewalls is lower than that for holes injected from the flat MQW [44–46]. As a result, the injection of holes into the MQW via the sidewalls of the V-pits is easier than via the flat region. On the other hand, the surface coverage ratio of V-pits in the InGaN/GaN MQW was calculated to be about 0.02 and 0.12 for green LED without and with an InGaN/GaN SL, respectively. The higher surface coverage ratio of V-pits was favorable for the injection of holes into MQWs from the sidewall of the V-pits. Consequently, the green LED with an InGaN/GaN SL exhibits a lower forward voltage than that of the green LED without an InGaN/GaN SL.

Figures 6d and 7c show temperature-dependent current-voltage characteristics of green LEDs under the reverse bias condition for temperatures ranging from 100 K to 400 K. We found that a higher reverse voltage tended to enhance the reverse leakage current for both of the green LEDs without and with an InGaN/GaN SL, which was attributed to the thermal activation of carriers from deep centers enhanced by an electric field. However, the reverse leakage current of the green LED with an InGaN/GaN SL was much lower than that of the green LED without an InGaN/GaN SL under the same reverse bias. For example, the leakage current of green LEDs without and with an InGaN/GaN SL were 18.8 and 0.109 μA, respectively, at −10 V and 300 K. It has been reported in our previous work that hopping conduction, including variable-range hopping (VRH) conduction and nearest-neighbor hopping (NNH) conduction, which is defined as carrier transport via electron hopping among localized states within the bandgap, is believed to the main mechanism causing the reverse leakage current [47]. The localized states are believed to be associated with TDs. Therefore, screening of TDs by the V-pits is considered to be an effective method for reducing the reverse leakage current. Additionally, the larger V-pits were also more effective in reducing the reverse leakage current of LEDs due to their higher Poole–Frenkel barrier height [24,48,49]. As a result, the reverse leakage current of the green LEDs decreased with increasing V-pit diameter.

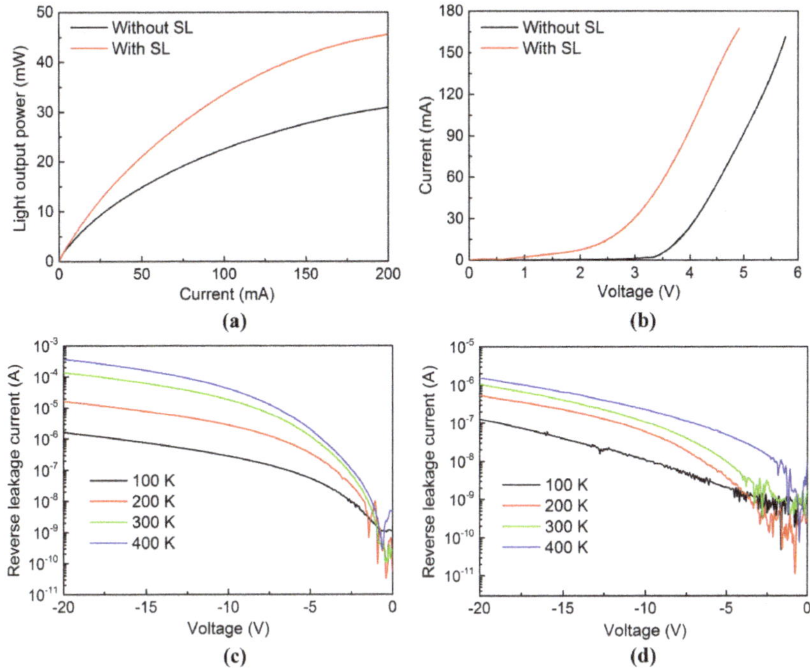

Figure 7. (**a**) Light output power versus current and (**b**) current versus voltage characteristics of green LEDs without and with an InGaN/GaN SL. Temperature-dependent current-voltage characteristics of green LEDs (**c**) without and (**d**) with an InGaN/GaN SL under the reverse bias condition.

4. Conclusions

We systematically investigated the effect of a V-pits-embedded InGaN/GaN SL on the optical and electrical properties of green LEDs. By performing a CL measurement, we have demonstrated that a potential barrier formed by a V-pit occurs around the TDs, and that the V-pit potential barrier height rises as the size of the V-pit increases, suggesting that a larger V-pit could more effectively hinder the non-radiative recombination of carrier at TDs. As a result, the EQE of green LEDs is improved by 29.6% at an injection current of 20 mA after implementing the V-pits-embedded InGaN/GaN SL. We observed that the forward voltage of green LEDs decreased with increasing V-pit diameter due to the enhanced holes injection. In addition, we found that the reverse leakage current of green LEDs decreased with increasing V-pit diameter, which was attributed to a more effective screening of TDs and also to a higher Poole–Frenkel barrier height.

Author Contributions: S.Z. conceived and designed the experiment. S.Z. led the project. S.Z. and M.L. wrote the manuscript. J.Z., J.H. and X.D. contributed to manuscript. Y.G. carried out the PL measurement. S.Z., M.L. and X.L. carried out the AFM, SEM, STEM, XRD, and CL measurements. All authors discussed the progress of the research and reviewed the manuscript.

Funding: This work was funded by the National Key Research and Development Program of China (No. 2017YFB1104900), the Hubei Province Science Fund for Distinguished Yong Scholars (No. 2018CFA091), and the National Natural Science Foundation of China (Grant Nos. 51675386, 51305266). We acknowledge the nanofabrication assistance from the Center for Nanoscience and Nanotechnology at Wuhan University.

Conflicts of Interest: The authors declare no conflict of interest.

References

1. Mukai, T.; Yamada, M.; Nakamura, S. Characteristics of InGaN-based UV/blue/green/amber/red light-emitting diodes. *Jpn. J. Appl. Phys.* **1999**, *38*, 3976–3981. [CrossRef]
2. Takano, T.; Mino, T.; Sakai, J.; Noguchi, N.; Tsubaki, K.; Hirayama, H. Deep-ultraviolet light-emitting diodes with external quantum efficiency higher than 20% at 275 nm achieved by improving light-extraction efficiency. *Appl. Phys. Express* **2017**, *10*, 031002. [CrossRef]
3. Zhou, S.; Liu, S. Study on sapphire removal for thin-film LEDs fabrication using CMP and dry etching. *Appl. Surf. Sci.* **2009**, *255*, 9469–9473. [CrossRef]
4. Massabuau, F.C.; Chen, P.; Horton, M.K.; Rhode, S.L.; Ren, C.X.; O'Hanlon, T.J.; Kovacs, A.; Kappers, M.J.; Humphreys, C.J.; Dunin-Borkowski, R.E.; et al. Carrier localization in the vicinity of dislocations in InGaN. *J. Appl. Phys.* **2017**, *121*, 013104. [CrossRef]
5. Zhou, S.; Liu, X.; Gao, Y.; Liu, Y.; Liu, M.; Liu, Z.; Gui, C.; Liu, S. Numerical and experimental investigation of GaN-based flip-chip light-emitting diodes with highly reflective Ag/TiW and ITO/DBR Ohmic contacts. *Opt. Express* **2017**, *25*, 26615–26627. [CrossRef] [PubMed]
6. Feng, Y.; Chen, Z.; Jiang, S.; Li, C.; Chen, Y.; Zhan, J.; Chen, Y.; Nie, J.; Jiao, F.; Kang, X.; et al. Study on the coupling mechanism of the orthogonal dipoles with surface plasmon in green LED by cathodoluminescence. *Nanomaterials* **2018**, *8*, 244. [CrossRef] [PubMed]
7. Hu, H.; Zhou, S.; Liu, X.; Gao, Y.; Gui, C.; Liu, S. Effects of GaN/AlGaN/Sputtered AlN nucleation layers on performance of GaN-based ultraviolet light-emitting diodes. *Sci. Rep.* **2017**, *7*, 44627. [CrossRef] [PubMed]
8. Maur, M.A.; Pecchi, A.; Penazzi, G.; Rodrigues, W.; Carlo, A.D. Efficiency drop in green InGaN/GaN light emitting diodes: The role of random alloy fluctuations. *Phys. Rev. Lett.* **2016**, *116*, 027401. [CrossRef] [PubMed]
9. Liu, J.; Jia, Z.; Ma, S.; Dong, H.; Zhai, G.; Xu, B. Enhancement of carrier localization effect and internal quantum efficiency through In-rich InGaN quantum dots. *Superlattice Microstruct.* **2018**, *113*, 497–501. [CrossRef]
10. Karpov, S.Y. Carrier localization in InGaN by composition fluctuations: Implication to the "green gap". *Photonics Res.* **2017**, *5*, A7–A12. [CrossRef]
11. Sano, T.; Doi, T.; Inada, S.A.; Sugiyama, T.; Honda, Y.; Amano, H.; Yoshino, T. High internal quantum efficiency blue-green light-emitting diode with small efficiency droop fabricated on low dislocation density GaN substrate. *Jpn. J. Appl. Phys.* **2013**, *52*, 08JK09. [CrossRef]
12. Liu, L.; Ling, M.; Yang, J.; Xiong, W.; Jia, W.; Wang, G. Efficiency degradation behaviors of current/thermal co-stressed GaN-based blue light emitting diodes with vertical-structure. *J. Appl. Phys.* **2012**, *111*, 093110. [CrossRef]
13. Zhou, S.; Yuan, S.; Liu, Y.; Guo, L.J.; Liu, S.; Ding, H. Highly efficient and reliable high power LEDs with patterned sapphire substrate and strip-shaped distributed current blocking layer. *Appl. Surf. Sci.* **2015**, *355*, 1013–1019. [CrossRef]
14. Chang, C.Y.; Li, H.; Shih, Y.T.; Lu, T.C. Manipulation of nanoscale V-pits to optimize internal quantum efficiency of InGaN multiple quantum wells. *Appl. Phys. Lett.* **2015**, *106*, 091104. [CrossRef]
15. Takeuchi, T.; Wetzel, C.; Yamaguchi, S.; Sakai, H.; Amano, H.; Akasaki, I. Determination of piezoelectric fields in strained GaInN quantum wells using the quantum-confined Stark effect. *Appl. Phys. Lett.* **1998**, *73*, 1691–1693. [CrossRef]
16. Leem, S.J.; Shin, Y.C.; Kim, K.C.; Kim, E.H.; Sung, Y.M.; Moon, Y.; Hwang, S.M.; Kim, T.G. The effect of the low-mole InGaN structure and InGaN/GaN strained layer superlattices on optical performance of multiple quantum well active layers. *J. Cryst. Growth* **2008**, *311*, 103–106. [CrossRef]
17. Jeong, H.; Jeong, H.J.; Oh, H.M.; Hong, C.H.; Suh, E.K.; Lerondel, G.; Jeong, M.S. Carrier localization in In-rich InGaN/GaN multiple quantum wells for green light-emitting diodes. *Sci. Rep.* **2015**, *5*, 9373. [CrossRef] [PubMed]
18. Chang, S.P.; Wang, C.H.; Chiu, C.H.; Li, J.C.; Lu, Y.S.; Li, Z.Y.; Yang, H.C.; Kuo, H.C.; Lu, T.C.; Wang, S.C. Characteristics of efficiency droop in GaN-based light emitting diodes with an insertion layer between the multiple quantum wells and GaN layer. *Appl. Phys. Lett.* **2010**, *97*, 251114. [CrossRef]

19. Tsai, Y.L.; Liu, C.Y.; Krishnan, C.; Lin, D.W.; Chu, Y.C.; Chen, T.P.; Shen, T.L.; Kao, T.S.; Charlton, M.D.B.; Yu, P.; et al. Bridging the "green gap" of LEDs: Giant light output enhancement and directional control of LEDs via embedded nano-void photonic crystals. *Nanoscale* **2016**, *8*, 1192–1199. [CrossRef] [PubMed]

20. Tomiya, S.; Kanitani, Y.; Tanaka, S.; Ohkubo, T.; Hono, K. Atomic scale characterization of GaInN/GaN multiple quantum wells in V-shaped pits. *Appl. Phys. Lett.* **2011**, *98*, 181904. [CrossRef]

21. Cho, H.K.; Lee, J.Y.; Yang, G.M.; Kim, C.S. Formation mechanism of V defects in the InGaN/GaN multiple quantum wells grown on GaN layers with low threading dislocation density. *Appl. Phys. Lett.* **2001**, *79*, 215–217. [CrossRef]

22. Wu, X.H.; Elsass, C.R.; Abare, A.; Mack, M.; Keller, S.; Petroff, P.M.; DenBaars, S.P.; Rosner, S.J. Structural origin of V-defects and correlation with localized excitonic centers in InGaN/GaN multiple quantum wells. *Appl. Phys. Lett.* **1998**, *72*, 692–694. [CrossRef]

23. Shiojiri, M.; Chuo, C.C.; Hsu, J.T.; Yang, J.R.; Saijo, H. Structure and formation mechanism of V defects in multiple InGaN/GaN quantum well layers. *J. Appl. Phys.* **2006**, *99*, 073505. [CrossRef]

24. Kim, J.; Kim, J.; Tak, Y.; Chae, S.; Kim, J.Y.; Park, Y. Effect of V-shaped pit size on the reverse leakage current of InGaN/GaN light-emitting diodes. *IEEE Electron Device Lett.* **2013**, *34*, 1409–1411. [CrossRef]

25. Hangleiter, A.; Hitzel, F.; Netzel, C.; Fuhrmann, D.; Rossow, U.; Ade, G.; Hinze, P. Suppression of nonradiative recombination by V-shaped pits in GaInN/GaN quantum wells produces a large increase in the light emission efficiency. *Phys. Rev. Lett.* **2005**, *95*, 127402. [CrossRef] [PubMed]

26. Sugahara, T.; Sato, H.; Hao, M.; Naoi, Y.; Kurai, S.; Tottori, S.; Yamashita, K.; Nishino, K.; Romano, L.T.; Sakai, S. Direct evidence that dislocations are non-radiative recombination centers in GaN. *Jpn. J. Appl. Phys.* **1998**, *37*, L398–L400. [CrossRef]

27. Kurai, S.; Higaki, S.; Imura, N.; Okawa, K.; Makio, R.; Okada, N.; Tadatomo, K.; Yamada, Y. Potential Barrier formed around dislocations in InGaN quantum well structures by spot cathodoluminescence measurements. *Phys. Status Solidi B* **2018**, *255*, 1700258. [CrossRef]

28. Liu, Y.J.; Tsai, T.Y.; Yen, C.H.; Chen, L.Y.; Tsai, T.H.; Liu, W.C. Characteristics of a GaN-based light-emitting diode with an inserted p-GaN/i-InGaN superlattice structure. *IEEE J. Quantum Electron.* **2010**, *46*, 492–498. [CrossRef]

29. Armstronga, A.M.; Bryant, B.N.; Crawford, M.H.; Koleske, D.D.; Lee, S.R.; Wierer, J.J. Defect-reduction mechanism for improving radiative efficiency in InGaN/GaN light-emitting diodes using InGaN underlayers. *J. Appl. Phys.* **2015**, *117*, 134501. [CrossRef]

30. Lee, K.; Lee, C.R.; Lee, J.H.; Chung, T.H.; Ryu, M.Y.; Jeong, K.U.; Leem, J.Y.; Kim, J.S. Influences of Si-doped graded short-period superlattice on green InGaN/GaN light-emitting diodes. *Opt. Express* **2016**, *24*, 7743–7751. [CrossRef] [PubMed]

31. Lee, K.; Lee, C.R.; Chung, T.H.; Park, J.; Leem, J.Y.; Jeong, K.U.; Kim, J.S. Influences of graded superlattice on the electrostatic discharge characteristics of green InGaN/GaN light-emitting diodes. *J. Cryst. Growth* **2017**, *464*, 138–142. [CrossRef]

32. Okadaa, N.; Kashiharab, H.; Sugimoto, K.; Yamada, Y.; Tadatomo, K. Controlling potential barrier height by changing V-shaped pit size and the effect on optical and electrical properties for InGaN/GaN based light-emitting diodes. *J. Appl. Phys.* **2015**, *117*, 025708. [CrossRef]

33. Zhou, S.; Liu, X. Effect of V-pits embedded InGaN/GaN superlattices on optical and electrical properties of GaN-based green light-emitting diodes. *Phys. Status Solidi A* **2017**, *214*, 1600782. [CrossRef]

34. Wu, X.; Liu, J.; Jiang, F. Hole injection from the sidewall of V-shaped pits into *c*-plane multiple quantum wells in InGaN light emitting diodes. *J. Appl. Phys.* **2015**, *118*, 164504. [CrossRef]

35. Kim, J.; Cho, Y.H.; Ko, D.S.; Li, X.S.; Won, J.Y.; Lee, E.; Park, S.H.; Kim, J.Y.; Kim, S. Influence of V-pits on the efficiency droop in InGaN/GaN quantum wells. *Opt. Express* **2014**, *22*, A857–A866. [CrossRef] [PubMed]

36. Zhou, S.; Liu, S. Transient measurement of light-emitting diode characteristic parameters for production lines. *Rev. Sci. Instrum.* **2009**, *80*, 095102. [CrossRef] [PubMed]

37. Pereira, S.; Correia, M.R.; Pereira, E.; O'Donnell, K.P.; Trager-Cowan, C.; Sweeney, F.; Alves, E. Compositional pulling effects in InxGa1-xN/GaN layers: A combined depth-resolved cathodoluminescence and Rutherford backscattering/channeling study. *Phys. Rev. B* **2001**, *64*, 205311. [CrossRef]

38. Hiramatsu, K.; Kawaguchi, Y.; Shimizu, M.; Sawaki, N.; Zheleva, T.; Davis, R.F.; Tsuda, H.; Taki, W.; Kuwano, N.; Oki, K. The composition pulling effect in MOVPE grown InGaN on GaN and AlGaN and its TEM characterization. *Mater. Res. Soc. Internet J. Nitride Semicond. Res.* **1997**, *2*. [CrossRef]

39. Pozina, G.; Ciechonski, R.; Bi, Z.; Samuelson, L.; Monemar, B. Dislocation related droop in InGaN/GaN light emitting diodes investigated via cathodoluminescence. *Appl. Phys. Lett.* **2015**, *107*, 251106. [CrossRef]

40. Cho, Y.H.; Gainer, G.H.; Fischer, A.J.; Song, J.J.; Keller, S.; Mishra, U.K.; DenBaars, S.P. "S-shaped" temperature-dependent emission shift and carrier dynamics in InGaN/GaN multiple quantum wells. *Appl. Phys. Lett.* **1998**, *73*, 1370–1372. [CrossRef]

41. Yasan, A.; McClintock, R.; Mayes, K.; Kim, D.H.; Kung, P.; Razeghi, M. Photoluminescence study of AlGaN-based 280 nm ultraviolet light-emitting diodes. *Appl. Phys. Lett.* **2003**, *83*, 4083–4085. [CrossRef]

42. Watanabe, S.; Yamada, N.; Nagashima, M.; Ueki, Y.; Sasaki, C.; Yamada, Y.; Taguchi, T.; Tadatomo, K.; Okagawa, H.; Kudo, H. Internal quantum efficiency of highly-efficient $In_xGa_{1-x}N$-based near-ultraviolet light-emitting diodes. *Appl. Phys. Lett.* **2003**, *83*, 4906–4908. [CrossRef]

43. Zhou, S.; Hu, H.; Liu, X.; Liu, M.; Ding, X.; Gui, C.; Liu, S.; Guo, L.J. Comparative study of GaN-based ultraviolet LEDs grown on different-sized patterned sapphire substrates with sputtered AlN nucleation layer. *Jpn. J. Appl. Phys.* **2017**, *56*, 111001. [CrossRef]

44. Nishizuka, K.; Funato, M.; Kawakami, Y.; Fujita, S.; Narukawa, Y.; Mukai, T. Efficient radiative recombination from $\langle 11\bar{2}2 \rangle$-oriented $In_xGa_{1-x}N$ multiple quantum wells fabricated by the regrowth technique. *Appl. Phys. Lett.* **2004**, *85*, 3122–3124. [CrossRef]

45. Li, Y.; Yun, F.; Su, X.; Liu, S.; Ding, W.; Hou, X. Deep hole injection assisted by large V-shape pits in InGaN/GaN multiple-quantum-wells blue light-emitting diodes. *J. Appl. Phys.* **2014**, *116*, 123101. [CrossRef]

46. Quan, Z.; Wang, L.; Zheng, C.; Liu, J.; Jiang, F. Roles of V-shaped pits on the improvement of quantum efficiency in InGaN/GaN multiple quantum well light-emitting diodes. *J. Appl. Phys.* **2014**, *116*, 183107. [CrossRef]

47. Zhou, S.; Lv, J.; Wu, Y.; Zhang, Y.; Zheng, C.; Liu, S. Reverse leakage current characteristics of InGaN/GaN multiple quantum well ultraviolet/blue/green light-emitting diodes. *Jpn. J. Appl. Phys.* **2018**, *57*, 051003. [CrossRef]

48. Shan, Q.; Meyaard, D.S.; Dai, Q.; Cho, J.; Schubert, E.F.; Son, J.K.; Sone, C. Transport-mechanism analysis of the reverse leakage current in GaInN light-emitting diodes. *Appl. Phys. Lett.* **2011**, *99*, 253506. [CrossRef]

49. Liu, S.; Zheng, C.; Lv, J.; Liu, M.; Zhou, S. Effect of high-temperature/current stress on the forward tunneling current of InGaN/GaN high-power blue-light-emitting diodes. *Jpn. J. Appl. Phys.* **2017**, *56*, 081001. [CrossRef]

Article

Improved Performance of Perovskite Light-Emitting Diodes by Quantum Confinement Effect in Perovskite Nanocrystals

Lung-Chien Chen [1,*], Zong-Liang Tseng [1,*], Dai-Wei Lin [1], Yu-Shiang Lin [1] and Sheng-Hui Chen [2]

[1] Department of Electro-optical Engineering, National Taipei University of Technology, 1, 3 Section, Chung-Hsiao East Road, Taipei 106, Taiwan; t104658068@ntut.org.tw (D.-W.L.); t104658044@ntut.org.tw (Y.-S.L.)

[2] Department of Optics and Photonics, National Central University, 300 Chung-Da Rd., Chung-Li 32001, Taiwan; ericchen@dop.ncu.edu.tw

* Correspondence: ocean@ntut.edu.tw (L.-C.C.); tw78787788@yahoo.com.tw (Z.-L.T.); Tel.: +886-2-27712171 (L.-C.C.)

Received: 25 May 2018; Accepted: 19 June 2018; Published: 25 June 2018

Abstract: In this study, we demonstrate an easy and reliable solution-processed technique using an extra adductive in the perovskite precursor solution. Using this method, a dense and uniform morphology with full surface coverage and highly fluorescent films with nanoscale crystal grains can be obtained. The high exciton binding energy in the resulting films employing octylammonium bromide (OAB) adductives proved that high fluorescence originated from the quantum confinement effect. The corresponding perovskite light-emitting diodes (PeLEDs) that were based on this technique also exhibited excellent device performance.

Keywords: PeLEDs; OAB; perovskite; quantum confinement effect

1. Introduction

Since the report on the perovskite light-emitting diodes (PeLEDs) in 2012 [1] expanded the research range of lead halide perovskites from their photovoltaic applications, a series of studies on device structures and deposition methods [2–10] have been presented. As a result, excellent electroluminescence efficiency of 42.9 cd/A has been achieved by employing an additive-based nanocrystal pinning technique [11].

The active layer of PeLEDs and $CH_3NH_3PbBr_3$ (MAPbBr$_3$) can be prepared by using a simple solution-processed coating, where a precursor solution containing CH_3NH_3Br (MAB) and $PbBr_2$ is spin-coated on the substrates and washed using toluene to rapidly crystalize while spinning [12,13]. The resulting film exhibited better optoelectronic properties, such as high mobility [14–17], long and balanced electron-hole diffusion lengths [18,19], low bulk defect densities, and slow Auger recombination [20] compared to the previous solution-processed semiconductors. On the other hand, Pérez-Prieto et al. [21,22] first reported that MAPbBr$_3$ perovskite quantum dots (QDs) were achievable and they synthesized them using octylammonium bromide ($CH_3(CH_2)_7NH_3Br$; OAB) with a long alkyl chain as a capping ligand to stabilize them. The longer alkyl chain cations are embedded in the MAPbBr$_3$ lattice to replace the methyl ammonium cations and their long chains dangling outside the lattice, as illustrated in Figure 1a. Because of the repulsion forces between long alkyl chains, the growth of the perovskite array is suppressed in three dimensions, resulting in perovskite QDs being produced. Although it is well known that colloidal QD dispersions exhibit narrow-band emission and high photoluminescence (PL) efficiency [21–27], the uniform, smooth, and large-area films are difficult to prepare by directly using colloidal dispersions. In addition, some solution-processed methods for highly efficient PeLEDs with high-quality perovskite films, such as the additive-based nanocrystal pinning technique, require great skill for processing controls. Therefore, it is necessary to develop an easy and reliable method for obtaining high-quality perovskite films with nanoscale crystal grains for high-performance PeLEDs [28–30].

In this study, we demonstrate the preparation of MAPbBr$_3$ thin films with highly uniform and dense nanoscale grains through a simple spin-coating method using OAB as an additive (Figure 1a). Highly fluorescent thin films with full-surface coverage were achieved by optimizing the amount of OAB in the perovskite precursors. Highly efficient PeLEDs were also prepared using the resulting MAPbBr$_3$ thin films as active layers. The champion device based on the OAB adductive method exhibited a maximum luminance of 310 cd/m^2 (at 4.5 V) and a maximum luminous current efficiency of 1.21 cd/A (at 4.5 V). We further show that the improved performance of PeLEDs and the enhanced fluorescence of MAPbBr$_3$ thin films are due to high exciton binding energy in nanometer-sized crystal grains, which leads to reduced nonradiative recombination and increased emission efficiency.

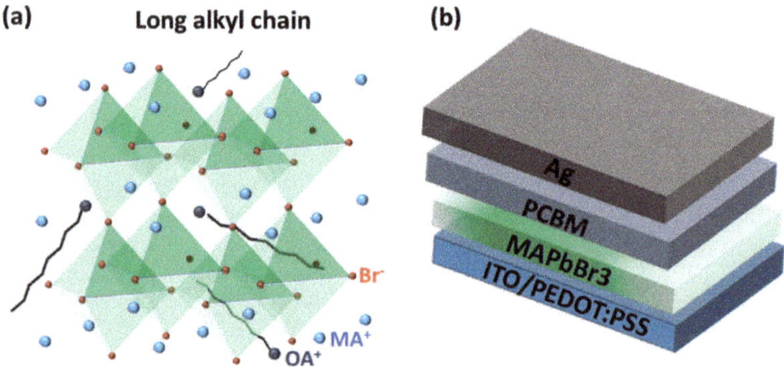

Figure 1. (a) Illustration for substituting octylammonium (OA$^+$) for methylammonium (MA$^+$) in the MAPbBr$_3$ array; (b) Schematic of the device structure of perovskite light-emitting diodes (PeLEDs) in this study. PCBM, [6,6]-phenyl-C61-butyric acid methyl ester. PEDOT:PSS, poly(3,4-ethylenedioxythiophene) polystyrene sulfonate. ITO, indium tin oxide.

2. Materials and Methods

PeLEDs with a device structure of glass/indium tin oxide (ITO)/poly(3,4-ethylenedioxythiophene) polystyrene sulfonate (PEDOT:PSS)/MAPbBr3/[6,6]-phenyl-C61-butyric acid methyl ester (PCBM)/Ag were used in this study, as illustrated in Figure 1b. The PEDOT:PSS layers using AI-4083 (Heraeus Clevios) were spin-coated on a cleaned ITO substrate at 5000 rpm for 30 s and were post-annealed at 120 °C for 10 min. PbBr$_2$ (99.999%; Sigma-Aldrich, St. Louis, MO, USA), methylammonium iodide (MAB; Lumtec, Hsinchu, Taiwan), and octylammonium bromide (OAB; Lumtec, Hsinchu, Taiwan) were dissolved in a dimethyl sulfoxide (DMSO)/dimethylformamide (DMF) mixture (7:3, v/v) as the precursor, with an MAPbBr$_3$ concentration of 0.5 M. Different OAB ratios were prepared with the weight ratio of MAB:OAB. The perovskite precursors were then spin-coated onto the PEDOT:PSS layers at 5000 rpm for 30 s, with 2 mL of anhydrous toluene dropped at 27 s during spin-coating. The large amount of toluene ensured the removal of OAB due to the long alkyl chain approaching nonpolar character. The as-deposited films were post-annealed at 90 °C for 5 min, and after being cooled to room temperature, [6,6]-phenyl-C61-butyric acid methyl ester (PCBM) dissolved in chlorobenzene (20 mg/mL) was then spin-coated on them at 1200 rpm for 30 s. Ag electrodes (100 nm) were evaporated through a metal mask to define the device area (0.1 cm^2).

The crystalline microstructure, absorbance spectra, and surface morphology of the films were determined by X-ray diffraction with Cu-Kα radiation (D8 Discover, Bruker, Karlsruhe, Germany), UV-visible spectroscopy (U-4100, Hitachi High-Technologies Co., Tokyo, Japan), and a field-emission scanning electron microscope (GeminiSEM, Zeiss, Oberkochen, Germany), respectively. The photoluminescence (PL) spectra were measured using an optical microscope-based system (UniRAM, Protrustech, New Taipei, Taiwan) with an excitation of 405 nm. The temperature-dependent photoluminescence (PL) spectra was measured under a nitrogen-filled atmosphere. The current density-voltage-luminesce (J-V-L) characteristics were measured using a Keithley 2400 combined with a SpectraScan Spectroradiometer (PR-670, Photo Research, New York, NY, USA).

3. Results and Discussion

To understand how the ratio of OAB to MAPbBr$_3$ in the precursor solution affects the performance of the resulting perovskite light-emitting diodes, X-ray diffraction patterns of MAPbBr$_3$ films grown with different OAB ratios were collected, and these are shown in Figure 2 (the corresponding 2D-XRD is shown in Supplementary Figure S1). Pure MAPbBr$_3$ cubic phase with (001), (011), (002), (021), (211), and (220) at 2θ range from 5–45° were identified [23] in all diffractograms. The thicknesses of the MAPbBr$_3$ films that were obtained by α-step had no significant differences at different OAB ratios (~200 nm; Supplementary Figure S2). Using a higher OAB ratio, the intensity of the (001) peak decreased and full width at half maximum (FWHM) increased, implying a smaller MAPbBr$_3$ grain size. Also, when the OAB ratio was increased to 6%, the intensity and FWHM ratio of the (001) to (110) peak decreased. Reduced crystallinity of MAPbBr$_3$ with an OAB adductive compared to pure MAPbBr$_3$ films suggests that the long alkyl chain of OAB indeed acts as a better capping ligand to limit MAPbBr$_3$ grain growth. Therefore, the OAB ratio in the precursor solution was confirmed to suppress the grain growth in the MAPbBr$_3$ films. Interestingly, when applying a higher OAB ratio (8%), MAPbBr$_3$ became very weak and produced a new diffraction peak at a low angle (~6°). Most reported XRD patterns for MAPbBr$_3$ were only detected from 10° (2θ > 10°), therefore it was difficult to know the exact components and structures of this new phase by only depending on XRD. Nevertheless, according to the Scherrer equation [30], we can obtain that the low-angle peak corresponded to a larger lateral ordered spacing.

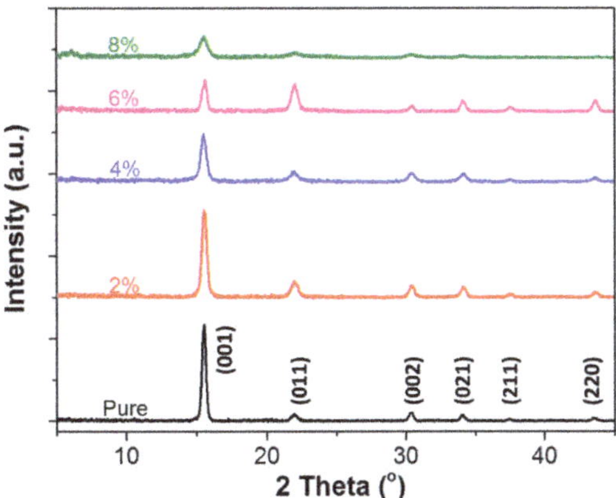

Figure 2. XRD patterns of the MAPbBr$_3$ films with different octylammonium bromide (OAB) ratios deposited on the glass.

Furthermore, the OAB ratio also affected the surface morphology of the MAPbBr$_3$ films, as revealed by scanning electron microscopy (SEM) micrographs, shown in Figure 3. The morphology of the pure MAPbBr$_3$ films showed large and nonuniform cubic grains from 200 to 1000 nm, but some interspace existed between the micrograins, which may have increased leakage due to the direct contact between the upper and bottom layers without passing through the perovskite films. The unwanted broad size distribution was due to the rapid crystallization process in the perovskite thin films [31,32], leading to not enough time for the thermodynamically spontaneous process, i.e., the well-known Ostwald ripening process [33,34], to form small-size crystals that recrystallized to large grains. Several methods have been suggested to improve this phenomenon in perovskite thin films, such as the solvent-annealing process [35], mixed halide treatment [34], and HBr/DMF cosolvent [10]. Applying OAB as an additive, the grain size was significantly reduced, which was in accord with XRD results. Smooth, uniform, and dense perovskite films (grain size of about 20 nm) were observed when the OAB ratio was equal to 2% and 4%, and furthermore, the grain size of the 2% sample was more uniform than that of the 4% sample (see insets in Figure 3). This suggested that the OAB additive is the main factor in the formation of nanograins in perovskite film. This dense MAPbBr$_3$ film was also reported using the HBr/DMF cosolvent method [10], which originated from slow crystallization rates during deposition. Unlike those, the reason in our case for morphological control was the grain growth being limited by the long alkyl chain of OAB. Therefore, our grain sizes were much smaller than theirs and were much like those using an additive-based nanocrystal pinning technique [11]. Under a higher OAB ratio, inhomogeneous humped structures could be seen, resulting in rough films with some cracks and defects, which negatively affected the device performance. The humped structure may be attributed to large-size ordered packing that was formed from the grain aggregation. This may be why the low angle phase was found in the XRD pattern. These results indicate that the film morphology of MAPbBr$_3$ films is strongly influenced by the OAB ratio in the precursor solution. This easy method, using an OAB adductive, provides a general way to control the morphology and the surface coverage of MAPbBr$_3$ films.

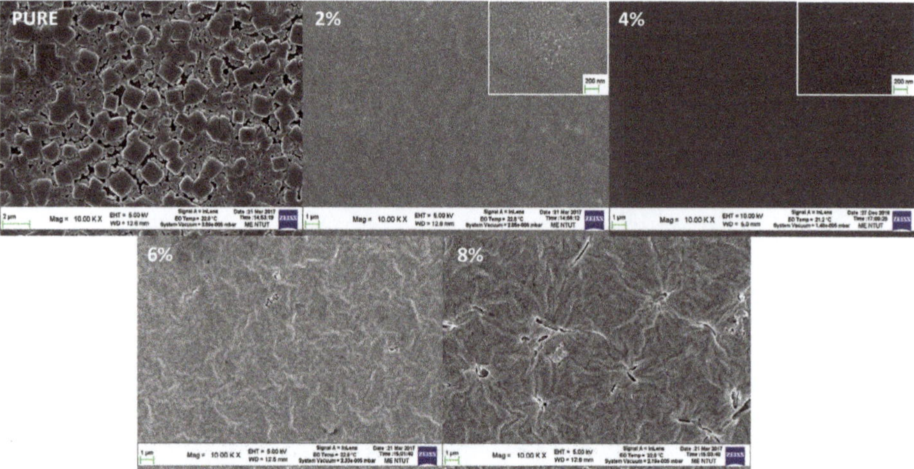

Figure 3. Scanning electron microscopy (SEM) images of the MAPbBr$_3$ films with different OAB ratios deposited on the glass. The insets for 2% and 4% show images with higher magnification.

Figure 4a shows the PL spectra of samples with different OAB ratios in the precursor solution. The intensity increased using lower OAB ratios (2% and 4%), but decreased using low OAB ratios (6% and 8%). The PL intensity in the 4% sample is better than that in the 2% sample, which can be attributed to more homogeneous grain size (Figure 3). The weak PL in the 6 and 8% samples may be due to the poor crystallinity and grain aggregation, which may increase the dissociation rate. However, the PL intensity in thin films compared to that in dispersions is much more complicated [36,37] due to exciton dissociation between grains or bottom/upper layers, leading to radiative loss. Therefore, the smaller grains in the 2% and 4% films had more grain boundaries (more grain package) to provide PL quenching sites, but the 2% and 4% samples exhibited better PL intensities. The reason is that excitons are confined in the nanometer-sized grains [21,22], leading to strong PL emission. These results are similar to those using colloidal perovskite QD dispersions to directly deposit thin films [38–41]. An exciton diffusion length of 67 nm in MAPbBr$_3$ films with nanograins was reported, which is much smaller than that of solution-processed perovskite films [11]. Moreover, a significant blue shift can be observed in PL spectra, in that emission wavelength decreased with an increase in the OAB ratio, from 540.1 nm (pure film) to 531.7 nm (8%). Similarly, the absorption edges in ultraviolet-visible (UV-vis) spectra (Figure 3b) also showed the same trend. The fitting bandgap (the corresponding Tauc plots shown in Supplementary Figure S3) from UV-vis spectra increased with an increase in the OAB ratio. The blue shift can typically be attributed to the quantum confinement effect in the nanocrystal [21,22]. Besides, nanocrystal materials have sharper density of states than higher dimensional materials. Therefore, they permit more electrons to occupy the states in conduction band, such that opportunity of spontaneous emission increases.

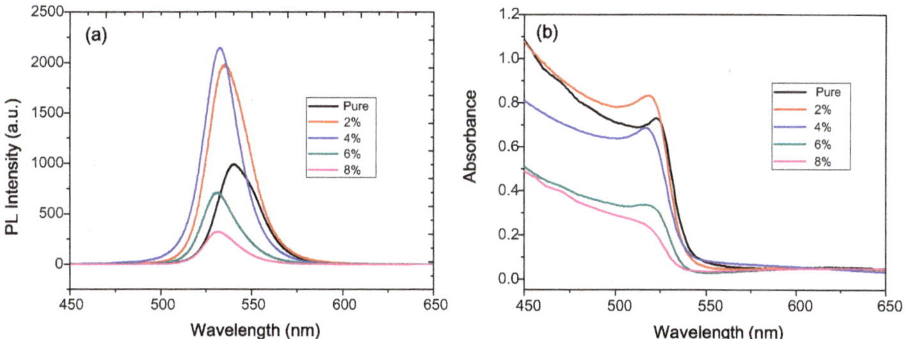

Figure 4. (a) PL spectra and (b) the absorbance of the MAPbBr$_3$ films with different OAB ratios deposited on the PEDOT:PSS glass.

Figure 5a shows the current–density vs. voltage (J-V) curves for PeLEDs using different OAB ratios. All of the curves revealed diode behavior. The inserted image in Figure 5a shows that our PeLED displayed a text patterned by a metal mask for Ag evaporation. Figure 5b,c shows the luminance vs. voltage (L-V) and the current efficiency vs. voltage (CE-V) of our PeLEDs with different OAB ratios. The optimized PeLED that was prepared with the OAB ratio of 6% exhibited a maximum luminance of 310 cd/m^2 (at 4.5 V) and a maximum luminous current efficiency of 1.21 cd/A (at 4.5 V). The PeLED that was based on the pure MAPbBr$_3$ film without OAB adductives showed poor luminance characteristics (maximum CE = 0.32 cd/A), mainly due to high leakage current, as mentioned in Figure 3, that was induced from the lateral space between the MAPbBr$_3$ grains or the pinholes on the film surface. Maximum current efficiency was achieved (1.21 cd/A) when the OAB ratio was increased to 4%. The excellent performance at the OAB ratio of 4% was mainly due to the dense and uniform MAPbBr$_3$ layer with full coverage avoiding leakage, the smooth surface providing good contact with the electrode transporting layer (PCBM), and the MAPbBr$_3$ nanograins having an enhanced radiative recombination rate for injected carriers. In contrast, poor device performance at higher OAB ratios (6% and 8%) may have been from the grain aggregation, leading to an increased dissociation path and a rough film surface. Furthermore, the electroluminescence (EL) spectra also exhibited a blue shift (Figure 5d), which was consistent with the PL observation. It is worth mentioning that the maximum luminance and current efficiency values of our PeLED are comparable to those of the previously reported PeLEDs based on MAPbBr$_3$ active layers [2–10]. Figure 5e shows the energy band structure to explain the mechanism of the radiative recombination caused by the quantum confinement effect, as results of the blue shift and enhance the spontaneous emission of the electroluminescence (EL) spectra. In Figure 5, the phenomenon of luminance decay at high applied voltage is caused either by heat due to the series resistance of the devices or by damage due to the high electrical field.

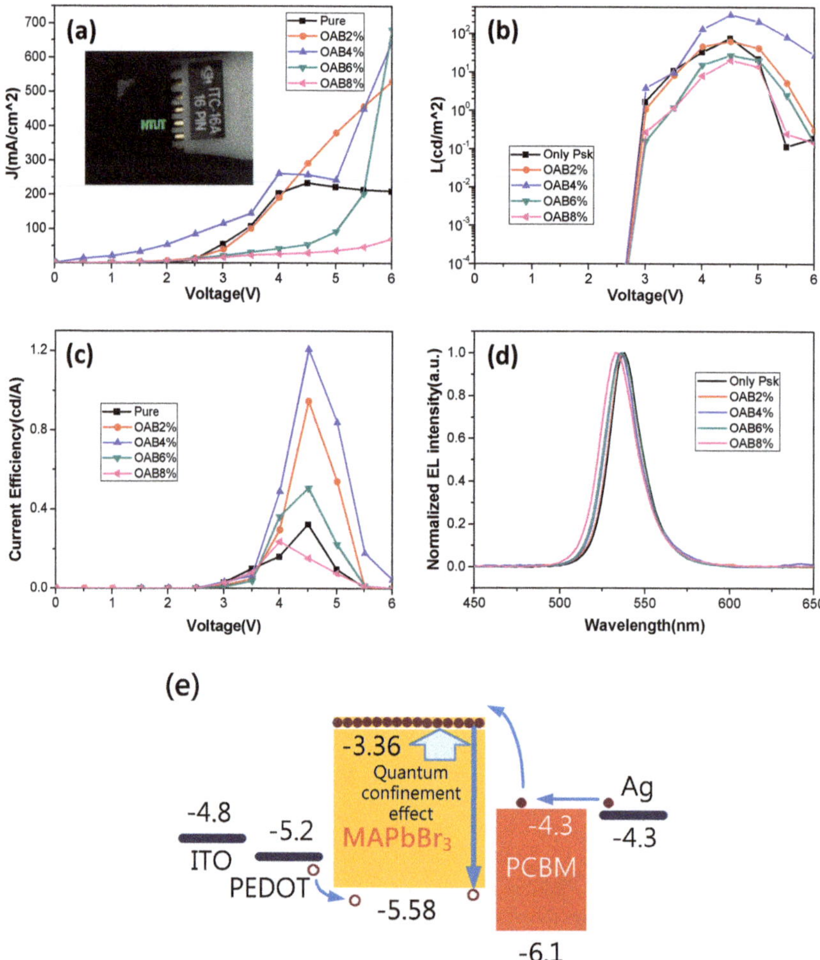

Figure 5. (**a**) Current–density vs. voltage, (**b**) luminance vs. voltage, (**c**) current efficiency vs. voltage, and (**d**) electroluminescence of PeLEDs based on MAPbBr$_3$ films with different OAB ratios deposited on the glass. (**e**) a diagram of the energy band structure of the PeLEDs. The inset in (**a**) shows a photo image of our PeLED.

To better realize the nature of strong PL emission in the MAPbBr$_3$ nanograins formed by the OAB adductive, the excitonic characteristics of the MAPbBr$_3$ films should be considered. As we reported previously [42], temperature-dependent photoluminescence (TDPL) can be used to determine the exciton binding energy by linear fitting with a PL spectral broadening equation [43] at different temperatures:

$$\ln(h\Delta v - h\Delta v_0) = \ln(hv_T) - E_b/K_B T \tag{1}$$

where $h\Delta v$ is the full width at half maximum (FWHM) of the PL spectrum at some temperature, $h\Delta v_0$ is the FWHM of the PL spectrum at the initial temperature, hv_T is related to the thermal dissociation rate, E_b is the exciton binding energy, K_B is the Boltzmann constant, and T is the temperature. The PL spectra at temperatures from 100 to 300 K for the MAPbBr$_3$ films deposited with (4%) and without (pure) the OAB adductive are shown in Figure 6. The peak of both samples significantly broadened with increased temperature due to exciton–phonon interaction [17]. Grätzel et al. suggested that the dual PL emissions in MAPbBr$_3$ at temperatures below 175 K were due to the coexistence of MA-ordered and MA-disordered domains in the MAPbBr$_3$ array [44]. However, only one peak of both samples can be found in Figure 6. Indeed, in our case, linear fitting could not be estimated (very low R-square) if the temperature selected was from 100 to 300 K. Therefore, the temperature was set from 180 to 300 K for linear fitting in Supplementary Figure S4. R-square over 0.99 reveals a good fit, and exciton binding energies of 48 and 85 meV were determined for pure MAPbBr$_3$ film and film using the OAB adductive, respectively. The higher exciton binding energy of the MAPbBr$_3$ film with OAB compared to that of the pure MAPbBr$_3$ film indicates strong exciton localization, which can block the exciton dissociation and increase the radiative recombination rate. Therefore, this provides evidence for the quantum confinement effect in nanometer-sized grains of MAPbBr$_3$ film using the OAB adductive.

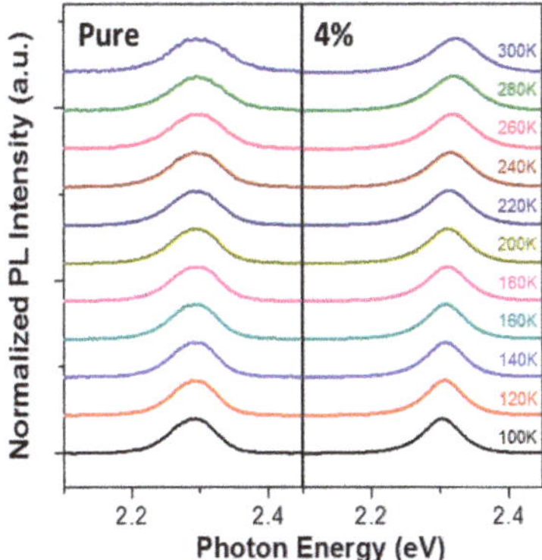

Figure 6. Temperature-dependent photoluminescence spectra for MAPbBr$_3$ film with an OAB ratio of 4% and without OAB.

4. Conclusions

In conclusion, we have demonstrated that the preparation method employing OAB adductive can obtain high-quality MAPbBr$_3$ thin films. The addition of a small amount of OAB in the MAPbBr$_3$ precursor solvent produces a dense and uniform film morphology with full surface coverage. Also, it enables the formation of nanoscale MAPbBr$_3$ grains, which enhance PL emission due to the quantum confinement effect. The high exciton binding energy in the MAPbBr$_3$ films that are formed by the OAB adductive is evidence for the quantum confinement effect in the nanometer-sized grains, which reduces exciton dissociation and enhances exciton radiation. The quantum confinement effect also affects bandgap, which shifts to a short wavelength with an increasing OAB ratio. A blue shift is found in absorbance, PL, and EL spectra at different OAB ratios. The optimized PeLED with an OAB ratio of 4% exhibited a maximum luminance of 310 cd/m^2 (at 4.5 V) and a maximum luminous current efficiency of 1.21 cd/A (at 4.5 V). This study offers a promising approach for reliably depositing high-quality MAPbBr$_3$ thin films for application to efficient PeLEDs.

Supplementary Materials: The following are available online at http://www.mdpi.com/2079-4991/8/7/459/s1, Figure S1: Original 2D XRD data of perovskite films with different OAB ratios, Figure S2: Film thicknesses of perovskite films with different OAB ratios. The error bars are standard deviations obtained from five samples, Figure S3: The corresponding Tauc plots from UV-vis spectra for linear fitting to determine bandgaps of perovskite films with different OAB ratios. The solid lines are fittings, Figure S4: Temperature-dependent data for linear fitting exciton binding energy. The solid lines are fittings.

Author Contributions: L.-C.C. wrote the paper, designed the experiments, and analyzed the data. Z.-L.T., D.-W.L., and Y.-S.L. prepared the samples and performed all of the measurements. S.-H.C. designed the experiments and discussed the results. All authors read and approved the final manuscript.

Funding: Financial support for this paper was provided by the Ministry of Science and Technology of the Republic of China under contract no. MOST 106-2221-E-027-091.

Acknowledgments: The authors want to thank Yu-Tai Tao, Institute of Chemistry, Academia Sinica, Taiwan, for his supportive chemistry knowledge and for providing J-V-L measurements. Financial support for this paper was provided by the Ministry of Science and Technology of the Republic of China under contract no. MOST 106-2221-E-027-091.

Conflicts of Interest: The authors declare no conflict of interest.

References

1. Tan, Z.-K.; Moghaddam, R.S.; Lai, M.L.; Docampo, P.; Higler, R.; Deschler, F.; Price, M.; Sadhanala, A.; Pazos, L.M.; Credgington, D.; et al. Bright light-emitting diodes based on organometal halide perovskite. *Nat. Nano* **2014**, *9*, 687–692. [CrossRef] [PubMed]

2. Kim, Y.-H.; Cho, H.; Heo, J.H.; Kim, T.-S.; Myoung, N.S.; Lee, C.-L.; Im, S.H.; Lee, T.-W. Multicolored Organic/Inorganic Hybrid Perovskite Light-Emitting Diodes. *Adv. Mater.* **2015**, *27*, 1248–1254. [CrossRef] [PubMed]

3. Hoye, R.L.Z.; Chua, M.R.; Musselman, K.P.; Li, G.; Lai, M.-L.; Tan, Z.-K.; Greenham, N.C.; MacManus-Driscoll, J.L.; Friend, R.H.; Credgington, D. Enhanced Performance in Fluorene-Free Organometal Halide Perovskite Light-Emitting Diodes using Tunable, Low Electron Affinity Oxide Electron Injectors. *Adv. Mater.* **2015**, *27*, 1414–1419. [CrossRef] [PubMed]

4. Li, G.; Tan, Z.-K.; Di, D.; Lai, M.L.; Jiang, L.; Lim, J.H.-W.; Friend, R.H.; Greenham, N.C. Efficient Light-Emitting Diodes Based on Nanocrystalline Perovskite in a Dielectric Polymer Matrix. *Nano Lett.* **2015**, *15*, 2640–2644. [CrossRef] [PubMed]

5. Jaramillo-Quintero, O.A.; Sanchez, R.S.; Rincon, M.; Mora-Sero, I. Bright Visible-Infrared Light Emitting Diodes Based on Hybrid Halide Perovskite with Spiro-OMeTAD as a Hole-Injecting Layer. *J. Phys. Chem. Lett.* **2015**, *6*, 1883–1890. [CrossRef] [PubMed]

6. Wong, A.B.; Lai, M.; Eaton, S.W.; Yu, Y.; Lin, E.; Dou, L.; Fu, A.; Yang, P. Growth and Anion Exchange Conversion of CH3NH3PbX3 Nanorod Arrays for Light-Emitting Diodes. *Nano Lett.* **2015**, *15*, 5519–5524. [CrossRef] [PubMed]

7. Sadhanala, A.; Ahmad, S.; Zhao, B.; Giesbrecht, N.; Pearce, P.M.; Deschler, F.; Hoye, R.L.Z.; Gödel, K.C.; Bein, T.; Docampo, P.; Dutton, S.E.; et al. Blue-Green Color Tunable Solution Processable Organolead Chloride–Bromide Mixed Halide Perovskites for Optoelectronic Applications. *Nano Lett.* **2015**, *15*, 6095–6101. [CrossRef] [PubMed]

8. Wang, J.; Wang, N.; Jin, Y.; Si, J.; Tan, Z.-K.; Du, H.; Cheng, L.; Dai, X.; Bai, S.; He, H.; et al. Interfacial Control Toward Efficient and Low-Voltage Perovskite Light-Emitting Diodes. *Adv. Mater.* **2015**, *27*, 2311–2316. [CrossRef] [PubMed]

9. Yu, J.C.; Kim, D.B.; Baek, G.; Lee, B.R.; Jung, E.D.; Lee, S.; Chu, J.H.; Lee, D.-K.; Choi, K.J.; Cho, S.; et al. High-Performance Planar Perovskite Optoelectronic Devices: A Morphological and Interfacial Control by Polar Solvent Treatment. *Adv. Mater.* **2015**, *27*, 3492–3500. [CrossRef] [PubMed]

10. Yu, J.C.; Kim, D.B.; Jung, E.D.; Lee, B.R.; Song, M.H. High-performance perovskite light-emitting diodes via morphological control of perovskite films. *Nanoscale* **2016**, *8*, 7036–7042. [CrossRef] [PubMed]

11. Cho, H.; Jeong, S.-H.; Park, M.-H.; Kim, Y.-H.; Wolf, C.; Lee, C.-L.; Heo, J.H.; Sadhanala, A.; Myoung, N.S.; Yoo, S.; et al. Overcoming the electroluminescence efficiency limitations of perovskite light-emitting diodes. *Science* **2015**, *350*, 1222–1225. [CrossRef] [PubMed]

12. Jeon, N.J.; Noh, J.H.; Kim, Y.C.; Yang, W.S.; Ryu, S.; Seok, S.I. Solvent engineering for high-performance inorganic–organic hybrid perovskite solar cells. *Nat. Mater.* **2014**, *13*, 897–903. [CrossRef] [PubMed]

13. Ahn, N.; Son, D.-Y.; Jang, I.-H.; Kang, S.M.; Choi, M.; Park, N.-G. Highly Reproducible Perovskite Solar Cells with Average Efficiency of 18.3% and Best Efficiency of 19.7% Fabricated via Lewis Base Adduct of Lead(II) Iodide. *J. Am. Chem. Soc.* **2015**, *137*, 8696–8699. [CrossRef] [PubMed]

14. Motta, C.; El-Mellouhi, F.; Sanvito, S. Charge carrier mobility in hybrid halide perovskites. *Sci. Rep.* **2015**, *5*, 12746. [CrossRef] [PubMed]

15. Wehrenfennig, C.; Eperon, G.E.; Johnston, M.B.; Snaith, H.J.; Herz, L.M. High Charge Carrier Mobilities and Lifetimes in Organolead Trihalide Perovskites. *Adv. Mater.* **2014**, *26*, 1584–1589. [CrossRef] [PubMed]

16. Stoumpos, C.C.; Malliakas, C.D.; Kanatzidis, M.G. Semiconducting Tin and Lead Iodide Perovskites with Organic Cations: Phase Transitions, High Mobilities, and Near-Infrared Photoluminescent Properties. *Inorg. Chem.* **2013**, *52*, 9019–9038. [CrossRef] [PubMed]

17. Wu, K.; Bera, A.; Ma, C.; Du, Y.; Yang, Y.; Li, L.; Wu, T. Temperature-dependent excitonic photoluminescence of hybrid organometal halide perovskite films. *Phys. Chem. Chem. Phys.* **2014**, *16*, 22476–22481. [CrossRef] [PubMed]

18. Stranks, S.D.; Eperon, G.E.; Grancini, G.; Menelaou, C.; Alcocer, M.J.P.; Leijtens, T.; Herz, L.M.; Petrozza, A.; Snaith, H.J. Electron-Hole Diffusion Lengths Exceeding 1 Micrometer in an Organometal Trihalide Perovskite Absorber. *Science* **2013**, *342*, 341–344. [CrossRef] [PubMed]

19. Xing, G.; Mathews, N.; Sun, S.; Lim, S.S.; Lam, Y.M.; Grätzel, M.; Mhaisalkar, S.; Sum, T.C. Long-Range Balanced Electron- and Hole-Transport Lengths in Organic-Inorganic $CH_3NH_3PbI_3$. *Science* **2013**, *342*, 344–347. [CrossRef] [PubMed]

20. Xing, G.; Mathews, N.; Lim, S.S.; Yantara, N.; Liu, X.; Sabba, D.; Gratzel, M.; Mhaisalkar, S.; Sum, T.C. Low-temperature solution-processed wavelength-tunable perovskites for lasing. *Nat. Mater.* **2014**, *13*, 476–480. [CrossRef] [PubMed]

21. Schmidt, L.C.; Pertegás, A.; González-Carrero, S.; Malinkiewicz, O.; Agouram, S.; Espallargas, G.M.; Bolink, H.J.; Galian, R.E.; Pérez-Prieto, J. Nontemplate Synthesis of $CH_3NH_3PbBr_3$ Perovskite Nanoparticles. *J. Am. Chem. Soc.* **2014**, *136*, 850–853. [CrossRef] [PubMed]

22. Gonzalez-Carrero, S.; Galian, R.E.; Perez-Prieto, J. Maximizing the emissive properties of $CH_3NH_3PbBr_3$ perovskite nanoparticles. *J. Mater. Chem. A* **2015**, *3*, 9187–9193. [CrossRef]

23. Luo, B.; Pu, Y.-C.; Yang, Y.; Lindley, S.A.; Abdelmageed, G.; Ashry, H.; Li, Y.; Li, X.; Zhang, J.Z. Synthesis, Optical Properties, and Exciton Dynamics of Organolead Bromide Perovskite Nanocrystals. *J. Phys. Chem. C* **2015**, *119*, 26672–26682. [CrossRef]

24. Jang, D.M.; Park, K.; Kim, D.H.; Park, J.; Shojaei, F.; Kang, H.S.; Ahn, J.-P.; Lee, J.W.; Song, J.K. Reversible Halide Exchange Reaction of Organometal Trihalide Perovskite Colloidal Nanocrystals for Full-Range Band Gap Tuning. *Nano Lett.* **2015**, *15*, 5191–5199. [CrossRef] [PubMed]

25. Zhang, F.; Zhong, H.; Chen, C.; Wu, X.-G.; Hu, X.; Huang, H.; Han, J.; Bingsuo Zou, B.; Dong, Y. Brightly Luminescent and Color-Tunable Colloidal $CH_3NH_3PbX_3$ (X = Br, I, Cl) Quantum Dots: Potential Alternatives for Display Technology. *ACS Nano* **2015**, *9*, 4533–4542. [CrossRef] [PubMed]

26. Protesescu, L.; Yakunin, S.; Bodnarchuk, M.I.; Krieg, F.; Caputo, R.; Hendon, C.H.; Yang, R.X.; Walsh, A.; Kovalenko, M.V. Nanocrystals of Cesium Lead Halide Perovskites (CsPbX3, X = Cl, Br, and I): Novel Optoelectronic Materials Showing Bright Emission with Wide Color Gamut. *Nano Lett.* **2015**, *15*, 3692–3696. [CrossRef] [PubMed]

27. Nedelcu, G.; Protesescu, L.; Yakunin, S.; Bodnarchuk, M.I.; Grotevent, M.J.; Kovalenko, M.V. Fast Anion-Exchange in Highly Luminescent Nanocrystals of Cesium Lead Halide Perovskites ($CsPbX_3$, X = Cl, Br, I). *Nano Lett.* **2015**, *15*, 5635–5640. [CrossRef] [PubMed]

28. Xiao, Z.; Kerner, R.A.; Zhao, L.; Tran, N.L.; Lee, K.M.; Koh, T.-W.; Scholes, G.D.; Rand, B.P. Efficient perovskite light-emitting diodes featuring nanometre-sized crystallites. *Nat Photonics* **2017**, *11*, 108–115. [CrossRef]

29. Zhao, L.; Yeh, Y.-W.; Tran, N.L.; Wu, F.; Xiao, Z.; Kerner, R.A.; Lin, Y.L.; Scholes, G.D.; Yao, N.; Rand, B.P. In Situ Preparation of Metal Halide Perovskite Nanocrystal Thin Films for Improved Light-Emitting Devices. *ACS Nano* **2017**, *11*, 3957–3964. [CrossRef] [PubMed]

30. Wu, Q.; Zhou, P.; Zhou, W.; Wei, X.; Chen, T.; Yang, S. Acetate Salts as Nonhalogen Additives to Improve Perovskite Film Morphology for High-Efficiency Solar Cells. *ACS Appl. Mater. Interfaces* **2016**, *8*, 15333–15340. [CrossRef] [PubMed]

31. Xiao, M.; Huang, F.; Huang, W.; Dkhissi, Y.; Zhu, Y.; Etheridge, J.; Gray-Weale, A.; Bach, U.; Cheng, Y.-B.; Spiccia, L. A Fast Deposition-Crystallization Procedure for Highly Efficient Lead Iodide Perovskite Thin-Film Solar Cells. *Angew. Chem.* **2014**, *126*, 10056–10061. [CrossRef]

32. Gaspera, E.D.; Peng, Y.; Hou, Q.; Spiccia, L.; Bach, U.; Jasieniak, J.J.; Cheng, T.-B. Ultra-thin high efficiency semitransparent perovskite solar cells. *Nano Energy* **2015**, *13*, 249–257. [CrossRef]

33. Baldan, A. Review Progress in Ostwald ripening theories and their applications to the γ'-precipitates in nickel-base superalloys Part II Nickel-base superalloys. *J. Mater. Sci.* **2002**, *37*, 2379–2405. [CrossRef]

34. Yang, M.; Zhang, T.; Schulz, P.; Li, Z.; Li, G.; Kim, D.H.; Guo, N.; Berry, J.J.; Zhu, K.; Zhao, Y. Facile fabrication of large-grain $CH_3NH_3PbI_{3-x}Br_x$ films for high-efficiency solar cells via CH_3NH_3Br-selective Ostwald ripening. *Nat. Commun.* **2016**, *7*, 12305. [CrossRef] [PubMed]

35. Xiao, Z.; Dong, Q.; Bi, C.; Shao, Y.; Yuan, Y.; Huang, J. Solvent Annealing of Perovskite-Induced Crystal Growth for Photovoltaic-Device Efficiency Enhancement. *Adv. Mater.* **2014**, *26*, 6503–6509. [CrossRef] [PubMed]

36. You, J.; Hong, Z.; Yang, Y.M.; Chen, Q.; Cai, M.; Song, T.-B.; Chen, C.-C.; Lu, S.; Liu, Y.; Zhou, H.; et al. Low-Temperature Solution-Processed Perovskite Solar Cells with High Efficiency and Flexibility. *ACS Nano* **2014**, *8*, 1674–1680. [CrossRef] [PubMed]

37. Chen, Q.; Zhou, H.; Song, T.-B.; Luo, S.; Hong, Z.; Duan, H.-S.; Dou, L.; Liu, Y.; Yang, Y. Controllable Self-Induced Passivation of Hybrid Lead Iodide Perovskites toward High Performance Solar Cells. *Nano Lett.* **2014**, *14*, 4158–4163. [CrossRef] [PubMed]

38. Bhaumik, S.; Veldhuis, S.A.; Ng, Y.F.; Li, M.; Muduli, S.K.; Sum, T.C.; Damodaran, B.; Mhaisalkar, S.; Mathews, N. Highly stable, luminescent core-shell type methylammonium-octylammonium lead bromide layered perovskite nanoparticles. *Chem. Commun.* **2016**, *52*, 7118–7121. [CrossRef] [PubMed]

39. Song, J.; Li, J.; Li, X.; Xu, L.; Dong, Y.; Zeng, H. Quantum Dot Light-Emitting Diodes Based on Inorganic Perovskite Cesium Lead Halides ($CsPbX_3$). *Adv. Mater.* **2015**, *27*, 7162–7167. [CrossRef] [PubMed]

40. Le, Q.V.; Park, M.; Sohn, W.; Jang, H.W.; Kim, S.Y. Investigation of Energy Levels and Crystal Structures of Cesium Lead Halides and Their Application in Full-Color Light-Emitting Diodes. *Adv. Electron. Mater.* **2017**, *3*, 1600448.

41. Deng, W.; Xu, X.; Zhang, X.; Zhang, Y.; Jin, X.; Wang, L.; Lee, S.-T.; Jie, J. Organometal Halide Perovskite Quantum Dot Light-Emitting Diodes. *Adv. Funct. Mater.* **2016**, *26*, 4797–4802. [CrossRef]

42. Chen, C.C.; Chang, S.H.; Chen, L.-C.; Tsai, C.-L.; Cheng, H.-M.; Huang, W.-C.; Chen, W.-N.; Lu, Y.-C.; Tseng, Z.-L.; Chiu, K.Y.; et al. Interplay between nucleation and crystal growth during the formation of $CH_3NH_3PbI_3$ thin films and their application in solar cells. *Sol. Energy Mater. Sol. Cells* **2017**, *159*, 583–589. [CrossRef]

43. D'Innocenzo, V.; Grancini, G.; Alcocer, M.J.P.; Kandada, A.R.S.; Stranks, S.D.; Lee, M.M.; Lanzani, G.; Snaith, H.J.; Petrozza, A. Excitons versus free charges in organo-lead tri-halide perovskites. *Nat. Commun.* **2014**, *5*, 3586. [CrossRef] [PubMed]

44. Dar, M.I.; Jacopin, G.; Meloni, S.; Mattoni, A.; Arora, N.; Boziki, A.; Zakeeruddin, S.M.; Rothlisberger, U.; Grätzel, M. Origin of unusual bandgap shift and dual emission in organic-inorganic lead halide perovskites. *Sci. Adv.* **2016**, *2*, e1601156. [CrossRef] [PubMed]

 *nanomaterials*

Article

Improvement in Color-Conversion Efficiency and Stability for Quantum-Dot-Based Light-Emitting Diodes Using a Blue Anti-Transmission Film

Jiasheng Li [1,2], Yong Tang [1], Zongtao Li [1,2,*], Xinrui Ding [1], Shudong Yu [1] and Binhai Yu [1]

[1] Engineering Research Center of Green Manufacturing for Energy-Saving and New-Energy Technology, South China University of Technology, Guangzhou 510640, China; jiasli@foxmail.com (J.L.); ytang@scut.edu.cn (Y.T.); ding.xinrui@mail.scut.edu.cn (X.D.); yu.shudong@mail.scut.edu.cn (S.Y.); mebhaiyu@scut.edu.cn (B.Y.)

[2] Foshan Nationstar Optoelectronics Company Ltd., Foshan 528000, China

* Correspondence: meztli@scut.edu.cn; Tel.: +86-138-2446-0886

Received: 8 June 2018; Accepted: 2 July 2018; Published: 9 July 2018

Abstract: In this report, a blue anti-transmission film (BATF) has been introduced to improve the color-conversion efficiency (CCE) and the stability of quantum dot (QD) films. The results indicate that the CCE can be increased by as much as 93% using 15 layers of BATFs under the same QD concentration. Therefore, the same CCE can be achieved using BATF-QD hybrid films with a lower QD concentration when compared with standard QD films. The hybrid and QD films with the same CCE of 60% were aged at an environmental temperature of 25°C and with a 10 mA injection current light-emitting diode source. The CCE and luminous efficacy that are gained by the hybrid film increased by 42.8% and 24.5%, respectively, when compared with that gained by the QD film after aging for the same time period of approximately 65 h. In addition, the hybrid film can effectively suppress the red-shift phenomenon of the QD light spectra, as well as an expansion of the full-width at half maximum. Consequently, these BATF-QD hybrid films with excellent optical performance and stability show great potential for illumination and display applications.

Keywords: light-emitting diodes; quantum dots; stability; color-conversion efficiency; photoluminescence

---

1. Introduction

Quantum dots (QDs) have several advantages for lighting applications, including high quantum yield, high color purity, and easy manufacture. They have demonstrated great potential in optoelectronic devices, such as light-emitting diodes [1]. Blue LED chips are generally utilized to excite QD films. In this process, the mixing of the blue light from the LEDs (chip light) and the conversion light from the QDs (QD light) can achieve different output colors [2]. Significant effort has been directed at attempting to improve the quantum yield (QY) of QDs by optimizing the core/shell structures [3] and the surface functional groups [4]. At present, the QY of CdSe/ZnS is already in excess of 90% [5]. This QD is widely regarded as one of the most promising candidates for replacing traditional rare-earth-based phosphor materials [6]. However, significant challenges are still encountered when these QDs are used in LEDs. Similar to other phosphor materials, such as yttrium aluminitum garnet (YAG) and nitride phosphor, it is necessary to disperse the QDs into a transparent matrix in order to achieve high light extraction [7] and to simultaneously prevent oxidation [8]. QDs are easy to aggregate in a matrix due to their small particle size of several nanometers. However, this results in aggregation induced quenching (AIQ) [9], which leads to high conversion losses. Some polymer matrixes, including polyvinyl alcohol (PVA) [10], polydimethylsiloxane (PDMS) [11], and gel glass [12] have been proposed to improve the dispersity of QDs, thereby decreasing this loss. In addition, QDs strongly absorb their emission light, which results in significant reabsorption loss [13–15]. This may lead to larger thermal power generation from the QDs and a reduced working stability [16].

The color-conversion efficiency (CCE) refers to the proportion of QD light radiant power to the total radiant power. It determines the output color coordinates of LEDs [17,18]. This is one of the most essential parameters when characterizing illuminations and displays. It is common to increase the CCE by increasing the QD concentration in the QD matrix [19]. However, a larger QD concentration can significantly increase AIQ and the reabsorption by the QDs [15], leading to a lower conversion efficiency of the QD films when compared to the QY of the QDs. Therefore, the luminous efficacy of the QD-based LEDs is still significantly less than that of traditional phosphor-based LEDs, especially for high CCE LEDs [20]. To solve this issue, the scattering effect that is introduced by particles with a high refractive index [21,22] or patterned structures [18] is used to improve the CCE by increasing the blue light path in the QD films. However, these methods also increase the light path of the QD light in the QD films, leading to higher reabsorption loss and heat power generation from the QDs [23]. Consequently, it is expected that any new solution for improving the CCE should simultaneously decrease reabsorption losses [20]. Solving this problem is an important aspect in the development of high efficiency and stable QD-based LEDs.

In this study, a blue anti-transmission film (BATF) is introduced in order to improve the CCE and stability of QD films. The absorption of QD films and BATF-QD hybrid films were investigated. Then, the influence of the BATF-QD hybrid films on the optical performance of LEDs was investigated. Finally, the optical performance and working stability of the BATF-QD hybrid films were compared with QD films under the same CCE.

2. Methods

The QDs with the CdSe/ZnS core/shell structure (purchased from China Beijing Beida Jubang Science & Technology Co., Ltd., Beijing, China, quantum yield larger than 80%) were used in this study. The emission and absorption properties of these QDs in chloroform solution are shown in Figure 1. PDMS (Sylgard-184, purchased from Dow Corning, Midland, MI, USA) was used as the transparent matrix to disperse the QDs. Since the blue light is necessary for the illumination and display applications, we have selected the BATF with a specific blue light transmittance instead of entirely absorbing the blue light. The commercial BATFs with approximately 70% blue light transmittance are widely used for eye-protection in screens of mobile phones and computers, which have been used in this study. These BATFs were purchased from Shen Zhen Fancy Package & Manufactory Co., Ltd. (Shenzhen, China). The blue blocking function is realized by the multilayer composites with gradient refractive index (the total thickness of the functional composites is 100 μm). The transparent substrate is PET polymer, as shown in Figure 2a. They strongly reflect blue light, as shown in Figure 2c, and are widely used as eye-protection films in screens. The LED source has a radius of 8 mm, and was packaged as 42 blue LED chips with an emission wavelength centered at 455 nm; all of the LED chips were sealed by the silicone-based encapsulant; the electrical injection power is 150 mW (injection current of 10 mA), which is a harsh condition for QD-based LED operation [24]. The QD films were fabricated using the evaporation-curing method [11,22,23]. Firstly, QDs were added into 2 mL of chloroform solution to uniformly disperse them in the solution systems with low viscosity; the total mass of the QDs were 10 mg, 15 mg, 20 mg, and 30 mg, in order to control the QD concentration in the QD films. After mixing for several seconds, 909 mg of pre-polymer A-PDMS was added to the QD-chloroform solution. This mixture was stirred for 40 mins until the chloroform solution completely evaporated. Then, 91 mg of curing agent of B-PDMS was added into the mixtures, which were stirred and evacuated for 15 mins. Finally, the QD-PDMS mixtures were injected into a mold and then cured at a temperature of 120 °C for 90 mins. QD films with a thickness of 0.5 mm and radius of 16 mm were produced. The BATF-QD hybrid films can be easily fabricated by placing the BATFs inside the mold during the curing process. Moreover, the BATFs were stamped to each other using PDMS in order to achieve multi-layer BATFs. The optical properties of the BATFs, QD films, and BATF-QD hybrid films were measured using a dual-beam UV-Vis spectrophotometer TU-1901 (Beijing Persee General Instrument Co., Ltd., Beijing, China). These films were also assembled on an LED source, and then their optical performance were measured using the Integrating Systems from Instrument Systems (Munich, Germany). The injection current of 10 mA is provided by a source from Keithley (Beaverton, OR, USA). During measurements, it should be noted that the BATFs are always on the upper side of the QD film.

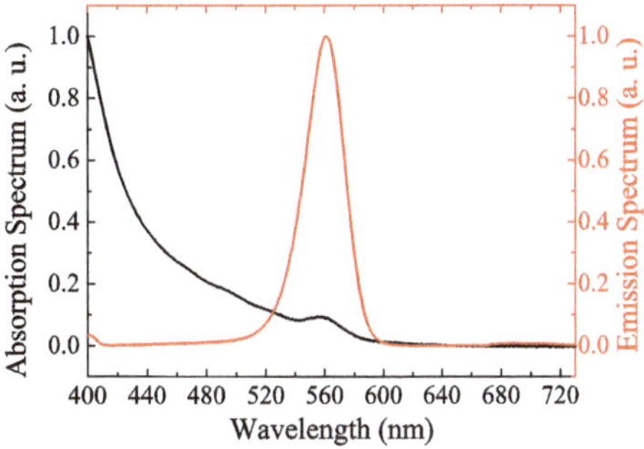

Figure 1. Absorption and emission spectra of CdSe/ZnS quantum dots (QDs) solution.

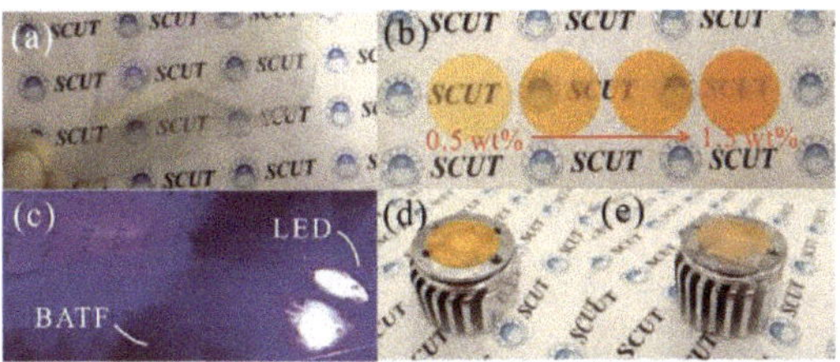

Figure 2. The (**a**) BATF; (**b**) QD films; (**c**) blue light spot reflection by BATFs; (**d**) LED assembled with a QD films; and (**e**) LED assembled with a BATF-QD hybrid film (5 layers of BTAFs).

The working stability was investigated according to the aging tests at an environmental temperature of 25 °C without heat sink for thermal managements [24]. The BATF-QD hybrid films and QD films were assembled to the same LED source mentioned above, respectively. The LED source was kept with an injection current of 10 mA (electrical injection power of 150 mW) to continuously excite these QDs films for degradations. The optical performance of LEDs assembled with BATF-QD hybrid films and QD films were measured at different aging time using the same injection current of 10 mA.

3. Results and Discussion

The optical properties including wavelength-dependent transmittance, reflection, and absorption of the commercial BATF were characterized, as shown in Figure 3a. The BATF has a low average transmittance of 69.1% for blue light with a wavelength less than 500 nm, which is due, in part, to its high reflection (average of 11.3%) and absorption (average 19.6%) for these wavelengths. Moreover, it exhibits almost no reflection for wavelengths that are longer than 500 nm. The average transmittance is as high as 91.5%, while the absorption is approximately 7% for these wavelengths. These values indicate that BATF is beneficial with respect to suppressing the transmittance of chip light (emission from LEDs), while also introducing some additional absorption loss for QD light (emission from QDs). Figure 3b shows the absorption spectra of the QD films and the BATF-QD hybrid films with different layers of BATFs. The concentration of the QD is kept as 0.75 wt. %. The BATF-QD hybrid films have a higher absorption when compared with the QD films, and the absorption becomes larger with additional layers of BATFs. Besides, it shows less increment in the absorption as the number of BATF layers increases. This is because much more reflection light can escapes from the lateral side of BATFs with a larger thickness instead of being absorbed by the QD film. The inset figures of Figure 3b show the absorption spectra when normalized to the absorption at the peak emission wavelength of the QDs. It can be observed that the BATF-QD hybrid films have a higher absorption for blue light (at 455 nm) when they have the same absorption for the peak wavelength of the QD light, especially for the hybrid films with several layers of BATFs. This implies that the BATF-QD hybrid films are useful in reducing the chip light with increasing layers of BATFs, while maintaining a large amount of QD light, which may be beneficial to increasing the CCE of the QDs.

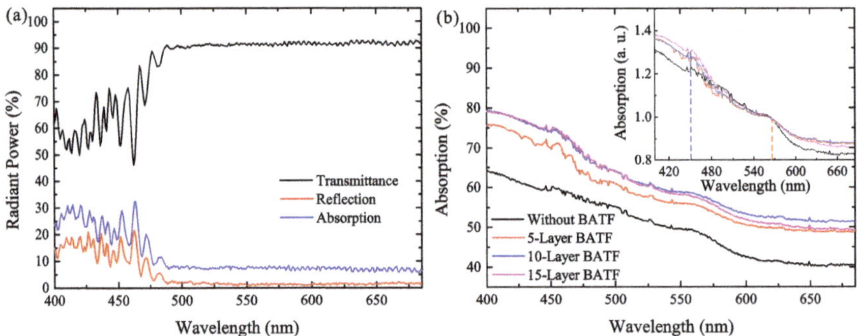

Figure 3. (a) The transmittance, reflection, and absorption spectrum of the BATF; and (b) the absorption spectra of the QD films and the BATF-QD hybrid films.

The QD film and BATF-QD hybrid film with different layers of BATFs and different concentrations of QDs were applied to a blue LED source. The radiant power, including the total radiant power, radiant power of chip light, and radiant power of QD light, of these QD-based LEDs is given in Figure 4a. It should be noted that the radiant power of the chip light and the QD light were obtained by integrating the emission radiant spectra of QD-based LEDs from 380 nm to 525 nm and from 525 nm to 730 nm, respectively. As the number of BATF layers was increased, the total radiant power significantly decreases. This demonstrates that the use of BATF can lead to a reduction in light extraction due to its high reflection and absorption. However, this decrease becomes smaller as the QD concentration increases. When the number of BATF layers increases from 0 to 15, the total radiant power that is gained for a QD concentration of 0 wt. % decreases by 62.2%. There is a decrease of 49.2% for a QD concentration of 1.5 wt. %. One reasonable explanation for this observation is that much more chip light is absorbed by the QD films prior to propagating into the BATFs as the QD concentration is increased. This results in more QD light with a higher transmittance. These results are also supported by the significant reduction in the radiant power of the chip light as the QD concentration increases. Although the BATF has a high absorption for QD light that is comparable with that of chip light, it is interesting that the radiant power shows a significant decrease with increasing BATF layers. Only a slight change is observed for the radiant power of the QD light. These results indicate that a portion of the chip light reflected by BATF has propagated into the QD films and it is absorbed by QDs, thereby increasing the excitation of the QD light. In other words, it is the reuse of the reflected chip light that prevents the reduction in the radiant power of the QD light.

Their luminous efficacy and CCE are given in Figure 4b. Similarly, the luminous efficacy also decreases with increasing BATF layers due to the absorption and reflection losses that are introduced by the BATFs. However, the reduction in the luminous efficacy is smaller than 37%, which is less than that of the radiant power. This is because the luminous efficacy is more sensitive to the QD light, which still maintains a high radiant power. Most importantly, the CCE shows a significant increase as the number of BATF layer increases. This increase is more significant at lower QD concentrations. An increase of 93% using 15 layers of BATFs is observed when the QD concentration is 0.5 wt. %. This is because when the QD film has a lower concentration, it more readily facilitates the escape of QD light due to the lower probability of reabsorption events.

Moreover, it is interesting that LEDs using BATF-QD hybrid films with a lower QD concentration can attain the same CCE as QD films with larger QD concentrations. To further investigate this issue, the results for the QD concentration for the same CCE when a different number of BATF layers were used, are given in Figure 4c. It is evident that more BATF layers are beneficial to decreasing the QD concentration at the same CCE. For example, when 15 layers of BATFs are used, a QD concentration of only 0.75 wt. % can result in a CCE that is as high as that of 1.5 wt. % without using BATFs. The conversion loss of the QDs is also given in Figure 4c, which was calculated using the equation (PCabs−PQDemit)/PCabs. PCabs and PQDemit are the QD absorption power of the chip light and the radiant power of the QD light, respectively. It is obvious that a larger QD concentration can cause higher conversion losses due to significantly more reabsorption events. The conversion loss gained with a 0.75 wt. % QD concentration can decrease by 11.3% when compared with that gained using 1.5 wt. % QD concentration. Therefore, the BATF-QD hybrid films can achieve a high CCE using a low QD concentration, which is beneficial to decreasing conversion losses inside QD films.

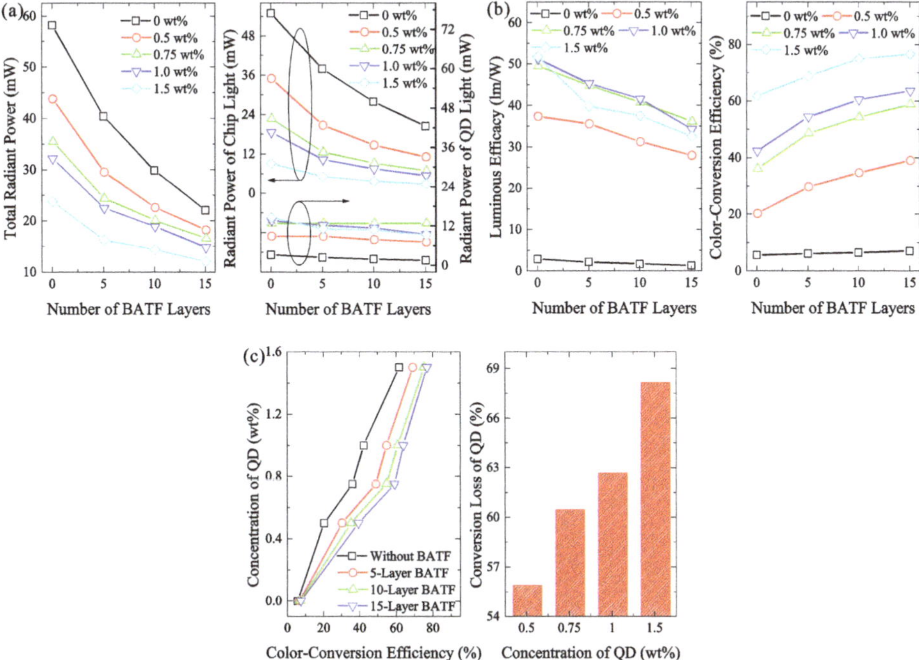

Figure 4. The (**a**) total radiant power, radiant power of the chip light, radiant power of the QD light; (**b**) luminous efficacy, and color-conversion efficiency of LEDs with different QD concentrations and BATF layers; (**c**) the QD concentration verse color-conversion efficiency with different BATF layers and the conversion loss with different QD concentration. The QD concentration is various from 0 wt. % to 1.5 wt. %.

The aging test was performed to evaluate this issue, the optical performances of LEDs using hybrid films (QD concentration of 0.75 wt. %, 15-layer BATFs, initial CCE of 60%) and QD films (QD concentration of 1.5 wt. %, initial CCE of 60%) with different aging time are shown in Figure 5a,b, respectively. The optical performances including the radiant power, luminous flux, and CCE of these LEDs dramatically decrease at the initial stage (within 200 min), which is similar to previous investigations on their stability [16,24]. One possible explanation is that much more QD light at the initial stage leads to much more reabsorption events between QDs, resulting in greater thermal power generation from QDs [20,25]. However, it is evident that the luminous efficacy that is gained by the QD film significantly decreases when compared with the BATF-QD hybrid film, although they have the same initial CCE of 60%. As discussed above, this is because that the BATF significantly reduces the chip light, while maintaining a large amount of QD light, which realizes the same CCE for LEDs using the hybrid film with a lower QD concentration compared with that only using the QD film with a higher QD concentration. A lower QD concentration leads to less reabsorption loss [20,26], thereby decreasing the thermal power generating from QDs and improving the stability. Their stability have been compared in detail. After aging for approximately 65 h, the CCE, total radiant power, and luminous efficacy gained by the QD film decrease by 66.5%, 62.4%, and 86.6%, respectively; while, that gained by the hybrid film only decreased by 39.7%, 19.4%, and 48%, respectively. Their spectra after aging for approximately 65 h are also given in Figure 6. It is interesting that the radiant power of the spectra of the QD light that are acquired from the hybrid film are even higher than that acquired from the QD film. As such, the hybrid film can have a higher CCE and luminous efficacy when compared to the QD film after aging. These values are increased by 42.8% and 24.5%, respectively, as shown in the inset figure. It should be noted that the initial luminous efficacy (before aging) of this hybrid film is 30% lower than that of the QD film at the same CCE of 60%, as shown in Figure 4. However, this property is significantly increased by 24.5% after aging. This indicates that the hybrid film has a lower conversion loss and generates less thermal power, which avoids the destruction of the QDs and silicone carbonization [25]. Therefore, the BATF-QD hybrid film appears to result in a better working stability and better optical performances after aging. In addition, the initial luminous efficacy using these hybrid films is far lower than that using the QD films at the same CCE. This is mainly because the absorption losses that are introduced by the BATFs are much larger than the reduction in the conversion loss. Consequently, it is necessary to modify the BATF structures in order to simultaneously improve light extraction of the LEDs. This will help to further improve the optical performances of these hybrid films. To avoid misinterpretation, it should be noticed that the percentage improvements given above are arbitrary and not absolute.

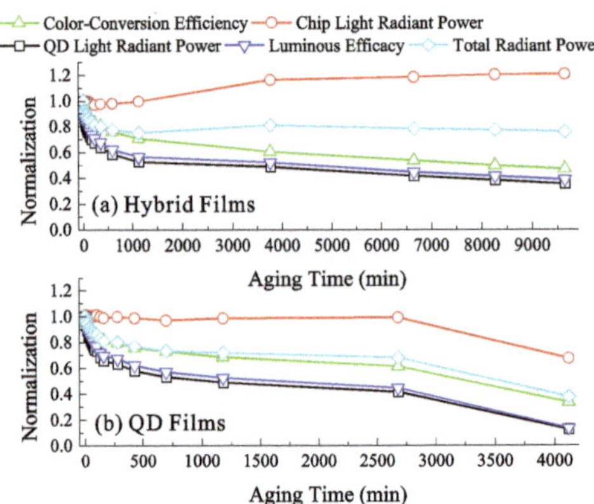

Figure 5. The optical performances of LEDs using (**a**) hybrid films (0.75 wt. % QD concentration and 15-layer BATFs) and (**b**) QD films (1.5 wt. % QD concentration) with different aging time, respectively.

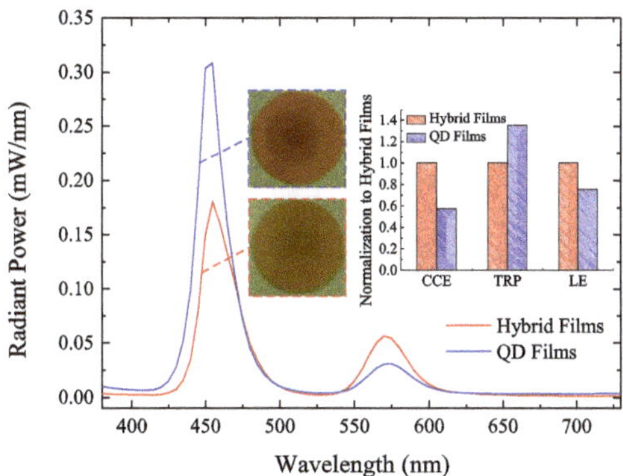

Figure 6. The emission spectra of LEDs using hybrid films (0.75 wt. % QD concentration and 15-layer BATFs) and QD films (1.5 wt. % QD concentration) after aging for approximately 65 h. The inset figure represents their color-conversion efficiency (CCE), total radiant power (TRP), and luminous efficacy (LE) normalized to that of the BATF.

The spectra of LEDs using QD films and BATF-QD hybrid films with different QD concentrations and BATF layers are further investigated, as given in Figure 7. Figure 7a shows the spectra that are acquired from the QD films. The radiant power of the spectra of the QD light increase with an increasing QD concentration, while it should be noted that their increment is significantly low when compared with the reduction in the radiant power of the spectra of the chip light, especially when the QD concentration is sufficiently large. This is because of the severe reabsorption losses of the QD films with high QD concentration, as previously indicated. Besides, the reabsorption effect leads to a red-shift of the spectra, as shown in the normalized spectra in the inset figures. When the QD concentration is 1.5 wt. %, the peak wavelength in the spectra of QD light shifts to 580 nm, and its full-width at half maximum (FWHM) is increased to 36 nm. As a result, although the increase in the QD concentration can result in an increase of the CCE, it also causes the spectra to be unpredictable. The spectra that are acquired from the BATF-QD hybrid films for a QD concentration of 0.75 wt. % are shown in Figure 7b. The spectra of the chip light is significantly decreased, while that of the QD light shows a slight decrease with an increasing number of BATF layers. This is beneficial to increasing the CCE, as previously discussed. However, it is interesting that their normalized spectra show almost no change. It is noteworthy that the CCE gained using 0.75 wt. % QD concentration and 15 layers of BATFs is almost the same as that gained by QD films using 1.5 wt. % QD concentration. The peak wavelength and the FWHM of the QD light are 575 nm and 34 nm, respectively. These values are the same as the values that are acquired from the QD films using 0.75 wt. % QD concentration and smaller than the values acquired from the QD films for 1.5 wt. % QD concentration. Therefore, compared with QD films with a higher QD concentration under the same CCE, the BATF-QD hybrid films with a lower QD concentration are also beneficial for suppressing the red-shift phenomenon. This is helpful for accurately obtaining color coordinates, which is critically important in illuminations and display applications.

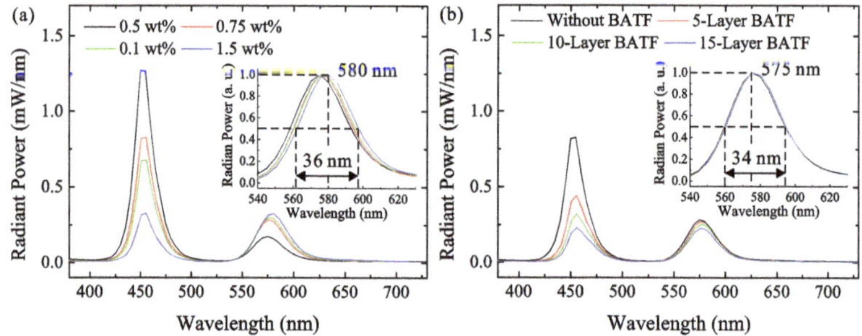

Figure 7. The emission spectra of LEDs using (**a**) QD films with different QD concentration and (**b**) hybrid films with 0.75 wt. % QD concentration and different BATFs layers.

4. Conclusions

In this report, BATF was introduced to improve the CCE and stability of QD films. The results show that the BATF-QD hybrid films significantly increase the CCE by 93% using 15 layers of BATFs with the same QD concentration. This is caused by the stronger absorption and reflection of chip light. The same CCE (as high as 60%) can be achieved using hybrid films with a QD concentration half that of QD films, although the initial luminous efficacy (before aging) acquired by the hybrid film was substantially lower than that acquired by the QD film with the same CCE of 60%. After aging for approximately 65 h, the CCE and luminous efficacy gained by the hybrid film increased by 42.8% and 24.5% when compared with that of the QD film. These results demonstrate that a hybrid film with a lower QD concentration can significantly decrease conversion losses due to the reduced probability of reabsorption events. As a result, the stability and optical performance are significantly increased after aging. Because of the lower QD concentration usage, the hybrid film can also effectively suppress the red-shift phenomenon of the spectra, which simultaneously leads to a smaller FWHM of the QD light. Therefore, the BATF-QD hybrid films show great potential in the illumination and display applications.

In future investigations, we intend to optimize the BATF structure to decrease its absorption while increasing its reflection. We believe that this will further improve the light extraction of LEDs by avoiding the absorption losses introduce by the BATFs.

Author Contributions: Formal analysis, J.L., Z.L., X.D. and S.Y.; Funding acquisition, Y.T., Z.L. and B.Y.; Investigation, Y.T., S.Y. and B.Y.; Methodology, J.L.; Resources, Y.T., Z.L. and B.Y.; Writing—original draft, J.L.; Writing—review & editing, Z.L. and X.D.

Funding: This work was supported by the National Natural Science Foundation of China (51775199, 51735004); the Natural Science Foundation of Guangdong Province (2014A030312017); the Science & Technology Program of Guangdong Province (2015B010132002); the Project of Science and Technology New Star in Zhujiang Guangzhou City (201806010102).

References

1. Shirasaki, Y.; Supran, G.J.; Bawendi, M.G.; Bulović, V. Emergence of colloidal quantum-dot light-emitting technologies. *Nat. Photonics* **2013**, *7*, 13–23. [CrossRef]

2. Xie, B.; Hu, R.; Luo, X. Quantum dots-converted light-emitting diodes packaging for lighting and display: Status and perspectives. *J. Electron. Packag.* **2016**, *138*, 020803. [CrossRef]

3. Dabbousi, B.O.; Rodriguez-Viejo, J.; Mikulec, F.V.; Heine, J.R.; Mattoussi, H.; Ober, R.; Jensen, K.F.; Bawendi, M.G. (CdSe) ZnS core— shell quantum dots: Synthesis and characterization of a size series of highly luminescent nanocrystallites. *J. Phys. Chem. B* **1997**, *101*, 9463–9475. [CrossRef]

4. Hou, J.; Wang, L.; Zhang, P.; Xu, Y.; Ding, L. Facile synthesis of carbon dots in an immiscible system with excitation-independent emission and thermally activated delayed fluorescence. *Chem. Commun.* **2015**, *51*, 17768–17771. [CrossRef] [PubMed]

5. Greytak, A.B.; Allen, P.M.; Liu, W.; Zhao, J.; Young, E.R.; Walker, B.; Nocera, D.G.; Bawendi, M.G. Alternating layer addition approach to CdSe/CdS core/shell quantum dots with near-unity quantum yield and high on-time fractions. *Chem. Sci.* **2012**, *3*, 2028–2034. [CrossRef] [PubMed]

6. Zhang, R.; Lin, H.; Yu, Y.; Chen, D.; Xu, J.; Wang, Y. A new-generation color converter for high-power white LED: Transparent Ce^{3+}:YAG phosphor-in-glass. *Laser Photonics Rev.* **2014**, *8*, 158–164. [CrossRef]

7. Tsai, Y.L.; Lai, K.Y.; Lee, M.J.; Liao, Y.K.; Ooi, B.S.; Kuo, H.C.; He, J.H. Photon management of GaN-based optoelectronic devices via nanoscaled phenomenon. *Prog. Quantum Electron.* **2016**, *49*, 1–25. [CrossRef]

8. Tsai, P.-Y.; Huang, H.-K.; Sung, J.-M.; Kan, M.-C.; Wang, Y.-H. High Thermal stability and wide angle of white light chip-on-board package using a remote phosphor structure. *IEEE Electron Device Lett.* **2015**, *36*, 250–252. [CrossRef]

9. Noh, M.; Kim, T.; Lee, H.; Kim, C.K.; Joo, S.W.; Lee, K. Fluorescence quenching caused by aggregation of water-soluble CdSe quantum dots. *Colloids Surf. A Physicochem. Eng. Asp.* **2010**, *359*, 39–44. [CrossRef]

10. Li, X.; Liu, Y.; Song, X.; Wang, H.; Gu, H.; Zeng, H. Intercrossed Carbon Nanorings with Pure Surface States as Low-Cost and Environment-Friendly Phosphors for White-Light-Emitting Diodes. *Angew. Chem. Int. Ed.* **2014**, *54*, 1759–1764. [CrossRef] [PubMed]

11. Yu, S.D.; Tang, Y.; Li, Z.T.; Chen, K.H.; Ding, X.R.; Yu, B.H. Enhanced optical and thermal performance of white light emitting diodes with horizontally-layered quantum dots phosphor nanocomposites. *Photonics Res.* **2018**, *6*, 90–98. [CrossRef]

12. Ma, L.; Xiang, W.; Gao, H.; Pei, L.; Ma, X.; Huang, Y.; Liang, X. Carbon dot-doped sodium borosilicate gel glasses with emission tunability and their application in white light emitting diodes. *J. Mater. Chem. C* **2015**, *3*, 6764–6770. [CrossRef]

13. Coropceanu, I.; Bawendi, M.G. Core/shell quantum dot based luminescent solar concentrators with reduced reabsorption and enhanced efficiency. *Nano Lett.* **2014**, *14*, 4097–4101. [CrossRef] [PubMed]

14. Li, J.; Tang, Y.; Li, Z.; Ding, X.; Yuan, D.; Yu, B. Study on Scattering and Absorption Properties of Quantum-Dot-Converted Elements for Light-Emitting Diodes Using Finite-Difference Time-Domain Method. *Materials* **2017**, *10*, 1264. [CrossRef] [PubMed]

15. Wang, X.; Yan, X.; Li, W.; Sun, K. Doped quantum dots for white-light-emitting diodes without reabsorption of multiphase phosphors. *Adv. Mater.* **2012**, *24*, 2742–2747. [CrossRef] [PubMed]

16. Woo, J.Y.; Kim, K.N.; Jeong, S.; Han, C.-S. Thermal behavior of a quantum dot nanocomposite as a color converting material and its application to white LED. *Nanotechnology* **2010**, *21*, 495704. [CrossRef] [PubMed]

17. Fulmek, P.; Sommer, C.; Hartmann, P.; Pachler, P.; Hoschopf, H.; Langer, G.; Nicolics, J.; Wenzl, F.P. On the Thermal Load of the Color-Conversion Elements in Phosphor-Based White Light-Emitting Diodes. *Adv. Opt. Mater.* **2013**, *1*, 753–762. [CrossRef]

18. Yu, S.; Zhuang, B.; Chen, J.; Li, Z.; Rao, L.; Yu, B.; Tang, Y. Butterfly-inspired micro-concavity array film for color conversion efficiency improvement of quantum-dot-based light-emitting diodes. *Opt. Lett.* **2017**, *42*, 4962. [CrossRef] [PubMed]

19. Zhang, W.; Yu, S.F.; Fei, L.; Jin, L.; Pan, S.; Lin, P. Large-area color controllable remote carbon white-light light-emitting diodes. *Carbon* **2015**, *85*, 344–350. [CrossRef]

20. Li, J.S.; Tang, Y.; Li, Z.T.; Ding, X.R.; Rao, L.S.; Yu, B.H. Effect of Quantum Dot Scattering and Absorption on the Optical Performance of White Light-Emitting Diodes. *IEEE Trans. Electron Devices* **2018**, *65*, 2877–2884. [CrossRef]

21. Chen, K.-J.; Han, H.-V.; Chen, H.-C.; Lin, C.-C.; Chien, S.-H.; Huang, C.-C.; Chen, T.-M.; Shih, M.-H.; Kuo, H.-C. White light emitting diodes with enhanced CCT uniformity and luminous flux using ZrO_2 nanoparticles. *Nanoscale* **2014**, *6*, 5378–5383. [CrossRef] [PubMed]

22. Li, J.S.; Tang, Y.; Li, Z.T.; Li, Z.; Ding, X.R.; Rao, L.S. Investigation of the Emission Spectral Properties of Carbon Dots in Packaged LEDs using TiO_2 Nanoparticles. *IEEE J. Sel. Top. Quantum Electron.* **2017**, *23*. [CrossRef]

23. Tang, Y.; Li, Z.; Li, Z.T.; Li, J.S.; Yu, S.D.; Rao, L.S. Enhancement of luminous efficiency and uniformity of CCT for quantum dot-converted LEDs by incorporating with ZnO nanoparticles. *IEEE Trans. Electron Devices* **2018**, *65*, 158–164. [CrossRef]

24. Hsu, S.-C.; Chen, Y.-H.; Tu, Z.-Y.; Han, H.-V.; Lin, S.-L.; Chen, T.-M.; Kuo, H.-C.; Lin, C.-C. Highly stable and efficient hybrid quantum dot light-emitting diodes. *IEEE Photonics J.* **2015**, *7*. [CrossRef]
25. Luo, X.; Fu, X.; Chen, F.; Zheng, H. Phosphor self-heating in phosphor converted light emitting diode packaging. *Int. J. Heat Mass Transf.* **2013**, *58*, 276–281. [CrossRef]
26. Li, J.; Tang, Y.; Li, Z.; Cao, K.; Yan, C.; Ding, X. Full spectral optical modeling of quantum-dot-converted elements for light-emitting diodes considering reabsorption and reemission effect. *Nanotechnology* **2018**, *29*, 295707. [CrossRef] [PubMed]

 *nanomaterials*

Article

Quantum Efficiency Enhancement of a GaN-Based Green Light-Emitting Diode by a Graded Indium Composition p-Type InGaN Layer

Quanbin Zhou [1], Hong Wang [1,2,*], Mingsheng Xu [1,2] and Xi-Chun Zhang [1,*]

[1] Engineering Research Center for Optoelectronics of Guangdong Province, School of Electronics and Information Engineering, South China University of Technology, Guangzhou 510640, China; zhouquanbin86@163.com (Q.Z.); mshxu@163.com (M.X.)

[2] School of Physics and Optoelectronics, South China University of Technology, Guangzhou 510640, China

* Correspondence: phhwang@scut.edu.cn (H.W.); xchzhang@scut.edu.cn (X.-C.Z.); Tel.: +86-136-0006-6193 (H.W.)

Received: 27 May 2018; Accepted: 7 July 2018; Published: 9 July 2018

Abstract: We propose a graded indium composition p-type InGaN (p-InGaN) conduction layer to replace the p-type AlGaN electron blocking layer and a p-GaN layer in order to enhance the light output power of a GaN-based green light-emitting diode (LED). The indium composition of the p-InGaN layer decreased from 10.4% to 0% along the growth direction. The light intensity of the LED with a graded indium composition p-InGaN layer is 13.7% higher than that of conventional LEDs according to the experimental result. The calculated data further confirmed that the graded indium composition p-InGaN layer can effectively improve the light power of green LEDs. According to the simulation, the increase in light output power of green LEDs with a graded indium composition p-InGaN layer was mainly attributed to the enhancement of hole injection and the improvement of the radiative recombination rate.

Keywords: p-type InGaN; graded indium composition; hole injection; quantum efficiency; green LED

---

1. Introduction

GaN-based light-emitting diodes (LEDs) have attracted considerable attention and have been seen as a promising replacement for conventional light sources in the last few decades [1,2]. The efficiency of blue LEDs is very high, and blue LEDs have been commercially used in many fields, such as lighting [3–6], display [7,8], light communication [9,10], back lighting [11,12], and so on. However, the internal quantum efficiency (IQE) of GaN-based green LEDs is still lower than that of blue LEDs, which is called the "Green Gap" [13]. It obstructs the green LED to be applied in Red-Green-Blue (RGB) lighting, full-color displays, and visible-light communication. A large polarization field [14–16] and poor crystal quality [17,18] are the main reasons for the low IQE of green LEDs with a high indium composition. In fact, the poor hole injection also plays an important role in the low quantum efficiency of GaN-based LEDs. Many researchers have proposed various methods to solve this problem based on band engineering of the electron blocking layer (EBL). Kim et al. employed an active-layer-friendly lattice-matched InAlN EBL to improve the quantum efficiency of green LEDs [19]. A graded superlattice AlGaN/GaN inserting layer was proposed by J. Kang et al. to enhance the efficiency of hole injection and performance of green LEDs [20]. An InAlGaN/GaN superlattice [21], an AlGaN/InGaN superlattice [22,23], and a composition-graded AlGaN EBL [24–26] were also employed to reduce the potential barrier of holes without damaging the electron confinement. A recently proposed method to improve the properties of p-type GaN is polarization doping [27]. It uses the internal polarization of the structures and material composition grading to induce free electrons or holes [28]. However, the growth temperature of AlGaN is always high in order to improve the crystal quality. The high indium content InGaN/GaN multiple quantum well (MQW) of green LEDs will be damaged during the high temperature process [29–32]. There are few reports about the p-type layer structure designed to improve the hole injection of GaN-based LEDs [33].

In this paper, we designed a new structure of p-type InGaN (p-InGaN) conduction layer with a graded indium composition to replace the conventional p-type AlGaN (p-AlGaN) EBL and p-type GaN (p-GaN) conduction layer of GaN-based green LEDs. The effect of the graded indium composition p-InGaN conduction layer on the light output power of green LEDs is studied by experiments and simulations.

2. Experimental Details

The LED samples were grown on (0001)-oriented sapphire substrates by an AIXTRON close-coupled showerhead metal-organic chemical vapor deposition (MOCVD) reactor (MOCVD, AIXTRON Inc., Herzogenrath, Germany). The trimethylgallium (TMGa), trimethylaluminum (TMAl), trimethylindium (TMIn), and ammonia (NH$_3$) were used as sources of gallium, aluminum, indium, and nitrogen, respectively. Silane (SiH$_4$) and bicyclopentadienyl magnesium (Cp$_2$Mg) were used as n-type and p-type doping sources, respectively. The epitaxial structure of conventional LEDs consisted of a 30 nm thick GaN nucleation layer grown at 530 °C, a 3 µm thick undoped GaN (u-GaN) buffer layer grown at 1100 °C, a 4 µm thick Si-doped n-type GaN (n-GaN) layer with 8×10^{18} cm^{-3} doping concentration grown at 1080 °C, five pairs of 3 nm and 10 nm thick In$_{0.22}$Ga$_{0.78}$N/GaN MQWs active layers, a 20 nm thick p-type Al$_{0.15}$Ga$_{0.85}$N (p-AlGaN) electron blocking layer grown at 1040 °C, and a 180 nm thick p-GaN layer grown at 940 °C. The doping concentration of the p-AlGaN and p-GaN layers was 5×10^{18} cm^{-3}. The conventional LEDs were denoted as sample A. Figure 1a is the profile of sample A. We also prepared green LEDs with a graded indium composition p-InGaN conduction layer, which are denoted as sample B. The growth conditions of sample B were similar to that of sample A except for the p-type layers. As shown in Figure 1b, the p-AlGaN and p-GaN layers were replaced by a p-InGaN single layer with a thickness of 200 nm. The growth temperature of p-InGaN layer was 860 °C, and the flow of TMIn changed from 175 sccm to 0 sccm along the growth direction in order to obtain a p-InGaN layer with a graded indium composition. Furthermore, we grew another sample with only a p-InGaN film on u-GaN in order to determine the indium composition of the p-InGaN layer. The TMIn flow of the p-InGaN film remained at 175 sccm, and the growth temperature was 860 °C. Figure 1c shows the schematic diagram of the p-InGaN film, which is denoted as sample C. The thickness of the p-InGaN layer in sample C was 80 nm.

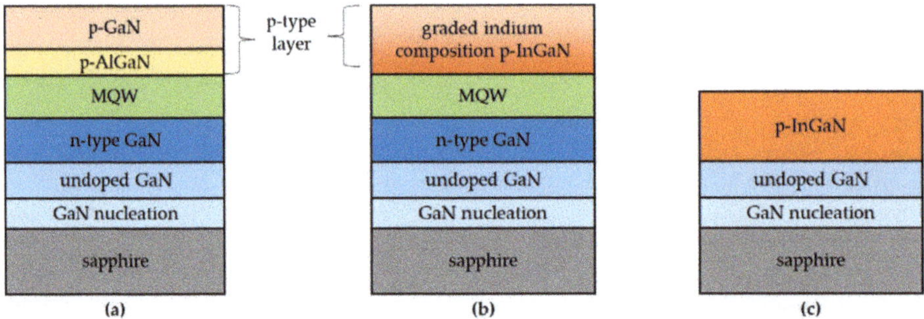

Figure 1. Schematic diagrams for the LED and p-InGaN samples: (**a**) sample A; (**b**) sample B; and (**c**) sample C.

After the growth of the LED structure of sample A and sample B, the epitaxial wafers were treated together to LED chips. The epitaxial wafers were first cleaned in acetone, isopropanol, deionized water, and partly etched to n-GaN by an inductively coupled plasma (ICP) system. Next, these samples were cleaned by a sulfuric acid peroxide mixture at 60 °C, ammonia water at 35 °C, deionized water at room temperature, and dried by nitrogen gas. Sequentially, the 80 nm indium tin oxide (ITO) transparent conductive electrodes (TCEs) were evaporated by electron beam evaporation and then annealed at 600 °C for 3 min in a mixture of ambient N_2/O_2 (200:35). The ITO TCEs were selectively removed by wet chemical etching. Then, the SiO_2 passivation layer was deposited by a plasma enhanced chemical vapor deposition (PECVD) system. Finally, Cr (50 nm)/Al (200 nm)/Ti (100 nm)/Au (100 nm) electrode layers were deposited by electron beam evaporation. The size of the LED chip was 1.14 mm × 1.14 mm.

3. Results and Discussion

We characterized the crystallography of the p-InGaN film in sample C by a Rigaku high-resolution X-ray diffraction (XRD, Rigaku Inc., Tokyo, Japan) with Cu K_α irradiation at 40 kV and 100 mA. Figure 2 is the HRXRD ω-2θ scan of p-InGaN film in sample C. The main peak and the secondary peak are GaN and InGaN, respectively. Chen et al. and Zhou et al. calculated the indium composition of the $In_xGa_{1-x}N$ film using Vegard's law and the XRD data [34–36]. We evaluated the indium content of the p-InGaN film by the separation between GaN and InGaN peaks in the XRD spectra. The indium content of p-InGaN in sample C was 10.4%. No evidence of phase separation could be found in the XRD spectrum. When the indium content is not too high, it almost linearly increases as the TMIn flow increasing [37]. Because the p-InGaN in sample B had similar growth conditions to sample C, except for the TMIn flow changing form 175 sccm to 0 sccm, the indium content in p-InGaN of sample B was from 10.4% to 0% along the growth direction. The expected indium content profiles of samples B and C are shown in the inset of Figure 2.

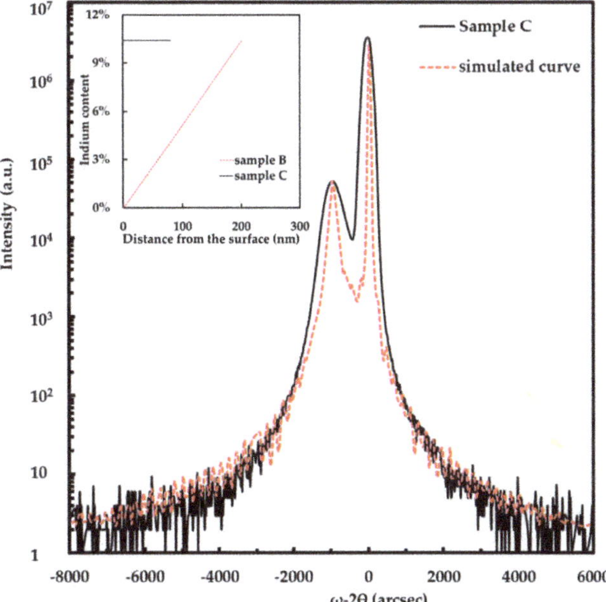

Figure 2. High-resolution X-ray diffraction ω-2θ scan of sample C. The dashed line in red is the simulated curve. The inset shows the indium content profiles of samples B and C.

The light output properties of the two LED samples are shown in Figure 3. Figure 3a shows the light output power as a function of injection current. Sample B had a great enhancement of light output power for the whole injection current range compared with the sample A. The light output power of sample B was 13.7% larger than that of sample A at a 300 mA injection current. The result revealed that the graded indium composition p-InGaN conduction layer was beneficial for enhancing the light output power of a GaN-based green LED. It is clear that the peak intensity of electroluminescence (EL) spectra at 300 mA of sample B was stronger than that of sample A, as shown in Figure 3b. Furthermore, the peak wavelengths of samples A and B were 529 nm and 534 nm respectively.

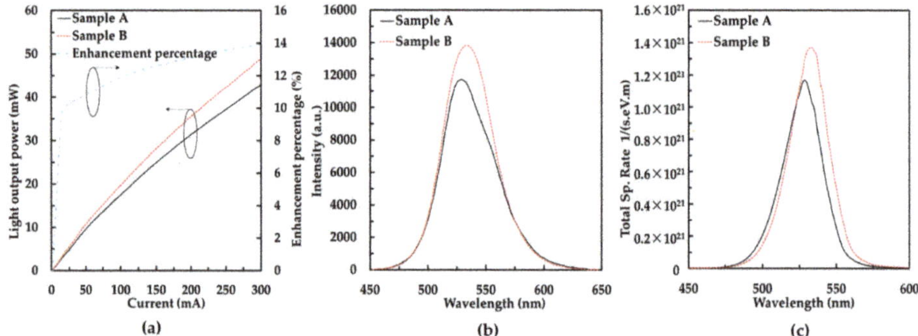

Figure 3. Light output properties of samples A and B: (**a**) measured light output power and enhancement percentage as a function of the injection current; (**b**) measured EL spectra at 300 mA; and (**c**) calculated EL spectra at 300 mA.

In order to better understand the influence of the graded indium composition p-InGaN conduction layer on the performance of GaN-based green LEDs, we performed a numerical simulation using APSYS software (version 2010, Crosslight Software Inc., Burnaby, BC, Canada). The software self-consistently solves the Poisson equation, continuity equation, and Schrödinger equation with boundary conditions [38,39]. Here, the band-offset ratio between the conduction band and the valence band for InGaN/GaN MQWs was 70% [40]. In addition, the values of the Shockley-Read-Hall (SRH) recombination lifetime and Auger coefficients assumed in this simulation were 50 ns and 1×10^{-30} cm^6/s, respectively [41–43]. Figure 3c displays the calculated EL spectra at 300 mA, which shows a similar trend of enhancement in light intensity with measured results. The simulated energy band diagrams at 300 mA of sample A and sample B are shown in Figure 4. The solid lines are conduction bands and valence bands and the dashed lines are the quasi-fermi level. In sample A, the height of the barrier, which obstructs holes injecting into the MQW, is 330 meV. However, in sample B, the height of the barrier is 285 meV when the p-AlGaN EBL and p-GaN conduction layer are replaced by the graded indium composition p-InGaN conduction layer. The lower potential barrier is beneficial for holes injecting into the MQW.

Figure 5 plots the carrier concentration distribution and radiation recombination rate distribution in the MQW of sample A and sample B from n-side to p-side. The holes of samples A and B both accumulate in the quantum well near the p-type layer. As shown in Figure 5a, the hole concentration in the quantum well near the p-type layer of sample B was much larger than that of sample A. The function of the EBL is to reduce the electron overflow leakage. Therefore, sample A had a little more electron concentration in the MQW, as shown in Figure 5b. The higher carrier concentration led to a larger radiation recombination rate of GaN-based LED. From Figure 5c, we could find that the radiation recombination rate of sample B was much bigger than that of sample A in the quantum well near the p-type layer. As a result, sample B had a higher total radiation recombination rate compared to sample A. The simulation data demonstrated that the graded indium composition p-InGaN conduction layer could enhance the hole injection and radiation recombination rate of GaN-based green LEDs.

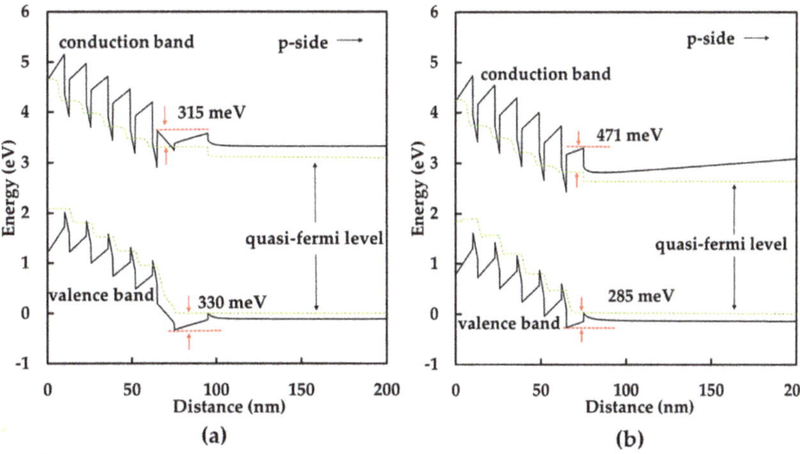

Figure 4. Calculated energy band diagrams at 300 mA of (**a**) sample A; and (**b**) sample B.

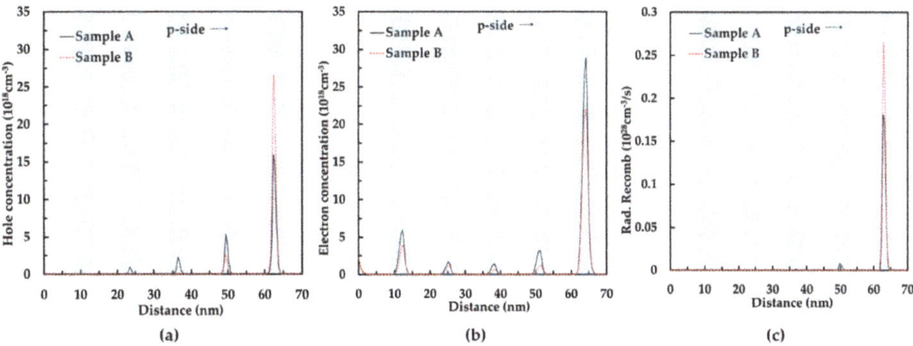

Figure 5. (**a**) Hole concentration distribution; (**b**) electron concentration distribution; and (**c**) radiation recombination rate distribution of samples A and B.

4. Conclusions

In conclusion, the light output properties of GaN-based green LEDs with and without a graded indium composition p-InGaN layer were numerically and experimentally investigated. Both the experimental results and simulated data revealed that the graded indium composition p-InGaN conduction layer can promote the light output power of green LEDs. The light output power of green LEDs with a p-InGaN conduction layer was enhanced by 13.7% compared to the conventional LED, according to the experimental data. The simulation results demonstrated that the improvement in light output property was mainly due to the increase of hole injection and the enhancement of the radiative recombination rate.

Author Contributions: Conceptualization, Q.Z. and H.W.; Data curation, Q.Z.; Formal analysis, Q.Z. and M.X.; Investigation, Q.Z.; Methodology, Q.Z. and H.W.; Supervision, X.-C.Z.; Writing—original draft, Q.Z. and M.X., Q.Z. designed the experiment, prepared the samples, performed the measurements and the simulation, and wrote the manuscript. H.W. contributed to the conception of the study and designed the experiment. M.X. contributed to the data analysis and wrote the manuscript. X.-C.Z. supervised the study and reviewed the manuscript.

Funding: This research was funded by the National High Technology Research and Development Program of China (No. 2014AA032609), in part by the National Natural Science Foundation of China (No. 61504044), the Key Technologies R&D Programs of Guangdong Province (Nos. 2014B0101119002, 2017B010112003), the Applied technologies R&D Projects of Guangdong Province (Nos. 2015B010127013, 2016B010123004), the technologies R&D Programs of Guangzhou City (Nos. 201504291502518, 201604046021), and the Science and Technology Development Special Fund Projects of Zhangshan City (Nos. 2017F2FC0002, 2017A1009).

Conflicts of Interest: The authors declare no conflict of interest.

References

1. Wang, L.; Liu, Z.; Zhang, Z.H.; Tian, Y.D.; Yi, X.; Wang, J.; Li, J.; Wang, G. Interface and photoluminescence characteristics of graphene-(GaN/InGaN)$_n$ multiple quantum wells hybrid structure. *J. Appl. Phys.* **2016**, *119*, 611–622. [CrossRef]

2. Wang, Q.; Ji, Z.; Zhou, Y.; Wang, X.; Liu, B.; Xu, X.; Gao, X.; Leng, J. Diameter-dependent photoluminescence properties of strong phase-separated dual-wavelength InGaN/GaN nanopillar LEDs. *Appl. Surf. Sci.* **2017**, *410*, 196–200. [CrossRef]

3. Pimputkar, S.; Speck, J.S.; Denbaars, S.P.; Nakamura, S. Prospects for LED lighting. *Nat. Photonics* **2009**, *3*, 180–182. [CrossRef]

4. Sinnadurai, R.; Khan, M.K.A.A.; Azri, M.; Vikneswaran, V. Development of White LED down Light for Indoor Lighting. In Proceedings of the IEEE Conference on Sustainable Utilization and Development in Engineering and Technology (STUDENT), Kuala Lumpur, Malaysia, 6–9 October 2012; pp. 242–247.

5. Moon, S.; Koo, G.B.; Moon, G.W. A new control method of interleaved single-stage flyback AC–DC converter for outdoor LED lighting systems. *IEEE Trans. Power Electron.* **2013**, *28*, 4051–4062. [CrossRef]

6. Poulet, L.; Massa, G.D.; Morrow, R.C.; Bourget, C.M.; Wheeler, R.M.; Mitchell, C.A. Significant reduction in energy for plant-growth lighting in space using targeted LED lighting and spectral manipulation. *Life Sci. Space Res.* **2014**, *2*, 43–53. [CrossRef]

7. Sanchot, A.; Consonni, M.; Calvez, S.L.; Robin, I.C.; Templier, F. Color conversion using quantum dots on high-brightness GaN LED arrays for display application. *MRS Proc.* **2015**, *1788*, 19–21. [CrossRef]

8. Lee, K.J. Flexible GaN LED on a polyimide substrate for display applications. *Proc. SPIE* **2012**, *8268*, 52.

9. Ferreira, R.; Xie, E.; Mckendry, J.; Rajbhandari, S. High bandwidth GaN-based micro-LEDs for multi-Gb/s visible light communications. *IEEE Photonics Technol. Lett.* **2016**, *28*, 2023–2026. [CrossRef]

10. Du, C.; Huang, X.; Jiang, C.; Pu, X.; Zhao, Z.; Jing, L.; Hu, W.; Wang, Z.L. Tuning carrier lifetime in InGaN/GaN LEDs via strain compensation for high-speed visible light communication. *Sci. Rep.* **2016**, *6*, 37132. [CrossRef] [PubMed]

11. Anandan, M. Progress of LED backlights for LCDS. *J. Soc. Inf. Displ.* **2008**, *16*, 287–310. [CrossRef]

12. Soon, C.M. *White Light Emitting Diode as Liquid Crystal Display Backlight*; Massachusetts Institute of Technology: Cambridge, MA, USA, 2007.

13. Zhou, Q.; Xu, M.; Wang, H. Internal quantum efficiency improvement of InGaN/GaN multiple quantum well green light-emitting diodes. *Opto-Electron. Rev.* **2016**, *24*, 1–9. [CrossRef]

14. Young, N.G.; Farrell, R.M.; Iza, M.; Nakamura, S.; DenBaars, S.P.; Weisbuch, C.; Speck, J.S. Germanium doping of GaN by metalorganic chemical vapor deposition for polarization screening applications. *J. Cryst. Growth* **2016**, *455*, 105–110. [CrossRef]

15. Prajoon, P.; Nirmal, D.; Menokey, M.A.; Pravin, J.C. Efficiency enhancement of InGaN MQW LED using compositionally step graded InGaN barrier on SiC substrate. *J. Disp. Technol.* **2016**, *12*, 1117–1121. [CrossRef]

16. Xu, M.; Yu, W.; Zhou, Q.; Zhang, H.; Wang, H. Efficiency enhancement of GaN-based green light-emitting diode with PN-doped quantum barriers. *Mater. Express* **2016**, *6*, 533–537. [CrossRef]

17. Ren, P.; Zhang, N.; Xue, B.; Liu, Z.; Wang, J.; Li, J. A novel usage of hydrogen treatment to improve the indium incorporation and internal quantum efficiency of green InGaN/GaN multiple quantum wells simultaneously. *J. Phys. D Appl. Phys.* **2016**, *49*, 175101. [CrossRef]

18. Qiao, L.; Ma, Z.-G.; Chen, H.; Wu, H.-Y.; Chen, X.-F.; Yang, H.-J.; Zhao, B.; He, M.; Zheng, S.-W.; Li, S.-T. Effects of multiple interruptions with trimethylindium-treatment in the InGaN/GaN quantum well on green light emitting diodes. *Chin. Phys. B* **2016**, *25*, 107803. [CrossRef]

19. Kim, H.J.; Choi, S.; Kim, S.-S.; Ryou, J.-H.; Yoder, P.D.; Dupuis, R.D.; Fischer, A.M.; Sun, K.; Ponce, F.A. Improvement of quantum efficiency by employing active-layer-friendly lattice-matched InAlN electron blocking layer in green light-emitting diodes. *Appl. Phys. Lett.* **2010**, *96*, 101102. [CrossRef]

20. Kang, J.; Li, H.; Li, Z.; Liu, Z.; Ma, P.; Yi, X.; Wang, G. Enhancing the performance of green GaN-based light-emitting diodes with graded superlattice AlGaN/GaN inserting layer. *Appl. Phys. Lett.* **2013**, *103*, 102104. [CrossRef]

21. Lin, D.-W.; Tzou, A.-J.; Huang, J.-K.; Lin, B.-C.; Chang, C.-Y.; Kuo, H.-C. Greatly improved efficiency droop for InGaN-based green light emitting diodes by quaternary content superlattice electron blocking layer. In Proceedings of the International Conference on Numerical Simulation of Optoelectronic Devices (NUSOD), Taipei, Taiwan, 7–11 September 2015; pp. 15–16.

22. Yu, C.-T.; Lai, W.-C.; Yen, C.-H.; Chang, S.-J. Effects of ingan layer thickness of AlGaN/InGaN superlattice electron blocking layer on the overall efficiency and efficiency droops of GaN-based light emitting diodes. *Opt. Express* **2014**, *22*, A663–A670. [CrossRef] [PubMed]

23. Chen, F.-M.; Liou, B.-T.; Chang, Y.-A.; Chang, J.-Y.; Kuo, Y.-T.; Kuo, Y.-K. *Numerical Analysis of Using Superlattice-AlGaN/InGaN as Electron Blocking Layer in Green InGaN Light-Emitting Diodes*; International Society for Optics and Photonics: Bellingham, WA, USA, 2013; pp. 862526–862527.

24. Lei, Y.; Liu, Z.; He, M.; Yi, X.; Wang, J.; Li, J.; Zheng, S.; Li, S. Enhancement of blue InGaN light-emitting diodes by using AlGaN increased composition-graded barriers. *J. Semicond.* **2015**, *36*, 054006. [CrossRef]

25. Wang, C.H.; Ke, C.C.; Lee, C.Y.; Chang, S.P. Hole injection and efficiency droop improvement in InGaN/GaN light-emitting diodes by band-engineered electron blocking layer. *Appl. Phys. Lett.* **2010**, *97*, 261103. [CrossRef]

26. Kuo, Y.K.; Chang, J.Y.; Tsai, M.C. Enhancement in hole-injection efficiency of blue InGaN light-emitting diodes from reduced polarization by some specific designs for the electron blocking layer. *Opt. Lett.* **2010**, *35*, 3285–3287. [CrossRef] [PubMed]

27. Kivisaari, P.; Oksanen, J.; Tulkki, J. Polarization doping and the efficiency of III-nitride optoelectronic devices. *Appl. Phys. Lett.* **2013**, *103*, 1029. [CrossRef]

28. Li, S.; Zhang, T.; Wu, J.; Yang, Y.; Wang, Z.; Wu, Z.; Chen, Z.; Jiang, Y. Polarization induced hole doping in graded $Al_xGa_{1-x}N$ ($x = 0.7{\sim}1$) layer grown by molecular beam epitaxy. *Appl. Phys. Lett.* **2013**, *102*, 132103. [CrossRef]

29. Oh, M.S.; Kwon, M.K.; Park, I.K.; Baek, S.H.; Park, S.J.; Lee, S.H.; Jung, J.J. Improvement of green LED by growing p-GaN on $In_{0.25}GaN/GaN$ MQWs at low temperature. *J. Cryst. Growth* **2006**, *289*, 107–112. [CrossRef]

30. Lee, W.; Limb, J.; Ryou, J.H.; Yoo, D.; Chung, T.; Dupuis, R.D. Influence of growth temperature and growth rate of p-GaN layers on the characteristics of green light emitting diodes. *J. Electron. Mater.* **2006**, *35*, 587–591. [CrossRef]

31. Lee, W.; Limb, J.; Ryou, J.H.; Yoo, D.; Chung, T.; Dupuis, R.D. Effect of thermal annealing induced by p-type layer growth on blue and green LED performance. *J. Cryst. Growth* **2006**, *287*, 577–581. [CrossRef]

32. Ju, J.W.; Zhu, J.; Kim, H.S.; Lee, C.R.; Lee, I.H. Effects of p-GaN growth temperature on a green InGaN/GaN multiple quantum well. *J. Korean Phys. Soc.* **2007**, *50*, 810. [CrossRef]

33. Lin, Z.; Wang, H.; Lin, Y.; Yang, M.; Li, G.; Xu, B. A new structure of p-GaN/InGaN heterojunction to enhance hole injection for blue GaN-based LEDs. *J. Phys. D Appl. Phys.* **2016**, *49*, 285106. [CrossRef]

34. Qin, Z.; Chen, Z.; Tong, Y.; Lu, S.; Zhang, G. Estimation of InN phase inclusion in InGaN films grown by MOVPE. *Appl. Phys. A* **2002**, *74*, 655–658. [CrossRef]

35. Chen, Z.Z.; Qin, Z.X.; Hu, X.D.; Yu, T.J.; Yang, Z.J.; Tong, Y.Z.; Ding, X.M.; Zhang, G.Y. Study of photoluminescence and absorption in phase-separation InGaN films. *Phys. B Condens. Matter* **2004**, *344*, 292–296. [CrossRef]

36. Zhou, S.Q.; Wu, M.F.; Hou, L.N.; Yao, S.D.; Ma, H.J.; Nie, R.; Tong, Y.Z.; Yang, Z.J.; Yu, T.J.; Zhang, G.Y. An approach to determine the chemical composition in InGaN/GaN multiple quantum wells. *J. Cryst. Growth* **2004**, *263*, 35–39. [CrossRef]

37. Guo, Y.; Liu, X.L.; Song, H.P.; Yang, A.L.; Xu, X.Q.; Zheng, G.L.; Wei, H.Y.; Yang, S.Y.; Zhu, Q.S.; Wang, Z.G. A study of indium incorporation in In-rich InGaN grown by MOVPE. *Appl. Surf. Sci.* **2010**, *256*, 3352–3356. [CrossRef]

38. *Apsys, Version 2010 Software for Electrical, Optical and Thermal Properties of Compound Semiconductor Devices*; Crosslight Software Inc.: Burnaby, BC, Canada, 2010.

39. Zhao, H.; Arif, R.A.; Ee, Y.K.; Tansu, N. Self-consistent analysis of strain-compensated InGaN-AlGaN quantum wells for lasers and light-emitting diodes. *IEEE J. Quantum Electron.* **2008**, *45*, 66–78. [CrossRef]

40. Li, J.; Guo, Z.; Li, F.; Lin, H.; Li, C.; Xiang, S.; Zhou, T.; Wan, N.; Liu, Y. Performance enhancement of blue light-emitting diodes by using special designed n and p-type doped barriers. *Superlattices Microstruct.* **2015**, *85*, 454–460. [CrossRef]

41. Zhang, M.; Yun, F.; Li, Y.; Ding, W.; Wang, H.; Zhao, Y.; Zhang, W.; Zheng, M.; Tian, Z.; Su, X. Luminescence properties of InGaN-based dual-wavelength light-emitting diodes with different quantum-well arrangements. *Phys. Status Solidi* **2015**, *212*, 954–959. [CrossRef]

42. Cheng, L.; Wu, S.; Chen, H.; Xia, C.; Kong, Q. Investigation of whether uniform carrier distribution in quantum wells can lead to higher performance in InGaN light-emitting diodes. *Opt. Quantum Electron.* **2016**, *48*, 1–9. [CrossRef]

43. Piprek, J. Efficiency droop in nitride-based light-emitting diodes. *Phys. Status Solidi* **2010**, *207*, 2217–2225. [CrossRef]

Article

Enhancing GaN LED Efficiency through Nano-Gratings and Standing Wave Analysis

Xiaomin Jin *, Simeon Trieu, Gregory James Chavoor and Gabriel Michael Halpin

Electrical Engineering Department, California Polytechnic State University, San Luis Obispo, CA 93407, USA; simeon.trieu@gmail.com (S.T.); gjchavoor@calpoly.edu (G.J.C.); gabriel.halpin@gmail.com (G.M.H.)
* Correspondence: xjin@calpoly.edu; Tel.: +01-805-756-7046

Received: 16 November 2018; Accepted: 10 December 2018; Published: 13 December 2018

Abstract: Based on our recent work, this paper reviews our theoretical study on gallium nitride (GaN) light-emitting-diode (LED). The focus of the paper is to improve LED light extraction efficiency through various nano-grating designs. The gratings can be designed at different locations, such as at the top, the middle, and the bottom, on the LED. They also can be made of different materials. In this study, we first present a GaN LED error-grating simulation model. Second, nano Indium Tin Oxide (ITO) top gratings are studied and compared with conventional LED (CLED) using standing wave analysis. Third, we present results related to a patterned sapphire substrate (PSS), SiO_2 Nanorod array (NR), and Ag bottom reflection layer. Finally, we investigate the nano-top ITO grating performance over different wavelengths to validate our design simulation, which focusing on a single wavelength of 460 nm.

Keywords: GaN; LED; nano-grating

1. Introduction

Gallium nitride (GaN) based light-emitting-diodes (LEDs) continue to prove themselves increasingly useful in the world of solid-state lighting. Although highly efficient, scientists continue to investigate ways to increase their internal and external quantum efficiencies. In general, the efficiency of an LED is limited by the maximum angle that light can escape from the surface as defined by Snell's law. Since GaN has a high refractive index compared to air, light can only escape the LED if it approaches the surface within $+/- 23°$ of the normal incidence. The large difference in the refractive index of air and GaN results in a low critical angle. Therefore, the low critical angle traps light within the LED, reduces light output, and increases the device temperature. Improving the light extraction efficiency of GaN LEDs has been approached by several different ways. Early approaches for improvement of light efficiency include surface roughness or texturing [1]. Then, adding a material with a refractive index between the indices of GaN and air to the LED surface increases the limiting angle [2]. Photonic crystal top grating [3] and nano pillar Multiple-quantum-well (MQW) [4] have also been proposed. Recently, several research groups have used various gratings as methods for increasing light output and placed these gratings either on the surface [5], the bottom [5,6], or the sidewalls [7] of the LED.

Fabrication techniques have improved in the recent years. They have allowed design structures, such as periodic top gratings [8–13], patterned sapphire substrates (PSS) [14–18], and reflection layers (R) [19–23], to improve the external quantum efficiency by enhancing light extraction. One study showed that moving an Ag-based reflective mirror from below to above the sapphire substrate increases light extraction by 21% [24]. Recently, a SiO_2 nano-rod array (NR) was also used in GaN LED [25–28]. As technology progresses, more complex structures are designed using a multi-nano structure in a single GaN LED. For example, in 2018, GaN LEDs with a hybrid structure were fabricated and studied, which were a combination of sidewalls and microhole arrays [26]. In general, a cost-effective method to design high efficiency GaN LEDs is still highly desired. A more comprehensive GaN LED design simulation or theoretical study, including several nano-stuctures, is very valuable in this case before the fabrication.

Currently, no study fully examines or directly compares the above several light extraction efficiency (LEE) improvement methods theoretically. For example, which method is more efficient, and whether they can be combined. How does an Ag reflector affect light extraction in those structures? In this paper, we present comprehensive simulation results of Indium Tin Oxide (ITO) top nano-grating, and a direct comparison of GaN LED with a patterned sapphire substrate, SiO_2 nano-rod array, and Ag bottom reflection layer. Additionally, we find that among all of the above technologies, NR has more potential of improving the light extraction efficiency.

2. GaN LED Grating Simulation with the Error Grating Model

2.1. Basic Structure of GaN LED

The GaN LED models used in this paper were built through 2D finite difference time domain (FDTD) analysis. We used FDTD to provide a solution to Maxwell's equations by using Yee's mesh, a system in which the E- and H-field components are solved based on the previous spatial E- and H-field components. FDTD also solves Maxwell's equations on a point by point basis and can accurately simulate the effects of nano-gratings in the LED, such as reflection due to linear dispersion or total internal reflection, transmission of escaping light from the LED, and scattering at the grating. Since FDTD decomposes space and time into separate components, the model is meshed into small cells whose side length must be much smaller than the wavelength of light to obtain accurate results. Using this method, we can simulate the average power emission of LEDs.

As shown in Figure 1, light is generated in the multiple quantum well (MQW) region between the positive and negatively doped GaN regions of the diode. Since the computer model simulates light propagation in the LED, we assign all light to emerge as a continuous wave (CW) from the middle of the MQW region. The model is built as a 2D representative section of the whole LED and implements a uniform light distribution across the MQW region. The four green bars around the edge of the LED in Figure 1 are the light monitors. They measure the intensity of light emerging from top, bottom, and sides of the LED. The top monitor always sits 560 nm above the top non-grating layer of the LED, 100 nm higher than the tallest grating triangles. The bottom and side monitors sit 100 nm from the LED body. These positions were held constant throughout this research to make sure that each set of simulations has only limit variables that can change the intensity of the light emitted from the LED.

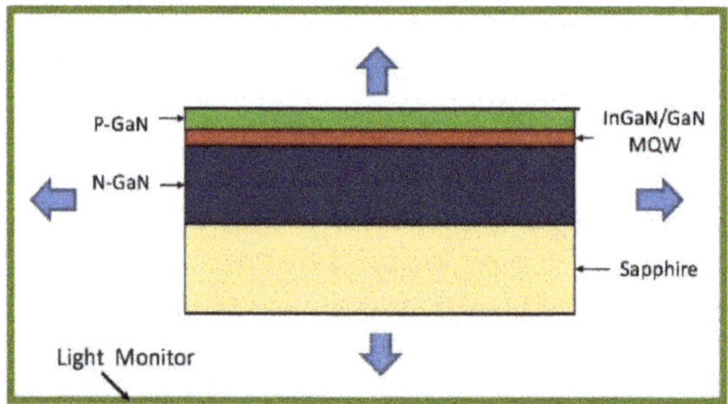

Figure 1. Structure and basic simulation model of conventional Gallium nitride (GaN) based light-emitting-diode (LED).

The conventional GaN LED model, as shown in Figure 1, consists of a p-GaN layer, InGaN/GaN MQW, n-GaN layer, and sapphire substrate, whose thicknesses and refractive indices are listed in Table 1. The sapphire substrate thickness was set to 80 μm.

Table 1. LED material thickness and refractive indices at λ = 460 nm.

Material	Sample Thickness (µm)	Refractive Index
ITO	0.23	2.1
p-GaN	0.2	2.5
InGaN/GaN MQWs	0.1	2.6
n-GaN	2	2.5
GaN	3	2.5
Sapphire	80	1.78
Ag	Reflection: 90%	

2.2. Error Grating Model

Usually, it is not very practical to fabricate all kinds of textures or patterns to select the optimized structure [29,30]. Therefore, we first simulated three typical gratings: Cylindrical pillar grating, conical pillar grating, and cylindrical nano-hole grating. Figure 2a–c are illustrations of the top grating implementing above three grating types. Additionally, we found that the conical pillar grating is more efficient compared to the cylindrical pillar grating, and a small grating period will yield a better light extraction efficiency [31,32]. In the grating simulation model, for top-grating, bottom grating, and nano-hole grating, there are three major parameters that affect the light extraction: The grating period (A), grating height (h), and bottom width (w).

We also proposed a top and bottom grating model with each cell randomly shifted a distance along the axis in varying degrees of randomization intensity to further understand the effects of fabrication defects on the top and bottom gratings, as shown in Figure 2d [5]. Usually, the widths of holes can be fabricated to great precision. Often the placement of holes causes concern, as it shifts the grating location and affects light extraction efficiency of otherwise ordered photonic crystal structures.

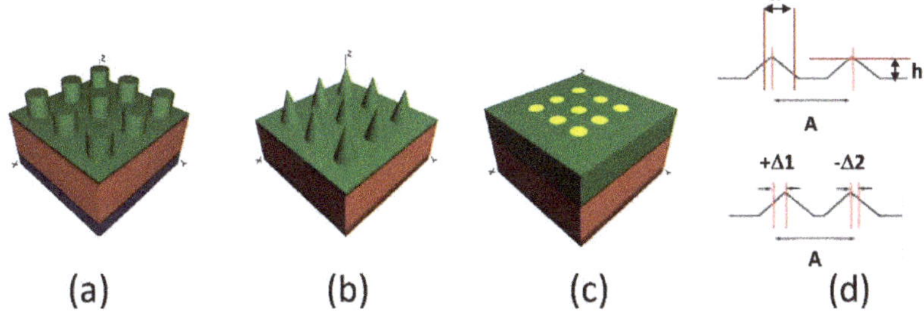

Figure 2. The schematic diagrams of the top grating simulation, (**a**) cylindrical pillar grating, (**b**) conical pillar grating, and (**c**) cylindrical nano-hole grating. (**d**) Error grating model: Normal reference grating model and error grating model with both positive and negative shifts.

The error grating model still makes use of the 2D FDTD method. Random displacements in position form the basis of the error grating model with a normal grating as a reference. Displacements can move either direction from the grating cell's original center point. The error grating model shows examples of a positive and negative Δx shift. This randomization then applies to all grating cells in the photonic crystal arrangement with Equation (1):

$$x_{pos} = N \times \text{period} + (2 \times \text{rand} -1) \times R \times \text{period} \tag{1}$$

where N is an integer index defining the original grating cell location, the period is the grating period (A), rand is a pseudo-randomly generated number from 0 to 1, and R is the randomization factor from 0 to 1. The quantity, Δx, in Figure 2d represents the $(2 \times \text{rand} - 1) \times R \times \text{period}$ in Equation (1). By varying R, which applies to all grating cells, from 0 to 1 in 40 steps, the individual rand factor can be emphasized or deemphasized. This process repeats for each of the grating models to calculate the light extraction efficiency variation.

The randomization creates local variation of the grating structure, *A* and *w*. Our simulation of several top and bottom grating shows that randomization in gratings appear to help the light extraction efficiency, peaking at about a randomization factor of R = 10% in most simulations [5]. In essence, a slight random variation or fabrication defect in grating cells would not only be beneficial, but also desirable for many top and bottom grating types up to a variation of 10–15% for most double grating cases. A double grating case, such as top and bottom gratings, usually optimize separately, and a small local perturbation could result in more matching gratings and introduce local light extraction improvement.

Randomization of grating cells increases the light extraction efficiency while having the added benefit of alleviating some of the fabrication complexities demanded by strict periodicities in photonic crystal LEDs. We believe the LED error grating model presents a unique model to analyze fabrication defects associated with laser positioning error and randomizations from chemical etching.

3. Results

3.1. Top ITO Layer with/without Grating Using Standing Wave Analysis

In our ITO nano-grating device design, it is very important to keep a layer of ITO at the bottom of the grating. This fixed thickness is used to prevent the P-GaN layer from being damaged in the etching process and protect the overall device charaterization. Furthermore, the ITO layer also acts as the current injection layer to protect the LED I-V charaterization from being affected by the nano-structure. The ITO top grating studied is shown in Figure 3. In this paper, we focus on the conical ITO study. More simulation results regrading cylindrical top ITO can be found in reference [33].

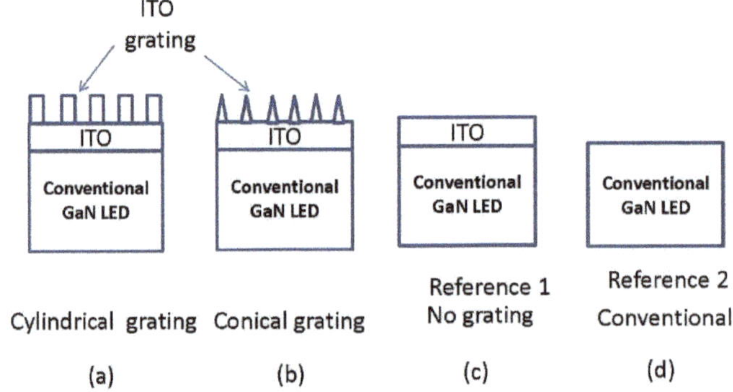

Figure 3. The schematic diagrams of the top grating simulation of (**a**) cylindrical pillar grating, (**b**) conical pillar grating, (**c**) conventional LED (CLED) with Indium Tin Oxide (ITO) layer, and (**d**) conventional LED.

Standing waves in an LED can either increase or decrease light output from an LED by constructively or destructively interfering at the top surface. By changing the thicknesses of the ITO and sapphire layers, we show that standing wave interference patterns exist in the LED. Studying this effect shows how much the standing wave pattern can either increase or decrease the light extraction from a chip. It is an important step in forming a complete light optimization study of a GaN LED, before any grating will be implemented on top of them, such as top ITO grating, and bottom nano Patterned Sapphire Substrates (PSS) [10].

We first chose to incrementally add ITO to the surface of the conventional LED (CLED) at a 460 nm wavelength, studying reference 1 in Figure 3c. If no standing wave is present, we should observe a linearly decreasing light output as the ITO is added. However, if a standing wave does exist, adding ITO should cause the output to sinusoidally fluctuate with a linear decrease in output intensity, as shown in Figure 4. As ITO is added, a roughly sinusoidal standing wave pattern emerges whose peak output intensity gradually decreases with increasing material thickness. From Figure 4, the standing wave period is 50 nm and one of the best-case ITO thicknesses occurs at 78 nm, the first constructive interference peak, while one of the worst-case ITO occurs at the 260 nm thickness. This yields a total light increase of 26.7% when the LED is changed from the worst- to best-case ITO thickness. The best-case values are fairly similar as well, from 1.26 a.u. to 1.23 a.u. as the ITO thickness increases. The more accurate values are summarized in Table 2, which presents the light output improvement by 9.6% over the conventional LED (reference 2). The 46 nm ITO thickness is a neutral case, with no improvement and no degradation.

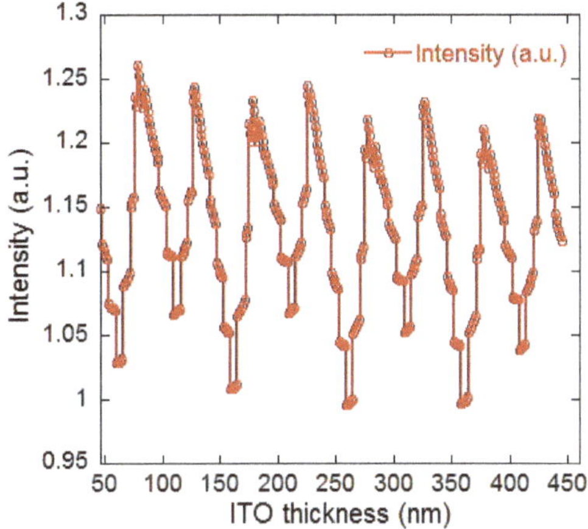

Figure 4. LED top output intensity as ITO thickness varies from 0 nm to 450 nm at 1 nm increments.

Table 2. Light output intensities at the best, worst, and neutral ITO thicknesses.

ITO Thickness	Output Intensity (a.u.)	% Improvement over CLED
46nm	1.1500	0.000
78nm	1.2606	9.617
260nm	0.9946	−13.513

Based on the ITO layer thickness study, the grating period is swept from 92 nm to 920 nm so we can find the grating period that maximizes the top light extraction. We chose this range for the grating because it sweeps the height and width of the cones in the grating from $\lambda/10$ to λ. The grating fill factor was held at 0.5 and the ratio of the cone height to the cone width was kept at 1 throughout the study to focus on the effect of the grating period and standing waves on the light output. Because the standing wave analysis showed material thicknesses for the ITO layer that both maximize and minimize light extraction before gratings are added, we studied the grating output at those key ITO thicknesses. As material thicknesses change, the grating output was compared to the instance of reference 1, that has the same ITO thickness. However, reference 2, the conventional LED, was used as an unchanging reference and its sapphire substrate thickness remains fixed at 10,000 nm in here to reduce the simulation time and space.

In our earlier study [31], we used a two-dimensional (2D) rigorous couple wave analysis (RCWA) GaN LED grating model to study top diffraction gratings, and compare none-grating, cylindrical-grating, and conical-grating cases. It showed that the cylindrical grating has better performance. Therefore, our study is focused on the conical nano-ITO grating case, and compares it with the none-grating (reference 1) and conventional LED (reference 2) [10].

Understanding where light is emitted is important, because some systems are designed to capture and direct light emitted from the sides and bottom of the LED. For these systems, maximizing total light output may be more important. Figure 5a shows that most of the light is emitting through the top, and is also most sensitive to the grating design, and the bottom is second. Side light emittance is very low and very insensitive to the grating implementation. Compared to the conventional LED (reference 2), top emittance also has the highest improvement, about 204% for both reference 1 and 2. Bottom light emittance for the best grating case can reach a 241% and 132% improvement according to reference 1 and reference 2, respectively. Left and right monitor simulations have some difference, which comes from slightly different monitor location placements. We also calculate reference 1 and reference 2's monitor outputs, as listed in Table 3 and calculated the total light extraction (top + bottom + left side + right side) according to the grating period variation (Figure 5b). It showed that gratings can significantly improve light output around the 500 nm grating period and increase light output at most periods. However, the results also show that the grating can reduce light output by more than 50% if a period of 400 nm or 900 nm is used. As expected, the grating on the best case ITO layer (78 nm) has the highest light output at 5.03 a.u., resulting in an approximately 199% improvement for reference 1 and 148% improvement for reference 2.

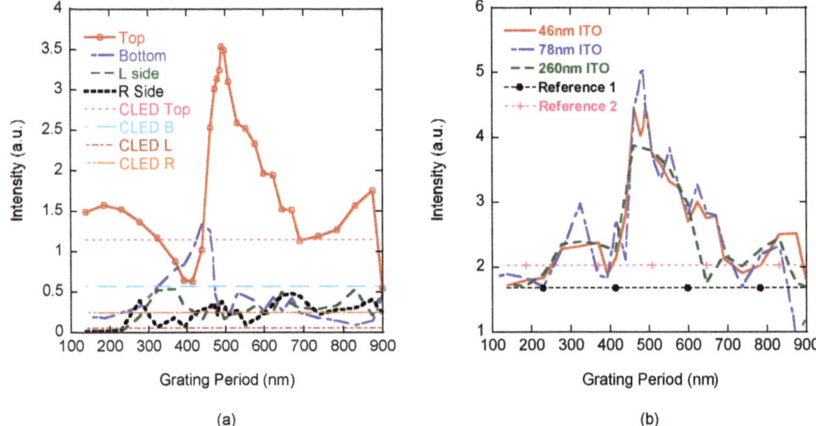

Figure 5. Light extracted from LED as the grating period is varied. (**a**) Light extraction from four monitors (top, bottom, left side, and right side) compared with conventional LED (reference 2) for an ITO thickness of 46 nm; (**b**) total light extraction for difference ITO thicknesses and compared with reference 1 with an ITO of 46 nm and reference 2.

Table 3. Light output intensities at difference monitors for ITO 46nm case and refrernces.

Monitor	Intensity (a.u.) reference 1	Intensity (a.u.) reference 2	Intensity (a.u.) for ITO 46 nm maximum value	Intensity (a.u.) for ITO 46 nm grating period 500 nm
Left	0.017	0.0605	0.533	0.16
Right	0.124	0.244	0.488	0.3
Bottom	0.3897	0.5726	1.322	0.20
Top	1.149	1.15	3.5778	3.54
Total	**1.680**	**2.027**	——	**4.20**

3.2. Nano-patterned Sapphire Substrates (PSS) Bottom Grating, SiO₂ Nano-rod Grating (NR), and Ag Reflector

We studied the conical bottom reflection grating and published top/bottom grating design simulation in reference [5]. The simulation results show that simple or direct combinations of the optimized top grating with the optimized bottom grating only produces a 42% light extraction improvement compared to the non-grating conventional LED, which is much lower than that of an optimized single grating case (about 165%). This is due to the mismatch of two grating parameters with the direct addition of the second grating structure, which changes optical modes in the LEDs. Therefore, it is very important to optimize both top and bottom gratings simultaneously for the double-grating design [5]. In this section, we optimized two or three structures simultaneously to achieve the final design comparison.

Most nano-bottom gratings are fabricated on the sapphire substrate, currently called a patterned sapphire substrate, as shown in Figure 6e. In this section, we simulated GaN LED structures with a patterned sapphire substrate and an embedded SiO_2 nanorod array. To confirm the accuracy of our software, we simulated a conventional LED (CLED) and PSS NR LED. Neither of these two structures contained an Ag reflection layer. Previous experiments have revealed that adding a patterned sapphire substrate and SiO_2 nanorod array increases the external quantum efficiency by 48% [34] or 56% [25] compared to CLED. Our corresponding simulation results show that the PSS NR structure increases light extraction by 52% compared to the CLED, from CLED 32.239 a.u. to PSS NR 48.925 a.u. These results agree very well with the published experimental data. Next, we optimized PSS by changing the period (d) from 2 μm to 3 μm in steps of 0.5 μm and width (w) from 1 μm to 2.5 μm in steps of 0.1 μm. A more stable point was obtained where $w = 2.5$ μm and $d = 3$ μm, which will be used in our later simulation. The more detailed results for various grating widths and periods for PSS are published in reference [35] and is not the focus of this paper. We also optimized the position of the SiO_2 NR array by moving it up and down in the z direction between 7 μm and 8 μm, and chose a maximum light output location of $z = 7.8$ μm for the final simulation.

Based on the basic simulation results, we fully investigated 12 state of the art LED structures, which include (Figure 6):

(a) Conventional LED (CLED);
(b) conventional LED with Ag between U-GaN and a sapphire substrate;
(c) conventional LED with Ag below a sapphire substrate;
(d) LED with an SiO_2 nanorod (NR)) array;
(e) LED with a patterned sapphire substrate (PSS);
(f) LED with a PSS and NR array (PSS NR);
(g) LED with PSS, Ag between U-GaN, and a sapphire substrate;
(h) LED with an NR array, Ag between U-GaN, and a sapphire substrate;
(i) LED with PSS, Ag below a sapphire substrate;
(j) LED with NR array, Ag below a sapphire substrate;
(k) LED with PSS, NR array, and Ag between U-GaN and a sapphire substrate; and
(l) LED with PSS, NR array, and Ag below a sapphire substrate.

Combinations of PSS and NR arrays have shown to increase crystal quality by reducing thread dislocations, and increase external quantum efficiency by scattering light [34]. We chose to simulate this structure because of its high efficiency. The conventional LEDs were simulated as a reference point. All simulations were compared with the conventional LED. If a structure cannot perform better than a conventional LED and is more difficult to fabricate, it may not be worth fabricating. Figure 6 does not show two key components of our simulation, the excitation source located at the MQW region and the time monitor located just above the p-GaN layer. The excitation source emits light at a wavelength of 460 nm, while the monitor records the average power extracted. Table 4 below provides a description of the conventional LED, including the substrate material, thickness, and refractive index, for this comparison simulation.

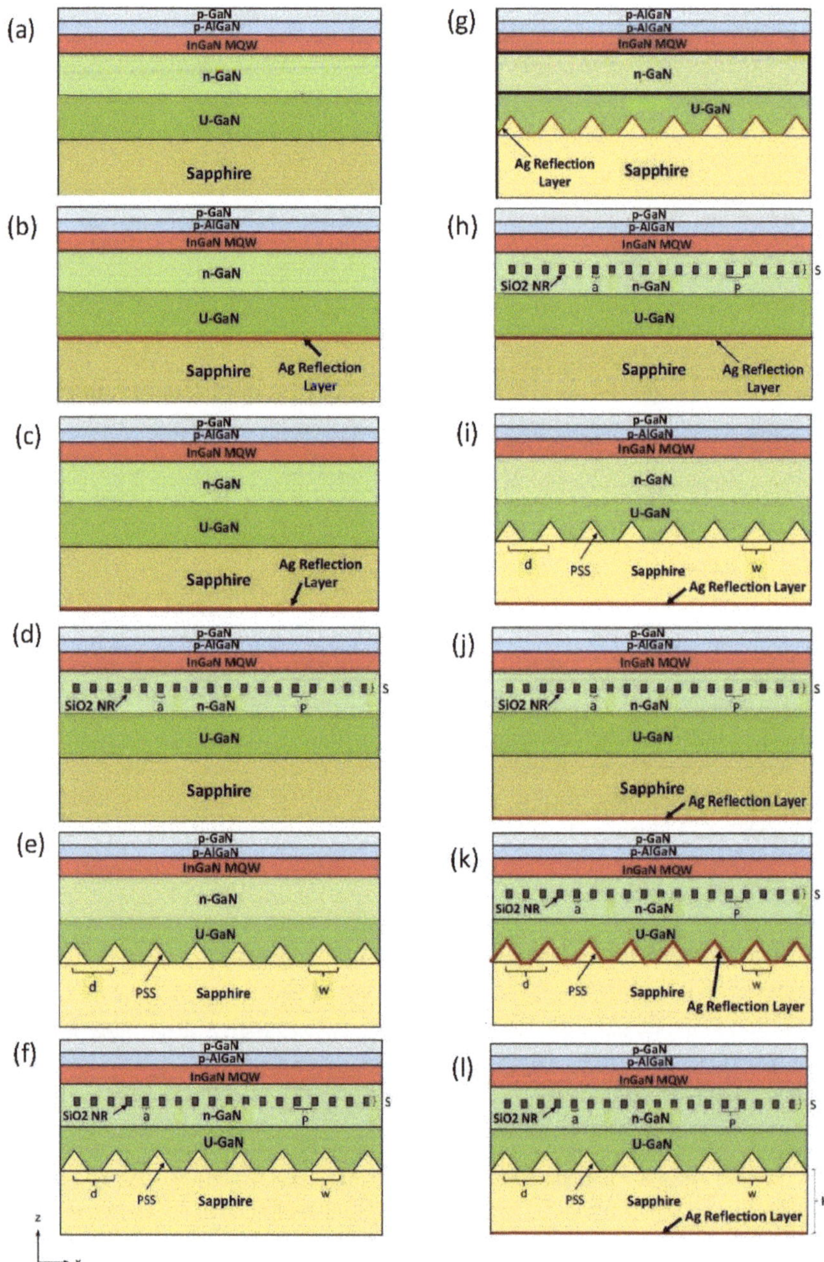

Figure 6. Diagram of LED with a patterned sapphire substrate (PSS), SiO$_2$ NR array, and *Ag Reflector*. The diagram defines the PSS period (d), PSS width (w), SiO$_2$ NR period (p), SiO$_2$ NR width (a), SiO$_2$ NR high (s), and shows the x and z direction of the LED. (**a**)–(**l**) are explained and listed in the paper.

Table 4. Conventional GaN LED Parameters.

Material	Thickness (μm)	Refractive Index
p-GaN	0.12	2.55
P-AlGaN	0.05	2.5
InGaN/GaN	0.115	2.6
n-GaN	2	2.55
GaN	3	2.55
Sapphire	80	1.77

The results in Figure 7 lead to several observations. First, Figure 7 show that both PSS and SiO_2 nano-rod structures enhance light extraction; a 15% improvement for the PSS structure and a 26% improvement for the PSS and SiO_2 NR array structure. The structure with both the PSS and SiO_2 nano-rod array increases light extraction more than the simulation only containing the PSS. This is because that SiO_2 NR structure alone can improve light extraction by 30%, which is higher than PSS only or PSS NR. Second, adding Ag reflection may not necessarily improve light extraction, as shown in the CLED Ag-middle and Ag-bottom cases. Inappropriate placement of the Ag layer may cause a destructive interference pattern or force minimum standing wave output at the top of the LED. Therefore, this may decrease the output light intensity. The third, NR with an Ag reflector improves light extraction by about 127% for Ag-middle and 116.38% with Ag-bottom, which are the best cases for the 12 designs. It is very worthwhile to fabricate NR with an Ag reflector structure. Fourth, the output power of the structures containing the bottom Ag reflection layer fluctuates with the sapphire substrate thickness because of the standing wave pattern inside the LED cavity [35], as shown in Figure 8. Our simulation also shows that the sapphire substrate height (H) does not affect structures without Ag reflectors or Ag reflectors above it. The conventional LED with the reflection layer shows the greatest change in output power. At a height of 40 μm, light extraction increases by 160% and at a height of 20 μm, light extraction decreases by 45% compared to the CLED structure, which can also be understood by the standing wave analysis.

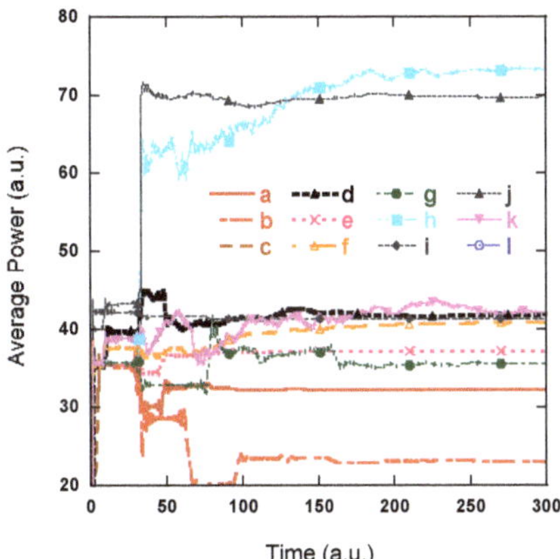

Figure 7. Average power of light extracted in 10 structures, while the PSS width, $w = 2.5$ μm, and period, $d = 3$ μm. The SiO_2 layer is centered at $z = 7.8$ μm. Sapphire layer thickness is 80 μm. (**a**) CLED, (**b**) CLED with Ag between U-GaN and the sapphire substrate, (**c**) CLED with Ag below the sapphire substrate, (**d**) NR, (**e**) PSS, (**f**) PSS NR, (**g**) PSS with Ag between U-GaN and the sapphire substrate, (**h**) NR with Ag between U-GaN and the sapphire substrate, (**i**) PSS, Ag below the sapphire substrate, (**j**) NR array, Ag below the sapphire substrate, (**k**) PSS NR array with Ag between U-GaN and the sapphire substrate, and (**l**) LED with PSS, NR array, and Ag below the sapphire substrate.

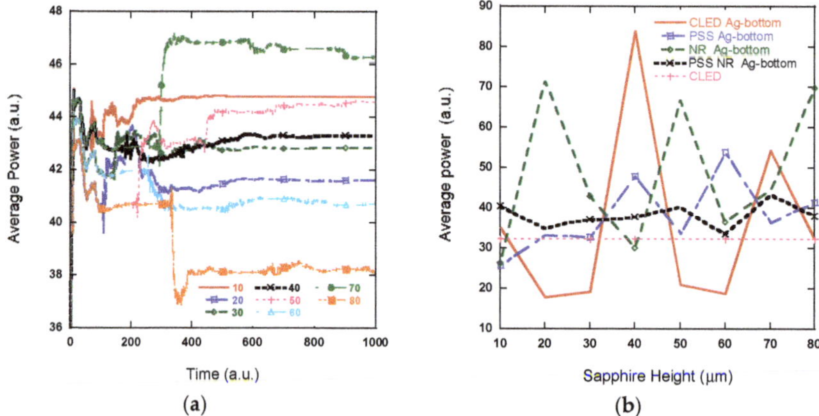

Figure 8. Average power vs. different sapphire substrate thicknesses with PSS NR and a bottom layer Ag reflector. PSS width of w = 2.5 µm and period of d = 3 µm. SiO$_2$ layer is centered at z = 7.8 µm. (**a**) Average power extracted against time according to different sapphire substrate thicknesses of 10 µm, 20 µm, 30 µm, 40 µm, 50 µm, 60 µm, 70 µm, and 80 µm, and (**b**) steady state average power.

A summary of the results, including the percent improvement compared to the conventional LED, are listed below in Table 5. For the three kinds of structures investigated, NR is the most efficient and intensively studied. PSS is a reasonable method to improve the light extraction efficiency. An Ag-reflector can improve or decrease light output, because it can change the standing wave pattern inside an LED dramatically and cause huge effects. When designing carefully, an NR plus Ag-reflector can be one of the best designs of GaN LEDs.

Table 5. Percent improvement of various LEDs compared to CLEDs.

	Structure	Average Power (a.u.)	Percent Improvement (%)
a	Conventional (CLED)	32.239	—
b	CLED Ag-middle	23.078	−28.416
c	CLED Ag-bottom	32.227	−0.037218
d	NR (only)	41.934	30.072
e	PSS (only)	37.159	15.261
f	PSS & NR	40.769	26.459
g	PSS Ag-middle	35.461	9.9941
h	NR Ag-middle	73.191	127.03
i	PSS Ag-bottom	41.358	28.29
j	NR Ag-bottom	69.759	116.38
k	PSS NR Ag-middle	42.221	30.963
l	PSS NR Ag-bottom	38.011	17.904

3.3. Nano-top Grating Performance over Different Wavelength

In reality, regarding LEDs' output light over a range of wavelengths surrounding the primary wavelength, we verified that the grating period that maximizes light extraction at the center wavelength is still effective at other wavelengths in the primary emission range. To better understand how the wavelength of light affects light output, the grating periods around the peak light emission were re-simulated for the ITO nano-top grating at free space wavelengths of 470 nm and 450 nm, as shown in Figure 9. These are wavelengths that are near the peak wavelength of 460 nm and are still strongly emitted by the LED. These simulations use an ITO thickness of 78 nm and a sapphire substrate thickness of 10,050 nm.

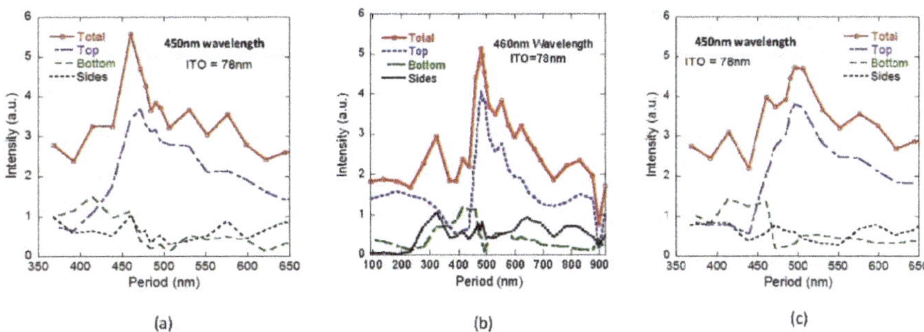

Figure 9. Total light output for ITO nano-top grating when the free space wavelength is at (**a**) 450 nm, (**b**) 460 nm, and (**c**) 470 nm.

It shows that the gratings are still highly effective at increasing the light output from the CLED for both 450 nm light and 470 nm light. Additionally, total light extraction, with a wavelength of λ_o = 450 nm, is substantially higher than any total light extraction that was achieved for either the 460 nm or 470 nm light. This is because the shoulder of the top emission peak overlaps with a sharp rise in light extraction from both the bottom and sides of the LED. The top light extraction efficiency is still highest for the 460 nm wavelength, which is our design wavelength. The most effective grating period in all three wavelengths is about 500 nm, which is our next simulation parameters.

The maximum output for each wavelength occurs at a different grating period, but all are a little smaller and close to 500 nm. It shows that for maximum top light extraction, shorter wavelengths correlate with smaller grating periods and longer wavelengths correlate with bigger grating periods [10]. Because the material layers were optimized for 460 nm light in the standing wave section, the larger grating periods, like 496 nm, enhance the longer wavelengths of light within the LED emitting spectrum and shorter periods around 472 nm enhance the shorter wavelengths, which is clearly shown in Figure 10 as well. The grating period that maximizes total light output for our design is a grating period near 484 nm.

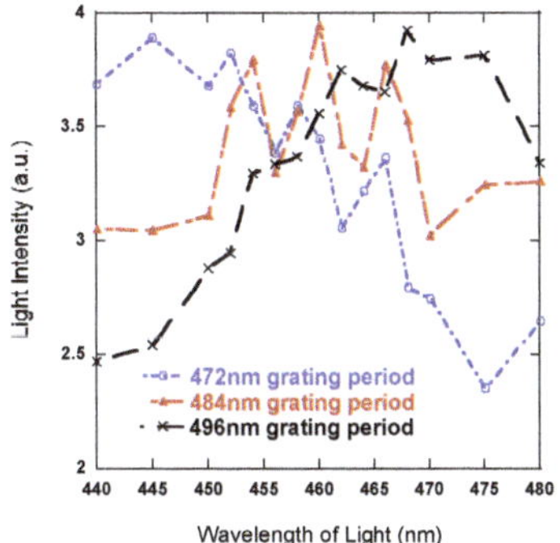

Figure 10. Light extraction intensity across the LED emitting spectrum.

4. Conclusions

We proposed an error grating model to simulate the fabrication variation and its effects. All simulations were based on the FDTD method. The scope of this research covers the effects of nano-scale top gratings, patterned sapphire substrates, SiO_2 nano-rod arrays, and Ag reflection layers.

We investigated the effect of nanoscale ITO transmission gratings on light emission from the top, the side, and the bottom of a GaN LED based on substrate standing wave analysis. Standing wave analysis of the LED showed that the light extraction efficiency can be improved by varying the ITO thickness, and optimized the ITO grating parameters, which can reach a 204% improvement compared to CLED. We also optimized light extraction for a structure containing a patterned sapphire substrate and an SiO_2 nano rod array. Both PSS and SiO_2 nano-rod structures enhanced light extraction. A 15% improvement for the PSS structure and a 30% improvement for NR was achieved. A combination of NR and PSS may not be a better design, with a 26.459% improvement here. A more careful design simulation is needed to obtain the overall optimizing results. We also found that the position of the reflection layer affected the light extraction. From these simulations, we found a maximum increase in light extraction of 127% for an SiO_2 nano-rod LED with an Ag reflection layer compared to a conventional LED. The LED with the NR and Ag reflector structure was the best design.

Finally, we studied the nano-top grating performance over different wavelengths to validate our design simulation and generate an LED emitting spectrum. The grating period had clear influence on the emitting spectrum. However, our design was still reasonable between 440 nm and 480 nm around the center wavelength of 460 nm.

Author Contributions: The paper was written by Xiaomin Jin based on three master thesis of her students. Simeon Trieu proposed the error grating model, Gregory James Chavoor investigated Nano-patterned Sapphire Substrates (PSS), SiO_2 Nano-rod Grating (NR), and Ag Reflector Bottom Grating. Gabriel Michael Halpin studied LED performance for different wavelength and LED ITO Nano-Gratings with Standing Wave Analysis.

Funding: This research was funded by National Science Foundation (NSF) International Research Experiences for Students (IRES) award number 1029153, PI: Xiaomin Jin, Co-PI: Xiao-hua Yu, from September 2010 to August 2014.

Conflicts of Interest: The authors declare no conflict of interest.

References

1. Wei, T.; Kong, Q.; Wang, J.; Jing, L. Improving light extraction of InGaN-based light emitting diodes with a roughened p-GaN surface using CsCl nano-islands. *Opt. Express* **2011**, *19*, 1065–1071. [CrossRef] [PubMed]
2. Bao, K.; Kang, X.N.; Zhang, B.; Dai, T.; Sun, Y.J.; Fu, Q.; Lian, G.J.; Xiang, G.C.; Zhang, G.Y.; Chen, Y. Improvement of light extraction from GaN-based thin-film light-emitting diodes by patterning undoped GaN using modified laser lift-off. *Appl. Phys. Lett.* **2008**, *92*, 141104. [CrossRef]
3. Cho, C.Y.; Kang, S.E.; Kim, K.S.; Lee, S.J.; Choi, Y.S.; Han, S.H.; Jung, G.Y.; Park, S.J. Enhanced light extraction in light-emitting diodes with photonic crystal structure selectively grown on p-GaN. *Appl. Phys. Lett.* **2010**, *96*, 181110. [CrossRef]
4. Yang, G.F.; Xie, F.; Tong, Y.Y.; Chen, P.; Yu, Z.G.; Yan, D.W.; Xue, J.J.; Zhu, H.X.; Guo, Y.; Li, G.H.; et al. Formation of nanorod InGaN/GaN multiple quantum wells using nickel nano-masks and dry etching for InGaN-based light emitting diodes. *Mat. Sci. Semicon. Proc.* **2015**, *30*, 694–706. [CrossRef]
5. Trieu, S.; Jin, X. Study of Top and Bottom Photonic Gratings on GaN LED with Error Grating Models. *IEEE J. Quantum Electron.* **2010**, *46*, 896–901. [CrossRef]
6. Lin, W.Y.; Wuu, D.S.; Huang, S.C.; Horng, R.H. Enhanced output power of near-ultraviolet InGaN/AlGaN LEDs with patterned distributed Bragg reflectors. *IEEE Trans. Electron Dev.* **2011**, *58*, 173–179. [CrossRef]
7. Kuo, C.W.; Lee, Y.C.; Fu, Y.K.; Tsai, C.H.; Wu, M.L.; Chi, G.C.; Kuo, C.H.; Tun, C.J. Optical Simulation and Fabrication of Nitride-Based LEDs with the Inverted Pyramid Sidewalls. *IEEE J. Sel. Top. Quant.* **2009**, *15*, 1264–1268. [CrossRef]
8. Bao, K.; Kang, X.N. Improvement of Light Extraction from Patterned Polymer Encapsulated GaN-Based Flip-Chip Light-Emitting Diodes by Imprinting. *IEEE Photonics Technol. Lett.* **2007**, *19*, 1840–1842. [CrossRef]
9. Ryu, H.; Shim, J. Structural Parameter Dependence of Light Extraction Efficiency in Photonic Crystal InGaN Vertical Light-Emitting Diode Structures. *IEEE J. Quantum Electron.* **2010**, *46*, 714–720.
10. Halpin, G.; Robinson, T.; Jin, X.; Kang, X.N.; Zhang, G.Y. Study of GaN LED ITO Nano-Gratings With Standing Wave Analysis. *IEEE Photonics J.* **2014**, *6*, 4500210. [CrossRef]
11. Shei, S.C. Multiple Nanostructures on Full Surface of GZO/GaN-Based LED to Enhance Light-Extraction Efficiency Using a Solution-Based Method. *IEEE J. Quantum Electron.* **2014**, *50*, 629–632. [CrossRef]
12. Chen, Z.; Liu, W.; Wan, W.; Chen, G.; Zhang, B.; Jin, C. Improving the Extraction Efficiency of Planar GaN-Based Blue Light-Emitting Diodes via Optimizing Indium Tin Oxide Nanodisc Arrays. *J. Disp. Technol.* **2016**, *12*, 1588–1593. [CrossRef]

13. Ding, Q.A.; Li, K.; Kong, F.; Zhao, J.; Yue, Q. Improving the Vertical Light Extraction Efficiency of GaN-Based Thin-Film Flip-Chip LED with Double Embedded Photonic Crystals. *IEEE J. Quantum Electron.* **2015**, *51*, 3300109. [CrossRef]

14. Dai, T.; Kang, X.; Zhang, B.; Xu, J.; Bao, K.; Xiong, C.; Gan, Z. Study and formation of 2D microstructures of sapphire by focused ion beam milling. *Microelectron. Eng.* **2008**, *85*, 640–645. [CrossRef]

15. Huang, X.H.; Liu, J.P.; Fan, Y.Y.; Kong, J.J.; Yang, H.; Wang, H.B. Effect of Patterned Sapphire Substrate Shape on Light Output Power of GaN-Based LEDs. *IEEE Photonics Technol. Lett.* **2011**, *23*, 944–946. [CrossRef]

16. Cui, H.; Park, S.H. Numerical simulations of light-extraction efficiencies of light-emitting diodes on micro and nanopatterned sapphire substrates. *IET Micro Nano Lett.* **2014**, *9*, 841–844. [CrossRef]

17. Che, Z.; Zhang, J.; Yu, X.; Xie, M.; Yu, J.; Lu, H.; Luo, Y.; Guan, H.; Chen, Z. Improvement of light extraction efficiency of GaN-based flip-chip LEDs by a double-sided spherical cap-shaped patterned sapphire substrate. In Proceedings of the 2016 International Conference on Numerical Simulation of Optoelectronic Devices (NUSOD), Sydney, Australia, 11–15 July 2016.

18. Xia, C.S.; Sheng, Y.; Li, Z.M.; Cheng, L. Simulation of GaN-Based Light-Emitting Diodes with Hemisphere Patterned Sapphire Substrate Based on Poynting Vector Analysis. *IEEE J. Quantum Electron.* **2015**, *51*, 3300105. [CrossRef]

19. Kim, B.J.; Jung, H.; Kim, S.H.; Bang, J.; Kim, J. GaN-Based Light-Emitting Diode with Three-Dimensional Silver Reflectors. *IEEE Photonics Technol. Lett.* **2009**, *21*, 700–702. [CrossRef]

20. Jeong, T.; Kim, K.H.; Lee, H.H.; Lee, S.J.; Lee, S.H.; Baek, J.H.; Lee, J.K. Enhance Light Output Power of GaN-Based Vertical Light-Emitting Diodes by Using Highly Reflective ITO-Ag-Pt Refelctors. *IEEE Photonics Technol. Lett.* **2008**, *20*, 1932–1934. [CrossRef]

21. Liou, J.K.; Chen, W.C.; Chang, C.H.; Chang, Y.C.; Tsai, J.H.; Liu, W.C. Enhanced Light Extraction of a High-Power GaN-Based Light-Emitting Diode with a Nanohemispherical Hybrid Backside Reflector. *IEEE Trans. Electron Dev.* **2015**, *62*, 3296–3301. [CrossRef]

22. Jin, X.; Chavoor, G. Position of Ag reflection layer and its effect on GaN LED light extraction efficiency. In Proceedings of the 2015 IEEE 15th International Conference on Nanotechnology (IEEE-NANO), Roma, Italy, 27–30 July 2015.

23. Tien, C.H.; Zhang, C.H.; Chung, S.H.; Ou, S.L.; Horng, R.H.; Wuu, D.S. ITO/nano-Ag plasmonic window applied for efficiency improvement of near-ultraviolet light emitting diodes. In Proceedings of the 2016 Compound Semiconductor Week (CSW) [Includes 28th International Conference on Indium Phosphide & Related Materials (IPRM) & 43rd International Symposium on Compound Semiconductors (ISCS)], Toyama, Japan, 26–30 June 2016.

24. Peng, W.C.; Wu, Y.S. Enhance performance of an InGaN-GaN light emitting diode by roughening the undoped-GaN surface and applying a mirror coating to the sapphire substrate. *Appl. Phys. Lett.* **2006**, *88*, 181117. [CrossRef]

25. Chiu, C.H.; Yen, H.H.; Chao, C.L.; Li, Z.Y.; Yu, P.C. Nanoscale epitaxial lateral overgrowth of GaN-based light-emitting diodes on a SiO_2 nanorod-array patterned sapphire template. *Appl. Phys. Lett.* **2008**, *93*, 081108. [CrossRef]

26. Lee, Y.L.; Liu, W.C. Enhanced Light Extraction of GaN-Based Light-Emitting Diodes with a Hybrid Structure Incorporating Microhole Arrays and Textured Sidewalls. *IEEE Trans. Electron Dev.* **2018**, *65*, 3305–3310. [CrossRef]

27. Liu, H.Y.; Yang, Y.C.; Liu, G.J.; Huang, R.C. Improved Light Extraction Efficiency of GaN-Based Ultraviolet Light-Emitting Diodes by Self-Assembled MgO Nanorod Arrays. *IEEE Trans. on Electron Dev.* **2017**, *64*, 5006–5011. [CrossRef]

28. Huang, J.K.; Liu, C.Y.; Chen, T.P.; Huang, H.W.; Lai, F.I.; Lee, P.T.; Lin, C.H.; Chang, C.Y.; Kao, T.S.; Kuo, H.C. Enhanced Light Extraction Efficiency of GaN-Based Hybrid Nanorods Light-Emitting Diodes. *IEEE J. Sel. Top. Quant.* **2015**, *21*, 6000107. [CrossRef]

29. Kim, S.H.; Lee, K.D.; Kim, J.Y.; Kwon, M.K.; Park, S.J. Fabrication of photonic crystal structures on light emitting diodes by nanoimprint lithography. *TOP Nanotechnol.* **2007**, *18*, 055306. [CrossRef]

30. Hong, H.G.; Kim, S.S.; Kim, D.Y.; Lee, T.; Song, J.O.; Cho, J.H.; Sone, C.; Park, Y.; Seonga, T.Y. Enhancement of the light output of GaN-based ultraviolet light-emitting diodes by a one-dimensional nanopatterning process. *Appl. Phys. Lett.* **2006**, *88*, 103505–103507. [CrossRef]

31. Jin, X.; Zhang, B.; Dai, T.; Wei, W.; Kang, X.-N.; Zhang, G.-Y.; Trieu, S.; Wang, F. Optimization of Top Polymer Gratings to Improve GaN LEDs Light Transmission. *Chin. Opt. Lett.* **2008**, *6*, 788–790.

32. Jin, X.; Trieu, S.; Wang, F.; Zhang, B.; Dai, T.; Kang, X.N.; Zhang, G.Y. Design Simulation of Top ITO Gratings to Improve Light Transmission for Gallium Nitride LEDs. In Proceedings of the 2009 Sixth International Conference on Information Technology: New Generations (ITNG2009), Las Vegas, NV, USA, 27–29 April 2009.

33. Halpin, G.; Jin, X.; Fu, X.X.; Kang, X.N.; Zhang, G.Y. Study of Top ITO Nano-gratings on GaN LEDs. In Proceedings of the 13th IEEE International Conference on Nanotechnology, Beijing, China, 5–8 August 2013.

34. Huang, H.W.; Huange, J.K.; Lin, C.H.; Lee, K.Y.; Hsu, H.W.; Yu, C.C.; Kuo, H.C. Efficiency Improvement of GaN-Based LEDs with a SiO_2 Nanorod Array and a Patterned Sapphire Substrate. *IEEE Electron Device Lett.* **2010**, *31*, 582–584. [CrossRef]

35. Jin, X.; Chavoor, G.; Liu, G. Patterned sapphire substrate and SiO_2 array in GaN LED. In Proceedings of the SPIE Photonic West 2018, San Francisco, CA, USA, 31 January 2018.

Article

Fabrication and Characterization of AlGaN-Based UV LEDs with a ITO/Ga$_2$O$_3$/Ag/Ga$_2$O$_3$ Transparent Conductive Electrode

Hong Wang [1,2,3,*], Quanbin Zhou [1], Siwei Liang [1,2] and Rulian Wen [1,2]

[1] Engineering Research Center for Optoelectronics of Guangdong Province, School of Electronics and Information Engineering, South China University of Technology, Guangzhou 510640, China; zhouquanbin86@163.com (Q.Z.); 201620122572@mail.scut.edu.cn (S.L.); wrl_789@163.com (R.W.)

[2] School of Physics and Optoelectronics, South China university of Technology, Guangzhou 510640, China

[3] Zhongshan Institute of Modern Industrial Technology, South China University of Technology, Zhongshan 528437, China

* Correspondence: phhwang@scut.edu.cn; Tel.: +86-136-0006-6193

Received: 30 November 2018; Accepted: 31 December 2018; Published: 5 January 2019

Abstract: We fabricated a complex transparent conductive electrode (TCE) based on Ga$_2$O$_3$ for AlGaN-based ultraviolet light-emitting diodes. The complex TCE consists of a 10 nm ITO, a 15 nm Ga$_2$O$_3$, a 7 nm Ag, and a 15 nm Ga$_2$O$_3$, forming a ITO/Ga$_2$O$_3$/Ag/Ga$_2$O$_3$ multilayer. The metal layer embedded into Ga$_2$O$_3$ and the thin ITO contact layer improves current spreading and electrode contact properties. It is found that the ITO/Ga$_2$O$_3$/Ag/Ga$_2$O$_3$ multilayer can reach a 92.8% transmittance at 365 nm and a specific contact resistance of 10^{-3} $\Omega \cdot cm^2$ with suitable annealing conditions.

Keywords: transparent conductive electrode; Ga$_2$O$_3$; AlGaN-based ultraviolet light-emitting diode; transmittance; sheet resistance

---

1. Introduction

AlGaN-based ultraviolet (UV) light-emitting diodes (LEDs) can achieve the full wavelength coverage of UVA (400–320 nm), UVB (320–280 nm) and UVC (280–200 nm) by changing Al content. As a result, AlGaN-based UV LEDs have attracted considerable attention and are seen as a promising lighting source for different applications in environmental cleaning, medicine, printing, microscopy and lighting [1–6]. However, the external quantum efficiency (EQE) of AlGaN-based UV LEDs is still much lower than that of the commercially available blue LEDs with an EQE close to 20% for UVA and <1% for UVC devices [7–9]. This phenomenon obstructs commercial applications of the AlGaN-based UV LEDs. Indium tin oxide (ITO) is widely used as transparent contact layers in traditional GaN-based blue and green LEDs. However, there is serious light absorption in the ITO in the ultraviolet band due to the band gap of ITO ranging from 3.5 eV to 4.3 eV [10,11]. Previous studies reported that doping metals in ITO would reduce the light absorption in near UV LEDs. The transmittance of ITO at wavelengths above 380 nm can reach about 90% by optimizing the thickness of metal and the annealing temperature [12–15]. But the transmittance of ITO still decreases rapidly when the wavelength becomes shorter. Thus, it is very urgent for a layer with higher transmittance in ultraviolet band to be able to replace the traditional ITO transparent conductive electrode (TCE) in UV LEDs.

Ga$_2$O$_3$, which has a bandgap from 4.9 eV to 5 eV, is an attractive alternative for TCE in UV LEDs because of its high transmittance in UV band [16–18]. In addition, a large size and high quality Ga$_2$O$_3$ thin film can be fabricated by single crystals synthesized by the melt growth method [19]. This material has been studied in the fields of metal semiconductor field effect transistors, metal oxide semiconductor field effect transistors and Schottky barrier diodes. However, the conductivity of Ga_2O_3 is very poor. Many approaches have been developed to promote the conductivity of Ga_2O_3. Orita Mi Hiramatsu H et al. improved the conductivity of β-Ga_2O_3 by doping In or Sn into Ga_2O_3 [16]. The (201)-oriented Sn-doped β-Ga_2O_3 films obtained a maximum conductivity of 8.2 S/cm (about 1.22×10^4 Ω/sq). But it is still too low to be used as TCE in UV LED. Liu JJ et al. grew ITO thin films in Ga_2O_3 films and improved the sheet resistance and transmittance of Ga_2O_3/ITO films by adjusting the growth temperature and the thickness of ITO [17]. A sheet resistance of 323 Ω/sq and a transmittance at 280 nm of 77.6% can be achieved. Jae-kwan Kim et al. realized that the transmittance at 380 nm is 80.944% and the sheet resistance is 58.6 Ω/sq [20]. The Kie Young Woo group in Korea prepared the Ag/Ga_2O_3 model by learning the ITO/Ag/ITO model [21]. The contact characteristics and conductivity of the Ga_2O_3 films were improved by the Ag intercalation layer, and the transmittance at 380 nm and specific contact resistivity of the Ag/Ga_2O_3 thin film were 91% and 3.06×10^{-2} $\Omega \cdot cm^2$ respectively.

In this paper, a complex TCE based on Ga_2O_3 is proposed to enhance the efficiency of UV LEDs. We prepared the complex Ga_2O_3-based TCE by depositing an ITO contact layer, a Ga_2O_3 layer, an Ag metal intercalation layer and another Ga_2O_3 layer in sequence, forming an ITO/Ga_2O_3/Ag/Ga_2O_3 multilayer. The resistance and transmittance ITO/Ga_2O_3/Ag/Ga_2O_3 multilayer with a different annealing temperature were studied and analyzed systematically. The sheet resistance of the ITO/Ga_2O_3/Ag/Ga_2O_3 multilayer was detected by four-point probe methods. The optical transmittance was measured by a UV/visible spectrophotometer. The surface roughness of these ITO/Ga_2O_3/Ag/Ga_2O_3 multilayer were measured by atomic force microscope (AFM). The X-ray photoelectron spectroscopy (XPS) and Auger electron spectroscopy (AES) measurements were also used to analyze the ITO/Ga_2O_3/Ag/Ga_2O_3 multilayer. Furthermore, we employed the ITO/Ga_2O_3/Ag/Ga_2O_3 multilayer as TCEs on 365 nm UV epitaxy in comparison to those with conventional ITO.

2. Materials and Methods

To investigate the influence of a Ag intercalation layer on the Ga_2O_3 layer, a Ga_2O_3/Ag/Ga_2O_3 (15 nm/7 nm/15 nm) multilayer was deposited on quartz substrates and then annealed at different conditions. The quartz substrates were first washed in acetone, isopropanol and deionized water and dried by nitrogen. After that, Ga_2O_3, Ag and Ga_2O_3 were sequentially deposited on the quartz substrates in magnetron sputtering equipment. In order to reduce the resistivity of the Ga_2O_3 layer but not affect its transmittance, the thickness of the Ag embedding interlayer and Ga_2O_3 were set to be 7 nm and 15 nm respectively. The Ga_2O_3 thin films were all deposited by RF magnetron sputtering of Ga_2O_3 (purity 99.99%) ceramic target, and the Ag thin film was deposited by direct current magnetron sputtering of the Ag target. The sputtering cavity was pumped to 5×10^{-6} Pa before the sputtering begin. The sputtering atmosphere was pure argon with the pressure of 5 mtorr. The rotation speed of the cavity substrate is 20 rpm. The temperature was controlled at about 35 °C $\pm$ 1 °C by feedback control heater during deposition. Afterwards, all the Ga_2O_3/Ag/Ga_2O_3 multilayer samples were annealed by a rapid thermal annealing (RTA) system at a different temperature and ambient. We used X-ray photoelectron spectroscopy (XPS) and Auger electron spectroscopy (AES) to analyze the element diffusion effect of the Ga_2O_3/Ag/Ga_2O_3 multilayer.

To further improve the contact between Ga_2O_3 and AlGaN-based UV epitaxy, we insert an ITO thin film below the Ga_2O_3/Ag/Ga_2O_3 multilayer as a contact layer. We prepared ITO/Ga_2O_3/Ag/Ga_2O_3 multilayer on quartz substrates. Before the deposition of Ga_2O_3/Ag/Ga_2O_3 multilayer, a 10 nm ITO was deposited on quartz substrates by RF magnetron sputtering of ITO (In_2O_3: 90 wt%, SnO_2: 10 wt%) and then annealed by RTA. Subsequently, Ga_2O_3/Ag/Ga_2O_3 multilayer was deposited on the annealed ITO thin films and the whole ITO/Ga_2O_3/Ag/Ga_2O_3 multilayer was annealed again.

Finally, we prepared ITO/Ga_2O_3/Ag/Ga_2O_3 multilayer on AlGaN-based UV epitaxy in the same method to study the specific contact resistance through the CTLM model. A 47 nm ITO thin film on quartz substrates and epitaxy was also prepared as reference, which was annealed at 600 °C for

1 min in a mixture of N_2/O_2 (200 sccm:35 sccm) ambient. The procedures of $ITO/Ga_2O_3/Ag/Ga_2O_3$ multilayer and optical micrograph of contact surface on CTLM patterns are shown in Figure 1.

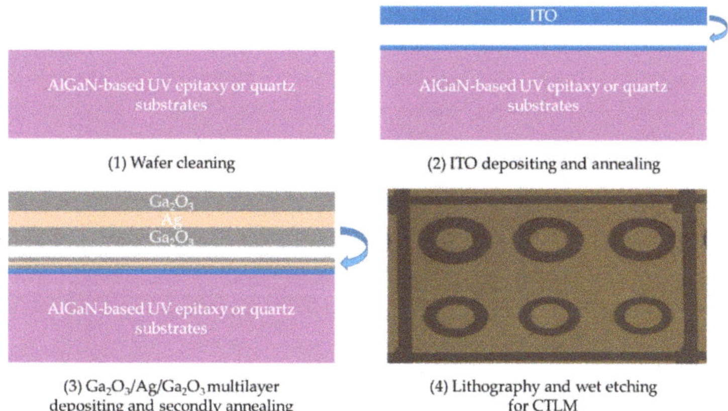

Figure 1. The procedures of $ITO/Ga_2O_3/Ag/Ga_2O_3$ multilayer and optical micrograph of contact surface on CTLM patterns.

3. Results

In order to study the influence of annealing conditions on the sheet resistance of $Ga_2O_3/Ag/Ga_2O_3$ multilayer, a series of $Ga_2O_3/Ag/Ga_2O_3$ multilayer on quartz substrates were annealed at different temperature and ambient. The annealing temperature changed from 400 °C to 600 °C with the annealing ambient changing from N_2/O_2 mixture and pure O_2 ambient. As shown in Table 1, the sheet resistance increases with the decrease of annealing temperature. It is found that the $Ga_2O_3/Ag/Ga_2O_3$ multilayer could reach the lowest sheet resistance of 16.45 Ω/sq after being annealed at 600 °C for 1 min in an N_2/O_2 mixture ambient. The result means that the effect of Ag as the insertion layer is not obvious at low temperature, and the metal diffusion reaction is not sufficient. The metal insertion layer in the film can fully diffuse to the Ga_2O_3 layer and decrease the resistance value of the $Ga_2O_3/Ag/Ga_2O_3$ multilayer when the temperature reaches 600 °C. The resistance of the multilayer annealed at 600 °C in pure oxygen ambient is higher than that of the multilayer in an N_2/O_2 mixture annealing ambient. Besides, the higher the oxygen ratio in the annealing atmosphere, the higher the multilayer resistance value becomes. The reason for this is that metal oxides form and then affect the resistance of film [22,23].

Table 1. Sheet resistance of a $Ga_2O_3/Ag/Ga_2O_3$ multilayer on quartz substrates at different annealing conditions.

No.	Annealing Temperature	Annealing Ambient	Annealing Time	Sheet Resistance (Ω/sq)
1	As deposited	As deposited	As deposited	23.86
2	400 °C	N_2 200 sccm:O_2 35 sccm	1 min	32.1
3	500 °C	N_2 200 sccm:O_2 35 sccm	1 min	27.74
4	600 °C	N_2 200 sccm:O_2 35 sccm	1 min	16.45
5	600 °C	O_2 35 sccm	3 min	30.6
6	600 °C	O_2 100 sccm	1 min	40.93

Figure 2 and Table 2 show the XPS energy spectral of $Ga_2O_3/Ag/Ga_2O_3$ multilayer on quartz substrate before and after annealing at 600 °C for 1 min in N_2/O_2 mixture ambient. The energy intensity, peak value quantum-number vertex, high half-width and atomic fraction content of Ag3d, O1s, Ga2p$_3$ were measured at the depth of about 10 nm of the multilayer. The open symbol and solid symbol in Figure 2 represent the energy intensity of the elements before and after annealing respectively. We can see that the energy value of element Ag3d is high, the quantum number per second is 13581 states/s, and the atomic fraction is 0.47 before annealing. After annealing, the energy intensity and the atomic ratio of Ag atom decrease relatively, which is 6651.8 counts/s and 0.29%, respectively. The energy and atomic ratio of Ga and O increase a little after annealing. The decrease of atomic ratio of Ag atom means that the process of annealing results in diffusion of Ag in the multilayer. Therefore, the sheet resistance of annealed $Ga_2O_3/Ag/Ga_2O_3$ multilayer decreases compared to that of the as-deposited sample due to the diffusion of internal elements.

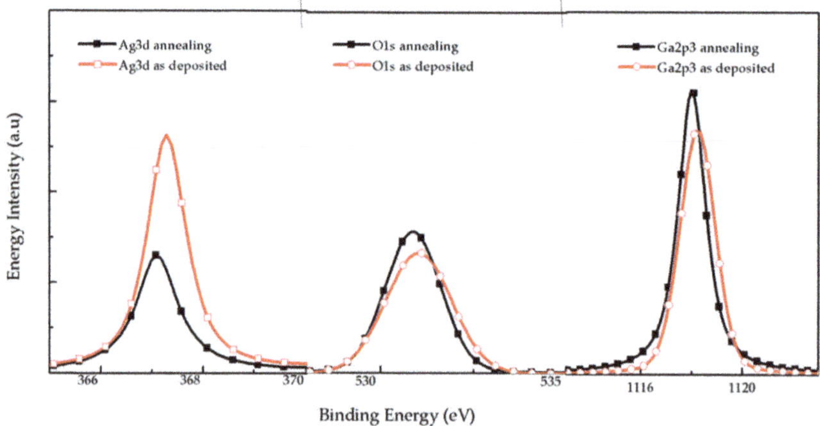

Figure 2. XPS spectrum for Ag3d, O1s and Ga2p3 of $Ga_2O_3/Ag/Ga_2O_3$ multilayer on quartz substrates.

Table 2. XPS data of $Ga_2O_3/Ag/Ga_2O_3$ multilayer on quartz substrates.

Name	As Deposited	Annealing	As Deposited	Annealing	As Deposited	Annealing	As Deposited	Annealing
	Peak BE		Height CPS		FWHM eV		Atomic %	
Ag3d	367.24	367.06	13,581.03	6651.81	0.95	0.95	0.47	0.29
O1s	530.89	530.7	224,669.16	265,423	2.04	1.7	50.67	50.97
Ga2p3	1118.21	1117.93	705,296.57	772,390.77	1.66	1.62	34.31	35.62

In addition, to further identify the distribution of composition in $Ga_2O_3/Ag/Ga_2O_3$ multilayer, we analyzed the $Ga_2O_3/Ag/Ga_2O_3$ multilayer on quartz substrate using AES measurement. Figure 3 shows the AES depth profiles of the $Ga_2O_3/Ag/Ga_2O_3$ multilayer before and after annealing at 600 °C for 1 min in N_2/O_2 mixture ambient. For the multilayer before annealing, the atomic percent of Ag is low in the surface and increases after a specific sputter time, which means that the Ag do not diffuse into the multilayer. Since the Ga_2O_3 and quartz substrates have poor conductivity, the atomic percent will become random and fluctuant due to the charge accumulation effect when the sputter time increases. By contrast, the atomic percent of Ag increases at the beginning of sputtering and the Ag atoms distribute more evenly in the whole multilayer after annealing as shown in Figure 3b. This result demonstrates that the Ag will diffuses into the Ga_2O_3 layer during the annealing process, leading to the reduction of sheet resistance of the $Ga_2O_3/Ag/Ga_2O_3$ multilayer.

Because of the bad contact property between Ga_2O_3 and p-GaN on epitaxial wafer, we insert a 10 nm ITO thin film below Ga_2O_3 as the contact layer. In order to optimal the transmittance and sheet resistance of the $ITO/Ga_2O_3/Ag/Ga_2O_3$ multilayer, we prepared five $ITO/Ga_2O_3/Ag/Ga_2O_3$ multilayer samples on quartz substrates and changed the annealing temperature as shown in Table 3. Among the five samples, sample 1 was not annealed. Sample 2 was annealed at 600 °C as a whole. For sample 3 to sample 5, the 10 nm ITO layer was firstly annealed at 550/600/650 °C respectively and then the whole $ITO/Ga_2O_3/Ag/Ga_2O_3$ multilayer were annealed at 600 °C. The annealing process of ITO and $ITO/Ga_2O_3/Ag/Ga_2O_3$ multilayer both maintained in N_2/O_2 (200 sccm:35 sccm) mixture ambient for 1 min. Figure 4 is the transmittance curves of five samples at range of 300 nm to 450 nm. It is obvious that sample 4, which was annealed at 600 °C at first and then at 600 °C again, has the highest transmittance of 92.68% at 365 nm and the lowest sheet resistance of 20.1 Ω/sq.

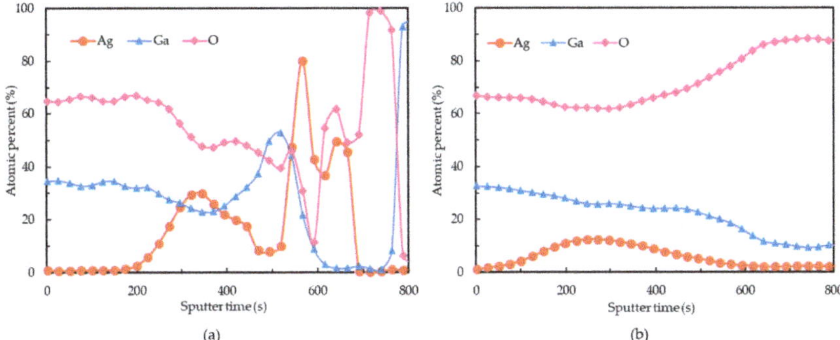

Figure 3. AES depth profiles of the Ga_2O_3/Ag/Ga_2O_3 multilayer on quartz substrates (**a**) before annealing and (**b**) after annealing.

Table 3. The transmittance and sheet resistance of ITO/Ga_2O_3/Ag/Ga_2O_3 multilayer on quartz substrates.

Sample	Annealing Temperature		Sheet Resistance	Transmittance at 365 nm
	10 nm ITO	**ITO/Ga$_2$O$_3$/Ag/Ga$_2$O$_3$**		
1	No annealing	No annealing	386.7 Ω/sq	48.04%
2	No annealing		164.0 Ω/sq	69.35%
3	550 °C	600 °C	36.9 Ω/sq	80.31%
4	600 °C		20.1 Ω/sq	92.68%
5	650 °C		48.1 Ω/sq	72.09%

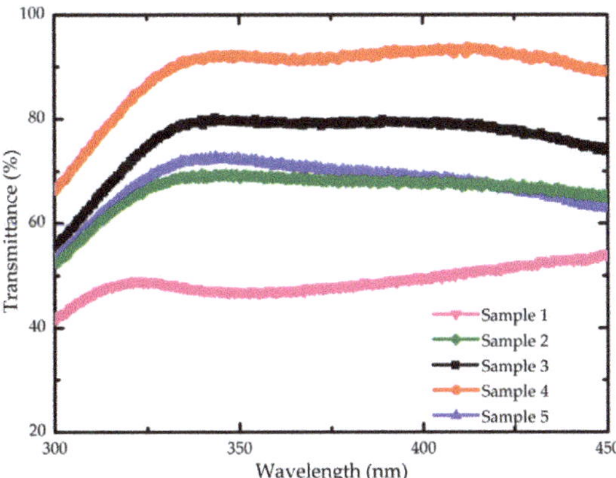

Figure 4. The transmittance curves of ITO/Ga_2O_3/Ag/Ga_2O_3 multilayer on quartz substrates after annealing.

In addition, we compared the transmittance and sheet resistance of sample 4 and a 47 nm ITO thin film on quartz substrate. The 47 nm ITO sample was annealed at 600 °C in N_2/O_2 (200 sccm:35 sccm) mixture ambient for 1 min. Figure 5a plots the transmittance curves of sample 4 and 47 nm ITO.

The ITO/Ga$_2$O$_3$/Ag/Ga$_2$O$_3$ multilayer demonstrates better transmittance property than 47 nm ITO especially in UV range. To further understand the origin of this result, the optical bandgap Energy *Eg* of sample 4 and 47 nm ITO was calculated. The *Eg* can be extracted from the relation between $(\alpha h v)^2$ and *hv* according to the Equations (1) and (2), as follow:

$$\alpha h v = C\left(h v - E_g\right)^{1/2} \tag{1}$$

$$h v = \frac{h c}{\lambda_i} \tag{2}$$

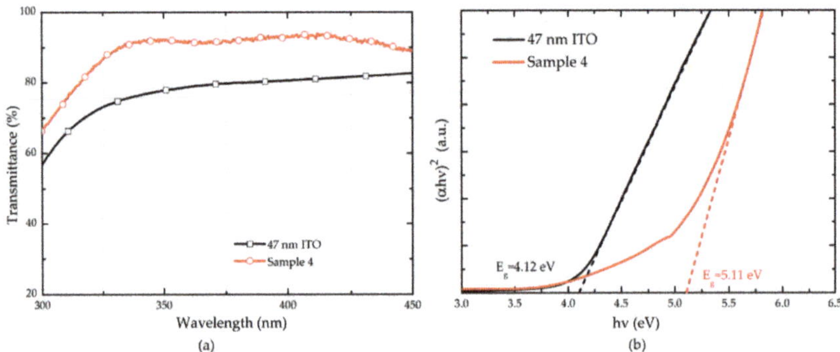

Figure 5. (a) Transmittance and (b) Energy bandgap of sample 4 and 47 nm ITO on quartz substrates.

The *Eg* can be obtained by extrapolating the linear $(\alpha h v)^2$ versus *hv* plots to the horizontal axis [24,25]. In Equations (1) and (2), *C* is a constant of direct transition, α is the light absorption coefficient, *hv* is the photon energy, *h* is Planck constant bright, *c* is the light speed, and λ_i is the wavelength [24,26,27].

If the transmittance *T* at each λ_i is known, the value of α at each λ_i can be obtained by Equations (3) and (4), as follow:

$$T = exp(-\alpha d) \tag{3}$$

$$\alpha = \frac{Ln\left(\frac{1}{T}\right)}{d} \tag{4}$$

where *d* is the thickness of films. Since we have measured the transmittance *T* of sample 4 and 47 nm ITO, the curves of $(\alpha h v)^2$ as a function of *hv* can be obtained as shown in Figure 5b. The optical Energy bandgap *Eg* of sample 4 is determined to be 4.12 eV, and that of ITO layer is 5.11 eV, by extrapolating the linear section of $(\alpha h v)^2$ to the *hv* axis. The large band gap means that the absorption of ITO/Ga$_2$O$_3$/Ag/Ga$_2$O$_3$ multilayer in UV range is smaller than that of 47 nm ITO layer. Table 4 shows the transmittance at 365 nm and sheet resistance of sample 4 and 47 nm ITO. The sample 4 has a reduction in sheet resistance compared to the 47 nm ITO sample. The transmittance of sample 4 is higher than that of 47 nm ITO and other reported metal-doped ITO [12–15]. These results reveal that the ITO/Ga$_2$O$_3$/Ag/Ga$_2$O$_3$ multilayer exhibits an advantage of transmittance at UV range and conductivity.

Table 4. Transmittance at 365 nm and sheet resistance of sample 4 and 47 nm ITO on quartz substrates.

Sample	47 nm ITO	Sample 4
Transmittance at 365 nm	79.15%	92.68%
Sheet resistance	57.63 Ω/sq	20.1 Ω/sq

Finally, we prepared a series of ITO/Ga$_2$O$_3$/Ag/Ga$_2$O$_3$ multilayers on AlGaN-based UV epitaxy to study the specific contact resistance through the CTLM model. These samples were fabricated in the same process as sample 1 and sample 3, 4, 5. The 10 nm ITO contact layer was annealed at 550/600/650 °C respectively, and then the whole ITO/Ga$_2$O$_3$/Ag/Ga$_2$O$_3$ multilayer was annealed at 600 °C. The annealing process of ITO and the ITO/Ga$_2$O$_3$/Ag/Ga$_2$O$_3$ multilayer both maintained in N$_2$/O$_2$ (200 sccm:35 sccm) mixture ambient for 1 min. As reference, a 47 nm ITO was also deposited on epitaxy and annealed at 600 °C in N$_2$/O$_2$ (200 sccm:35 sccm) mixture ambient for 1 min. Figure 6 shows the Ohmic contact characteristics of annealed ITO/Ga$_2$O$_3$/Ag/Ga$_2$O$_3$ multilayer with p-GaN measured by Electroluminescence system and CTLM mode. The I–V characteristics of the as-deposited ITO/Ga$_2$O$_3$/Ag/Ga$_2$O$_3$ multilayer are insulated, because the Ga$_2$O$_3$ films have the properties of non-diffusion of metals, poor conductivity and insulation on the p-GaN surface. However, the multilayer whose ITO was annealed in nitrogen-oxygen atmosphere at 550 °C /600 °C /650 °C shows linear I–V characteristics on the surface of p-GaN. Also, all the annealed ITO/Ga$_2$O$_3$/Ag/Ga$_2$O$_3$ multilayers exhibit higher current compared to the 47 nm ITO on p-GaN. The slope of 600 °C annealed I-V curve is highest. The specific contact resistance of 600 °C annealed sample could reach 2.36×10^{-3} Ω·cm^2. In contrast, the specific contact resistance of 47 nm ITO on AlGaN-based UV epitaxy is 5.68×10^{-3} Ω·cm^2.

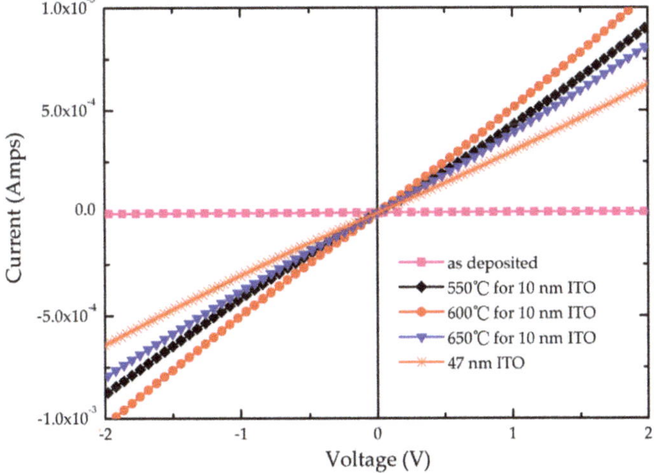

Figure 6. Ohmic contact characteristics of ITO/Ga$_2$O$_3$/Ag/Ga$_2$O$_3$ multilayer with different annealing temperature for ITO contact layer.

To further compare the differences between ITO/Ga$_2$O$_3$/Ag/Ga$_2$O$_3$ multilayer and the 47 nm ITO, we measured the surface morphology using scanning electron microscope (SEM) and AFM. The Figure 7a,b show SEM micrographs of the ITO/Ga$_2$O$_3$/Ag/Ga$_2$O$_3$ multilayer and 47 nm ITO on the AlGaN-based UV epitaxy. The surface of 47 nm ITO is smoother than that of the ITO/Ga$_2$O$_3$/Ag/Ga$_2$O$_3$ multilayer. Besides, the thickness of the multilayer is about 48 nm measured by SEM cross-section micrograph. The root-mean-square (RMS) surface roughness of the ITO/Ga$_2$O$_3$/Ag/Ga$_2$O$_3$ multilayer and 47 nm ITO on a 10 × 10 μm^2 area are 6.92 nm and 2.36 nm respectively measured by AFM. A rough surface is beneficial for light emitting form chips to external. The rougher surface of the ITO/Ga$_2$O$_3$/Ag/Ga$_2$O$_3$ multilayer may be another reason for its higher transmittance.

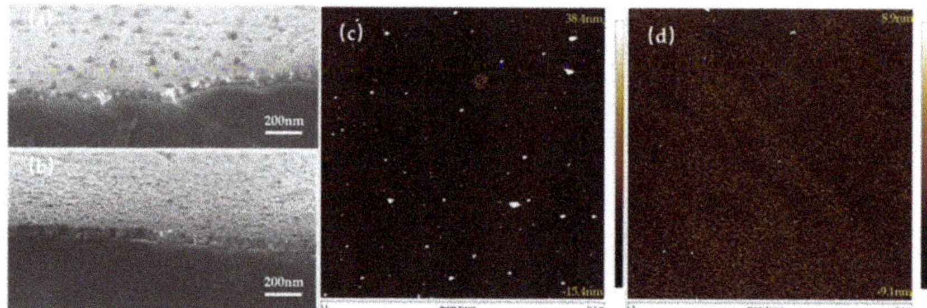

Figure 7. The surface morphology measured by SEM and AFM. (**a,c**) are for ITO/Ga$_2$O$_3$/Ag/Ga$_2$O$_3$ multilayer on AlGaN-based UV epitaxy after annealing at 600 °C. (**b,d**) are for 47 nm ITO on AlGaN-based UV epitaxy after annealing at 600 °C.

4. Conclusions

In this paper, a complex transparent conductive electrode based on Ga$_2$O$_3$ for AlGaN-based UV LEDs is proposed. The complex transparent conductive electrode consists of a 10 nm ITO, a 15 nm Ga$_2$O$_3$, a 7 nm Ag, and a 15 nm Ga$_2$O$_3$, forming a ITO/Ga$_2$O$_3$/Ag/Ga$_2$O$_3$ multilayer. The ITO/Ga$_2$O$_3$/Ag/Ga$_2$O$_3$ multilayer was grown by magnetron sputtering. The resistance and transmittance ITO/Ga$_2$O$_3$/Ag/Ga$_2$O$_3$ multilayer with a different annealing temperature was studied and analyzed systematically. With suitable annealing conditions, the ITO/Ga$_2$O$_3$/Ag/Ga$_2$O$_3$ multilayer reaches a 92.8% transmittance at 365 nm and a specific contact resistance of 2.36×10^{-3} $\Omega \cdot$cm^2. The XPS and AES results show that the diffusion of Ag in the multilayer leads to a low sheet resistance of the ITO/Ga$_2$O$_3$/Ag/Ga$_2$O$_3$ multilayer. The reason for the high transmittance of the ITO/Ga$_2$O$_3$/Ag/Ga$_2$O$_3$ multilayer in the UV range is the 5.11 eV band gap. These situations provide the improvement in optical characteristics of 365 nm UV LEDs. These results indicate that the proposed ITO/Ga$_2$O$_3$/Ag/Ga$_2$O$_3$ multilayer is a promising alternative for TCE to further improve the optical and electrical performances of AlGaN-based UV LED.

Author Contributions: These authors contributed equally.

Funding: This work was supported by Science and Technologies plan Projects of Guangdong Province (Nos. 2017B010112003, 2017A050506013), and by Applied Technologies Research and Development Projects of Guangdong Province (Nos. 2015B010127013, 2016B010123004), and by Science and Technologies plan Projects of Guangzhou City (Nos. 201504291502518, 201604046021, 201704030139), and by Science and Technology Development Special Fund Projects of Zhongshan City (Nos. 2017F2FC0002, 2017A1009).

Conflicts of Interest: The authors declare no conflict of interest.

References

1. Maitland, J. UV curing of inks and coatings in digital printing applications. In Proceedings of the NIP & Digital Fabrication Conference, Baltimore, MD, USA, 18–23 September 2005.
2. Yu, C.C.; Chuah, E.C.; Ng, Y.T.; Seah, Y.S.; Tan, P.P.; Chiu, T.H.; Hsieh, T.T. Neonatal status in cesarean section under epidural anesthesia with supplementary oxygen. *Ma Zui Xue Za Zhi = Anaesthesiologica Sinica* **1992**, *30*, 229. [PubMed]
3. Hockberger, P.E. A history of ultraviolet photobiology for humans, animals and microorganisms. *Photochem. Photobiol.* **2010**, *76*, 561–579. [CrossRef]
4. Schreiner, M.; Martínez-Abaigar, J.; Glaab, J.; Jansen, M. Uv-b induced secondary plant metabolites. *Opt. Photonik* **2014**, *9*, 34–37. [CrossRef]
5. Lui, G.Y.; Roser, D.; Corkish, R.; Ashbolt, N.; Jagals, P.; Stuetz, R. Photovoltaic powered ultraviolet and visible light-emitting diodes for sustainable point-of-use disinfection of drinking waters. *Sci. Total Environ.* **2014**, *493*, 185–196. [CrossRef]
6. Hodgkinson, J.; Tatam, R.P. Optical gas sensing: A review. *Meas. Sci. Technol.* **2013**, *24*, 012004. [CrossRef]

7. Takano, T.; Mino, T.; Sakai, J.; Noguchi, N.; Tsubaki, K.; Hirayama, H. Deep-ultraviolet light-emitting diodes with external quantum efficiency higher than 20% at 275 nm achieved by improving light-extraction efficiency. *Appl. Phys. Express* **2017**, *10*, 031002. [CrossRef]

8. Inoue, S.; Tamari, N.; Taniguchi, M. 150 mw deep-ultraviolet light-emitting diodes with large-area aln nanophotonic light-extraction structure emitting at 265 nm. *Appl. Phys. Lett.* **2017**, *110*, 141106. [CrossRef]

9. Shatalov, M.; Sun, W.; Lunev, A.; Hu, X.; Dobrinsky, A.; Bilenko, Y.; Yang, J.; Shur, M.; Gaska, R.; Moe, C. Algan deep-ultraviolet light-emitting diodes with external quantum efficiency above 10%. *Appl. Phys. Express* **2012**, *5*, 2101. [CrossRef]

10. Toyota, A.; Nakashima, N.; Sagara, T. Uv–visible transmission–absorption spectral study of au nanoparticles on a modified ito electrode at constant potentials and under potential modulation. *J. Electroanal. Chem.* **2004**, *565*, 335–342. [CrossRef]

11. Lin, Y.H.; Liu, C.Y. Reflectivity and abnormal absorption at ito/al interface. *J. Electron. Mater.* **2009**, *38*, 108–112. [CrossRef]

12. Lin, Y.H.; Liu, Y.S.; Liu, C.Y. Light output enhancement of near uv-led by using ti-doped ito transparent conducting layer. *IEEE Photonics Technol. Lett.* **2010**, *22*, 1443–1445. [CrossRef]

13. Jae Hoon, L.; Kie Young, W.; Kyeong Heon, K.; Hee-Dong, K.; Tae Geun, K. Ito/ag/ito multilayer-based transparent conductive electrodes for ultraviolet light-emitting diodes. *Opt. Lett.* **2013**, *38*, 5055–5058.

14. Kim, M.J.; Kim, T.G. Fabrication of metal-deposited indium tin oxides: Its applications to 385 nm light-emitting diodes. *ACS Appl. Mater. Interfaces* **2016**, *8*, 5453–5457. [CrossRef] [PubMed]

15. Xu, J.; Zhang, W.; Peng, M.; Dai, J.; Chen, C. Light-extraction enhancement of gan-based 395 nm flip-chip light-emitting diodes by an al-doped ito transparent conductive electrode. *Opt. Lett.* **2018**, *43*, 2684–2687. [CrossRef] [PubMed]

16. Orita, M.; Hiramatsu, H.; Ohta, H.; Hirano, M.; Hosono, H. Preparation of highly conductive, deep ultraviolet transparent β-ga 2 o 3 thin film at low deposition temperatures. *Thin Solid Films* **2002**, *411*, 134–139. [CrossRef]

17. Liu, J.; Yan, J.; Shi, L.; Li, T. Semiconductor materials electrical and optical properties of deep ultraviolet transparent conductive Ga_2O_3/ito films by magnetron sputtering. *J. Semicond.* **2010**, *31*, 5–9.

18. Passlack, M.; Schubert, E.F.; Hobson, W.S.; Hong, M.; Moriya, N.; Chu, S.N.G.; Konstadinidis, K.; Mannaerts, J.P.; Schnoes, M.L.; Zydzik, G.J. Ga_2O_3 films for electronic and optoelectronic applications. *J. Appl. Phys.* **1995**, *77*, 686–693. [CrossRef]

19. Higashiwaki, M.; Sasaki, K.; Murakami, H.; Kumagai, Y.; Koukitu, A.; Kuramata, A.; Masui, T.; Yamakoshi, S. Recent progress in Ga_2O_3 power devices. *Semicond. Sci. Technol.* **2016**, *31*, 034001. [CrossRef]

20. Kim, J.K.; Lee, J.M. Electrical and optical properties of near uv transparent conductive ito/Ga_2O_3 multilayer films deposited by rf magnetron sputtering. *Appl. Phys. Lett.* **2016**, *109*, 4166. [CrossRef]

21. Woo, K.Y.; Lee, J.H.; Kim, K.H.; Kim, S.J.; Kim, T.G. Highly transparent conductive ag/Ga_2O_3 electrode for near-ultraviolet light-emitting diodes. *Phys. Status Solidi* **2015**, *211*, 1760–1763. [CrossRef]

22. Wang, G.; Liu, H.; Zhao, C.; Yang, B.; Huang, X. Optical and electrical properties of annealed ito films. *Acta scientiarium Naturalium Universitatis Jilinensis* **1999**, *4*, 61–65.

23. Cai, Q.; Cao, C.; Jiang, X.; Song, X.; Sun, Z. Microstructures and stoichiometrics of indium-tin-oxide films. *Zhenkong Kexue Yu Jishu Xuebao/J. Vac. Sci. Technol.* **2007**, *27*, 195–199.

24. He, J.; Sun, L.; Chen, S.; Chen, Y.; Yang, P.; Chu, J. Composition dependence of structure and optical properties of cu2znsn (s, se) 4 solid solutions: An experimental study. *J. Alloy. Compd.* **2012**, *511*, 129–132. [CrossRef]

25. Washizu, E.; Yamamoto, A.; Abe, Y.; Kawamura, M.; Sasaki, K. Optical and electrochromic properties of rf reactively sputtered wo 3 films. *Solid State Ionics* **2003**, *165*, 175–180. [CrossRef]

26. Chandramohan, S.; Kanjilal, A.; Tripathi, J.K.; Sarangi, S.N.; Sathyamoorthy, R.; Som, T. Structural and optical properties of mn-doped cds thin films prepared by ion implantation. *J. Appl. Phys.* **2009**, *105*, 3272. [CrossRef]

27. Lu, X.; Zhuang, Z.; Peng, Q.; Li, Y. Wurtzite cu 2 znsns 4 nanocrystals: A novel quaternary semiconductor. *Chem. Commun.* **2011**, *47*, 3141–3143. [CrossRef] [PubMed]

 *nanomaterials* 

Article

Improvement in Light Output of Ultraviolet Light-Emitting Diodes with Patterned Double-Layer ITO by Laser Direct Writing

Jie Zhao [1,2], Xinghuo Ding [1,2], Jiahao Miao [1,2], Jinfeng Hu [1,2], Hui Wan [1,2] and Shengjun Zhou [1,2,3,*]

1   Key Laboratory of Hydraulic Machinery Transients (Wuhan University), Ministry of Education, Wuhan 430072, China; zjie1994@whu.edu.cn (J.Z.); xinghuo@whu.edu.cn (X.D.); miaojh@whu.edu.cn (J.M.); 2017282080101@whu.edu.cn (J.H.); wanhui_hb@whu.edu.cn (H.W.)
2   Center for Photonics and Semiconductors, School of Power and Mechanical Engineering, Wuhan University, Wuhan 430072, China
3   State Key Laboratory of Applied Optics, Changchun Institute of Optics, Fine Mechanics and Physics, Chinese Academy of Sciences, Changchun 130033, China
*   Correspondence: zhousj@whu.edu.cn; Tel.: +86-027-5085-3293

Received: 7 January 2019; Accepted: 30 January 2019; Published: 4 February 2019

Abstract: A patterned double-layer indium-tin oxide (ITO), including the first unpatterned ITO layer serving as current spreading and the second patterned ITO layer serving as light extracting, was applied to obtain uniform current spreading and high light extraction efficiency (LEE) of GaN-based ultraviolet (UV) light-emitting diodes (LEDs). Periodic pinhole patterns were formed on the second ITO layer by laser direct writing to increase the LEE of UV LED. Effects of interval of pinhole patterns on optical and electrical properties of UV LED with patterned double-layer ITO were studied by numerical simulations and experimental investigations. Due to scattering out of waveguided light trapped inside the GaN film, LEE of UV LED with patterned double-layer ITO was improved as compared to UV LED with planar double-layer ITO. As interval of pinhole patterns decreased, the light output power (LOP) of UV LED with patterned double-layer ITO increased. In addition, UV LED with patterned double-layer ITO exhibited a slight degradation of current spreading as compared to the UV LED with a planar double-layer ITO. The forward voltage of UV LED with patterned double-layer ITO increased as the interval of pinhole patterns decreased.

Keywords: UV LEDs; double-layer ITO; pinhole pattern; current spreading; light output power

---

1. Introduction

Ultraviolet (UV) light-emitting diodes (LEDs) have attracted considerable attention due to their potential applications in sterilization, water purification, photocatalyst, and solid-state lighting [1–8]. The improvement of light output is crucial for the realization of high-efficiency UV LEDs. The uniform current spreading is strongly correlated with light output and device reliability [9–12]. Thus, p-type electrodes possessing low contact resistivity as well as high transmittance have been extensively studied. Several kinds of transparent conductive layer such as metal nanowires [13–16], graphene [17–19], carbon nanotubes [20,21], and conductive polymers [22–24] have been reported to improve the current spreading of LEDs. For top-emitting LEDs, indium-tin-oxide (ITO) has been widely used to increase current spreading uniformity over the entire active region due to its high optical transmittance and excellent electrical conductivity [25–27]. In addition, thermal annealing process has been used to improve ohmic contact performance between ITO and p-GaN [28,29]. However, the ITO exhibits optical degradation at UV wavelength region during the thermal annealing process [30].

On the other hand, the light extraction efficiency (LEE) is limited by total internal reflection (TIR) because of large difference in refractive index between sapphire substrate, GaN film, ITO and the air. To overcome TIR at the ITO-air interface, microstructures formed on ITO have been reported to redirect the orientation of photons to favor emission into the air, leading to increasing scattering probability of photons at the ITO-air interface [31–37]. Moreover, compared to randomly distributed ITO pattern, the escape probability of photons emitting from the active region can be further increased by a regularly distributed ITO pattern [38]. However, patterned ITO has been shown to degrade the current spreading performance of UV LED due to an increase in sheet resistance [39–41]. Adding an annealed bottom ITO layer is an effective way to stabilize the current spreading and prevent electrical degradation.

In this study, we combined two single ITO layers into double-layer ITO to obtain excellent ohmic contact and high LEE. The first ITO layer was annealed to form low-resistance ohmic contact with p-GaN. The unannealed second ITO layer was patterned to increase the scattering probability of photons at the ITO-air interface. The periodic pinhole patterns with different intervals were fabricated on the unannealed second ITO layer using laser direct writing. Effects of interval of pinhole patterns on the optical and electrical characteristics of UV LED with patterned double-layer ITO were studied in detail using numerical simulations and experimental investigations.

2. Materials and Methods

The GaN-based UV LED was grown on patterned sapphire substrate (PSS) with a sputtered AlN nucleation layer (15 nm) using the metal-organic chemical vapor deposition (MOCVD) technique. Trimethylgallium (TMGa), trimethylindium (TMIn), trimethylaluminum (TMAl), and ammonia (NH$_3$) were used as precursors of Ga, In, Al, and N, respectively. Silane (SiH$_4$) and bis (cyclopentadienyl) magnesium (Cp$_2$Mg) were used as the n- and p-dopant sources, respectively. The epitaxial structure of UV LED includes a 15-nm-thick sputtered AlN nucleation layer, a 2.5-μm-thick undoped GaN layer, a 2-μm-thick n-GaN layer, a 12-pair In$_{0.02}$Ga$_{0.98}$N (2.1 nm)/GaN (2.3 nm) superlattice layer (SL), a 15-pair In$_{0.07}$Ga$_{0.93}$N (3 nm)/GaN (12 nm) multiple quantum well (MQW) active layer, a 28-nm-thick p-InGaN layer, a 5-pair p-Al$_{0.2}$Ga$_{0.8}$N (1 nm)/GaN(2 nm) SL, a 50-nm-thick p-GaN layer, and a 10-nm-thick p$^+$-GaN layer. The peak emission wavelength of UV LED is 375 nm. Figure 1a shows a schematic representation of the 375 nm UV LED epitaxial structure. Figure 1b shows the cross-sectional transmission electron microscope (TEM) image of the UV LED epitaxial structure grown on PSS. Figure 1c shows the magnified cross-sectional TEM image of In$_{0.02}$Ga$_{0.98}$N/GaN SL, In$_{0.07}$Ga$_{0.93}$N/GaN MQW, p-InGaN layer, and p-Al$_{0.2}$Ga$_{0.8}$N /GaN SL.

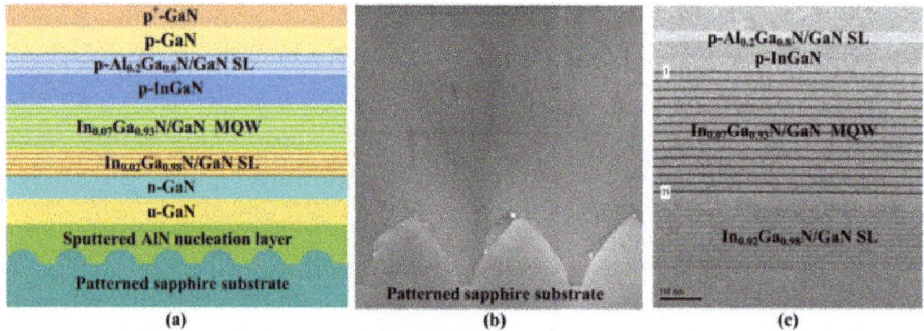

Figure 1. (a) Schematic representation of the UV LED epitaxial structure; (b) cross-sectional TEM image of the UV LED epitaxial structure; (c) magnified cross-sectional TEM image showing In$_{0.02}$Ga$_{0.98}$N/GaN SL, In$_{0.07}$Ga$_{0.93}$N/GaN MQW, and p-Al$_{0.2}$Ga$_{0.8}$N /GaN SL.

UV LEDs with a patterned double-layer ITO were fabricated. The detailed fabrication process is composed of the following steps. Firstly, inductively coupled plasma (ICP) etching based on BCl$_3$/Cl$_2$ mixture gas was used to form GaN mesa by removing a portion of p-GaN layer and MQW active layer to expose the n-GaN layer (Step (1)). Then, the first 30-nm-thick ITO layer was subsequently deposited on the p-GaN layer using an electron beam evaporator, followed by thermal annealing in N_2 ambient at 540 °C for 20 min to obtain good ohmic contact with p-GaN (Step (2)). Next, the second 120-nm-thick ITO layer was deposited on the first ITO layer, and the periodic pinhole patterns were then formed on the second ITO layer by laser direct writing (Step (3)). Finger-like Cr/Pt/Au (30 nm/30 nm/2000 nm) multilayers were adopted for both n-electrode and p-electrodes. The Cr/Pt/Au multilayers were deposited on the second patterned ITO layer for the formation of p-electrode. Meanwhile, the same Cr/Pt/Au multilayers were also deposited on the exposed n-GaN layer for the formation of the n-electrode (Step (4)). Finally, the UV LED wafers were thinned down to about 120 μm and diced into chips with dimension of 1132×1132 μm^2 (Step (5)). Except for different interval of pinhole patterns fabricated on the second ITO layer, the fabrication process of these UV LEDs is identical. The diameters of pinhole patterns were 2 μm in UV LED II, UV LED III, and UV LED IV, while intervals of pinhole patterns were 400 nm, 600 nm, and 800 nm, respectively. For comparison, the UV LED I with planar double-layer ITO was also prepared. The light output power (LOP)-current-voltage (L-I-V) characteristics of UV LEDs were measured using an integrating sphere and a semiconductor parameter analyzer (Keysight B2901A). Figure 2a shows a schematic representation of the UV LED with a patterned double-layer ITO. Figure 2b shows a cross-sectional schematic representation of the UV LED with a patterned double-layer ITO. Figure 2c shows the top-view scanning electron microscopy (SEM) image of the fabricated UV LED chip.

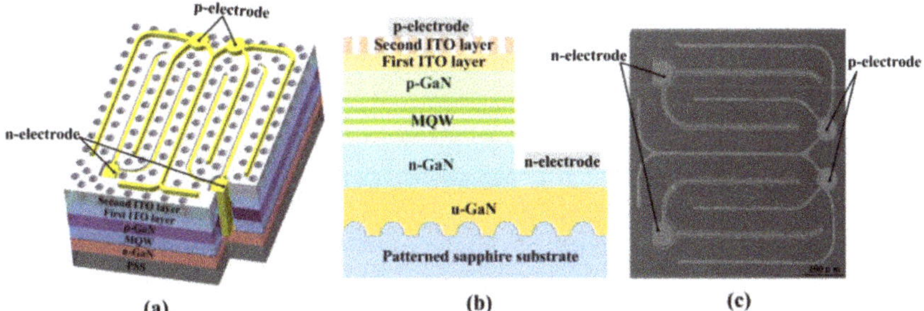

Figure 2. (a) Schematic representation of the UV LED with a patterned double-layer ITO; (b) cross-sectional schematic representation of the UV LED with a patterned double-layer ITO; (c) top-view SEM image of the fabricated UV LED chip.

Figure 3 shows an energy band diagram of the UV LED at forward bias. When a current is passed through the UV LED, electrons and holes are injected into the $In_{0.07}Ga_{0.93}N$/GaN MQW active layer from the n-type GaN and p-type GaN layers, respectively. The electrons and holes recombine radiatively in the $In_{0.07}Ga_{0.93}N$/GaN MQW active layer, thereby emitting photons of light, as shown in Figure 3.

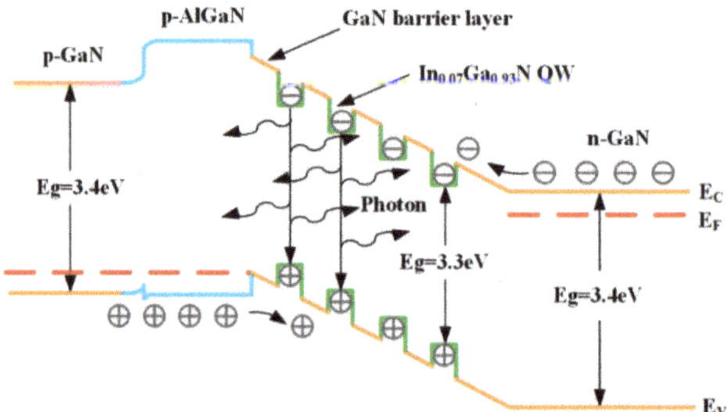

Figure 3. Energy band diagram of the UV LED at forward bias.

3. Results and Discussion

The pinhole patterns on the second ITO layer were fabricated by laser direct writing system (4PICO.BV, Netherlands). First, the photoresist (ma-p1275HV) was spin-coated on GaN-based wafer at 4000 rpm for 60 s. Then, laser direct writing was performed on the photoresist film to obtain pinhole patterns, which will be transferred to the ITO film. The irradiating red laser beam (λ = 635 nm) was focused onto the surface of spin-coated photoresist through a dedicated objective lens (numerical aperture (NA) = 0.85, focal length = 0.6 mm, and magnification = 100×) to correct for surface errors. Then an ultraviolet laser beam (λ = 405 nm) was normally directed to the surface of photoresist with a spot diameter of 300 nm and an energy density of 60 mJ/cm^2 through the same objective lens. The laser beam was scanned on the photoresist with a scan speed of 200 mm/s and stepping distance of 150 nm by a motorized stage. The wafer was then submersed into developer (ma-D 331) for 70 s to remove the area that has been exposed. Finally, pinhole patterns were transferred from photoresist onto the second ITO layer by wet chemical etching process. Figure 4 shows top-view SEM images of the pinhole patterns with different intervals. The diameter and intervals of the pinhole patterns of UV LED II, UV LED III, and UV LED IV were 2 µm/400 nm, 2 µm/600 nm, and 2 µm/800 nm, respectively.

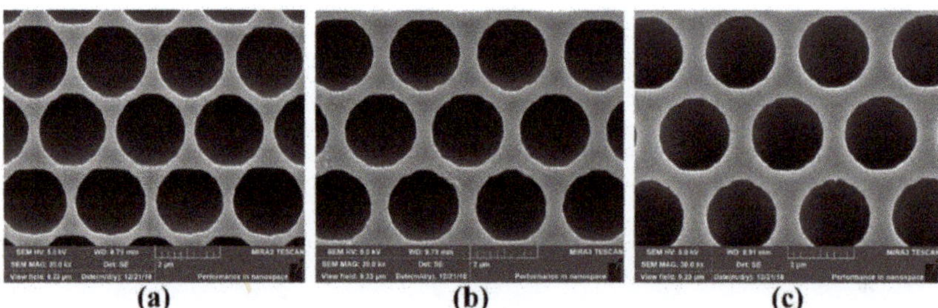

(a) **(b)** **(c)**

Figure 4. Top-view SEM images of the pinhole patterns with different intervals: (**a**) 400 nm; (**b**) 600 nm; (**c**) 800 nm.

We measured transmittance spectra of the evaporated ITO films before and after thermal annealing. Figure 5 shows the transmittance spectra of unannealed ITO and annealed ITO films in the wavelength range of 370–440 nm measured by a UV/VIS/NIR spectrometer. The transmittance of the unannealed ITO and annealed ITO were 68.81% and 66.88% at 375 nm, respectively. The transmittance of the unannealed ITO was 2.89% higher than that of the annealed ITO. It was indicated that the transmittance of ITO exhibited optical degradation at UV wavelength region after thermal annealing, which would lead to a decrease in LEE of UV LED. To avoid the decrease in transmittance of ITO at the UV wavelength region, the second ITO layer remains unannealed in our design. In addition, a combined method including electron beam evaporation and subsequent magnetron sputtering can further improve the optical transmittance of ITO [42].

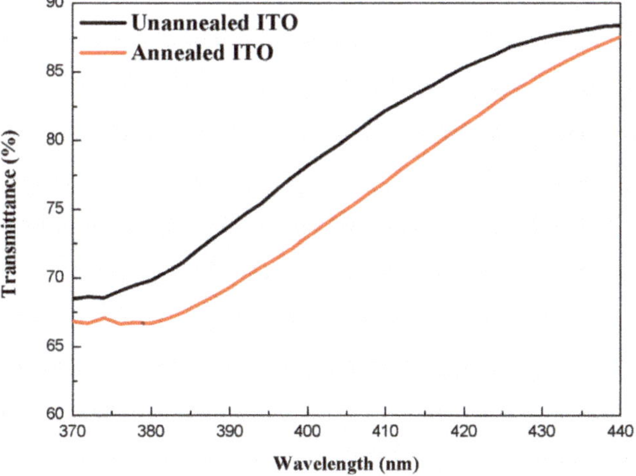

Figure 5. Transmission spectra of the unannealed ITO and annealed ITO films in the wavelength range of 370–440 nm.

We investigated the effect of interval of pinhole patterns on current spreading of UV LED with double-layer ITO using SimuLED commercial software package. Figure 6 shows simulation results of the current density distribution of UV LED I, UV LED II, UV LED III, and UV LED IV at 350 mA. The root mean square (RMS) of current density in the active region of UV LED I, UV LED II, UV LED III, and UV LED IV was 35.02 A/cm^2, 38.76 A/cm^2, 37.27 A/cm^2, and 37.09 A/cm^2, respectively. Compared with UV LED I, the RMS of current density in the active region of UV LED with patterned double-layer ITO was slightly increased. Meanwhile, the RMS of current density in the active region of UV LEDs increased with decreasing interval of the pinhole patterns on the second ITO layer. The slightly increased RMS of current density resulted from non-uniform current spreading.

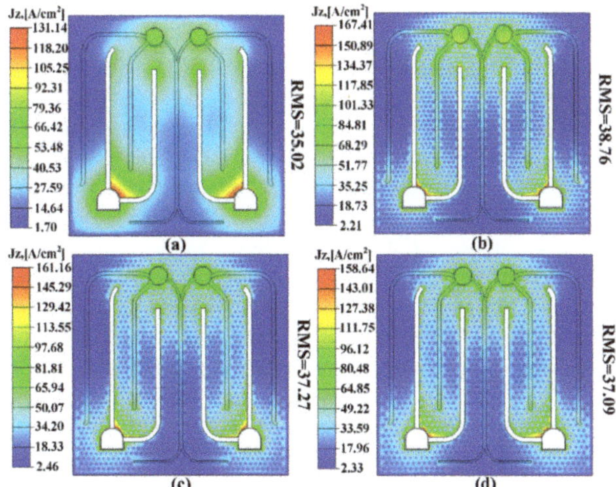

Figure 6. SimuLED simulation of the current density distribution in the active region of UV LED at 350 mA: (**a**) UV LED I; (**b**) UV LED II; (**c**) UV LED III; (**d**) UV LED IV.

The Finite-Difference Time-Domain (FDTD) method was applied for numerical simulation to clarify the effect of interval of pinhole patterns on the LEE of UV LED. To save computational memory and simulation time, the thickness of the sapphire substrate was fixed at 1 μm. Other structure parameters in the simulation were similar with the actual LED structure. The simplified UV LED structure model was depicted in Figure 7. The boundary conditions were a perfectly matched layer (PML) on top of the simulation region. The PML boundary absorbs incident light with zero reflection. Reflective metal boundaries were used at the bottom and four broad sidewalls. The top monitor was located 800 nm above the surface of the second patterned ITO to receive the optical waves extracted to air and calculate the LEE of the UV LED.

In order to present the random propagation of unpolarized photons within the active region, sufficient dipole sources with wavelength of 375 nm were regularly distributed throughout the middle of the MQW layer with an interval of 1 μm. Three orientations of dipole sources were along the x-, y-, and z-axis. Compared to UV LED I, the LEEs of UV LED II, UV LED III, and UV LED IV were improved by 13.06%, 12.04%, and 7.02%, respectively. The enhancement in LEE was attributed to the formation of pinhole patterns on the second ITO layer, which redirected the photon propagation into the air escape cone and provided the photons with multiple opportunities to escape from the LED surface.

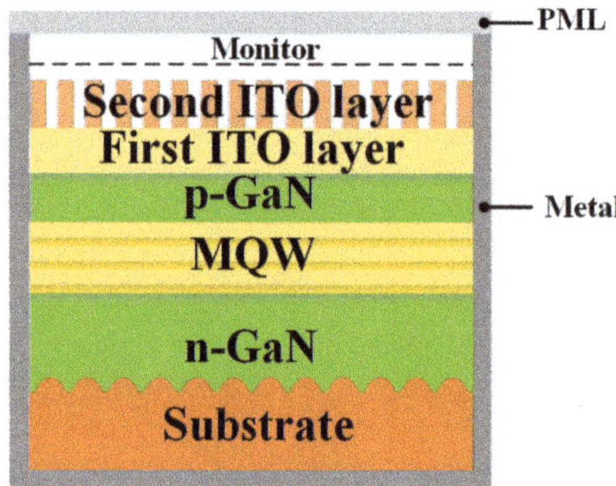

Figure 7. Schematic illustration of the FDTD simulation model for UV LED with a patterned double layer ITO.

The LOPs of UV LEDs versus injection current were shown in Figure 8a. At an injection current of 350 mA, the LOPs of UV LED I, UV LED II, UV LED III, and UV LED IV were 32.60 mW, 36.70 mW, 36.38 mW, and 34.75 mW, respectively. Compared with UV LED I, the LOPs of UV LED II, UV LED III, and UV LED IV were improved by 12.4%, 11.7%, and 6.5%, respectively. The improvement of LOP was attributed to increased LEE caused by formation of a pinhole pattern on the second ITO layer, as evidenced by FDTD simulation. Moreover, LOP of UV LED with a patterned double-layer ITO increased with the decrease of interval of pinhole patterns, owing to the higher density of pinhole pattern on the second ITO layer leading to an increased escape probability of photons. Figure 8b shows the I–V characteristic of UV LEDs. At 350 mA, the forward voltages of UV LED I, UV LED II, UV LED III, and UV LED IV were 3.88 V, 4.01 V, 3.99 V, and 3.97 V, respectively. The forward voltage of UV LED with patterned double-layer ITO was slightly higher than that of UV LED with planar double-layer ITO. The slightly higher forward voltage originated from the pinhole patterns on the second ITO that impeded lateral current spreading. Additionally, the forward voltage of UV LED with a patterned double-layer ITO increased with decreasing interval of the pinhole patterns.

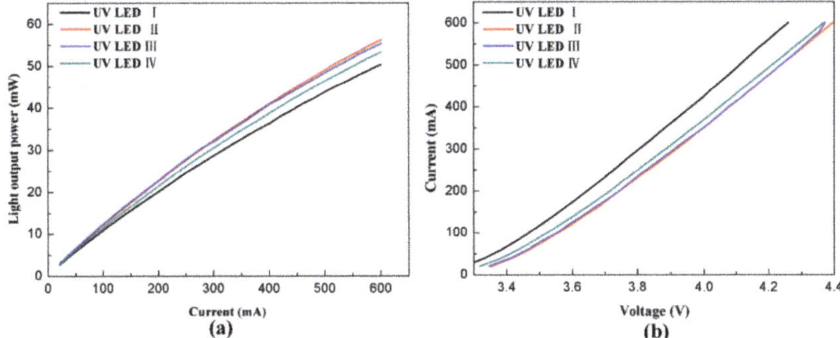

Figure 8. (**a**) L-I characteristic and (**b**) I–V characteristic of UV LED I, UV LED II, UV LED III, and UV LED IV.

4. Conclusions

The first unpatterned current spreading layer and the second patterned light extracting layer were combined into a patterned double-layer ITO to improve current spreading and LEE of UV LEDs. The interval of the pinhole patterns on the second ITO has a significant impact on the sheet resistance of patterned double-layer ITO and the way of light propagation at the ITO-air interface. Compared to UV LED with planar double-layer ITO, LEE of UV LED with a patterned double-layer ITO was improved due to the scattering out of the waveguided light trapped inside the GaN film. The LOP of UV LED with a patterned double-layer ITO increased with decreasing interval of the pinhole patterns due to the higher density of pinhole pattern formed on the second ITO. Owing to non-uniform current spreading caused by pinhole patterns on the second ITO, the forward voltage of UV LED with a patterned double-layer ITO was slightly increased.

Author Contributions: S.Z. originally conceived the idea. J.Z. and S.Z. wrote the manuscript. X.D. contributed to the manuscript preparation. J.M. carried out SEM, TEM, and transmittance measurements. J.H. and H.W. carried out the simulations. All authors discussed the progress of the research and reviewed the manuscript.

Funding: This work was supported by the National Key R&D Program of China (No. 2017YFB1104900), the National Natural Science Foundation of China (Grant Nos. 51675386, 51305266), and Hubei Province Science Fund for Distinguished Young Scholars (No. 2018CFA091). We acknowledge nanofabrication assistance from the Center for Nanoscience and Nanotechnology at Wuhan University.

Conflicts of Interest: The authors declare no conflict of interest.

References

1. Zou, X.; Lin, Y.; Xu, B.; Cao, T.; Tang, Y.; Pan, Y.; Gao, Z.; Gao, N. Enhanced inactivation of *E. coli* by pulsed UV-LED irradiation during water disinfection. *Sci. Total Environ.* **2019**, *650*, 210–215. [CrossRef] [PubMed]

2. Rattanakul, S.; Oguma, K. Inactivation kinetics and efficiencies of UV-LEDs against *Pseudomonas aeruginosa*, *Legionella pneumophila*, and surrogate microorganisms. *Water Res.* **2018**, *130*, 31–37. [CrossRef] [PubMed]

3. Zhou, S.; Xu, H.; Hu, H.; Gui, C.; Liu, S. High quality GaN buffer layer by isoelectronic doping and its application to 365 nm InGaN/AlGaN ultraviolet light-emitting diodes. *Appl. Surf. Sci.* **2019**, *471*, 231–238. [CrossRef]

4. Oguma, K.; Kanazawa, K.; Kasuga, I.; Takizawa, S. Effects of UV Irradiation by Light Emitting Diodes on Heterotrophic Bacteria in Tap Water. *Photochem. Photobiol.* **2018**, *94*, 570–576. [CrossRef] [PubMed]

5. Hu, H.; Zhou, S.; Liu, X.; Gao, Y.; Gui, C.; Liu, S. Effects of GaN/AlGaN/Sputtered AlN nucleation layers on performance of GaN-based ultraviolet light-emitting diodes. *Sci. Rep.* **2017**, *7*, 44627. [CrossRef] [PubMed]

6. Guo, H.; Huang, X.; Zeng, Y. Synthesis and photoluminescence properties of novel highly thermal-stable red-emitting Na$_3$Sc$_2$(PO$_4$)$_3$:Eu^{3+} phosphors for UV-excited white-light-emitting diodes. *J. Alloys Compd.* **2018**, *741*, 300–306. [CrossRef]

7.   Fernandes, R.A.; Sampaio, M.J.; Da Silva, E.S.; Serp, P.; Faria, J.L.; Silva, C. Synthesis of selected aromatic aldehydes under UV-LED irradiation over a hybrid photocatalyst of carbon nanofibers and zinc oxide. *Catal. Today* **2018**. [CrossRef]

8.   Zhou, S.; Hu, H.; Liu, X.; Liu, M.; Ding, X.; Gui, C.; Liu, S.; Guo, L.J. Comparative study of GaN-based ultraviolet LEDs grown on different-sized patterned sapphire substrates with sputtered AlN nucleation layer. *Jpn. J. Appl. Phys.* **2017**, *56*, 111001. [CrossRef]

9.   Qian, C.; Li, Y.; Fan, J.; Fan, X.; Fu, J.; Zhao, L.; Zhang, G. Studies of the light output properties for a GaN based blue LED using an electro-optical simulation method. *Microelectron. Reliab.* **2017**, *74*, 173–178. [CrossRef]

10.   Hao, G.D.; Taniguchi, M.; Tamari, N.; Inoue, S. Enhanced wall-plug efficiency in AlGaN-based deep-ultraviolet light-emitting diodes with uniform current spreading p-electrode structures. *J. Phys. D Appl.* **2016**, *49*, 235101. [CrossRef]

11.   Kudryk, Y.Y.; Zinovchuk, A.V. Efficiency droop in InGaN/GaN multiple quantum well light-emitting diodes with nonuniform current spreading. *Semicond. Sci. Technol.* **2011**, *26*, 095007. [CrossRef]

12.   Huang, S.; Fan, B.; Chen, Z.; Zheng, Z.; Luo, H.; Wu, Z.; Wang, G.; Jiang, H. Lateral current spreading effect on the efficiency droop in GaN based light-emitting diodes. *J. Disp. Technol.* **2013**, *9*, 266–271. [CrossRef]

13.   Djavid, M.; Choudhary, D.D.; Philip, M.R.; Bui, T.H.Q.; Akinnuoye, O.; Pham, T.T.; Nguyen, H.P.T. Effects of optical absorption in deep ultraviolet nanowire light-emitting diodes. *Photonics Nanostruct.* **2018**, *28*, 106–110. [CrossRef]

14.   Mock, J.; Bobinger, M.; Bogner, C.; Lugli, P.; Becherer, M. Aqueous Synthesis, Degradation, and Encapsulation of Copper Nanowires for Transparent Electrodes. *Nanomaterials* **2018**, *8*, 767. [CrossRef]

15.   Park, J.S.; Kim, J.H.; Na, J.Y.; Kim, D.H.; Kang, D.; Kim, S.K.; Seong, T.Y. Ag nanowire-based electrodes for improving the output power of ultraviolet AlGaN-based light-emitting diodes. *J. Alloys Compd.* **2017**, *703*, 198–203. [CrossRef]

16.   Gui, C.; Ding, X.; Zhou, S.; Gao, Y.; Liu, X.; Liu, S. Nanoscale Ni/Au wire grids as transparent conductive electrodes in ultraviolet light-emitting diodes by laser direct writing. *Opt. Laser Technol.* **2018**, *104*, 112–117. [CrossRef]

17.   Alonso, E.T.; Rodrigues, D.P.; Khetani, M.; Shin, D.W.; Sanctis, A.D.; Joulie, H.; Schrijver, I.; Baldycheva, A.; Alves, H.; Neves, A.I.S.; et al. Graphene electronic fibres with touch-sensing and light-emitting functionalities for smart textiles. *NPJ Flexible Electron.* **2018**, *2*, 25. [CrossRef]

18.   Bointon, T.H.; Russo, S.; Craciun, M.F. Is graphene a good transparent electrode for photovoltaics and display a pplications? *IET Circ. Devices Syst.* **2015**, *9*, 403–412. [CrossRef]

19.   Alonso, E.T.; Karkera, G.; Jones, G.F.; Craciun, M.F.; Russo, S. Homogeneously bright, flexible, and foldable lighting devices with functionalized graphene electrodes. *ACS Appl. Mater. Interfaces* **2016**, *8*, 16541–16545. [CrossRef]

20.   Wu, Z.; Chen, Z.; Du, X.; Logan, J.M.; Sippel, J.; Nikolou, M.; Kamaras, K.; Reynolds, J.R.; Tanner, D.B.; Hebard, A.F.; et al. Transparent, Conductive Carbon Nanotube Films. *Science* **2004**, *305*, 1273–1276. [CrossRef]

21.   Hu, L.; Hecht, D.S.; Grüner, G. Percolation in Transparent and Conducting Carbon Nanotube Networks. *Nano Lett.* **2004**, *4*, 2513–2517. [CrossRef]

22.   Park, J.S.; Kim, J.H.; Kim, J.Y.; Kim, D.H.; Kang, D.; Sung, J.S.; Seong, T.Y. Hybrid indium tin oxide/Ag nanowire electrodes for improving the light output power of near ultraviolet AlGaN-based light-emitting diode. *Curr. Appl. Phys.* **2016**, *16*, 545–548. [CrossRef]

23.   Lai, W.C.; Lin, C.N.; Lai, Y.C.; Yu, P.; Chi, G.C.; Chang, S.J. GaN-based light-emitting diodes with graphene/indium tin oxide transparent layer. *Opt. Express* **2014**, *22*, A396–A401. [CrossRef] [PubMed]

24.   Kim, Y.H.; Lee, J.; Hofmann, S.; Gather, M.C.; Müller-Meskamp, L.; Leo, K. Achieving High Efficiency and Improved Stability in ITO-Free Transparent Organic Light-Emitting Diodes with Conductive Polymer Electrodes. *Adv. Funct. Mater.* **2013**, *23*, 3763–3769. [CrossRef]

25.   Kim, H.; Lee, S.N. Theoretical considerations on current spreading in GaN-based light emitting diodes fabricated with top-emission geometry. *J. Electrochem. Soc.* **2010**, *157*, H562–H564. [CrossRef]

26.   Jo, Y.J.; Hong, C.H.; Kwak, J.S. Improved electrical and optical properties of ITO thin films by using electron beam irradiation and their application to UV-LED as highly transparent p-type electrodes. *Curr. Appl. Phys.* **2011**, *11*, S143–S146. [CrossRef]

27. Liu, Y.J.; Huang, C.C.; Chen, T.Y.; Hsu, C.S.; Liou, J.K.; Tsai, T.Y.; Liu, W.C. Implementation of an indium-tin-oxide (ITO) direct-Ohmic contact structure on a GaN-based light emitting diode. *Opt. Express* **2011**, *19*, 14662–14670. [CrossRef]

28. Waki, I.; Fujioka, H.; Oshima, M.; Miki, H.; Fukizawa, A. Low-temperature activation of Mg-doped GaN using Ni films. *Appl. Phys. Lett.* **2001**, *78*, 2899. [CrossRef]

29. Nakamura, S.; Mukai, T.; Senoh, M.; Iwasa, N. Thermal annealing effects on p-type Mg-doped GaN films. *Jpn. J. Appl. Phys.* **1992**, *31*, L139. [CrossRef]

30. Zhou, S.; Yuan, S.; Liu, Y.; Guo, L.J.; Liu, S.; Ding, H. Highly efficient and reliable high power LEDs with patterned sapphire substrate and strip-shaped distributed current blocking layer. *Appl. Surf. Sci.* **2015**, *355*, 1013–1019. [CrossRef]

31. Jung, S.H.; Song, K.M.; Choi, Y.S.; Park, H.H.; Shin, H.B.; Kang, H.K.; Lee, J. Light output enhancement of InGaN/GaN light-emitting diodes with contrasting indium tin-oxide nanopatterned structures. *J. Nanomater.* **2013**, *2013*, 7. [CrossRef]

32. Tsai, C.F.; Su, Y.K.; Lin, C.L. Improvement in External Quantum Efficiency of InGaN-Based LEDs by Micro-Textured Surface with Different Geometric Patterns. *J. Electrochem. Soc.* **2011**, *159*, H151–H156. [CrossRef]

33. Zhong, Y.; Shin, Y.C.; Kim, C.M.; Lee, B.G.; Kim, E.H.; Park, Y.J.; Sobahan, K.M.A.; Hwangbo, C.K.; Lee, Y.P.; Kim, T.G. Optical and electrical properties of indium tin oxide thin films with tilted and spiral microstructures prepared by oblique angle deposition. *J. Mater. Res.* **2008**, *23*, 2500–2505. [CrossRef]

34. Chang, S.J.; Shen, C.F.; Chen, W.S.; Kuo, C.T.; Ko, T.K.; Shei, S.C.; Sheu, J.K. Nitride-based light emitting diodes with indium tin oxide electrode patterned by imprint lithography. *Appl. Phys. Lett.* **2007**, *91*, 013504. [CrossRef]

35. Horng, R.H.; Yang, C.C.; Wu, J.Y.; Huang, S.H.; Lee, C.E.; Wuu, D.S. GaN-based light-emitting diodes with indium tin oxide texturing window layers using natural lithography. *Appl. Phys. Lett.* **2005**, *86*, 221101. [CrossRef]

36. Leem, D.S.; Cho, J.; Sone, C.; Park, Y.; Seong, T.Y. Light-output enhancement of GaN-based light-emitting diodes by using hole-patterned transparent indium tin oxide electrodes. *J. Appl. Phys.* **2005**, *98*, 076107. [CrossRef]

37. Kang, J.H.; Ryu, J.H.; Kim, H.K.; Kim, H.Y.; Han, N.; Park, Y.J.; Uthirakumar, P.; Hong, C.H. Comparison of various surface textured layer in InGaN LEDs for high light extraction efficiency. *Opt. Express* **2011**, *19*, 3637–3646. [CrossRef]

38. Zhou, S.; Cao, B.; Liu, S.; Ding, H. Improved light extraction efficiency of GaN-based LEDs with patterned sapphire substrate and patterned ITO. *Opt. Laser Technol.* **2012**, *44*, 2302–2305. [CrossRef]

39. Huang, H.; Hu, J.; Wang, H. GaN-based light-emitting diodes with hybrid micro/nano-textured indium-tin-oxide layer. *J. Semicond.* **2014**, *35*, 084006. [CrossRef]

40. Oh, S.; Su, P.C.; Yoon, Y.J.; Cho, S.; Oh, J.H.; Seong, T.Y.; Kim, K.K. Nano-patterned dual-layer ITO electrode of high brightness blue light emitting diodes using maskless wet etching. *Opt. Express* **2013**, *21*, A970–A976. [CrossRef]

41. Leem, D.S.; Lee, T.; Seong, T.Y. Enhancement of the light output of GaN-based light-emitting diodes with surface-patterned ITO electrodes by maskless wet-etching. *Solid-State Electron.* **2007**, *51*, 793–796. [CrossRef]

42. Markov, L.K.; Smirnova, I.P.; Pavluchenko, A.S.; Kukushkin, M.V.; Zakheim, D.A.; Pavlov, S.I. Use of double-layer ITO films in reflective contacts for blue and near-UV LEDs. *Semiconductors* **2014**, *48*, 1674–1679. [CrossRef]

 *nanomaterials*

Article

Enhanced Light Extraction of Flip-Chip Mini-LEDs with Prism-Structured Sidewall

Bin Tang [1], Jia Miao [1], Yingce Liu [2], Hui Wan [3], Ning Li [1], Shengjun Zhou [1,4,5,*] and Chengqun Gui [3]

1   Key Laboratory of Hydraulic Machinery Transients (Wuhan University), Ministry of Education, Wuhan 430072, China; bintang@whu.edu.cn (B.T.); miaojia789@whu.edu.cn (J.M.); liningmick@whu.edu.cn (N.L.)
2   Xiamen Changelight Co. Ltd., Xiamen 361000, China; tigerlyc@163.com
3   The Institute of Technological Sciences, Wuhan University, Wuhan 430072, China; wanhui_hb@whu.edu.cn (H.W.); cheng.gui.2000@gmail.com (C.G.)
4   Center for Photonics and Semiconductors, School of Power and Mechanical Engineering, Wuhan University, Wuhan 430072, China
5   State Key Laboratory of Applied Optics, Changchun Institute of Optics, Fine Mechanics and Physics, Chinese Academy of Sciences, Changchun 130033, China
*   Correspondence: zhousj@whu.edu.cn

Received: 22 January 2019; Accepted: 26 February 2019; Published: 28 February 2019

Abstract: Current solutions for improving the light extraction efficiency of flip-chip light-emitting diodes (LEDs) mainly focus on relieving the total internal reflection at sapphire/air interface, but such methods hardly affect the epilayer mode photons. We demonstrated that the prism-structured sidewall based on tetramethylammonium hydroxide (TMAH) etching is a cost-effective solution for promoting light extraction efficiency of flip-chip mini-LEDs. The anisotropic TMAH etching created hierarchical prism structure on sidewall of mini-LEDs for coupling out photons into air without deteriorating the electrical property. Prism-structured sidewall effectively improved light output power of mini-LEDs by 10.3%, owing to the scattering out of waveguided light trapped in the gallium nitride (GaN) epilayer.

Keywords: flip-chip mini-LED; prism-structured sidewall; waveguide photons; light extraction

1. Introduction

The developments in GaN-based light-emitting diodes (LEDs) have promoted the liquid crystal display (LCD) as a highly competitive display technology in the past few decades [1,2]. Recently, the application of mini-LEDs with size below 200 microns gains new advantages for LCDs in market competition owing to their prominent merits as backlight unit, such as long life span, low energy consumption and high resolution [3,4]. However, to meet the high dynamic range (HDR) requirements of next generation displays, the luminance of the LCD bright state should be over 1000 nits, which requests the mini-LED backlight unit to be much more energy efficient [5].

Tremendous efforts have been done to improve the efficiency of GaN-based LEDs, which can be principally divided into two categories: improving the crystal quality of epilayer [6–10] and boosting the light extraction efficiency (LEE) [11–15]. Since mini-LEDs are obtained from the identical epilayer as broad-area LEDs, the fruitful methods for high crystal quality epilayer are universal in fabrication of the two kinds of LEDs. Since the ratio of top emitting area to sidewall emitting area is greatly different for mini-LEDs and broad-area LEDs, the methods applicable in broad-area LEDs for improved LEE need to be reassessed in mini-LEDs.

Since the top surface area of mini-LED is much reduced, flip-chip structure is the preferable choice for mini-LEDs to ensure enough top emitting area as well as p-contact area. Owing to the index mismatch of GaN epilayer, sapphire substrate and air, the majority of photons are trapped in the high-index epilayer and sapphire substrate by total internal reflection (TIR) and guided laterally as waveguide modes, which finally dissipate in the lossy epilayer. Several methods have been proposed to extract the waveguide photons of flip-chip LEDs, such as texturing the sapphire subtract surface [16–18], shaping the sapphire substrate [19,20], plasmonic structure [21], and depositing nanoparticles or scattering layers [22,23]. Such approaches are effective in extraction of the sapphire substrate mode photons, while additional protection procedures are generally needed owing to the more inert property of sapphire substrate relative to the epilayer. More importantly, these approaches have no effect on epilayer mode photons. Simulation and experimental results have shown that reducing the pattern size of patterned sapphire substrate (PSS) is effective in improving the LEE by scattering out epilayer mode photons [24,25]. However, small pattern size is disadvantage for crystalline quality of epilayer [26] and the minimal spacing of PSS achieved in practice is far from the best outcoupling spacing for light extraction [27]. Thus, further works are still needed to extract the epilayer mode photons of flip-chip LEDs.

In this study, we demonstrated a simple and reliable method to extract the epilayer mode photons. Prism-structured sidewall generated by tetramethylammonium hydroxide (TMAH)-based crystallographic etching was introduced to scatter out epilayer mode photons. The size of prism structure on the sidewall of mini-LED could be manipulated from nanoscale to a few microns by adjusting TMAH etching time to achieve the best outcoupling efficiency. The anisotropic TMAH etching is damage-free and practical in mass-production, which makes the prism-structured sidewall based on TMAH etching a promising solution for highly efficient mini-LEDs.

2. Materials and Methods

The GaN-based LEDs were grown on c-plane PSS using metal–organic chemical vapor deposition (MOCVD) method. The LED epitaxial structure consisted of a 25-nm-thick low temperature GaN nucleation layer, a 3.0-μm-thick undoped GaN buffer layer, a 2.5-μm-thick Si-doped n-GaN layer, a 12-pair of InGaN (3 nm)/GaN (12 nm) multiple quantum well (MQW), a 40-nm-thick p-AlGaN electron blocking layer, and a 112-nm-thick Mg-doped p-GaN layer. The LED wafer was subsequently annealed at 750 °C at N_2 atmosphere to activate Mg acceptor in the p-GaN. Then, the photolithography and inductively coupled plasma etching (ICP) process based on BCl_3/Cl_2 mixture gas were performed to form the mesa structure and deep isolation trench. Afterwards, the TMAH-based crystallographic etching procedure was applied. The LED wafers were dipped into the 15 wt% TMAH solution at 85 °C during the TMAH etching process. After rinsing with deionized water and drying under N_2 flow, a 60-nm-thick indium tin oxide (ITO) transparent conductive layer was evaporated on the p-GaN layer. Cr/Al/Ti/Pt/Au metallization schemed as ohmic contact layer was deposited on the ITO and n-GaN layer. Sixteen pairs of quarter-wavelength-thick TiO_2/SiO_2 stacks, as distributed Bragg reflectors, were sputtered by ion beam deposition followed by the opening of via through DBR using ICP etching based on $CHF_3/Ar/O_2$ mixture gas. Cr/Ti/Pt/Au metallization was evaporated as contact pads subsequently. Finally, the LED wafer was thinned down to about 150 μm and diced into chips with dimensions of 101 μm × 200 μm. The peak emission wavelength of the fabricated mini-LEDs was 456 nm. The light output power–current–voltage (*L-I-V*) characteristics of mini-LEDs were measured using a semiconductor parameter analyzer (Keysight B2901A, Santa Rosa, CA, USA) with an integrating sphere. In this work, two types of mini-LED chips with different sidewall orientations on the same LED wafer (as shown in Figure 1) were investigated by the scanning electron microscope (SEM) owing to the anisotropic etching behavior of TMAH-based crystallographic etching. The two types of mini-LEDs were named as mini-LED I and mini-LED II according to the sidewall orientation. The larger sidewalls of mini-LED I were set to be orientated along [10-10] direction, while the larger sidewalls of mini-LED II were set to be orientated along [1-210] direction.

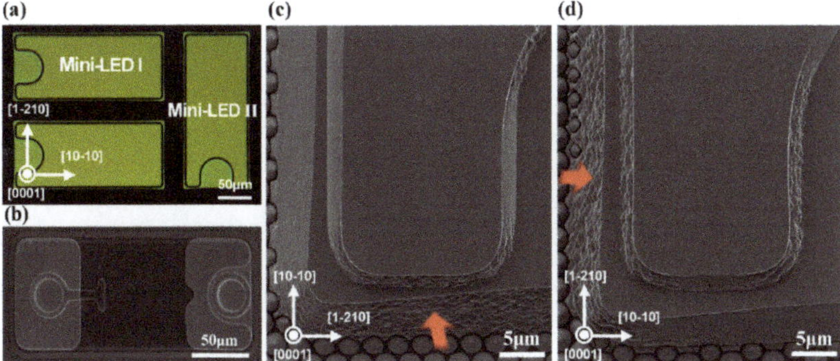

Figure 1. (**a**) Optical microscope image of the epilayer after ICP procedure, showing the orthogonal arrangements of mini-LED I and mini-LED II. (**b**) SEM image of the fabricated flip-chip mini-LED with a bird's eye view, showing the dimension of the flip-chip mini-LED. (**c**) SEM image of the mini-LED I after TMAH etching treatment, the red arrow in the image points to the prism-structured sidewall. (**d**) SEM image of the mini-LED II after TMAH etching treatment, the red arrow in the image points to the prism-structured sidewall.

3. Results and Discussion

We took advantage of the anisotropic etching behavior of TMAH-based crystallographic etching to obtain the textured sidewall structure. No additional protection procedure was incorporated owing to the damage-free and anisotropic etching properties of TMAH-based crystallographic etching, making it a convenient and cost effective solution to texture the sidewall of GaN epilayer. The selective etching ability of TMAH solution arises from the difference in density of N dangling bonds on different GaN lattice planes [28]. Surfaces with high density of N dangling bonds possess large repulsion to the hydroxide ions, which stops the crystallographic etching at such surface [29]. The density of N dangling bonds on different GaN lattice planes can be ranked as follows: (0001) plane > (1-210) plane > (10-10) plane > (000-1) plane [30]. Thus, the (0001) plane and (1-210) plane have larger repulsion force to the hydroxide ions, and the TMAH etching did not proceed on the top surface and the sidewall along [10-10] direction under our experiment condition. As shown in Figure 1c,d, no prism structure appeared on the top surface and the sidewall along [10-10] direction while the hierarchical prism structure appeared throughout the sidewall along [1-210] direction. Moreover, the textured sidewall surface area was different for the two types of mini-LEDs investigated owing to their orthogonal arrangements, which resulted in discrepant light output power as demonstrated by the *L-I-V* characteristics. The surface morphology of the sidewall along [1-210] direction with different TMAH etching time was characterized by SEM, as shown in Figure 2. Owing to the anisotropic etching property, the smooth surface was left with hierarchical prism structure after TMAH treatment. Trigonal prisms close to the PSS presented larger size than that in other regions, suggesting a larger TMAH etching rate near the interface of PSS and GaN. The larger etching rate may arise from that the TMAH etching started from the (000-1) plane at the interface. Within the etching time investigated, the TMAH etching proceeded at selected lattice planes and trigonal prism structures varied from nanoscale to a few microns as the TMAH etching time increased. The influence of prism size on light extraction is discussed below.

To verify the TMAH etching is a damage-free process, the *I-V* characteristics of mini-LEDs with and without TMAH etching treatment were investigated, as shown in Figure 3a. A 7.5 min TMAH etching procedure only caused slight variation in forward voltages of mini-LEDs, suggesting no electrical degradation was brought in by the TMAH etching process. Previous reports on improving the sidewall surface area by ICP etching generally incorporate plasma damage, which leads to deteriorated electrical property [31,32]. The TMAH etching demonstrated here provided an alternative solution without reducing the top emitting area and deteriorating the electrical property.

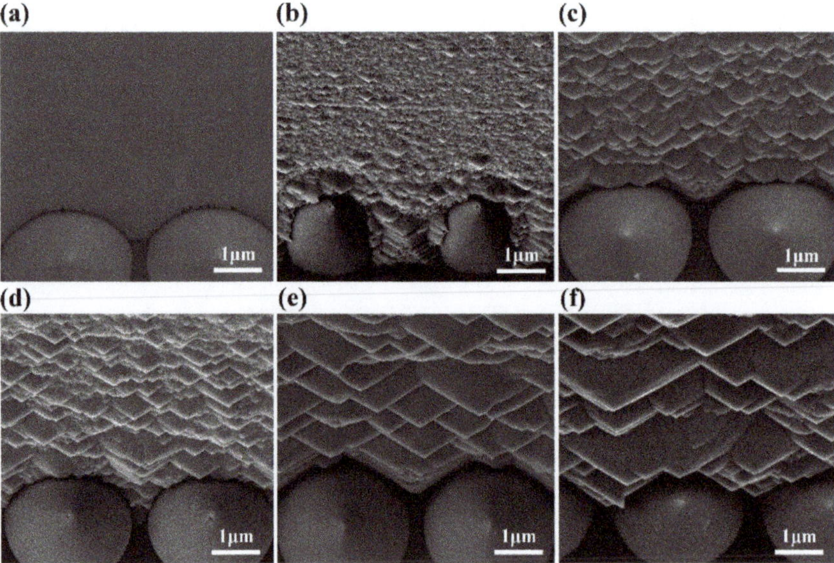

Figure 2. (**a–f**) SEM images of the chip sidewall along [1-210] direction with various TMAH etching time: (**a**) with 0 min TMAH etching treatment; (**b**) with 2.5 min TMAH etching treatment; (**c**) with 5 min TMAH etching treatment; (**d**) with 7.5 min TMAH etching treatment; (**e**) with 10 min TMAH etching treatment; and (**f**) with 20 min TMAH etching treatment.

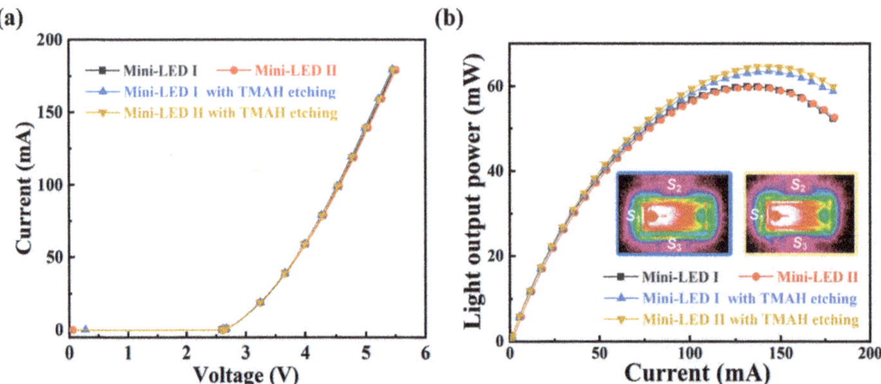

Figure 3. (**a**) *I-V* curves of the investigated two types of mini-LEDs with and without TMAH etching treatment. (**b**) *L-I* curves of the fabricated mini-LEDs with and without 7.5 min TMAH etching treatment. The insets show the photographs of TMAH treated LEDs under 10 mA injection current: mini-LED I with TMAH etching (**left**) and mini-LED II with TMAH etching (**right**).

Figure 3b shows the *L-I* curves of the investigated mini-LEDs, and the insets are photographs of TMAH treated mini-LEDs under 10 mA injection current. The *L-I* characteristics for mini-LED I and mini-LED II almost overlapped since they were fabricated from the same wafer. With injection current of 120 mA, the light output powers of the TMAH treated mini-LED I and TMAH treated mini-LED II were 62.5 mW and 65.6 mW, which were improved by 4.5% and 10.3% as compared to the mini-LEDs without TMAH treatment. The inset photographs show obvious brightness difference at the sidewall regions for the two types of mini-LEDs with TMAH etching. Mini-LED I with TMAH etching showed brighter S_1 sidewall while mini-LED II with TMAH etching showed brighter S_2 and S_3 sidewall, corresponding to their prism-structured sidewalls. According to the far-field radiation patterns shown in Figure 4, the light emission of TMAH treated mini-LEDs from the side direction was significantly improved as compared to the mini-LEDs without TMAH treatment, while only slight variation along the surface normal direction was observed for the investigated mini-LEDs. These results suggest that the light output power enhancement of mini-LEDs with TMAH treatment can be mainly attributed to increased light extraction from the prism-structured sidewall surfaces.

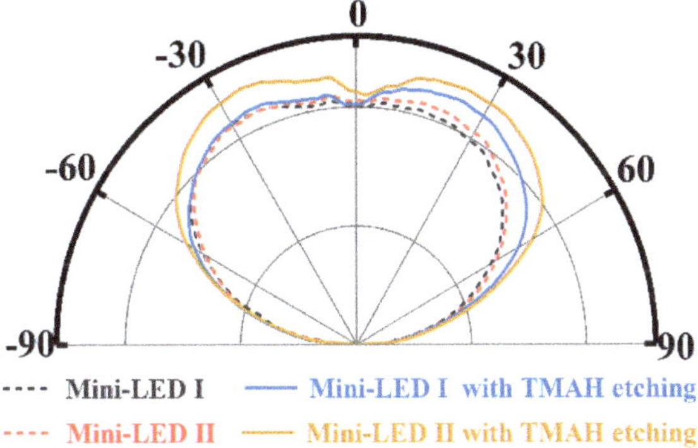

Figure 4. Far-field radiation patterns of the flip-chip mini-LEDs without TMAH etching treatment and with 7.5 min TMAH etching treatment.

To reveal the fundamental principle of prism structure on light extraction, the finite-difference time-domain (FDTD) simulation was conducted. The simulation model was built based on the above-described device structure and scaled to the size of 15 μm × 30 μm considering the computational capacity. Perfectly matched layers (PML) was adopted as boundaries to avoid unnecessary reflected light. The grid size in the simulation domain was 10 nm for accuracy with the limitation of computer memory. Transverse electric (TE) and transverse magnetic (TM) polarized point sources with a ratio of 1.8:1 were positioned in the center region of MQW [33] and the emission wavelength was set to be 456 nm. Figure 5a shows the simulated electric field intensity distribution nearby the smooth and prism-structured sidewalls of epilayer. The electric field emitting out from the smooth sidewall is mainly confined at the center region, while an intensified electric field emitted out from the prism-structured sidewall with a broader distribution in air. The broader and stronger electric field emitting out from the prism-structured sidewall suggested that the prism-structured sidewall acted more effectively in extracting light out than the smooth sidewall. The relationship between sidewall light extraction efficiency and prism size is presented in Figure 5b. The strong dependence of sidewall light extraction efficiency on the prism size indicating that the enhancement mainly arose from more light scattering out from the epilayer mode rather than only randomizing of light rays [34].

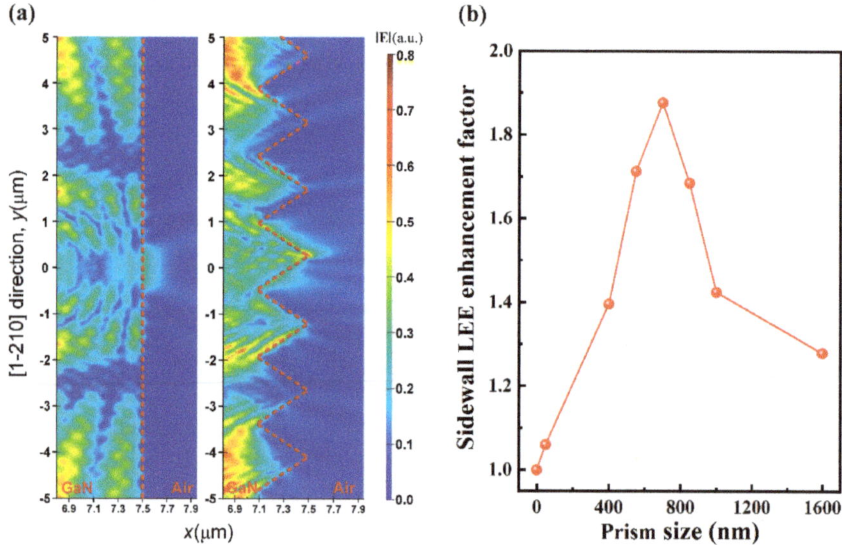

Figure 5. (a) Normalized electric field intensity distribution nearby the smooth (**left**) and prism-structured sidewalls (**right**) of LED chips from FDTD simulations. The color scale measures the electric field intensity E. (**b**) Simulated dependence of single sidewall light extraction efficiency on prism size parameters.

4. Conclusions

In summary, we demonstrated the prism-structured sidewall based on TMAH etching as an effective approach for scattering out waveguided light from the GaN epilayer. After TMAH etching procedure, hierarchical prism structure generated on the sidewall along [1-210] direction without bringing in damages on other surfaces owing to the anisotropic property of TMAH etching. Compared to the control mini-LEDs, the light output power of mini-LEDs with prism-structured sidewall improved by 4.5% or 10.3%, respectively, according to the different arrangements of LED chips on the wafer. We suggest the cost-effective sidewall texturing approach proposed in this work is a promising way to realize high-efficiency flip-chip mini-LEDs.

Author Contributions: S.Z. conceived and designed the experiment. S.Z. led the project. S.Z. and B.T. wrote the manuscript. J.M. and C.G. contributed to manuscript. Y.L. carried out *L-I-V* measurement. B.T., H.W. and N.L. carried out SEM measurement. B.T. and H.W. carried out the FDTD simulation. All authors discussed the progress of research and reviewed the manuscript.

Funding: This research was funded by the National Natural Science Foundation of China (Grant Nos. 51675386, U1501241 and 51775387), National Key Research and Development Program of China (No. 2017YFB1104900) and Natural Science Foundation of Hubei Province (No. 2018CFA091).

Conflicts of Interest: The authors declare no conflict of interest.

References

1. Tan, G.; Huang, Y.; Li, M.-C.; Lee, S.-L.; Wu, S.-T. High dynamic range liquid crystal displays with a mini-LED backlight. *Opt. Express* **2018**, *26*, 16572–16584. [CrossRef] [PubMed]
2. Cai, Y.; Zou, X.; Liu, C.; Lau, K.M. Voltage-Controlled GaN HEMT-LED Devices as Fast-Switching and Dimmable Light Emitters. *IEEE Electron Device Lett.* **2018**, *39*, 224–227. [CrossRef]
3. Wu, T.; Sher, C.-W.; Lin, Y.; Lee, C.-F.; Liang, S.; Lu, Y.; Huang Chen, S.-W.; Guo, W.; Kuo, H.-C.; Chen, Z. Mini-LED and Micro-LED: Promising Candidates for the Next Generation Display Technology. *Appl. Sci.* **2018**, *8*, 1557. [CrossRef]

4. Son, K.R.; Lee, T.H.; Lee, B.R.; Im, H.S.; Kim, T.G. Nitride-Based Microlight-Emitting Diodes Using AlN Thin-Film Electrodes with Nanoscale Indium/Tin Conducting Filaments. *Small* **2018**, *14*, 1801032. [CrossRef] [PubMed]

5. Daly, S.; Kunkel, T.; Sun, X.; Farrell, S.; Crum, P. 41.1: Distinguished Paper: Viewer Preferences for Shadow, Diffuse, Specular, and Emissive Luminance Limits of High Dynamic Range Displays. In *SID Symposium Digest of Technical Papers*; Blackwell Publishing Ltd.: Oxford, UK, 2013; Volume 44, pp. 563–566.

6. Hu, H.; Zhou, S.; Liu, X.; Gao, Y.; Gui, C.; Liu, S. Effects of GaN/AlGaN/Sputtered AlN nucleation layers on performance of GaN-based ultraviolet light-emitting diodes. *Sci. Rep.* **2017**, *7*, 44627. [CrossRef] [PubMed]

7. Alhassan, A.I.; Young, N.G.; Farrell, R.M.; Pynn, C.; Wu, F.; Alyamani, A.Y.; Nakamura, S.; DenBaars, S.P.; Speck, J.S. Development of high performance green c-plane III-nitride light-emitting diodes. *Opt. Express* **2018**, *26*, 5591–5601. [CrossRef] [PubMed]

8. Zhou, S.; Yuan, S.; Liu, Y.; Guo, L.J.; Liu, S.; Ding, H. Highly efficient and reliable high power LEDs with patterned sapphire substrate and strip-shaped distributed current blocking layer. *Appl. Surf. Sci.* **2015**, *355*, 1013–1019. [CrossRef]

9. Muhammed, M.M.; Alwadai, N.; Lopatin, S.; Kuramata, A.; Roqan, I.S. High-Efficiency InGaN/GaN Quantum Well-Based Vertical Light-Emitting Diodes Fabricated on β-Ga_2O_3 Substrate. *ACS Appl. Mater. Interfaces* **2017**, *9*, 34057–34063. [CrossRef] [PubMed]

10. Zhou, S.; Xu, H.; Hu, H.; Gui, C.; Liu, S. High quality GaN buffer layer by isoelectronic doping and its application to 365 nm InGaN/AlGaN ultraviolet light-emitting diodes. *Appl. Surf. Sci.* **2019**, *471*, 231–238. [CrossRef]

11. Zhao, J.; Ding, X.; Miao, J.; Hu, J.; Wan, H.; Zhou, S. Improvement in Light Output of Ultraviolet Light-Emitting Diodes with Patterned Double-Layer ITO by Laser Direct Writing. *Nanomaterials* **2019**, *9*, 203. [CrossRef] [PubMed]

12. Xu, J.; Zhang, W.; Peng, M.; Dai, J.; Chen, C. Light-extraction enhancement of GaN-based 395 nm flip-chip light-emitting diodes by an Al-doped ITO transparent conductive electrode. *Opt. Lett.* **2018**, *43*, 2684–2687. [CrossRef] [PubMed]

13. Zhou, S.; Liu, X.; Gao, Y.; Liu, Y.; Liu, M.; Liu, Z.; Gui, C.; Liu, S. Numerical and experimental investigation of GaN-based flip-chip light-emitting diodes with highly reflective Ag/TiW and ITO/DBR Ohmic contacts. *Opt. Express* **2017**, *25*, 26615–26627. [CrossRef] [PubMed]

14. Wei, T.B.; Ji, X.L.; Wu, K.; Zheng, H.Y.; Du, C.X.; Chen, Y.; Yan, Q.F.; Zhao, L.X.; Zhou, Z.; Wang, J.X.; et al. Efficiency improvement and droop behavior in nanospherical-lens lithographically patterned bottom and top photonic crystal InGaN/GaN light-emitting diodes. *Opt. Lett.* **2014**, *39*, 379–382. [CrossRef] [PubMed]

15. Kang, J.H.; Kim, H.G.; Chandramohan, S.; Kim, H.K.; Kim, H.Y.; Ryu, J.H.; Park, Y.J.; Beak, Y.S.; Lee, J.-S.; Park, J.S.; et al. Improving the optical performance of InGaN light-emitting diodes by altering light reflection and refraction with triangular air prism arrays. *Opt. Lett.* **2012**, *37*, 88–90. [CrossRef] [PubMed]

16. Shen, C.F.; Chang, S.J.; Chen, W.S.; Ko, T.K.; Kuo, C.T.; Shei, S.C. Nitride-Based High-Power Flip-Chip LED with Double-Side Patterned Sapphire Substrate. *IEEE Photonics Technol. Lett.* **2007**, *19*, 780–782. [CrossRef]

17. Huang, K.-C.; Huang, Y.-R.; Tseng, C.-M.; Tseng, S.H.; Huang, J.-E. Increased viewing angle and light extraction efficiency of flip-chip light-emitting diode using double-side patterned sapphire substrate. *Scr. Mater.* **2015**, *108*, 40–43. [CrossRef]

18. Lee, C.; Lee, Y.; Kuo, H.; Tsai, M.; Cheng, B.S.; Lu, T.; Wang, S.; Kuo, C. Enhancement of Flip-Chip Light-Emitting Diodes with Omni-Directional Reflector and Textured Micropillar Arrays. *IEEE Photonics Technol. Lett.* **2007**, *19*, 1200–1202. [CrossRef]

19. Lee, C.E.; Kuo, H.C.; Lee, Y.C.; Tsai, M.R.; Lu, T.C.; Wang, S.C.; Kuo, C.T. Luminance Enhancement of Flip-Chip Light-Emitting Diodes by Geometric Sapphire Shaping Structure. *IEEE Photonics Technol. Lett.* **2008**, *20*, 184–186. [CrossRef]

20. Huang, S.; Horng, R.; Wen, K.; Lin, Y.; Yen, K.; Wuu, D. Improved Light Extraction of Nitride-Based Flip-Chip Light-Emitting Diodes Via Sapphire Shaping and Texturing. *IEEE Photonics Technol. Lett.* **2006**, *18*, 2623–2625. [CrossRef]

21. Nami, M.; Feezell, D.F. Optical properties of plasmonic light-emitting diodes based on flip-chip III-nitride core-shell nanowires. *Opt. Express* **2014**, *22*, 29445–29455. [CrossRef] [PubMed]

22. Baek, S.; Kang, G.; Shin, D.; Bae, K.; Kim, Y.H.; Kim, K. Improvement of Light Extraction Efficiency in Flip-Chip Light Emitting Diodes on SiC Substrate via Transparent Haze Films with Morphology-Controlled Collapsed Alumina Nanorods. *ACS Appl. Mater. Interfaces* **2016**, *8*, 135–141. [CrossRef] [PubMed]

23. Chang, S.; Lin, N.; Shei, S. GaN-Based Power Flip-Chip LEDs with SILAR and Hydrothermal ZnO Nanorods. *IEEE J. Sel. Top. Quantum Electron.* **2015**, *21*, 431–435. [CrossRef]

24. Su, Y.K.; Chen, J.J.; Lin, C.L.; Kao, C.C. Structural analysis of nitride-based LEDs grown on micro- and nano-scale patterned sapphire substrates. *Phys. Status Solidi C* **2010**, *7*, 1784–1786. [CrossRef]

25. Lin, Z.; Yang, H.; Zhou, S.; Wang, H.; Hong, X.; Li, G. Pattern Design of and Epitaxial Growth on Patterned Sapphire Substrates for Highly Efficient GaN-Based LEDs. *Cryst. Growth Des.* **2012**, *12*, 2836–2841. [CrossRef]

26. Su, Y.K.; Chen, J.J.; Lin, C.L.; Chen, S.M.; Li, W.L.; Kao, C.C. Pattern-size dependence of characteristics of nitride-based LEDs grown on patterned sapphire substrates. *J. Cryst. Growth* **2009**, *311*, 2973–2976. [CrossRef]

27. Zhou, S.; Hu, H.; Liu, X.; Liu, M.; Ding, X.; Gui, C.; Liu, S.; Guo, L.J. Comparative study of GaN-based ultraviolet LEDs grown on different-sized patterned sapphire substrates with sputtered AlN nucleation layer. *Jpn. J. Appl. Phys.* **2017**, *56*, 111001. [CrossRef]

28. Zhuang, D.; Edgar, J.H. Wet etching of GaN, AlN, and SiC: A review. *Mat. Sci. Eng. R* **2005**, *48*, 1–46. [CrossRef]

29. Li, D.; Sumiya, M.; Fuke, S.; Yang, D.; Que, D.; Suzuki, Y.; Fukuda, Y. Selective etching of GaN polar surface in potassium hydroxide solution studied by x-ray photoelectron spectroscopy. *J. Appl. Phys.* **2001**, *90*, 4219–4223. [CrossRef]

30. Lai, Y.-Y.; Hsu, S.-C.; Chang, H.-S.; Wu, Y.S.; Chen, C.-H.; Chen, L.-Y.; Cheng, Y.-J. The study of wet etching on GaN surface by potassium hydroxide solution. *Res. Chem. Intermed.* **2017**, *43*, 3563–3572. [CrossRef]

31. Choi, H.W.; Dawson, M.D.; Edwards, P.R.; Martin, R.W. High extraction efficiency InGaN micro-ring light-emitting diodes. *Appl. Phys. Lett.* **2003**, *83*, 4483–4485. [CrossRef]

32. Yang, Y.; Cao, X.A. Removing plasma-induced sidewall damage in GaN-based light-emitting diodes by annealing and wet chemical treatments. *J. Vac. Sci. Technol. B* **2009**, *27*, 2337–2341. [CrossRef]

33. Shakya, J.; Knabe, K.; Kim, K.H.; Li, J.; Lin, J.Y.; Jiang, H.X. Polarization of III-nitride blue and ultraviolet light-emitting diodes. *Appl. Phys. Lett.* **2005**, *86*, 091107. [CrossRef]

34. Campbell, P.; Green, M.A. Light trapping properties of pyramidally textured surfaces. *J. Appl. Phys.* **1987**, *62*, 243–249. [CrossRef]

 *nanomaterials*

Article

Effects of Different Oxidation Degrees of Graphene Oxide on P-Type and N-Type Si Heterojunction Photodetectors

Ching-Kuei Shih, Yu-Tang Ciou, Chun-Wei Chiu, Yu-Ru Li, Jia-Syun Jheng, Yen-Chun Chen and Chu-Hsuan Lin *

Department of Opto-electronic Engineering, National Dong Hwa University, Hualien 97401, Taiwan; u9925044@ems.ndhu.edu.tw (C.-K.S.); u9925001@ems.ndhu.edu.tw (Y.-T.C.); 410025023@ems.ndhu.edu.tw (C.-W.C.); 410325022@gms.ndhu.edu.tw (Y.-R.L.); 410325003@gms.ndhu.edu.tw (J.-S.J.); 410225039@gms.ndhu.edu.tw (Y.-C.C.)
* Correspondence: chlin0109@gms.ndhu.edu.tw; Tel.: +886-38-903-188

Received: 25 May 2017; Accepted: 2 July 2018; Published: 4 July 2018

Abstract: Oxygen-containing functional groups in graphene oxide (GO), a derivative of graphene, can widen the bandgap of graphene. In this study, we varied the amount of hydrogen peroxide used to prepare GO samples with different degrees of oxidation. Transmittance measurement, Raman spectroscopy, and X-ray photoelectron spectroscopy were used to completely characterize the change in oxidation degree. The effects of oxidation degree on p-type and n-type Si heterojunction photodetectors were compared. Notably, GO with a lower oxidation degree led to a larger photoresponse of p-type Si, whereas that with a higher oxidation degree achieved a larger photoresponse of n-type Si.

Keywords: graphene oxide; oxidation; photodetector

---

1. Introduction

Heterojunction photodetectors combine two types of materials for light detection, and Si is the optimal choice for one of the two materials due to mature and cheap Si complementary metal–oxide–semiconductor (CMOS) technology with Si acting as the substrate [1–3]. Numerous materials have been investigated to be combined with the Si substrate in heterojunction photodetectors for the optimization of photodetector applications, and studies increasingly concentrate on graphene-based materials [4]. The common graphene-based materials in Si heterojunction photodetectors are composed of graphene/Si [5–7] or reduced-graphene-oxide (rGO)/Si [8–10] structures. In these studies, graphene or rGO is typically only used as the electrode in Schottky diodes. The absorption of graphene itself can also be utilized for certain structures [11] or certain wavelengths [12], but the gapless properties of graphene (and ideal rGO) still restrict its applications in optoelectronic devices [13]. Graphene, with attached oxygen-containing functional groups, is known as graphene oxide (GO) [14]. These oxygen-containing functional groups in GO open the bandgap of graphene. Even partially-reduced GO with suitable manipulation of O/C ratios can still exhibit a large bandgap for utilization in photodetection [15]. The opened bandgap yields more flexibility to graphene-based Si heterojunction photodetectors. For example, an opened bandgap in GO can contribute to light absorption, and enrich the possibility of band engineering. Recently, GO has been integrated on p-type Si substrates (p-Si) to fabricate heterojunction photodetectors [16,17], and GO has also been integrated with n-Si [18,19]. However, to our knowledge, the effects of GO oxidation degrees on the performance of GO/Si heterojunction photodetectors have not been investigated either on p-Si or n-Si. It was reported that the bandgap of GO can vary from 2.3 to 2.7 eV with increasing GO oxidation degrees [20]. Since GO has the advantage of an opened bandgap over graphene, the influence of oxidation degrees is a meaningful subject for GO/Si heterojunction photodetectors. Therefore, the influence of oxidation degrees of GO on Si heterojunction photodetectors is studied in this manuscript. Furthermore, the comparison between influence of oxidation degrees on GO/p-Si and the influence of oxidation degrees on GO/n-Si reveal different trends. We propose that high-oxidation-degree GO is suitable for GO/n-Si heterojunction photodetectors and low-oxidation-degree GO is suitable for GO/p-Si heterojunction photodetectors. The mechanisms are interpreted on the basis of band diagrams.

2. Experiments

We prepared GO powder using the modified Hummers method [21]. The main approach to varying the oxidation degree was changing the amount of hydrogen peroxide (H_2O_2) used. The details of the experimental procedures are as follows.

Graphite powder, potassium permanganate ($KMnO_4$), and sulfuric acid (H_2SO_4, 98%) were mixed at a ratio of 1:3:24 below 35 °C. The resulting solution was divided between two beakers. H_2O_2 (35%) was then added to both beakers until its volume fraction reached 1% and 4%. The corresponding solutions were used to prepare samples GO1 and GO4, respectively. The resultant solutions were both diluted with deionized (DI) water and continually stirred. Each solution was filtered, and the residual slurry was washed with HCl (3%) and DI water. The slurry was then dried and ground into powder. The powder was added to the DI water, and tiny flakes of GO could be exfoliated and collected through ultrasonication and centrifugation. GO aqueous suspension was then obtained from the supernatant.

To uniformly deposit the hydrophilic GO flakes on Si substrates, the Si substrates were immersed in SC1 solution (1:2:8 $NH_4OH/H_2O_2/H_2O$ ratio) for 30 min at a temperature of 70 °C [22]. Because of the abundant hydroxyl radicals (OH^-) in the SC1 solution, the Si substrates also became hydrophilic. Drops of the GO suspension were then applied to the Si surface to form a GO film. Uniform deposition of the GO flakes was possible because of the hydrophilicity of the GO due to its oxygen-containing groups. In our previous study [22], samples treated with SC1 for 0, 15, and 30 min were compared. The longest treatment (30 min) resulted in the lowest tunneling current because of the dense coverage. Because we wanted to form a uniform GO layer, we performed a 30-min SC1 treatment in this study. The morphology of GO1 and GO4 was obtained using atomic force microscopy, as shown in Figure 1a,b. Since both substrates were subjected to the same hydrophilic treatment for 30 min, GO flakes with dense coverage coated the entire surface. The difference in electrical characteristics (which is discussed later) may be attributed to the oxidation degree, but not to the coverage condition.

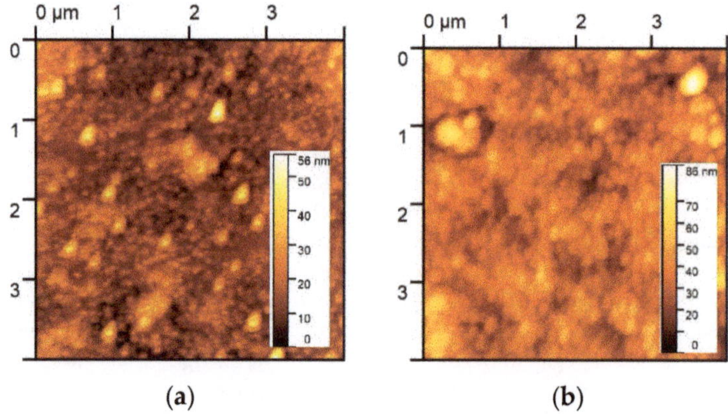

(a)   (b)

Figure 1. Surface morphology of (**a**) GO1 and (**b**) GO4. GO flakes densely cover the entire surface. Therefore, the difference in electrical characteristics discussed in the text was attributed to the oxidation degree and not to the coverage condition.

After GO deposition, a circular Al gate with an area of 5×10^{-4} cm^2 was evaporated on the top of the Si substrate, and a large area of Al was evaporated on the back side of the Si to form an ohmic contact. Both p-type and n-type Si substrates were used to fabricate GO-on-p-Si and GO-on-n-Si heterojunction photodetectors. Current-voltage (IV) measurement was performed using a semiconductor analyzer (Agilent B1500A, Santa Rosa, CA, USA).

3. Characterization of Oxidation Degree

A study found that a higher degree of oxidation resulted in a larger bandgap of GO [20]. To verify whether our procedure changed the oxidation degree, we first performed a spectral absorption measurement (Figure 2). For short-wavelength light, the energy of photons was larger than the GO bandgap, and GO absorbed the light, resulting in a low transmittance. Conversely, GO did not absorb long-wavelength light, which resulted in a high transmittance. Therefore, the onset of the transmittance increase was directly related to the GO bandgap. As shown in Figure 2, the onset for GO4 was shifted to wavelengths shorter than the wavelengths for GO1. This result implies that the bandgap of GO4 was larger than that of GO1. More hydrogen peroxide used in the modified Hummers method resulted in a higher degree of oxidation, which contributed to the larger bandgap.

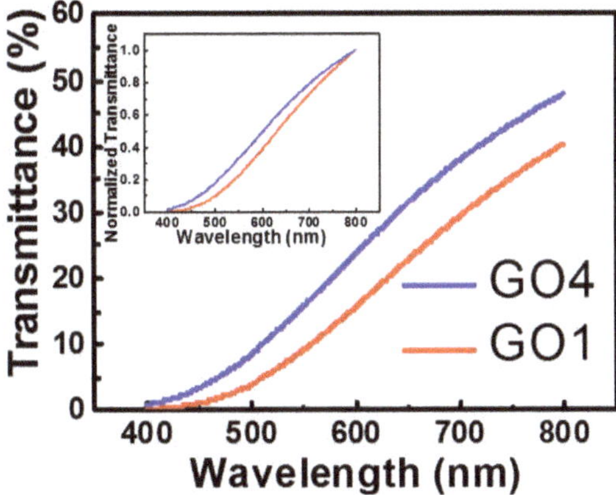

Figure 2. Transmittance of GO aqueous solutions with different oxidation degrees. The inset presents the normalized curves. The onset for GO4 is shifted to shorter wavelengths than those for GO1.

Raman spectroscopy is a powerful tool for analyzing the chemical composition and nanostructure of GO [23,24]. The Raman spectroscopy results of GO1 and GO4 are shown in Figure 3. Three bands are typically used to identify graphene-based materials. The vibration caused by defects in the material and disorder at the edge is called the D band, which is located at approximately 1350 cm^{-1} [23,25]. Another band at approximately 2650 cm^{-1}, known as the 2D band, is closely related to the number of layers of graphene flakes [23]. Undoped graphene exhibits a peak at approximately 1580 cm^{-1} [26], termed the G band. The position of this band is sensitive to the doping level. Thus, a G-band shift due to the introduction of oxygen, a form of doping, can be an index of the oxidation degree [27].

A magnified view of the G-band peaks of the two samples is shown in the inset of Figure 3. The peak shifted to a higher wavenumber from GO1 to GO4. As observed in a related study [23], the higher oxygen-to-carbon ratio led to a larger Raman shift of the G band. Therefore, it was suspected that the GO4 sample, which had a larger G-band Raman shift, corresponded to a higher oxygen-to-carbon ratio.

X-ray photoelectron spectroscopy (XPS) provided direct evidence of the difference in H$_2$O$_2$ treatment resulting in differences in oxygen content. Figure 4a,b show the C1s XPS results for GO1 and GO4, respectively. The well-known binding energy of sp^2 and sp^3 C (C–C peak in Figure 4) was at

284.5 eV [22,28]. In addition, the reported binding energies of C–O and O=C–OH bonds matched those from our peak fitting (~286.5 and 288.4 eV, respectively; Figure 4) [28,29]. The XPS intensity ratios of C–O (at 286.5 eV) to C–C (at 284.5 eV) bonds for GO1 and GO4 were 0.28 and 0.40, respectively. These indicated that the oxygen content in GO4 was higher than that of GO1.

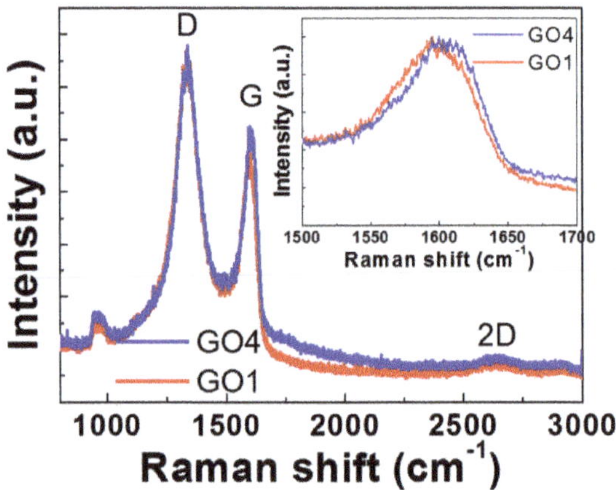

Figure 3. Raman spectra of the GO1 and GO4 samples. A magnified view of the G band is provided in the inset. The excitation wavelength in Raman measurements was located at 632.8 nm. The G peak of GO4 shifted to a higher wavenumber than that of GO1.

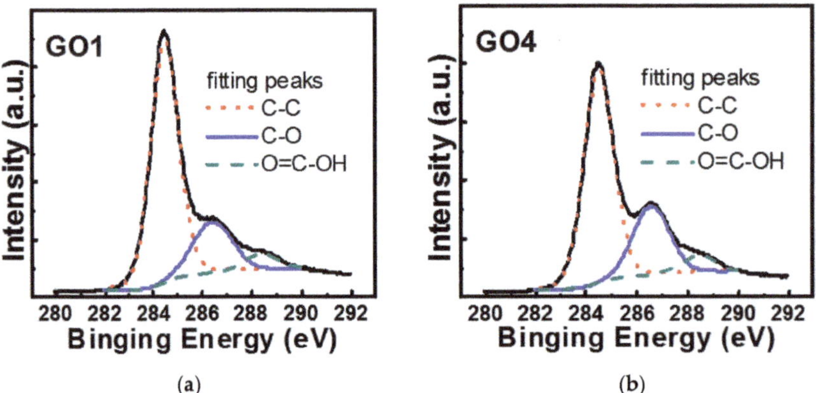

Figure 4. C1s XPS results for (**a**) GO1 and (**b**) GO4. The XPS intensity ratio of C–O (at 286.5 eV) to C–C (at 284.5 eV) bonds of GO4 was larger than that of GO1.

4. Electrical Characteristics

Si substrates with GO1 and GO4 deposited on the surface were used to fabricate GO1 and GO4 photodetectors, respectively. GO flakes and the native oxide comprised the oxide layers. The photocurrents and dark currents of p-Si and n-Si heterojunction photodetectors were measured, as shown in Figure 5a,b.

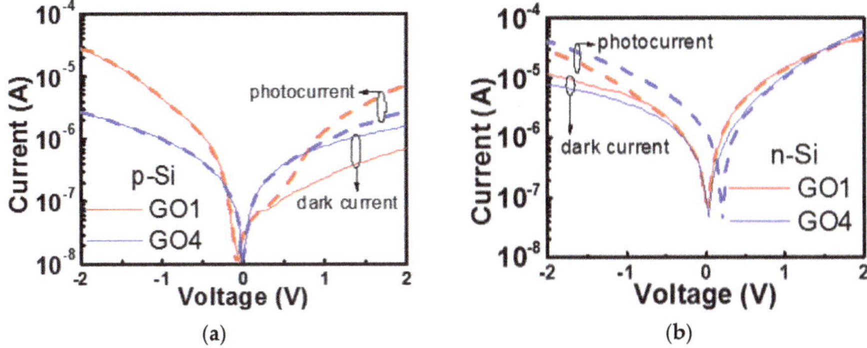

Figure 5. Dark currents and photocurrents of GO1 and GO4 heterojunction photodetectors on (**a**) p-Si and (**b**) n-Si. The photocurrents were measured under white light from white light emitting diodes with a power density of 0.5 mW/cm^2.

A higher oxidation degree corresponds to a larger bandgap [20]. Schematic band diagrams of our p-Si detectors with GO1 or GO4 (Figure 6) were constructed according to the data in [20]. A higher oxidation degree of GO4 resulted in a larger work function and bandgap compared with those of GO1. In [20], the conduction band edge (E_C) did not change but the valence band edge (E_V) was lowered as the oxidation degree increased. Therefore, a downward shift of E_V was present regarding GO4 relative to that in GO1 of our photodetectors (Figure 6d as compared with Figure 6c), which resulted in a large barrier for holes from Si to the Al gate at the accumulation (negative) bias. This is illustrated in the upper insets of Figure 6c,d. The hole flow from Si to Al might have been blocked by GO4 in the GO4 device (the upper inset of Figure 6d), whereas more hole flow from Si could reach the Al gate successfully in the GO1 device (the upper inset of Figure 6c). Therefore, the accumulation current of GO4-on-p-Si was smaller (Figure 5a). There might have been some electron flow from the Al gate to Si that overcame the E_C barrier at the accumulation bias. However, this component was small for the GO1 and GO4 cases because the E_C barrier of GO was large and thick. Therefore, the GO4-on-p-Si photodetector exhibited a smaller accumulation current than that of GO1-on-p-Si mainly due to the larger barrier for the holes from the Si to the Al gate. Photodetection should be performed at the inversion bias (positive bias in the p-substrate case) because the high electric-field in the depletion region at the inversion bias could help the carrier separation of photo-generated electron-hole pairs. The band alignment suggested that the larger work function of GO4 was detrimental to the formation of a depletion region in the p-Si. The Fermi level (E_F) of GO4 was lower than the E_F of p-Si, resulting in initial accumulation (the Si band at the surface was oriented upward at 0 V) (Figure 6d). At the positive bias, the initial accumulation of the GO4-on-p-Si resulted in a thinner depletion region (red solid line on the lower inset of Figure 6d) compared with the thicker depletion region of GO1-on-p-Si (red solid line of the lower inset of Figure 6c). Thus, the photocurrent of the GO4-on-p-Si was smaller than that of the GO1-on-p-Si. The photo-to-dark ratio of the GO4-on-p-Si was 1.75, which was smaller than that of the GO1-on-p-Si at 2 V (10.1). At 2 V, the absolute photoresponse (difference between the photocurrent and dark current) of the GO4-on-p-Si (1.2×10^{-6} A) was therefore smaller than that of GO1-on-p-Si (6.4×10^{-6} A). To validate these findings, more samples were prepared and measured. The cumulative probability plot of photo-to-dark ratio is shown in Figure 7. On p-Si, all the GO4 photodetectors exhibited smaller photo-to-dark ratios than those of the GO1 photodetectors.

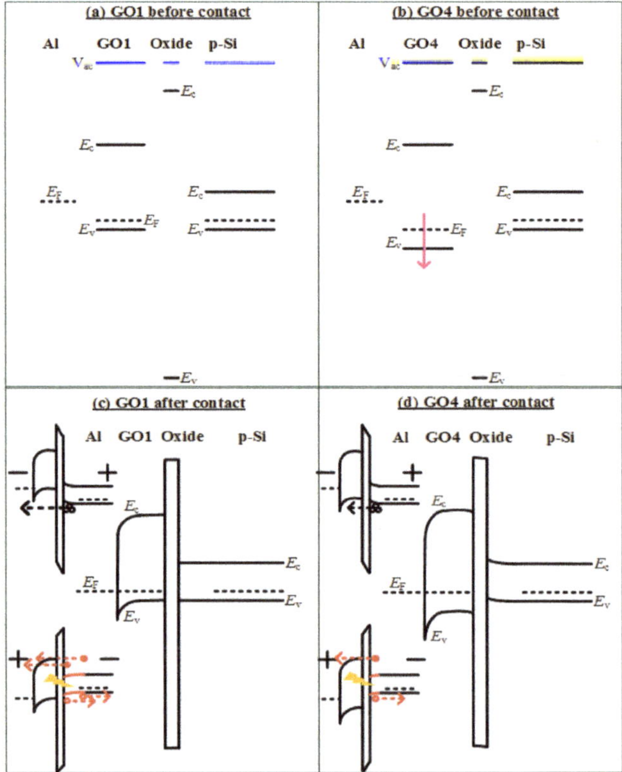

Figure 6. Band diagrams for (**a**) GO1 and (**b**) GO4 detectors on p-Si before and (**c**) GO1 and (**d**) GO4 detectors after contact. The larger work function of GO4 was detrimental to the formation of a depletion region in the p-Si.

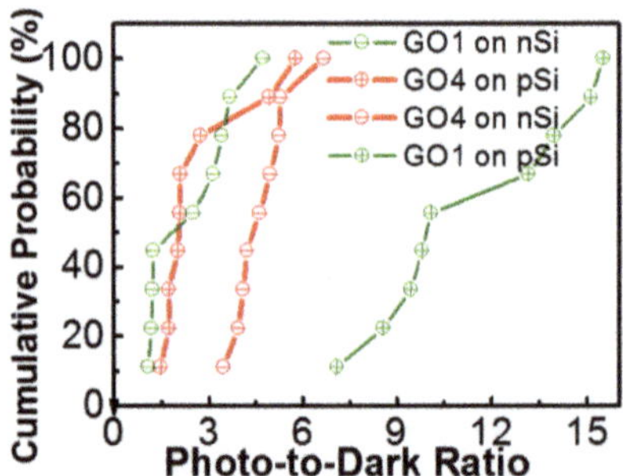

Figure 7. Cumulative probability plot of photo-to-dark ratios. On p-Si, GO1 results in larger ratios than those of GO4, whereas GO1 leads to smaller ratios on n-Si.

Figure 5b shows the IV behaviors of n-Si photodetectors. On n-Si, the accumulation bias (positive bias in the n-substrate case) drove electrons from the Si to the Al gate and the holes from the Al gate to the Si. In the n-Si photodetector, the E_C barrier for electrons from the n-Si to Al gate formed by GO became smaller than the E_C barrier from the Al gate to the p-Si in p-Si photodetectors. Therefore, the electron flow could not be ignored now. The electron flow will considerably contribute the total current in the n-Si photodetectors.

The accumulation current of the GO4-on-n-Si photodetector, therefore, displayed no obvious difference with that of the GO1-on-n-Si photodetector, because the barrier for electrons from n-Si to the Al gate in the GO4-on-n-Si photodetector (lower inset of Figure 8d) was similar to that in the GO1-on-n-Si photodetector (lower inset of Figure 8c). Oxidation degrees mainly influenced the E_V and E_F, but not the E_C [20] as shown in Figure 8a,b. However, the relation between the photoresponse at the inversion bias (negative bias in the n-substrate case) and oxidation degrees in the n-Si case was the inverse of that of the p-Si case. The photo-to-dark ratio of the GO4-on-n-Si was 5.3, which was larger than that of the GO1-on-n-Si at -2 V (2.5). The absolute photoresponse of the GO4-on-n-Si at -2 V (3.4×10^{-5} A) was also larger than that of the GO1-on-n-Si (1.8×10^{-5} A). Notably, GO4 provided a superior photoresponse than GO1 on n-Si, whereas GO4 led to an inferior photoresponse on p-Si. This substantial photoresponse of the GO4-on-n-Si occurred because the larger work function of GO4 contributed to a thicker depletion region (red solid line of the upper inset of Figure 8d) in n-Si compared with the thinner depletion region of the GO1-on-n-Si (red solid line of the upper inset of Figure 8c). Therefore, the cumulative probability plot demonstrated that the photo-to-dark ratios of the GO4-on-n-Si photodetector were significantly larger than those of GO1-on-n-Si (Figure 7).

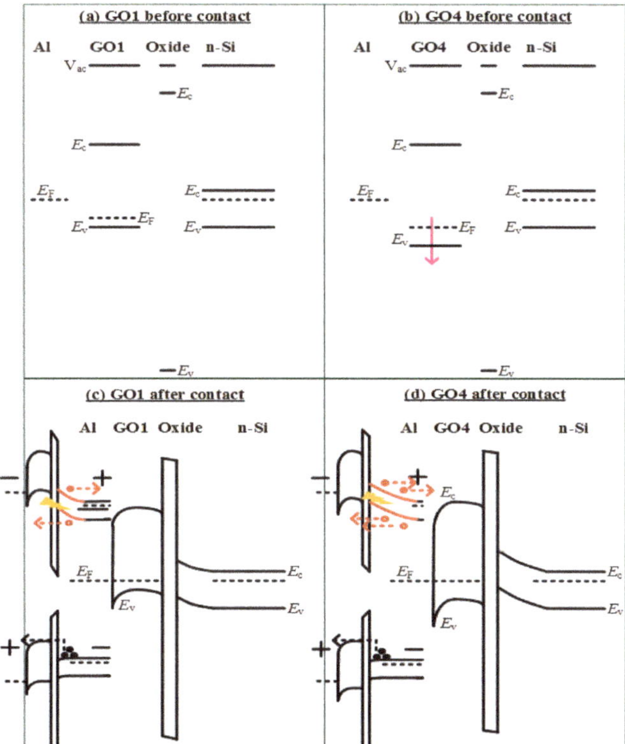

Figure 8. Band diagrams for (**a**) GO1 and (**b**) GO4 detectors on n-Si before and (**c**) GO1 and (**d**) GO4 detectors after contact. At the inversion bias (upper inset of (**c**,**d**)), the larger work function of GO4 contributed to a wider depletion region in the n-Si. At the accumulation bias (lower inset of (**c**,**d**)), the component of "electrons from Si to the Al gate" would predominate over the component of "holes from the Al gate to Si" due to the smaller SiO_2 oxide barrier.

5. Conclusions

GO-on-p-Si and GO-on-n-Si heterojunction photodetectors were comprehensively studied. Different oxidation degrees of GO were achieved by varying the amount of hydrogen peroxide used in the oxidation procedure of GO. Our results indicated that GO with lower oxidation degrees should be applied to p-Si heterojunction photodetectors to achieve superior photoresponses, whereas GO with higher oxidation degrees was applicable for n-Si heterojunction photodetectors. Oxidation degrees led to opposing influence on GO/p-Si and GO/n-Si photodetectors, and the corresponding influence on the depletion region provided the possibility of photodetector optimization through such band engineering.

Author Contributions: Investigation: Y.-R.L., J.-S.J., Y.-C.C., and C.-H.L.; Methodology: C.-K.S., Y.-T.C., and C.-W.C.; writing—Original draft: C.-H.L.

Funding: This work was funded by the Ministry of Science and Technology, R.O.C., under contract No. MOST 104-2221-E-259-030-MY3.

Acknowledgments: The authors are grateful to the National Center for High-Performance Computing for computer time and facilities.

References

1. Ismail, R.A.; Rashid, F.F.; Tariq, M.S. Preparation and characteristics study of $CuAlO_2$/Si heterojunction photodetector by pulsed laser deposition. *J. Mater. Sci. Mater. Electron.* **2017**, *28*, 6889–6896. [CrossRef]

2. Goossens, S.; Navickaite, G.; Monasterio, C.; Gupta, S.; Piqueras, J.J.; Pérez, R.; Burwell, G.; Nikitskiy, I.; Lasanta, T.; Galán, T.; et al. Broadband image sensor array based on graphene–CMOS integration. *Nat. Photonics* **2017**, *11*, 366–371. [CrossRef]

3. Khan, W.; Ajmal, H.M.S.; Khan, F.; Huda, N.U.; Kim, S.D. Induced Photonic Response of ZnO Nanorods Grown on Oxygen Plasma-Treated Seed Crystallites. *Nanomaterials* **2018**, *8*, 371. [CrossRef] [PubMed]

4. Koppens, F.H.L.; Mueller, T.; Avouris, P.; Ferrari, A.C.; Vitiello, M.S.; Polini, M. Photodetectors based on graphene, other two-dimensional materials and hybrid systems. *Nat. Nanotechnol.* **2014**, *9*, 780–793. [CrossRef]

5. Li, X.; Zhu, M.; Du, M.; Lv, Z.; Zhang, L.; Li, Y.; Yang, Y.; Yang, T.; Li, X.; Wang, K.; et al. High Detectivity Graphene-Silicon Heterojunction Photodetector. *Small* **2016**, *12*, 595–601. [CrossRef] [PubMed]

6. Casalino, M.; Sassi, U.; Goykhman, I.; Eiden, A.; Lidorikis, E.; Milana, S.; Fazio, D.D.; Tomarchio, F.; Iodice, M.; Coppola, G.; et al. Vertically Illuminated, Resonant Cavity Enhanced, Graphene–Silicon Schottky Photodetectors. *ACS Nano* **2017**, *11*, 10955–10963. [CrossRef] [PubMed]

7. Luongo, G.; Giubileo, F.; Genovese, L.; Iemmo, L.; Martucciello, N.; Di Bartolomeo, A. I-V and C-V Characterization of a High-Responsivity Graphene/Silicon Photodiode with Embedded MOS Capacitor. *Nanomaterials* **2017**, *7*, 158. [CrossRef] [PubMed]

8. Chandrakalavathi, T.; Peta, K.R.; Jeyalakshmi, R. Enhanced UV photoresponse with Au nanoparticles incorporated rGO/Si heterostructure. *Mater. Res. Express* **2018**, *5*, 025011. [CrossRef]

9. Azmi, S.N.C.; Rahman, S.F.A.; Nawabjan, A.; Hashim, A.M. Junction properties analysis of silicon back-to-back Schottky diode with reduced graphene oxide Schottky electrodes. *Microelectron. Eng.* **2018**, *196*, 32–37. [CrossRef]

10. Dehkharghani, M.N.; Rajabi, A.; Amirmazlaghani, A. Responsivity improvement of Si-based NIR photodetector using reduced Graphene Oxide. In Proceedings of the 2017 Iranian Conference on Electrical Engineering (ICEE), Tehran, Iran, 2–4 May 2017; pp. 523–526. [CrossRef]

11. Casalino, M. Design of Resonant Cavity-Enhanced Schottky Graphene/Silicon Photodetectors at 1550 nm. *J. Lightwave Technol.* **2018**, *36*, 1766–1774. [CrossRef]

12. Wan, X.; Xu, Y.; Guo, H.; Shehzad, K.; Ali, A.; Liu, Y.; Yang, J.; Dai, D.; Telin, C.; Liu, L.; et al. A self-powered high-performance graphene/silicon ultraviolet photodetector with ultra-shallow junction: Breaking the limit of silicon? *npj 2D Mater. Appl.* **2017**, *1*, 4. [CrossRef]

13. Wang, L.; Jie, J.; Shao, Z.; Zhang, Q.; Zhang, X.; Wang, Y.; Sun, Z.; Lee, S. MoS$_2$/Si Heterojunction with Vertically Standing Layered Structure for Ultrafast, High-Detectivity, Self-Driven Visible–Near Infrared Photodetectors. *Adv. Funct. Mater.* **2015**, *25*, 2910–2919. [CrossRef]

14. Han, K.I.; Kim, S.; Lee, I.G.; Kim, J.P.; Kim, J.H.; Hong, S.W.; Cho, B.J.; Hwang, W.S. Compliment Graphene Oxide Coating on Silk Fiber Surface via Electrostatic Force for Capacitive Humidity Sensor Applications. *Sensors* **2017**, *17*, 407. [CrossRef] [PubMed]

15. Singh, M.; Kumar, G.; Prakash, N.; Khanna, S.P.; Pal, P.; Singh, S.P. Large bandgap reduced graphene oxide (rGO) based n-p + heterojunction photodetector with improved NIR performance. *Semiconduct. Sci. Technol.* **2018**, *33*, 045012. [CrossRef]

16. Son, S.B.; Huang, F.; Bae, T.S.; Hong, W.K. Wettability effects of graphene oxide aqueous solution in photodetectors based on graphene oxide/silicon heterojunctions via ultraviolet ozone treatment. *J. Alloys Compd.* **2017**, *698*, 384–389. [CrossRef]

17. Ahmad, H.; Thandavan, T.M.K. Characterization of graphene oxide/silicon dioxide/p-type silicon heterojunction photodetector towards infrared 974 nm illumination. *Opt. Quantum Electron.* **2017**, *49*, 395. [CrossRef]

18. Maiti, R.; Manna, S.; Midya, A.; Ray, S.K. Broadband photoresponse and rectification of novel graphene oxide/n-Si heterojunctions. *Opt. Express* **2013**, *21*, 26034–26043. [CrossRef] [PubMed]

19. Kalita, G.; Wakita, K.; Umeno, M.; Tanemura, M. Fabrication and characteristics of solution-processed graphene oxide-silicon heterojunction. *Phys. Status Solidi Rapid Res. Lett.* **2013**, *7*, 340–343. [CrossRef]

20. Yeh, T.F.; Chan, F.F.; Hsieh, C.T.; Teng, H. Graphite Oxide with Different Oxygenated Levels for Hydrogen and Oxygen Production from Water under Illumination: The Band Positions of Graphite Oxide. *J. Phys. Chem. C* **2011**, *115*, 22587–22597. [CrossRef]

21. Lin, C.H.; Yeh, W.T.; Chan, C.H.; Lin, C.C. Influence of graphene oxide on metal-insulator-semiconductor tunneling diodes. *Nanoscale Res. Lett.* **2012**, *7*, 343. [CrossRef] [PubMed]

22. Lin, C.H.; Yeh, W.T.; Chen, M.H. Metal-Insulator-Semiconductor Photodetectors with Different Coverage Ratios of Graphene Oxide. *IEEE J. Sel. Top. Quantum Electron.* **2014**, *20*, 3800105. [CrossRef]

23. Haluška, M.; Obergfell, D.; Meyer, J.C.; Scalia, G.; Ulbricht, G.; Krauss, B.; Chae, D.H.; Lohmann, T.; Lebert, M.; Kaempgen, M.; et al. Investigation of the shift of Raman modes of graphene flakes. *Phys. Stat. Solidi B* **2007**, *244*, 4143–4146. [CrossRef]

24. Karthick, S.; Lee, H.S.; Kwon, S.J.; Natarajan, R.; Saraswathy, V. Standardization, Calibration, and Evaluation of Tantalum-Nano rGO-SnO$_2$ Composite as a Possible Candidate Material in Humidity Sensors. *Sensors* **2016**, *16*, 2079. [CrossRef] [PubMed]

25. Kudin, K.N.; Ozbas, B.; Schniepp, H.C.; Prud'homme, R.K.; Aksay, I.A.; Car, R. Raman Spectra of Graphite Oxide and Functionalized Graphene Sheets. *Nano Lett.* **2008**, *8*, 36–41. [CrossRef] [PubMed]

26. Kim, M.; Safron, N.S.; Han, E.; Arnold, M.S.; Gopalan, P. Electronic Transport and Raman Scattering in Size-Controlled Nanoperforated Graphene. *ACS Nano* **2012**, *6*, 9846–9854. [CrossRef] [PubMed]

27. Krishnamoorthy, K.; Veerapandian, M.; Yun, K.; Kim, S.J. The chemical and structural analysis of graphene oxide with different degrees of oxidation. *Carbon* **2013**, *53*, 38–49. [CrossRef]

28. Dreyer, D.R.; Park, S.; Bielawski, C.W.; Ruoff, R.S. The chemistry of graphene oxide. *Chem. Soc. Rev.* **2010**, *39*, 228–240. [CrossRef] [PubMed]

29. Yang, D.; Velamakanni, A.; Bozoklu, G.; Park, S.; Stoller, M.; Piner, R.D.; Stankovich, S.; Jung, I.; Field, D.A.; Ventrice, C.A., Jr.; et al. Chemical analysis of graphene oxide films after heat and chemical treatments by X-ray photoelectron and Micro-Raman spectroscopy. *Carbon* **2009**, *47*, 145–152. [CrossRef]

 *nanomaterials*

Article

Controllable Synthesis of 2D Perovskite on Different Substrates and Its Application as Photodetector

Yunzhou Xue [1], Jian Yuan [2], Jingying Liu [3] and Shaojuan Li [2,*]

[1] College of Chemistry and Environmental Engineering, Shenzhen University, Shenzhen 518000, China; xueyz@iccas.ac.cn
[2] Collaborative Innovation Center of Suzhou Nano Science and Technology, Jiangsu Key Laboratory for Carbon-Based Functional Materials and Devices, Institute of Functional Nano and Soft Materials (FUNSOM), Soochow University, Suzhou 215123, China; jianyuan_cq@126.com
[3] Department of Materials Science and Engineering, Monash University, Clayton, VIC 3800, Australia; jingying.liu@monash.edu
* Correspondence: sjli@suda.edu.cn; Tel.: +86-0512-6588-2337

Received: 14 June 2018; Accepted: 12 July 2018; Published: 3 August 2018

Abstract: Perovskites have recently attracted intense interests for optoelectronic devices application due to their excellent photovoltaic and photoelectric properties. The performance of perovskite-based devices highly depends on the perovskite material properties. However, the widely used spin-coating method can only prepare polycrystalline perovskite and physical vapor deposition (PVD) method requires a higher melting point (>350 °C) substrate due to the high growth temperature, which is not suitable for low melting point substrates, especially for flexible substrates. Here, we present the controlled synthesis of high quality two-dimensional (2D) perovskite platelets on random substrates, including SiO_2/Si, Si, mica, glass and flexible polydimethylsiloxane (PDMS) substrates, and our method is applicable to any substrate as long as its melting point is higher than 100 °C. We found that the photoluminescence (PL) characteristics of perovskite depend strongly on the platelets thickness, namely, thicker perovskite platelet has higher PL wavelength and stronger intensity, and thinner perovskite exhibits opposite results. Moreover, photodetectors based on the as-produced perovskite platelets show excellent photoelectric performance with a high photoresponsivity of 8.3 $A \cdot W^{-1}$, a high on/off ratio of ~10^3, and a small rise and decay time of 30 and 50 ms, respectively. Our approach in this work provides a feasible way for making 2D perovskite platelets for wide optoelectronic applications.

Keywords: 2D perovskite; controllable synthesis; flexible substrate; photodetector; photoelectric performance

---

1. Introduction

Organic-inorganic halide perovskite are materials described by AMX_3 formula, in which A is organic cation, M is metal cation and X is halogen anion [1]. Although perovskites have been discovered for more than one century, the use of perovskites in solar cells only happened in recent years [2–10]. Among the variety kinds of perovskites, methylammonium lead iodide perovskite ($CH_3NH_3PbI_3$) has attracted intensive interest due to its extraordinary optoelectronic properties such as long electron/hole diffusion lengths, high optical absorption coefficient, and optimal bandgap [11,12]. These advantages have generated widely growing interest for diverse applications such as photodetectors [13–16], light-emitting diodes (LEDs) [17–21], waveguides [22], field effect transistors (FET) [23], lasers [24,25], etc. These applications requires high performance devices which highly depends on the perovskite material properties. To date, there are several methods to prepare $CH_3NH_3PbI_3$ perovskite. Directly spin-coating perovskite compound solution to prepare perovskite films is the simplest method, which is suitable for use in solar cells with perovskite as the light harvester [18,20]. However, the material properties obtained by this method still cannot fully display the nature of perovskite, due to its inhomogeneous, polycrystalline structure and large surface roughness. Homogeneous and high crystalline perovskite film can be obtained by thermal evaporation, but this method needs dedicated equipment and can easily cause lead poisoning as the lead iodide (PbI_2) vapor cannot be avoided in the experiment process [8,26]. Chemical vapor deposition (CVD) is promising for the synthesis of high quality perovskite with well-defined structures and morphologies [27,28], especially for two-dimensional (2D) $CH_3NH_3PbI_3$ perovskite platelets. This method involves two steps: firstly, PbI_2 platelets were grown on mica substrate under 350–510 °C, and then the as-grown PbI_2 platelets were converted to $CH_3NH_3PbI_3$ perovskites through inserting the CH_3NH_3I molecules into the PbI_2 platelets under 120 °C [29]. Although high quality perovskite can be obtained by this method, it would still induce lead poisoning owing to the PbI_2 vapor produced during the first growth step. Moreover, the high temperature growth process requires substrates that have high melting point (>350 °C), which are not suitable for flexible substrates with low melting point. Therefore, the need for developing other alternative routes to produce high quality perovskite platelets without lead halide vapour during the process on diverse substrates, especially on flexible substrates remains a challenge.

Previously, we introduce a two-steps method to produce high quality 2D perovskite platelets on SiO_2/Si substrate [13]. The highest growth temperature during the process was around 180 °C. In this manuscript, we carefully control the growth temperature to below 100 °C, therefore high quality 2D perovskite platelets can be obtained on random substrates, especially on transparent, flexible and lower melting point substrates. Our method is applicable to any substrate as long as its melting point is higher than 100 °C. Moreover, photodetectors based on the as-produced 2D perovskite platelets were fabricated. The devices exhibit excellent photoresponse performance, including a high photoresponsivity of 8.3 AW^{-1}, a high on/off ratio of ~10^3, and a small rise and decay time of 30 and 50 ms, respectively. Considering the feasibility of preparing 2D perovskite platelets with different thickness on diverse substrates, especially on transparent, flexible and lower melting point substrates, the results in this paper would greatly extend the device applications of 2D perovskite.

2. Materials and Methods

The 2D $CH_3NH_3PbI_3$ platelets were prepared by two steps. Firstly, 0.2 mg PbI_2 powder was dissolved in 1 mL distilled water and heated at 100 °C for 1 h to get oversaturated PbI_2 solution. After this, the hot PbI_2 solution (around 100 °C) was dropped on the aimed heating substrates, including SiO_2/Si, Si, mica, glass and PDMS. During this process, the PbI_2 oversaturated solution would nucleate and form the 2D PbI_2 platelets on the corresponding substrates. Secondly, CH_3NH_3I powder was put into the center of the furnace and the above obtained 2D PbI_2 platelets were placed 10–15 cm downstream from the CH_3NH_3I powder in a CVD system (Hefei, Anhui, China) to convert to $CH_3NH_3PbI_3$ perovskite. After that, 500 sccm Ar was introduced into the CVD system for 30 min to clear the air in the quartz tube. The Ar flow rate was then kept at 30 sccm to maintain the system pressure to be lower than 1 Torr. Afterword, the furnace was heated to 100 °C under a heating rate of 2.5 °C/min and kept at 100 °C for 5–30 min. Finally, we stopped heating and opened the furnace quickly. During this process, the CH_3NH_3I molecules would insert into the 2D PbI_2 platelets to convert them into 2D $CH_3NH_3PbI_3$ perovskite platelets. Moreover, the heating rate should not be very quick as higher heating rate usually induces temperature overshoot (>100 °C), and there would be more CH_3NH_3I molecules inserted into the PbI_2 platelets than as needed.

The photodetectors were fabricated by picking up a single organic ribbon by a mechanical probe and placed over the 2D perovskite platelet as an organic ribbon mask. After this, 30 nm Au was deposited on the 2D perovskite platelet by thermal evaporation. Finally, the organic ribbon mask was peeled off by a mechanical probe to form the source and drain electrodes over the 2D perovskite platelet. The morphology and structure of the as-grown 2D perovskite platelets were characterized by field-emission scanning electron microscopy (FESEM, Model S-4800, Hitachi, Tokyo,

Japan), optical microscopy (Olympus BX51, Shinjuku, Tokyo, Japan), atomic force microscopy (AFM, Bruker, Dimension Icon SPM, Billerica, MA, USA) in the tapping model and X-ray diffraction (XRD, Bruker D8 advanced diffractometer, Cu-Kα radiation (λ = 1.54050 Å), Billerica, Massachusetts, United States) scanned from 10 to 60° with a step of 0.02°. Chemical composition and crystal orientation of the 2D perovskite platelets were analyzed by X-ray photoelectron spectroscopy (XPS) and transmission electron microscope (TEM, FEI, Hillsboro, OR, USA, Tecnai G2 F20). Photoluminescence (PL) spectrum and mapping measurements were performed using a confocal microscope system (WITec, Ulm, Germany, alpha 300R) with 532 nm wavelength laser to excite the samples. The photoresponse properties of the photodetectors were characterized using the probe station (Cascade, Kingsey Falls, QC, Canada, M150) and a semiconductor property analyzer (Keithley, Cleveland, OH, USA, 4200) under 405 nm laser excitation.

3. Results

Figure 1 shows the morphology of 2D PbI_2 and $CH_3NH_3PbI_3$ perovskite platelets characterized by optical microscopy, SEM and AFM. Figure 1a–d show the optical microscopy images of hexagonal and triangular 2D PbI_2 on Si, mica, glass and flexible PDMS substrates, respectively. We can observe that the surfaces of all the 2D PbI_2 platelets are smooth and uniform, no matter the substrate is rigid (Si, mica and glass)/flexible (PDMS) or smooth (Si and mica)/rough (PDMS and glass), which indicates that the toughness and roughness of the substrates do not play a key role during the PbI_2 single crystal growth process. Figure 1e–h show the optical microscope images of 2D perovskite platelets on Si, mica, glass and PDMS substrates, respectively, which were converted from the single crystal PbI_2 by inserting CH_3NH_3I molecules into them. After the conversion, the surfaces of perovskite platelets were found to be non-uniform and relatively rough compared to their counterparts before the conversion. The surface morphology change is induced by insertion of CH_3NH_3I molecules into the lattice of single crystal PbI_2. Notably, we can obtain $CH_3NH_3PbI_3$ perovskite on almost all kinds of substrates as long as its melting point is higher than 100 °C, which is the highest temperature during the whole process. Figure 1i–p show the optical and SEM images of perovskites with different thickness from hundreds of nanometers to a single-unit-cell thick (2 nm) on SiO_2/Si substrate. Clearly, the surface of the 2D perovskites is more rough for the hundreds-nanometer-thick (Figure 1i) and the single-unit-cell-thick samples (Figure 1l), but the ten-nanometer-thick sample exhibits a much smoother surface (Figure 1k). AFM was used to explore the morphology and thickness of the 2D perovskite platelets, as shown in Figure 1q–t. Samples with thickness of 450 nm, 175 nm, 150 nm and 60 nm were measured. AFM morphology indicates a surface roughness of 24 nm, 22 nm, 18 nm and 13 nm, corresponding to 450 nm, 175 nm, 150 nm and 60 nm samples, respectively. From Figure 1s,t, we can even find small particles on the surface of thinner samples, this proves again that the CH_3NH_3I molecules insert into the PbI_2 crystal and result in the surface morphology change.

Figure 2a shows the XRD patterns of the 2D PbI_2 and the corresponding $CH_3NH_3PbI_3$ perovskite platelets on glass substrates with a thickness of 10 nm. The strong (110) and (220) diffraction peaks with 2θ located at 13.9° and 28.17° indicate that the obtained $CH_3NH_3PbI_3$ perovskite is of tetragonal crystalline structure. Moreover, comparing the diffraction peaks of PbI_2 and $CH_3NH_3PbI_3$ perovskite, we can observe that the (001), (002), (003), (004) diffraction peaks of PbI_2 disappeared after the conversion process, indicating that the PbI_2 was completely converted to $CH_3NH_3PbI_3$ perovskite crystals. Note that the normally observed (112), (211), (310) and (224) diffraction peaks in perovskite synthesized via solution method are not observed in our samples, attesting the fine crystal orientation of the converted $CH_3NH_3PbI_3$ perovskite [30]. In order to characterize the optical properties of the converted perovskite, PL spectra were collected under 532 nm laser excitation at room temperature, as shown in Figure 2b and Figure S1a,b. Samples with different thickness that varies from 63 nm to single unit cell thick (2 nm) were marked by p1 to p6. The PL peak shifts towards shorter wavelength from 770 to 720 nm as the thickness of the perovskite decreases from 100 nm to single unit cell (2 nm), which is consistent with the previous report [13] and can be ascribed to the lattice expansion, namely, the structure relaxation of the in-plane crystal lattice could increase the optical band gap [31]. Moreover, we found that the PL intensity reduces dramatically by over 30 times as the film thickness decreases from 63 nm (marked by p1 in Figure 2b and Figure S1a,b) to single unit cell (2 nm, marked by p6 as shown in the upper left inset in Figure 2b and Figure S1a,b), owing to more excited charge carriers in the thicker perovskite under the laser irradiation. While for thinner perovskite, the carrier density is much less than that of thicker one, resulting in lower PL quantum yield efficiency. To further elucidate the relationship between the perovskite thickness and the corresponding PL intensity and peak position, PL mapping measurements were performed on 2D perovskite platelets, as shown in Figure 2c–f. When the perovskite thickness is larger than 100 nm, the boundary PL intensity is much higher than that in the central part (Figure 2c). In addition, this phenomenon still can be observed when the thickness decreases to 20 nm (inner triangular perovskite sheet in Figure 2d), although the intensity contrast is lower than that of the thicker ones. However, as the thickness decreases to thinner than 10 nm, the PL intensity is uniform over the whole crystal and no clear intensity contrast can be observed (inner triangular perovskite sheet in Figure 2e). While converting PbI_2 to perovskite by inserting CH_3NH_3I molecules into it, the crystal boundary region of PbI_2 is much easier to react with CH_3NH_3I molecules due to the large exposed edges than the central part during the conversion process, whereas, for the central part, the reaction first occurs at the surface and then towards inside the material. For thicker PbI_2, the CH_3NH_3I molecules cannot insert into it through the surface easily, and consequently there are not enough CH_3NH_3I molecules involved into the reaction, resulting in the lower PL intensity from the central part than that from the boundary part. As the PbI_2 thickness decreases, the insertion of CH_3NH_3I molecules becomes much easier though its surface, and thus the PL intensity difference between the central part and the boundary part is gradually reduced. As the thickness further decreases to 10 nm, sufficient number of CH_3NH_3I molecules can insert into PbI_2 though its surface, thereby there is no obvious intensity difference between the central part and the boundary part (Figure 2e). Figure 2f proved again that the boundary PL intensity is much higher than the central part and the thicker perovskite have higher PL intensity as clarified in Figure 2b. To more intuitively demonstrate the above effect, PbI_2 platelet is converted to $CH_3NH_3PbI_3$ perovskite without supplying sufficient CH_3NH_3I molecules, and the resultant PL spectrum of the $CH_3NH_3PbI_3$ perovskite is shown in Figure S2a,b, from which we can observe that the boundary exhibits higher PL intensity, and the intensity in the central part is very low, attesting the above analysis.

Figure 3a displays the TEM image of a hexagonal 2D perovskite platelet with a thickness of 10 nm. The corresponding high resolution TEM (HRTEM) image in Figure 3b shows clear lattice fringes with (200) and (0$\bar{2}$2) planes, further revealing the single crystalline structure of the converted perovskite platelet [13]. In order to understand the elemental arrangement of Pb in the perovskite platelets, scanning photoelectron microscopy (SPEM) was applied to obtain the XPS mapping image, as shown in Figure 3c. From the Pb mapping results, we can observe that the Pb arranged uniformly in the crystal after inserting the CH_3NH_3I molecules into the crystal. Moreover, single XPS spectra of Pb, I, C and N elements in the $CH_3NH_3PbI_3$ perovskite were also acquired and shown in Figure 3d–g, respectively. The prominent C_{1s} peak located at 285.3 eV corresponds to the carbon atoms in $CH_3NH_3PbI_3$ crystal, and a small amount of amorphous carbon located at 284.2 eV can also be seen, which may be due to the Si substrate contamination (Figure 3f). The I_{4d}, Pb_{4f} and N_{1s} peaks are located at (49.8 eV, 51.4 eV), (138.8 eV, 144.3 eV) and (399 eV, 401.8 eV) respectively, which are in good agreement with previous reports [32–34].

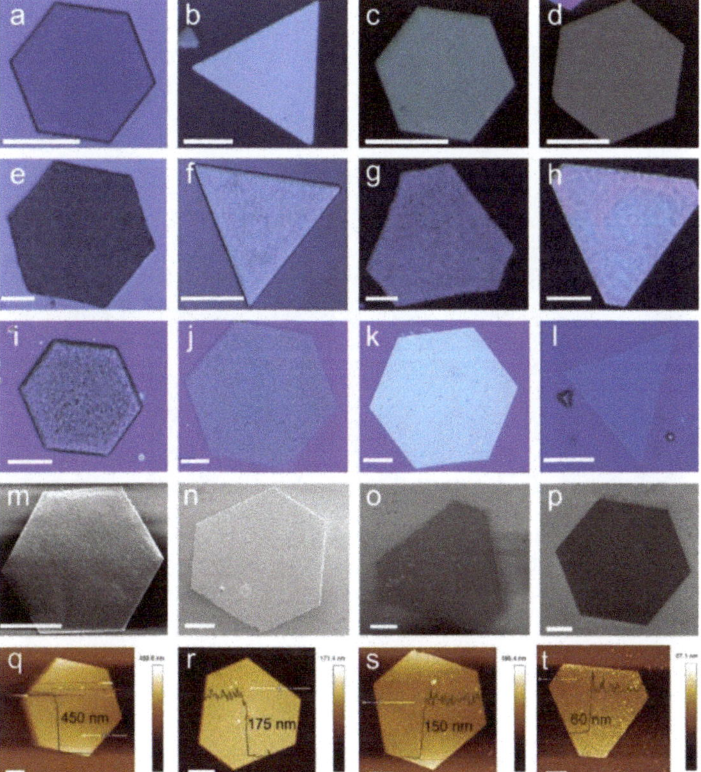

Figure 1. (**a–d**) Optical microscopy images of the hexagonal and triangular 2D PbI$_2$ platelets on Si, mica, glass and PDMS substrates, respectively; (**e–h**) Optical microscopy images of the hexagonal and triangular 2D CH$_3$NH$_3$PbI$_3$ perovskite on Si, mica, glass and PDMS substrates, respectively; (**i–l**) and (**m–p**) Optical microscopy and SEM images of 2D CH$_3$NH$_3$PbI$_3$ perovskite with different thicknesses from hundreds of nanometers to single unit cell thick (2 nm) on SiO$_2$/Si substrate; (**q–t**) AFM topography images of 2D CH$_3$NH$_3$PbI$_3$ perovskite platelets with different thicknesses from 450 nm to 60 nm, respectively. All the scale bars are 10 μm.

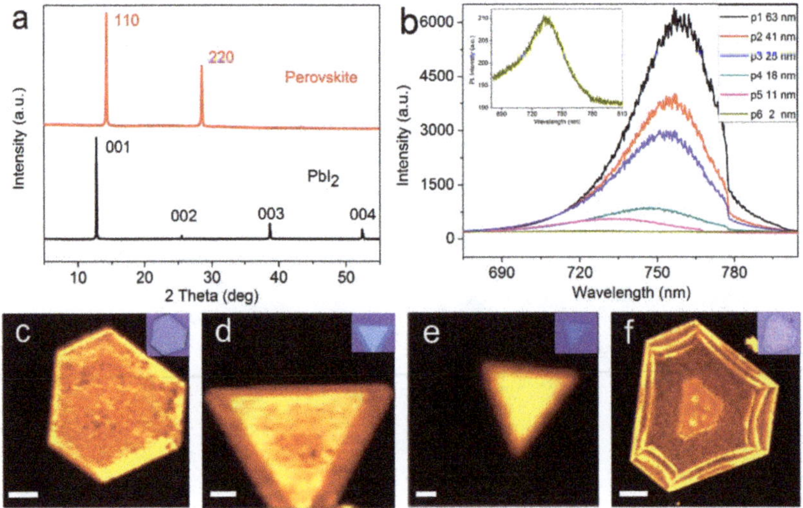

Figure 2. (**a**) XRD patterns of PbI_2 platelets and corresponding converted 2D $CH_3NH_3PbI_3$ platelets. The thickness of $CH_3NH_3PbI_3$ platelets is 10 nm; (**b**) PL spectra of perovskite platelets with different thicknesses. The upper left inset shows the PL spectra of 2 nm perovskite; (**c–f**) PL mapping images of 2D $CH_3NH_3PbI_3$ platelets with different thicknesses. As shown in (**c,d**), when the platelet is thicker than 10 nm, the PL intensities of the central part are lower than that of the boundary part. The upright insets in (**c–f**) show the corresponding optical microscopy images of 2D $CH_3NH_3PbI_3$ platelets. The scale bars are 4, 7, 1 and 10 μm, respectively.

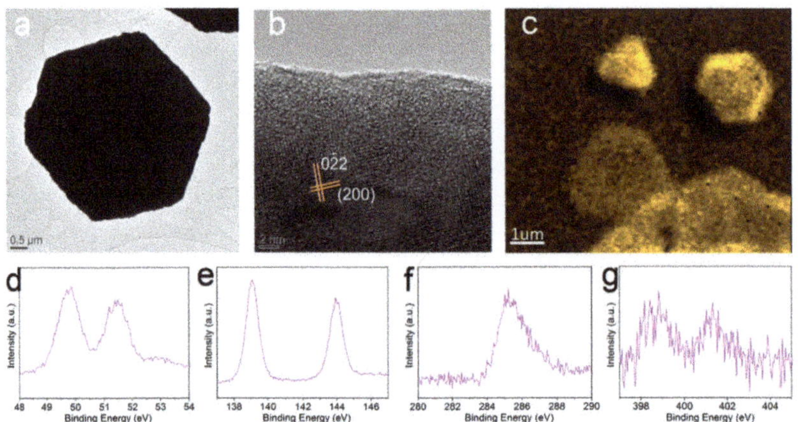

Figure 3. (**a**) TEM image of a 2D $CH_3NH_3PbI_3$ platelet; (**b**) High-resolution TEM image of the 2D $CH_3NH_3PbI_3$ platelet; (**c**) XPS mapping images of Pb element in 2D $CH_3NH_3PbI_3$ platelets acquired with SPEM; (**d–g**) XPS spectra of Pb, I, C and N elements in 2D $CH_3NH_3PbI_3$ platelets, respectively.

4. Discussion

Controlled growth of 2D perovskite platelets on different substrates enables us to probe their intrinsic optoelectrical properties. As an example, perovskite platelets were explored as the semiconducting channel of FETs on SiO_2/Si substrate with two gold electrodes as source/drain electrodes and Si as the back gate. Schematic and optical microscopy image of a 2D $CH_3NH_3PbI_3$ platelet phototransistor are shown in Figure 4a and Figure S3a. The results in Figure 4b display a nearly zero dark current and linear *I-V* curves under two different illumination powers with 405 nm laser excitation, indicating that the device has excellent photoresponse capability. The linear *I-V* curves indicate the Ohmic contact between perovskite and gold source/drain electrodes. Figure 4c shows the device response to pulsed light at different optical pumping power, from which we can observe that the device can be effectively switched "ON" and "OFF" while the laser source is turned on and off. The amplitude of the electrical signal is modulated by different light powers. Also from Figure 4c we can calculate the photocurrent to dark current ratio of our devices and the value can reach up to three orders of magnitude. Figure 4d shows the photoresponsivity and photocurrent as a function of the light power. It is found that both the photocurrent and photoresponsivity change nonlinearly with increasing the laser power. The photoresponsivity of our 2D perovskite based FET can reach up to 8.3 AW^{-1} under a bias voltage of 1 V, higher than the bulk perovskite film based devices (3 AW^{-1}) but lower than perovskite crystal (40 AW^{-1}) based ones with channel length reducing to 100 nm [15]. The response speed of our device is characterized by a rise time and a decay time of less than 30 and 50 ms, respectively, demonstrating a much faster response than the bulk perovskite film based devices. In addition, time-dependent photocurrents at different source-drain voltages are shown in Figure 4f and Figure S3b. The photocurrent can be significantly increased by increasing the source-drain voltage. As the laser is turned on and off, the photocurrent changes periodically, indicating a very good operation repeatability. The above results prove that our 2D perovskite platelets show excellent photoelectric properties and hold a potential for broader optoelectronics application, especially for applications that need low temperature processing.

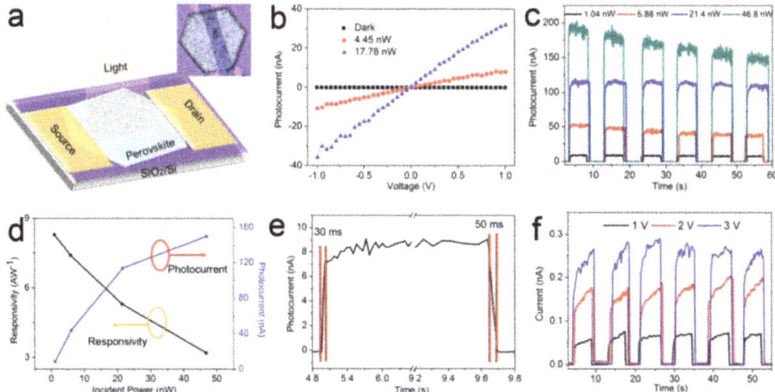

Figure 4. (a) Schematic of a 2D $CH_3NH_3PbI_3$ platelet phototransistor. The upright inset shows the optical microscopy image of the 2D $CH_3NH_3PbI_3$ platelet phototransistor; (b) *I-V* curves of the 2D perovskite-based device in dark and under light irradiation with different power; (c) Time-dependent photocurrent of the 2D $CH_3NH_3PbI_3$ platelet with different incident power; (d) Dependence of photocurrent and photoresponsivity on incident light power; the blue and black dots correspond to original data; (e) Time photocurrent response excited at 405 nm laser. The rise time and the decay time are 30 ms and 50 ms, respectively; (f) Time dependent photocurrent of the device based on the 2D $CH_3NH_3PbI_3$ platelet during the laser switching on/off process under positive source-drain voltage, V_{sd}. V_{sd} is from 1 to 3 V.

5. Conclusions

In summary, high quality 2D $CH_3NH_3PbI_3$ perovskite platelets were prepared by a two-steps method. By utilizing this method, we can produce 2D perovskites platelets with different thicknesses from hundreds of nanometers to single unit cell (2 nm) on different substrates, as long as the melting point of the substrate is higher than 100 °C. It was found that the PL characteristics of perovskite depends strongly on the platelet thickness, namely, thicker perovskite platelet has higher PL wavelength and stronger intensity, whereas, thinner perovskite exhibits opposite results. Moreover, photoelectric measurements confirm that our 2D perovskite platelets show excellent photoelectric properties. Phototransistors based on the 2D perovskite platelet exhibit a high photoresponsivity of 8.3 AW^{-1}, a high on/off ratio of ~10^3 with a small rise and a decay time of 30 and 50 ms, respectively. Considering the feasibility of preparing 2D perovskite platelets with different thickness on diverse substrates, especially on transparent, flexible and lower melting point substrates, our method would greatly extend the device applications of 2D perovskite.

Supplementary Materials: The following are available online at http://www.mdpi.com/2079-4991/8/8/591/s1, Figure S1: Optical microscopy and PL mapping images of 2D $CH_3NH_3PbI_3$ perovskite, Figure S2: Optical microscopy and PL mapping images of the converted 2D $CH_3NH_3PbI_3$ perovskite without supplying sufficient CH_3NH_3I molecules during the conversion process, Figure S3: supplementary photoelectrical performance of the 2D $CH_3NH_3PbI_3$ platelet phototransistor.

Author Contributions: Conceptualization, Y.X. and S.L.; Methodology, J.L. and J.Y.; Validation, Y.X.; Formal Analysis, Y.X., J.L., J.Y. and S.L.; Investigation, Y.X.; Resources, S.L.; Data Curation, Y.X.; Writing-Original Draft Preparation, Y.X., J.L., J.Y. and S.L.; Writing-Review & Editing, S.L.; Visualization, Y.X. and S.L.; Supervision, S.L.; Project Administration, S.L.; Funding Acquisition, S.L.

Funding: This research was funded by the National Natural Science Foundation of China (No. 61604102, 51222208, 51290273), the National Key Research & Development Program (No. 2016YFA0201902).

Acknowledgments: This work was performed in part at the Melbourne Centre for Nanofabrication (MCN) in the Victorian Node of the Australian National Fabrication Facility (ANFF).

Conflicts of Interest: The authors declare no conflict of interest. The founding sponsors had no role in the design of the study; in the collection, analyses, or interpretation of data; in the writing of the manuscript, and in the decision to publish the results.

References

1. Shi, S.; Li, Y.; Li, X.; Wang, H. Advancements in all-solid-state hybrid solar cells based on organometal halide perovskites. *Mater. Horiz.* **2015**, *2*, 378–405. [CrossRef]

2. Kojima, A.; Teshima, K.; Shirai, Y.; Miyasaka, T. Organometal Halide Perovskites as Visible-Light Sensitizers for Photovoltaic Cells. *J. Am. Chem. Soc.* **2009**, *131*, 6050–6051. [CrossRef] [PubMed]

3. Im, J.H.; Lee, C.R.; Lee, J.W.; Park, S.W.; Park, N.G. 6.5% Efficient perovskite quantum-dot-sensitized solar cell. *Nanoscale* **2011**, *3*, 4088–4093. [CrossRef] [PubMed]

4. Kim, H.S.; Lee, C.R.; Im, J.H.; Lee, K.B.; Moehl, T.; Marchioro, A.; Moon, S.J.; Humphry-Baker, R.; Yum, J.H.; Moser, J.E.; et al. Lead Iodide Perovskite Sensitized All-Solid-State Submicron Thin Film Mesoscopic Solar Cell with Efficiency Exceeding 9%. *Sci. Rep.* **2012**, *2*, 591. [CrossRef] [PubMed]

5. Lee, M.M.; Teuscher, J.; Miyasaka, T.; Murakami, T.N.; Snaith, H.J. Efficient Hybrid Solar Cells Based on Meso-Superstructured Organometal Halide Perovskites. *Science* **2012**, *338*, 643–647. [CrossRef] [PubMed]

6. Etgar, L.; Gao, P.; Xue, Z.; Peng, Q.; Chandiran, A.K.; Liu, B.; Nazeeruddin, M.K.; Grätzel, M. Mesoscopic $CH_3NH_3PbI_3$/TiO_2 Heterojunction Solar Cells. *J. Am. Chem. Soc.* **2012**, *134*, 17396–17399. [CrossRef] [PubMed]

7. Burschka, J.; Pellet, N.; Moon, S.J.; Humphry-Baker, R.; Gao, P.; Nazeeruddin, M.K.; Grätzel, M. Sequential deposition as a route to high-performance perovskite-sensitized solar cells. *Nature* **2013**, *499*, 316–319. [CrossRef] [PubMed]

8. Liu, M.; Johnston, M.B.; Snaith, H.J. Efficient planar heterojunction perovskite solar cells by vapour deposition. *Nature* **2013**, *501*, 395–398. [CrossRef] [PubMed]

9. Heo, J.H.; Im, S.H.; Noh, J.H.; Mandal, T.N.; Lim, C.-S.; Chang, J.A.; Lee, Y.H.; Kim, H.-J.; Sarkar, A.; Nazeeruddin, M.K.; et al. Efficient inorganic–organic hybrid heterojunction solar cells containing perovskite compound and polymeric hole conductors. *Nat. Photonics* **2013**, *7*, 486–491. [CrossRef]

10. Kamat, P.V. Organometal Halide Perovskites for Transformative Photovoltaics. *J. Am. Chem. Soc.* **2014**, *136*, 3713–3714. [CrossRef] [PubMed]

11. Xing, G.C.; Mathews, N.; Sun, S.Y.; Lim, S.S.; Lam, Y.M.; Grätzel, M.; Mhaisalkar, S.; Sum, T.C. Long-Range Balanced Electron- and Hole-Transport Lengths in Organic-Inorganic $CH_3NH_3PbI_3$. *Science* **2013**, *342*, 344–347. [CrossRef] [PubMed]

12. Stranks, S.D.; Eperon, G.E.; Grancini, G.; Menelaou, C.; Alcocer, M.J.P.; Leijtens, T.; Herz, L.M.; Petrozza, A.; Snaith, H.J. Electron-Hole Diffusion Lengths Exceeding 1 Micrometer in an Organometal Trihalide Perovskite Absorber. *Science* **2013**, *342*, 341–344. [CrossRef] [PubMed]

13. Liu, J.; Xue, Y.; Wang, Z.; Xu, Z.Q.; Zheng, C.; Weber, B.; Song, J.; Wang, Y.; Lu, Y.; Zhang, Y. Two-Dimensional $CH_3NH_3PbI_3$ Perovskite: Synthesis and Optoelectronic Application. *ACS Nano* **2016**, *10*, 3536–3542. [CrossRef] [PubMed]

14. Hu, X.; Zhang, X.; Liang, L.; Bao, J.; Li, S.; Yang, W.; Xie, Y. High-Performance Flexible Broadband Photodetector Based on Organolead Halide Perovskite. *Adv. Funct. Mater.* **2014**, *24*, 7373–7380. [CrossRef]

15. Wang, G.; Li, D.; Cheng, H.C.; Li, Y.; Chen, C.Y.; Yin, A.; Zhao, Z.; Lin, Z.; Wu, H.; He, Q.; et al. Wafer-scale growth of large arrays of perovskite microplate crystals for functional electronics and optoelectronics. *Sci. Adv.* **2015**, *1*, e1500613. [CrossRef] [PubMed]

16. Wang, Y.; Zhang, Y.; Lu, Y.; Xu, W.; Mu, H.; Chen, C.; Qiao, H.; Song, J.; Li, S.; Sun, B.; et al. Hybrid Graphene–Perovskite Phototransistors with Ultrahigh Responsivity and Gain. *Adv. Opt. Mater.* **2015**, *3*, 1389. [CrossRef]

17. Li, J.; Bade, S.G.R.; Shan, X.; Yu, Z. Single-Layer Light-Emitting Diodes Using Organometal Halide Perovskite/Poly(ethylene oxide) Composite Thin Films. *Adv. Mater.* **2015**, *27*, 5196–5202. [CrossRef] [PubMed]

18. Tan, Z.K.; Moghaddam, R.S.; Lai, M.L.; Docampo, P.; Higler, R.; Deschler, F.; Price, M.; Sadhanala, A.; Pazos, L.M.; Credgington, D.; et al. Bright light-emitting diodes based on organometal halide perovskite. *Nat. Nanotechnol.* **2014**, *9*, 687–692. [CrossRef] [PubMed]

19. Qasim, K.; Wang, B.; Zhang, Y.; Li, P.; Wang, Y.; Li, S.; Lee, S.T.; Liao, L.S.; Lei, W.; Bao, Q. Solution-Processed Extremely Efficient Multicolor Perovskite Light-Emitting Diodes Utilizing Doped Electron Transport Layer. *Adv. Funct. Mater.* **2017**, *27*, 1606874. [CrossRef]

20. Wang, J.; Wang, N.; Jin, Y.; Si, J.; Tan, Z.K.; Du, H.; Cheng, L.; Dai, X.; Bai, S.; He, H.; et al. Interfacial Control Toward Efficient and Low-Voltage Perovskite Light-Emitting Diodes. *Adv. Mater.* **2015**, *27*, 2311–2316. [CrossRef] [PubMed]

21. Xing, G.; Mathews, N.; Lim, S.S.; Yantara, N.; Liu, X.; Sabba, D.; Gratzel, M.; Mhaisalkar, S.; Sum, T.C. Low-temperature solution-processed wavelength-tunable perovskites for lasing. *Nat. Mater.* **2014**, *13*, 476–480. [CrossRef] [PubMed]

22. Wang, Z.; Liu, J.; Xu, Z.Q.; Xue, Y.; Jiang, L.; Song, J.; Huang, F.; Wang, Y.; Zhong, Y.L.; Zhang, Y.; et al. Wavelength-tunable waveguides based on polycrystalline organic–inorganic perovskite microwires. *Nanoscale* **2016**, *8*, 6258–6264. [CrossRef] [PubMed]

23. Chin, X.Y.; Cortecchia, D.; Yin, J.; Bruno, A.; Soci, C. Lead iodide perovskite light-emitting field-effect transistor. *Nat. Commun.* **2015**, *6*, 7383. [CrossRef] [PubMed]

24. Ha, S.T.; Shen, C.; Zhang, J.; Xiong, Q. Laser cooling of organic–inorganic lead halide perovskites. *Nat. Photonics* **2016**, *10*, 115–121. [CrossRef]

25. Zhang, Q.; Ha, S.T.; Liu, X.; Sum, T.C.; Xiong, Q. Room-Temperature Near-Infrared High-Q Perovskite Whispering-Gallery Planar Nanolasers. *Nano Lett.* **2014**, *14*, 5995–6001. [CrossRef] [PubMed]

26. Mitzi, D.B.; Prikas, M.T.; Chondroudis, K. Thin Film Deposition of Organic–Inorganic Hybrid Materials Using a Single Source Thermal Ablation Technique. *Chem. Mater.* **1999**, *11*, 542–544. [CrossRef]

27. Zhang, Y.; Wang, Y.; Xu, Z.Q.; Liu, J.; Song, J.; Xue, Y.; Wang, Z.; Zheng, J.; Jiang, L.; Zheng, C.; et al. Reversible Structural Swell–Shrink and Recoverable Optical Properties in Hybrid Inorganic–Organic Perovskite. *ACS Nano* **2016**, *10*, 7031–7038. [CrossRef] [PubMed]

28. Zhang, Y.; Liu, J.; Wang, Z.; Xue, Y.; Ou, Q.; Polavarapu, L.; Zheng, J.; Qi, X.; Bao, Q. Synthesis, properties, and optical applications of low-dimensional perovskites. *Chem. Commun.* **2016**, *53*, 13637–13655. [CrossRef] [PubMed]

29. Ha, S.T.; Liu, X.; Zhang, Q.; Giovanni, D.; Sum, T.C.; Xiong, Q. Synthesis of Organic–Inorganic Lead Halide Perovskite Nanoplatelets: Towards High-Performance Perovskite Solar Cells and Optoelectronic Devices. *Adv. Opt. Mater.* **2014**, *2*, 838–844. [CrossRef]

30. Noh, J.H.; Im, S.H.; Heo, J.H.; Mandal, T.N.; Seok, S.I. Chemical Management for Colorful, Efficient, and Stable Inorganic–Organic Hybrid Nanostructured Solar Cells. *Nano Lett.* **2013**, *13*, 1764–1769. [CrossRef] [PubMed]

31. Dou, L.; Wong, A.B.; Yu, Y.; Lai, M.; Kornienko, N.; Eaton, S.W.; Fu, A.; Bischak, C.G.; Ma, J.; Ding, T.; et al. Atomically thin two-dimensional organic-inorganic hybrid Perovskites. *Science* **2015**, *349*, 1518–1521. [CrossRef] [PubMed]

32. Li, Y.; Xu, X.; Wang, C.; Wang, C.; Xie, F.; Yang, J.; Gao, Y. Degradation by Exposure of Coevaporated $CH_3NH_3PbI_3$ Thin Films. *J. Phys. Chem. C* **2015**, *119*, 23996–24002. [CrossRef]

33. Wang, C.; Wang, C.; Liu, X.; Kauppi, J.; Shao, Y.; Xiao, Z.; Bi, C.; Huang, J.; Gao, Y. Electronic structure evolution of fullerene on $CH_3NH_3PbI_3$. *Appl. Phys. Lett.* **2015**, *106*, 111603. [CrossRef]

34. Quan, L.N.; Yuan, M.; Comin, R.; Voznyy, O.; Beauregard, E.M.; Hoogland, S.; Buin, A.; Kirmani, A.R.; Zhao, K.; Amassian, A.; et al. Ligand-Stabilized Reduced-Dimensionality Perovskites. *J. Am. Chem. Soc.* **2016**, *138*, 2649–2655. [CrossRef] [PubMed]

 *nanomaterials*

Article

Arrayed CdTeMicrodots and Their Enhanced Photodetectivity via Piezo-Phototronic Effect

Dong Jin Lee [1], G. Mohan Kumar [2], P. Ilanchezhiyan [2,*], Fu Xiao [2], Sh.U. Yuldashev [2], Yong Deuk Woo [3], Deuk Young Kim [1] and Tae Won Kang [2]

[1] Quantum-Functional Semiconductor Research Center, Dongguk University-Seoul, Seoul 04623, Korea; jin514rin@naver.com (D.J.L.); dykim@dgu.edu (D.Y.K.)
[2] Nano-Information Technology Academy (NITA), Dongguk University-Seoul, Seoul 04623, Korea; selvi1382@gmail.com (G.M.K.); xiaofu.04@foxmail.com (F.X.); shavkat@dongguk.edu (S.U.Y.); twkang@dongguk.edu (T.W.K.)
[3] Department of Mechanical and Automotive Engineering, Woosuk University, Chonbuk 55338, Korea; wooyongd@woosuk.ac.kr
* Correspondence: ilancheziyan@dongguk.edu

Received: 27 November 2018; Accepted: 23 January 2019; Published: 1 February 2019

Abstract: In this paper, a photodetector based on arrayed CdTe microdots was fabricated on Bi coated transparent conducting indium tin oxide (ITO)/glass substrates. Current-voltage characteristics of these photodetectors revealed an ultrahigh sensitivity under stress (in the form of force through press) while compared to normal condition. The devices exhibited excellent photosensing properties with photoinduced current increasing from 20 to 76 μA cm^{-2} under stress. Furthermore, the photoresponsivity of the devices also increased under stress from 3.2×10^{-4} A/W to 5.5×10^{-3} A/W at a bias of 5 V. The observed characteristics are attributed to the piezopotential induced change in Schottky barrier height, which actually results from the piezo-phototronic effect. The obtained results also demonstrate the feasibility in realization of a facile and promising CdTe microdots-based photodetector via piezo-phototronic effect.

Keywords: CdTe microdots; Schottky barrier; photodetector; piezo-phototronic effect

1. Introduction

Semiconductor nanostructures receive immense attention for their distinct physical properties and applications in high-performance nano devices, owing to their rationally designed surface and large surface to volume ratio [1–3]. Their unique morphological and crystalline characteristics make them very promising for designing and developing novel nanoscaled devices. In such cases, a control over their size and structure allowsus to tune their optical properties as well as their band gaps. Among many semiconductors, CdTe belonging to II–VI group is an optically active material with a band gap of 1.5 eV. This makes it a promising light-absorbing material for photovoltaics. Due to its excellent optical properties with high absorption coefficient and high specific power, CdTe is of great interest for application in optoelectronics devices such as photovoltaics, photodetectors, near-infrared detectors, gamma-ray detectors for medical imaging and lasers [4–10].

Due to its attractive bandgap and optical properties, CdTe in the nanostructures form is a potential candidate for the fabrication of a high-performance devices. Different morphologies of CdTe, such as nanoparticles, nanowires, nanorods, and nanotubes have been synthesized via a variety of methods including molecular beam epitaxy, chemical vapor deposition (CVD), physical vapor deposition, electrodeposition, closed spaced sublimation (CSS) method, and radio frequency(RF) magnetron sputtering [11–20]. CdTe solar cells on ultrathin glass substrates, yielding a high efficiency of 16.4%, have also been reported by Mahabaduge et al [21]. Recently, CdTe in the form of nanoribbons and nanowires were shown to exhibit significant photoresponse with high responsivity and gain [22,23].

In spite of the several advantageous optoelectronic properties of CdTe, their piezoelectric properties have seldomly been reported. Recently elastic and piezoelectric properties of zinc blende and wurtzite crystalline nanowire heterostructures were also reported [24–26]. Nonlinear piezoelectricity in CdTe was demonstrated by Corso et al by performing first principle calculation through density-functional theory (DFT) [27]. More recently, Hou et al. reported nanogenerator based on zinc blende CdTe micro/nanowires [28]. Inspired by these literatures, we showcase the piezopotential distribution on arrayed CdTe microdots under the application of external stress.

In this paper, we demonstrate the controlled growth of arrayed CdTe microdots using vapor phase epitaxy. The structural and morphology characterization are explained in detail. Additionally, investigations were also performed on the role of precursor to substrate distance over the controlled growth of CdTe microdots. Additionally, to showcase their potential for photoelectronic applications, we have investigated their electrical and photoelectronic properties by fabricating a photodetector using arrayed CdTe microdots. A strong photoelectric response was observed in the devices at room temperature without external power supply. The performance of this photodetector was significantly improved under the stress via piezo-phototronic effect. The photoresponsivity of the photodetector shows enhanced performance under stress than that of normal condition. The device displayed a detectivity of 1.68×10^{11} Jones under stress. This indicate that the performance of the arrayed CdTe microdots-based photodetector can effectively be enhanced through utilizing piezo-phototronic effect.

2. Materials and Methods

2.1. Synthesis of CdTe Films and Microdots

Figure 1 shows the schematic experimental setup involved in the deposition of arrayed CdTe microdots on Bi coated indium tin oxide (ITO)/glass.CdTe films and microdots were selectively grown on Bi coated ITO/glass substrates by vapor phase epitaxy (VPE) method in a temperature controlled three-zone furnace. Bi films were first predeposited on precleanedITO/glass substrate via e-beam evaporator system as the specific substrate for selective growth of CdTe (20 nm). For this a grain type Bi metal source of 99.999% purity (Alfa Aesar product, Haverhill, MA, USA) was used. The film was deposited by using a shadow metal mask at a deposition rate of 1 Å/s at 2×10^{-6} Torr. The prepared substrate and CdTe bulk source (3.8 g, 99.888% purity Alfa Aesar product, Haverhill, MA, USA) was placed in Zone 2 (growth zone) and Zone 3 (source zone) of the VPE. The source to substrate distance was varied from 5 to 15 cm in order to obtain a clear micro dot pattern. The quartz chamber was pumped to a pressure of 2.4×10^{-3} Torr via rotary pump and unreacted (Ar) carrier gas was flown out at 60 sccm for 10 min before growth. The flow was maintained until the temperature had completely lowered after the growth. The temperatures of Zone 1 and Zone 2 were kept the same (250, 350 and 450 °C), while the temperature of Zone 3 was held at 600 °C. The temperature rise rate of all the samples was 10 degrees per minute and the growth time was 1 h. Bi was additionally coated on Al_2O_3 (111) and Si (100) substrates under similar growth conditions to determine their role as catalyst that effectively promotes the selectively growth of CdTe.

2.2. Device Fabrication

We also fabricated devices using CdTe microdots to observe the current changes due to photoreaction and physical forces. ITO/glass plates were placed on top of selectively grown CdTe microdots/Bi/ITO/glass samples (for physical contact and as the top electrode by fixing with a tape). The top and bottom electrode were connected to a wire and silver paste, respectively. The current was then measured according to the voltage when light irradiation and physical force were applied. We further studied the reproducibility of changes in current due to optical response and physical force.

2.3. Characterization

The surface morphology/microstructure of CdTe samples were monitored through field-emission scanning electron microscopy (FESEM, Philips, Model: XL-30, Amsterdam, The Netherland). The microstructure of CdTe samples were inferred through X-ray diffraction (XRD, Bede scientific instruments, Model: Bede D1, Bowburn, UK) and Micro Raman spectrometer (DawoolAttonics, Model: Micro Raman System, Seongnam, Korea). Optical properties were obtained using a UV/VIS spectrophotometer (K LAB, Model: Optizen POP, Daejeon, Republic of Korea). A Keithley 617 semiconductor parameter analyzer (Tektronix, Model: Keithley 617, Beaverton, OR, USA) was employed to study the photoresponse of the device under solar simulator (Newport, AM1.5) (SERIC, Model: XIL-01B50KP, Tokyo, Japan). A specially designed spring-type soft stick was used to give a constant physical force to the device. The constant physical force was measured using a digital force gauge (Amittari, Model:FG-104, Guangdong, China) and the value was found to be 6.03 N/m^2. A UV cut-off filter was employed in our experiments to exclude the influence of ITO on the photocurrent values of our devices. Hall effect measurements for CdTe microdots were provided in the Table S1.

3. Results and Discussion

Figure 2 shows the SEM and cross-sectional images of the CdTe films grown on Bi catalyst under different growth temperature. As seen from Figure 1, the surface topography of the film changes under different growth temperature. The films deposited at 250 °C (Figure 2a) shows uniform distribution of small grains on the substrate with many pinholes. On increasing the temperature to 350 °C, the grain size increases and fills the pin holes and result in continuous film (Figure 2c). The increase in the size of the grains might be related to the fact that the critical grain radius increases due to disappearance of the smaller grains and enhanced growth of larger grains. For further increase in the temperature (450 °C), the grain size increases substantially, and gets turned into elongated grain-like structure (Figure 2e). Compared with films deposited at temperature of 250 °C and 350 °C, the film uniformity has been improved for 450 °C.

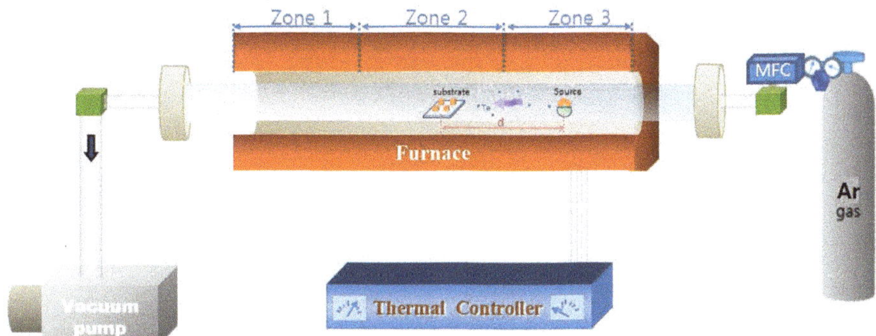

Figure 1. Apparatus for the deposition of the CdTe microdots arrays on Bi coated indium tin oxide (ITO)/glass.

To understand the effect of Bi film as seed layer for the formation of CdTe films, controlled experiments were conducted with and without Bi films on ITO substrates. The corresponding scanning electron microscopy (SEM) images of CdTe films grown under different growth temperatures are shown in Figure S1. In the early stages when the temperature was maintained at 250 °C nucleation of small grains of CdTe was observed on both with and without Bi film. When the temperature was increased to 450 °C, CdTe films was formed only on the Bi film. In this case no films were formed without Bi film, regardless of reaction condition. This indicates Bi film acts as a seed layer for the deposition of CdTe films. Additionally, different substrates such as Bi coated Al$_2$O$_3$ (111) and Si (100) substrate, were also employed to test the effect of substrate at the same growth temperature (Figure S2). No significant variation was observed in the results, as CdTe was formed only on the Bi coated area rather than uncoated area.

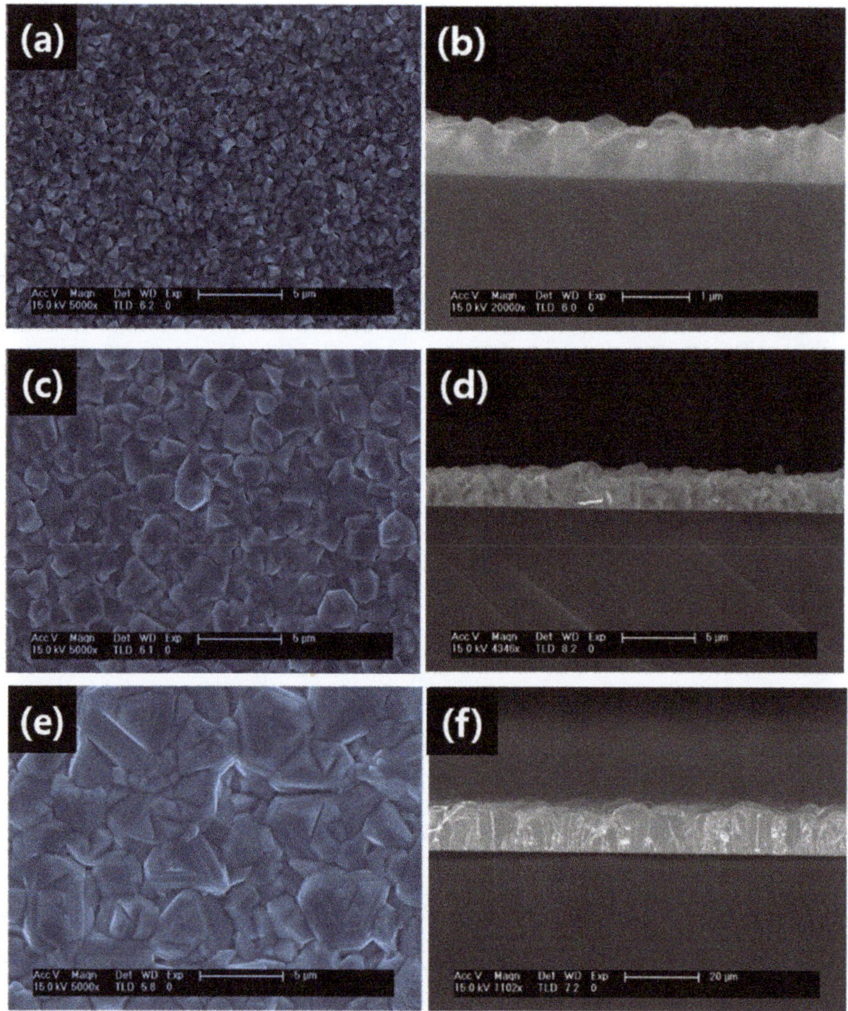

Figure 2. Scanning electron microscopy (SEM) images of CdTe grown at different substrate temperature (**a**) 250 °C (**c**) 350 °C (**e**) 450 °C along with their corresponding cross-sectional image (**b**,**d**,**f**).

Figure 3a displays XRD patterns of CdTe films grown under different temperatures. The diffraction peak observed at 23° corresponds to (111) plane of CdTe, suggesting that the crystal structure of CdTe films is zinc blende with a preferential orientation of the (111) plane, regardless of the growth temperature. The XRD pattern of the films deposited in the present study is consistent with that reported in the literature [29]. A rocking curve measurement for the $2\theta = 23°$ diffraction peak was carried out for films grown under different temperatures (Figure 3b). The width of the rocking curve remains fairly narrow with a full width at half maximum (FWHM) of 0.356° for 250 °C and 0.321°, 0.310° for 350 °C and 450 °C (Figure 3b). The decrease in full width at half maximum (FWHM) with increasing temperature indicates that the crystallinity is improved.

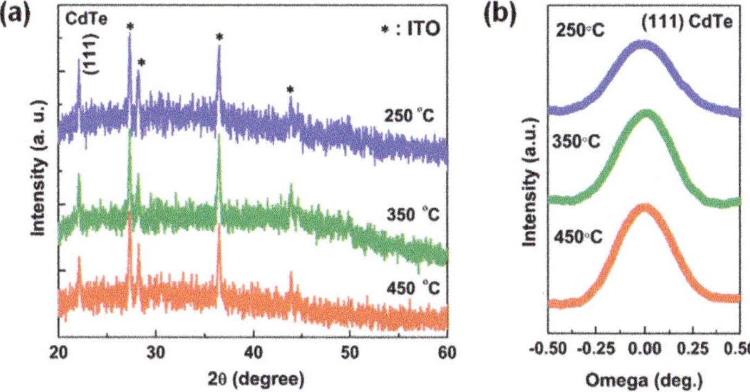

Figure 3. (**a**) XRD patterns of CdTe films grown under different substrate temperatures; (**b**) A rocking curve measurement for the 2θ = 23° diffraction peak.

The Raman spectrum of CdTe films under different substrate temperature is shown in Figure 4. The spectrum fitted with Lorentzian functions exhibits six peaks at 120, 140, 160, 210, 260 and 330 cm^{-1}. The vibration mode at 120 cm^{-1} and 260 cm^{-1} corresponds to elemental Tellurium phases [30]. The mode observed at 140 cm^{-1} could be attributed to combination of both transverse optical phonon (TO) and elemental Te [31]. The Raman modes at 162 and 330 cm^{-1} correspond to the longitudinal optical phonon (LO) and 2LO phonons of CdTe [32]. Similarly, a broad band located at 210 cm^{-1} which we believe probably originates from the combination bands and we tentatively assign them to the overtones of E and A1 modes in Te [32]. Interestingly, by increasing the substrate temperature to 450 °C, (LO) and 2LO phonons modes of CdTe improved with reduction of Te related peaks and indicates improvement in crystallinity of CdTe.

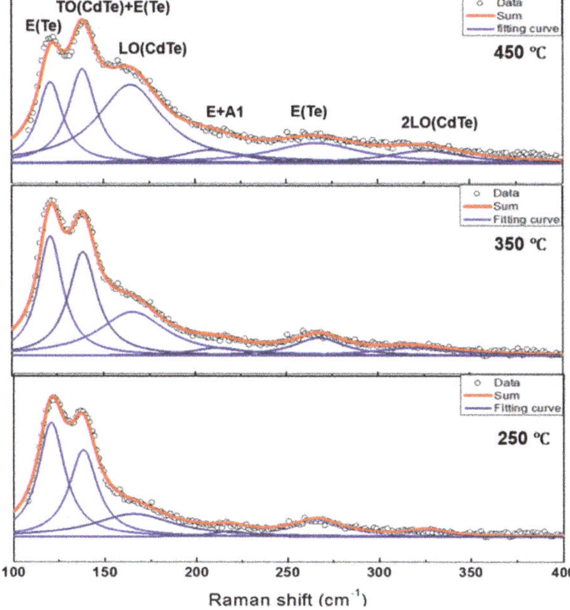

Figure 4. Raman spectrum (fitted using Lorentzian functions) of the CdTe films under different substrate temperature.

The fabrication procedures for the CdTe microdots arrays photodetectors is illustrated in Figure 5a. Initially, 20 nm thick Bi film was deposited onto precleaned ITO glass substrate through a metal shadow mask using e-beam evaporation. The Bi film serves as seed layer for the growth of CdTe microdots-arrayed pattern. The CdTe microdots were grown using VPE. Figure 5b shows the optical microscopic image of CdTe microdots arrays. It has been shown that precursor to substrate distance (D) plays an important role in controlling the quality of CdTe films [33,34]. Interestingly the synthesized CdTe microdots were significantly different for substrate positions (D1 = 15 cm, D2 =10 cm, and D3 = 5 cm) as shown in Figure 5b–d. In case of sample (D1 = 15 cm), isolated and irregular CdTe nanoparticle were formed on the surface of Bi film (Figure 5b). When the distance (D1 = 15 cm) was kept longer from the CdTe source the concentration of the reactant species decreased, which resulted in slow reaction rate leading to uneven growth. Hence, the particle integration into films was incomplete. For sample (D2 = 10 cm), the film with high quality and smooth surface was formed on the Bi film (Figure 5c). In this case the concentration of the reactant species, gas flow, and reaction temperature weremore suitable for nucleation, thereby resulting inthe growth of high-quality film. For sample (D3 = 5 cm) the surface seems to be non-uniform with larger grain size (Figure 5d). Here, the distance (D3 = 5 cm) was close enough to the CdTe source, so a large number of disordered particles was formed due to rapid nucleation.Therefore, CdTe microdots grown at (D2 = 10 cm) possess smooth surface with higher quality than samples D1 and D3. Based on the above results, the precursor to substrate distances seems to have a huge influence on the formation of well-arrayed CdTe microdots.

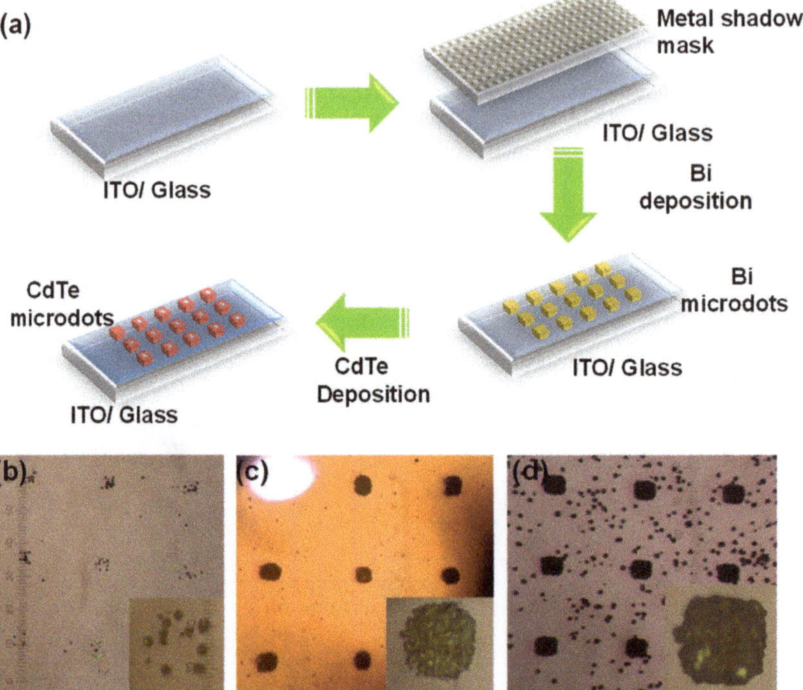

Figure 5. (**a**) Schematic of the fabrication process for the CdTe microdots arrays. Optical microscopy image of grown CdTe microdots arrays at different precursor to substrate distance (**b**) 15 cm, (**c**) 10 cm and (**d**) 5 cm.

Figure 6a shows an SEM image of CdTe microdots arrays grown on Bi coated ITO substrate. As seen from Figure 6a, microdots were assembled in a perfect array and the distance between each dot was measured to be 470 µm (Figure S3). The enlarged version of the single microdots shows the size to be approximately 100 µm in diameter and has a smooth surface with high quality (Figure 6b). This result demonstrates that microdots could be arranged with precise control in size and position. Figure 6c shows the photograph of the arrayed CdTe microdots films. The corresponding transmittance spectra of arrayed CdTe microdots film are shown in Figure 6d. Here, the ITO glass shows about 80% transmittance across the visible range, while it drops to 72% when the CdTe micro dot film is deposited on ITO substrate. This decrease in transmittance is due to the absorption characteristics of the CdTe microdots film. The inset in Figure 6d shows the internal absorption around 825 nm in the case of CdTe microdots.

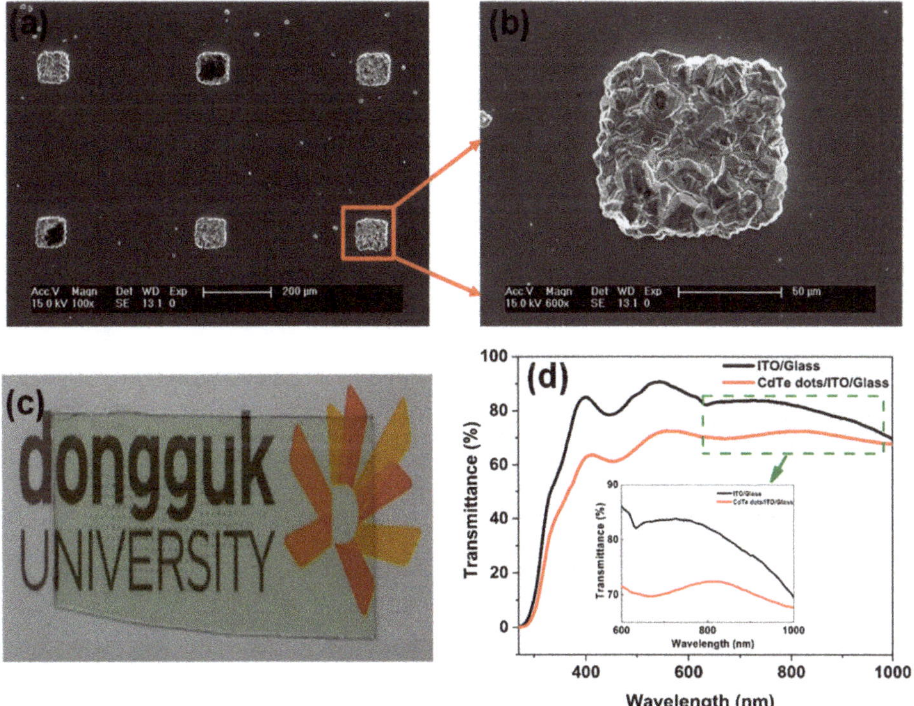

Figure 6. (**a**) SEM image of CdTe microdots arrays grown on Bi coated ITO substrate; (**b**) Enlarged version of the single microdots; (**c**) Photograph of the arrayed CdTe microdots films; (**d**) Transmittance spectra of CdTe microdots arrays film.

Encouraged by the arrayed pattern and good crystallinity, we constructed a photodetector device based on CdTe microdots arrays film to explore its potential for optoelectronic applications. For the fabrication of photodetectors, we adopted a simple contact method using a sandwich structure of ITO glass that serves as transparent electrode and CdTe microdots on Bi-ITO as bottom electrode (See Experimental Section for the detailed fabrication process). Figure 7a shows schematic representation of CdTe microdots-arrayed photodetector on Bi/ITO substrates. For photoresponse measurement, light was illuminated to the device through the top ITO electrode (as shown in Figure 7a). A digital photographical image of a typical device is displayed in Figure 7b. Current-voltage (I–V) curves of the CdTe microdots-arrayed photodetector under normal conditions in dark and illumination are shown in Figure 7c. Here, the curves display asymmetric nonlinear behavior, suggesting the formation of a Schottky-like junction at ITO/CdTe microdots and CdTe/Bi/ITO interface. In other words, the device corresponds to two back-to-back Schottky junctions, with possibly slightly different Schottky barriers [35]. In contrast, the device exhibits strong photoresponse under illumination, indicating the contribution from photogenerated carriers. The dark current was measured to be 0.8 µA at a bias of 5 V. However, the current reaches to 5.3 µA at same bias voltage under illumination, indicating excellent sensitivity of the CdTe microdots. Under illumination, the absorption could mainly take place in CdTe, therefore most of the photogenerated electron-hole pairs are created at the CdTe/ITO junction. The photogenerated electrons-holes are quickly separated by the strong built-in electric field and will be collected at nearby electrodes.

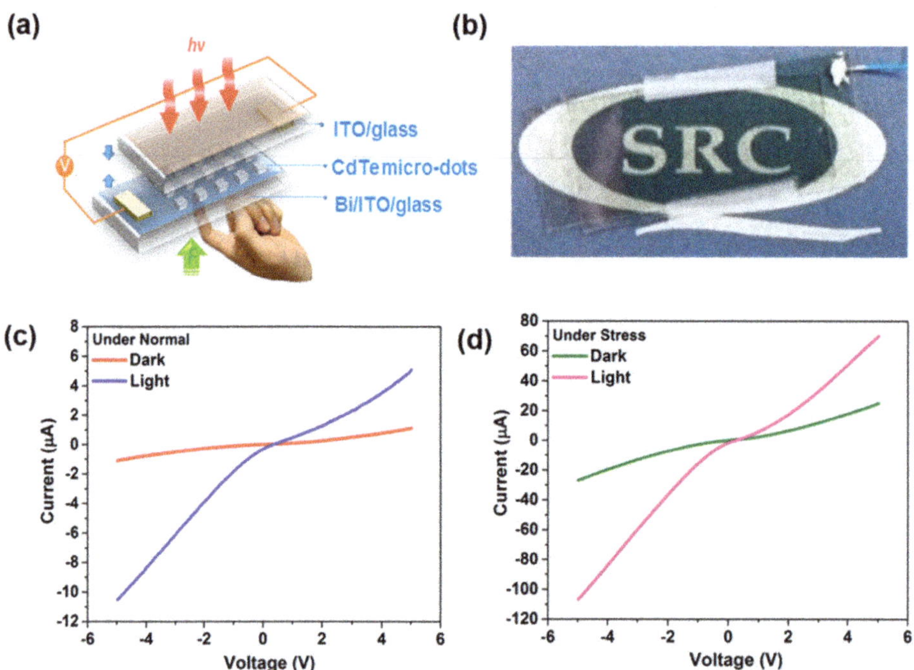

Figure 7. (**a**) Schematic of CdTe microdots array photodetectors; (**b**) Digital photographical image of a typical device; (**c**) Current-voltage (I–V) curve of the CdTe microdots array photodetectors under normal conditions in dark and illumination; (**d**) I–V characteristics of the device under stress (pressing condition) in dark and illumination.

To study the influence of piezoelectric effect on the performance of CdTe microdots array photodetectors, we applied an external stress (pressing the device from the back). We believe that, while pressing the sample, a stress could be completely applied to the CdTe microdots. The corresponding I–V characteristics of the device under such stress in dark and illumination is shown in Figure 7d. Here, the current flowing through the device increased compared to normal conditions. Such enhancement in currents under stress can be ascribed to the piezoelectric effects along CdTe microdots and ITO interface. The induced piezopotential at the CdTe microdots and ITO interface actually results in the change in the Schottky barrier height at CdTe/ITO interface and local electric field dominating the dark current. To confirm piezo-phototronic effect in CdTe microdots array photodetectors, I–V curves were measured by applying stress under illumination.Here, a notable enhancement in photocurrent (76 µA·cm^{-2}) under illumination was observed while compared with that of dark current (20 µA·cm^{-2}) under stress. It can be understood that the mechanically generated piezopotentials along the interface play a critical role in enhancing the performance of photodetector under illumination. When a stress is applied to the device, piezopotential is generated along the CdTe/ITO interface. The piezopotential at the CdTe/ITO interface, reduces the valence and conduction band energy level in CdTe. Meanwhile when the device is illuminated, the stronger electric field enhances the extraction and separation of photogenerated carriers, which results in enhanced photoresponse from the device (Figure 8b). The piezopotential induced under the stress condition plays a crucial role in increasing the performance of piezo-phototronic effect-based devices [36–39]. Similar experiments were also performed using CdTe thin films instead of CdTe microdots. However, we did not observe such a significant variation in the current values (Figure S4).

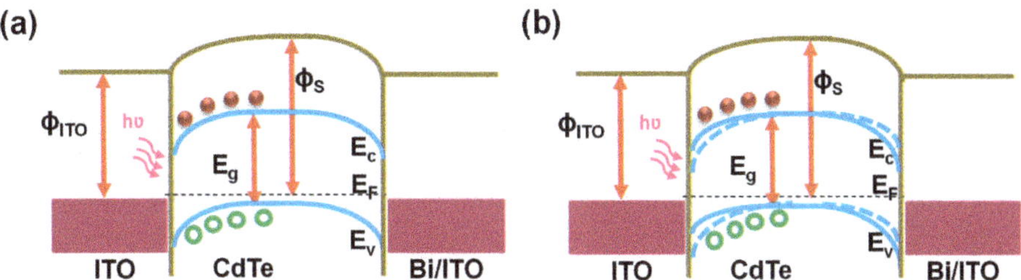

Figure 8. Schematic band diagrams of a CdTe microdots photodetector under (**a**) normal conditions and (**b**) with stress under illumination to illustrate the working mechanism of Piezo-phototronic effect-enhanced Photodetector performance.

Current–time (I–t) characteristics of the devices under normal and stress (pressing) conditions were recorded to investigate the piezo-phototronic effect on CdTe microdots devices (when external force wasapplied in a pulsed way to CdTe microdots device at 1 V). Figure S5 shows the current values of the device under normal and under stress condition. Here, in normal conditions the current values were found to be near zero and on applying stress to the device corresponding current of the CdTe device increased from 0.02 µA to 2.5 µA. It is worth mentioning that, the current always reached the same values at "press" state and recovered to the original current value when "no press" was applied to the substrate. As seen from the Figure 9a, under illumination on applying stress the photocurrent increases to 6.5 µA and then falls back to original current value. The time response of the device remains identical under normal and stress (pressing) conditions with no obvious degradation, indicating the excellent reproducibility and photocurrent stability of the CdTe microdots photodetector.

Responsivity (R), is a critical parameter to determine the photoresponse performance of a photodetector. Photoresponsivity (R) is generally expressed based on the equation [40,41]

$$R = (I_p - I_d)/P_{light} \tag{1}$$

where I_p and I_d represent the photocurrent and dark current respectively,and P_{light} is incident light intensity. Figure 9b displays the responsivity of the photodetector under normal and stress upon illumination. It can be seen that the responsivity of the photodetector increases, with increasing bias voltage. The high photocurrent can be generated with the application of higher bias under illumination, since more charge carriers can pass through the Schottky junction, thereby resulting in enhanced responsivity. The value of photoresponsivity was 3.2×10^{-4} A/W at (5 V). However, when a stress was applied to the device the photoresponsivity of the photodetector increases to one order on par with that of the normal condition. The value of photoresponsivity increases to 5.5×10^{-3} A/W at (5 V). The significantly higher photoresponsivity values of CdTe microdots-arrayed photodetectors under stress condition is attributed to piezopolarization charges induced along interface between the CdTe and ITO. The induced piezoelectric effects under stress results in the effective modulation of the Schottky barrier height at CdTe/ITO interface which results in enhanced transport of photogenerated electrons and holes with reduced recombination probability for charge carriers.

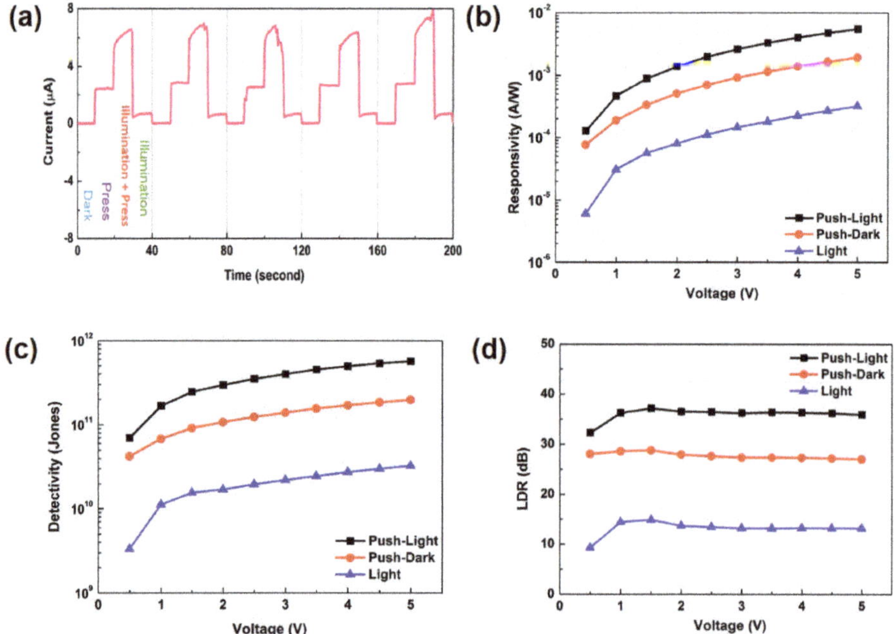

Figure 9. (**a**) Repeatable switching response of the device under 200 cycles with stress and illuminations. (**b**) Responsivity as a function of voltage under normal and stress condition. (**c**) Detectivity as a function of voltage under normal and stress condition. (**d**) Linear dynamic range LDR as a function of voltage under normal and stress condition.

Detectivity (D*) is another important parameter to determine the performance of photodetector. The detectivity is expressed based on the following equation [42]

$$D^* = RA^{0.5} / (2eI_d)^{0.5} \tag{2}$$

where R, A, e and I_d represents the responsivity, effective area of the photodetector, electron charge, and dark current respectively. Figure 9c shows the D* of the CdTe microdots array photodetector under normal and stress condition. As seen from the Figure 9c, the detectivity (D*) was estimated to be 1.12×10^{10} Jones at (1V) for normal conditions. However, under pressing condition it increases to 1.68×10^{11} Jones, which reflects excellent sensitivity detection ability and outstanding performance in the photodetector. The higher values of detectivity observed under stress condition are attributed to the piezo-phototronic effect on the CdTe microdots.

Another critical parameter to determine the performance of photodetector is the linear dynamic range (LDR). The LDR is expressed by

$$LDR = 20 \cdot \log(I_p/I_d) \tag{3}$$

where I_p and I_d is the photocurrent and dark current, respectively. The LDR for normal and stress condition of the device under illumination is shown in Figure 9d. The calculated LDR for the CdTe microdots device under normal and stress conditions are 13 dB and 36 dB, respectively. The observed larger LDR value for stress conditions indicates that the CdTe microdots device opens up a great route to the next generation photodetectors.

4. Conclusions

In conclusion, we successfully fabricated controlled growth of CdTe microdots array photodetectors on Bi coated ITO substrates. The significant enhancement in the electrical transport and photosensing properties of CdTe microdots array under stress condition can be ascribed to the piezo potentials induced along the CdTe/ITO interface. The induced piezopotential results in a stronger local electric field, resulting in enhanced separation and extraction of the photoexcited carriers at the CdTe/ITO interface. This results in the enhancement of photoresponse of the Schottky junction. Under the application of stress, the photoresponsivities of the CdTe microdots array photodetectors are increased from 3.2×10^{-4} A/W to 5.5×10^{-3} A/W, respectively, compared to that of normal condition. These results demonstrate that the photoresponse performance of CdTe microdots array photodetectors can be effectively enhanced using piezo-phototronic effect and provide a feasible approach to extend new design concepts using CdTe with different morphologies and broaden the scope of its potential in optoelectronics applications.

Supplementary Materials: The following are available online at http://www.mdpi.com/2079-4991/9/2/178/s1, Figure S1. SEM images of CdTe grown with and without Bi films under different substrate temperature (a) 250 °C; (b) 350 °C and (c) 450 °C, Figure S2. Photograph images of CdTe grown with and without Bi films on (a) Al2O3 (b) Si and (c) ITO substrate, Figure S3. SEM image of CdTe microdots arrays grown on Bi coated ITO substrate, Table S1: Hall effect measurements for CdTe microdots, Figure S4. Current vs. time characteristics of CdTe thin films under with press and without press conditions, Figure S5. Current values of the device under normal and under stress (pressing) condition.

Author Contributions: Conceptualization, D.J.L.; Data curation, F.X.; Supervision, T.W.K.; Validation, G.M.K. and S.U.Y.; Visualization, Y.D.W. and D.Y.K.; Writing—original draft, P.I. All authors read and approved the final manuscript.

Funding: This research was funded by the Basic Science Research Program through the National Research Foundation of Korea (NRF) grant funded by the Ministry of Education (No.2016R1A6A1A03012877), (No.2016R1D1A1B03935948), (No. 2018R1D1A1B07051461), (No.2018R1D1A1B07051474), (No.2018R1D1A1B07051406) and (No.2018R1D1A1B07050237).

Conflicts of Interest: The authors declare no conflicts of interest.

References

1. Law, M.; Greene, L.E.; Johnson, J.C.; Saykally, R.; Yang, P. Nanowire dye-sensitized solar cells. *Nat. Mater.* **2005**, *4*, 455–459. [CrossRef] [PubMed]

2. Duan, X.F.; Huang, Y.; Agarwal, R.; Lieber, C.M. Single-nanowire electrically driven lasers. *Nature* **2003**, *421*, 241–245. [CrossRef] [PubMed]

3. Yang, Q.; Guo, X.; Wang, W.; Zhang, Y.; Xu, S.; Lien, D.H.; Wang, Z.L. Enhancing sensitivity of a single ZnO micro-/nanowire photodetector by piezo-phototronic effect. *ACS Nano* **2010**, *4*, 6285–6291. [CrossRef] [PubMed]

4. Williams, B.L.; Halliday, D.P.; Mendis, B.G.; Durose, K. Microstructure and point defects in CdTe nanowires for photovoltaic applications. *Nanotechnology* **2013**, *24*, 135703. [CrossRef] [PubMed]

5. Kranz, L.; Gretener, C.; Perrenoud, J.; Schmitt, R.; Pianezzi, F.; La Mattina, F.; Blösch, P.; Cheah, E.; Chirilă, A.; Fella, C.M.; et al. Doping of polycrystalline CdTe for high-efficiency solar cells on flexible metal foil. *Nat. Commun.* **2013**, *4*, 2306. [CrossRef] [PubMed]

6. Shaygan, M.; Davami, K.; Kheirabi, N.; Baek, C.K.; Cuniberti, G.; Meyyappan, M.; Lee, J.-S. Single-crystalline CdTe nanowire field effect transistors as nanowire-based photodetector. *Phys. Chem. Chem. Phys.* **2014**, *16*, 22687–22693. [CrossRef]

7. Yang, G.; Kim, D.; Kim, J. Self-aligned growth of CdTe photodetectors using a graphene seed layer. *Opt. Express* **2015**, *23*, A1081–A1086. [CrossRef]

8. Xie, X.; Kwok, S.-Y.; Lu, Z.; Liu, Y.; Cao, Y.; Luo, L.; Zapien, J.A.; Bello, I.; Lee, C.-S.; Lee, S.-T.; et al. Visible-NIR photodetectors based on CdTe nanoribbons. *Nanoscale* **2012**, *4*, 2914–2919. [CrossRef]

9. Eisen, Y.; Shor, A. CdTe and CdZnTe materials for room-temperature X-ray and gamma ray detectors. *J. Cryst. Growth* **1998**, *184*, 1302–1312. [CrossRef]

10. Huang, M.H.; Mao, S.; Feick, H.; Yan, H.Q.; Wu, Y.Y.; Kind, H.; Weber, E.; Russo., R.; Yang, P.D. Room-Temperature Ultraviolet Nanowire Nanolasers. *Science* **2001**, *292*, 1897–1899. [CrossRef]

11. Olson, J.D.; Rodriguez, Y.W.; Yang, L.D.; Alers, G.B.; Carter, S.A. CdTe Schottky diodes from colloidal nanocrystals. *Appl. Phys. Lett.* **2010**, *96*, 242103. [CrossRef]

12. Baines, T.; Papageorgiou, G.; Hutter, O.S.; Bowen, L.; Durose, K.; Major, J.D. Self-Catalyzed CdTe Wires. *Nanomaterials* **2018**, *8*, 274. [CrossRef] [PubMed]

13. Wang, X.N.; Wang, J.; Zhou, M.J.; Wang, H.; Xiao, X.D.; Li, Q. CdTe nanorods formation via nanoparticle self-assembly by thermal chemistry method. *J. Cryst. Growth* **2010**, *312*, 2310–2314. [CrossRef]

14. Wang, X.; Xu, Y.; Zhu, H.; Liu, R.; Wang, H.; Li, Q. Crystalline Te nanotube and Te nanorods-on-CdTe nanotube arrays on ITO via a ZnO nanorod templating-reaction. *CrystEngComm* **2011**, *13*, 2955–2959. [CrossRef]

15. Kret, S.; Szuszkiewicz, W.; Dynowska, E.; Domagala, J.; Aleszkiewicz, M.; Baczewski, L.T.; Petroutchik, A. MBE Growth and Properties of ZnTe- and CdTe-Based Nanowires. *J. Korean Phys. Soc.* **2008**, *53*, 3055–3063.

16. Lee, S.K.C.; Yu, Y.; Perez, O.; Puscas, S.; Kosel, T.H.; Kuno, M. Bismuth-assisted CdSe and CdTe nanowire growth on plastics. *Chem. Mater.* **2010**, *22*, 77–84. [CrossRef]

17. Consonni, V.; Rey, G.; Bonaim, J.; Karst, N.; Doisneau, B.; Roussel, H.; Renet, S.; Bellet, D. Synthesis and physical properties of ZnO/CdTe core shell nanowires grown by low-cost deposition methods. *Appl. Phys. Lett.* **2011**, *98*, 96–99. [CrossRef]

18. Salim, H.I.; Patel, V.; Abbas, A.; Walls, J.M.; Dharmadasa, I.M. Electrodeposition of CdTe thin films using nitrate precursor for applications in solar cells. *J. Mater. Sci. Mater. Electron.* **2015**, *26*, 3119–3128. [CrossRef]

19. Yang, G.; Jung, Y.; Chun, S.; Kim, D.; Kim, J. Catalytic growth of CdTe nanowires by closed space sublimation method. *Thin Solid Films* **2013**, *546*, 375–378. [CrossRef]

20. Kulkarni, R.; Rondiya, S.; Pawbake, A.; Waykar, R.; Jadhavar, A.; Jadkar, V.; Bhorde, A.; Date, A.; Pathan, H.; Jadkar, S. Structural and optical properties of CdTe thin films deposited using RF magnetron sputtering. *Energy Procedia* **2017**, *110*, 188–195. [CrossRef]

21. Mahabaduge, H.P.; Rance, W.L.; Burst, J.M.; Reese, M.O.; Meysing, D.M.; Wolden, C.A.; Li, J.; Beach, J.D.; Gessert, T.A.; Metzger, W.K.; et al. High-efficiency, flexible CdTe solar cells on ultra-thin glass substrates. *Appl. Phys. Lett.* **2015**, *106*, 133501. [CrossRef]

22. Xie, C.; Luo, L.-B.; Zeng, L.-H.; Zhu, L.; Chen, J.-J.; Nie, B.; Hu, J.-G.; Li, Q.; Wu, C.-Y.; Wang, L.; et al. p-CdTe nanoribbon/n-silicon nanowires array heterojunctions: Photovoltaic devices and zero-power photodetectors. *CrystEngComm* **2012**, *14*, 7222–7228. [CrossRef]

23. Yang, G.; Kim, B.-J.; Kim, D.; Kim, J. Single CdTe microwire photodetectors grown by close-spaced sublimation method. *Opt. Express* **2014**, *22*, 18843–18848. [CrossRef] [PubMed]

24. Boxberg, F.; Søndergaard, N.; Xu, H.Q. Elastic and Piezoelectric properties of Zinc blende and Wurtzite crystalline nanowire heterostructures. *Adv. Mater.* **2012**, *24*, 4692–4706. [CrossRef] [PubMed]

25. Cibert, J.; Andre, R.; Dang, L.S. Piezoelectric effect in strained CdTe-based heterostructures. *Acta Phys. Pol. A* **1995**, *88*, 591–600. [CrossRef]

26. Xin, J.; Zheng, Y.; Shi, E. Piezoelectricity of zinc-blende and wurtzite structure binary compounds. *Appl. Phys. Lett.* **2007**, *91*, 112902.

27. Corso, A.D.; Resta., R.; Baroni, S. Nonlinear piezoelectricity in CdTe. *Phys. Rev. B* **1993**, *47*, 16252–16256. [CrossRef]

28. Hou, T.-C.; Yang, Y.; Lin, Z.-H.; Ding, Y.; Park, C.; Pradel, K.C.; Chen, L.-J.; Wang, Z.L. Nanogenerator based on zinc blende CdTe micro/nanowires. *Nano Energy* **2013**, *2*, 387–393. [CrossRef]

29. Matsune, K.; Oda, H.; Toyoma, T.; Okamoto, H.; Kudriavysevend, Y.; Asomoza, R. 15% efficiency CdS/CdTe thin film solar cells using CdS layers doped with metal organic compounds. *Sol. Energy Mater. Sol. Cells* **2006**, *90*, 3108–3114. [CrossRef]

30. Ilanchezhiyan, P.; Mohan Kumar, G.; Xiao, F.; Siva, C.; Yuldasheva, S.U.; Lee, D.J.; Jeon, H.C.; Kang, T.W. Surface induced charge transfer in CuxIn2-xS3 nanostructures and their enhanced photoelectronic and photocatalytic performance. *Sol. Energy Mater. Sol. Cells* **2019**, *191*, 100–107. [CrossRef]

31. Ilanchezhiyan, P.; Mohan, K.G.; Xiao, F.; Poongothai, S.; Madhan, K.A.; Siva, C.; Yuldashev, S.U.; Lee, D.J.; Kwon, Y.H.; Kang, T.W. Ultrasonic-assisted synthesis of ZnTe nanostructures and their structural, electrochemical and photoelectrical properties. *Ultrason. Sonochem.* **2017**, *39*, 414–419. [PubMed]

32. Xi, L.; Chua, K.H.; Zhao, Y.; Zhang, J.; Xiong, Q.; Lam, Y.M. Controlled synthesis of CdE (E = S, Se and Te) nanowires. *RSC Adv.* **2012**, *2*, 5243–5253. [CrossRef]

33. Nayak, T.R.; Wang, H.; Pant, A.; Zheng, M.; Junginger, H.; Goh, W.J.; Lee, C.K.; Zou, S.; Alonso, S.; Czarny, B.; et al. ZnO Nano-Rod Devices for Intradermal Delivery and Immunization. *Nanomaterials* **2017**, *7*, 147.

34. Cheng, H.; Li, J.; Wu, D.; Li, Y.; Wang, Z.; Wang, X.; Zheng, X. Effects of Precursor-Substrate Distances on the Growth of GaN Nanowires. *J. Nanomater.* **2015**, *2015*, 343541. [CrossRef]

35. Bartolomeo, A.D.; Grillo, A.; Urban, F.; Iemmo, L.; Giubileo, F.; Luongo, G.; Amato, G.; Croin, L.; Sun, L.; Liang, S.-J.; et al. Asymmetric schottky contacts in bilayer MoS_2 field effect transistors. *Adv. Funct. Mater.* **2018**, *28*, 1800657. [CrossRef]

36. Yang, R.; Qin, Y.; Dai, L.; Wang, Z.L. Power generation with laterally packaged piezoelectric fine wires. *Nat. Nanotechnol.* **2009**, *4*, 34–39. [CrossRef] [PubMed]

37. Jeong, S.; Kim, M.W.; Jo, Y.-R.; Kim, T.-Y.; Leem, Y.-C.; Kim, S.-W.; Kim, B.-J.; Park, S.-J. Crystal-structure-dependent piezotronic and piezo-phototronic effects of ZnO/ZnS core/shell nanowires for enhanced electrical transport and photosensing performance. *ACS Appl. Mater. Interfaces* **2018**, *10*, 28736–28744. [CrossRef] [PubMed]

38. Zhang, X.; Qiu, Y.; Yang, D.; Li, B.; Zhang, H.; Hu, L. Enhancing performance of Ag–ZnO–Ag UV photodetector by piezo-phototronic effect. *RSC Adv.* **2018**, *8*, 15290–15296. [CrossRef]

39. Lu, S.; Qi, J.; Liu, S.; Zhang, Z.; Wang, Z.; Lin, P.; Liao, Q.; Liang, Q.; Zhang, Y. Piezotronic interface engineering on ZnO/Au-based schottky junction for enhanced photoresponse of a flexible self-powered uv detector. *ACS Appl. Mater. Interfaces* **2014**, *6*, 14116–14122. [CrossRef]

40. Wang, Z.; Yu, R.; Pan, C.; Liu, Y.; Ding, Y.; Wang, Z.L. Piezo-phototronic uv/visible photosensing with optical-fiber-nanowire hybridized structures. *Adv. Mater.* **2015**, *27*, 1553–1560. [CrossRef] [PubMed]

41. Oh, S.; Kim, C.-K.; Kim, J. High responsivity β-Ga_2O_3 metal−semiconductor−metal solar-blind photodetectors with ultraviolet transparent graphene electrodes. *ACS Photonics* **2018**, *5*, 1123–1128. [CrossRef]

42. Kong, W.Y.; Wu, G.A.; Wang, K.Y.; Zhang, T.F.; Zou, Y.F.; Wang, D.D.; Luo, L.B. Graphene-β-Ga_2O_3 heterojunction for highly sensitive deep UV photodetector application. *Adv. Mater* **2016**, *28*, 10725–10731. [CrossRef] [PubMed]

 *nanomaterials*

Review

Research Progress in Organic Photomultiplication Photodetectors

Linlin Shi [1], Qiangbing Liang [1], Wenyan Wang [1], Ye Zhang [1], Guohui Li [1], Ting Ji [1], Yuying Hao [1,2] and Yanxia Cui [1,2,*]

1   Key Laboratory of Advanced Transducers and Intelligent Control System of Ministry of Education,
    College of Physics and Optoelectronics, Taiyuan University of Technology, Taiyuan 030024, China;
    shilinlinsll@sina.cn (L.S.); liangqiangbing@126.com (Q.L.); wangwenyan@tyut.edu.cn (W.W.);
    zhangye@tyut.edu.cn (Y.Z.); liguohui@tyut.edu.cn (G.L.); jiting@tyut.edu.cn (T.J.);
    haoyuyinghyy@sina.com (Y.H.)
2   Key Laboratory of Interface Science and Engineering in Advanced Materials,
    Taiyuan University of Technology, Taiyuan 030024, China
*   Correspondence: yanxiacui@gmail.com; Tel.: +86-35-1317-6639

Received: 31 July 2018; Accepted: 31 August 2018; Published: 11 September 2018

Abstract: Organic photomultiplication photodetectors have attracted considerable research interest due to their extremely high external quantum efficiency and corresponding high detectivity. Significant progress has been made in the aspects of their structural design and performance improvement in the past few years. There are two types of organic photomultiplication photodetectors, which are made of organic small molecular compounds and polymers. In this paper, the research progress in each type of organic photomultiplication photodetectors based on the trap assisted carrier tunneling effect is reviewed in detail. In addition, other mechanisms for the photomultiplication processes in organic devices are introduced. Finally, the paper is summarized and the prospects of future research into organic photomultiplication photodetectors are discussed.

Keywords: photodetector; organic; photomultiplication; tunneling; external quantum efficiency

1. Introduction

Photodetectors are optoelectronic devices which can absorb light energy and convert it into electrical energy, having found applications in wide areas of image sensing, missile guidance, environmental pollution monitoring, light communications, photometric metrology, industrial automation, and so on [1–8]. In particular applications, it is required that the sensitivity of photodetectors is sufficiently high to detect weak light signal, for example, bio-imaging sensing or long range light communication [9–12]. There are two routes to improve the sensitivity of photodetectors, one of which is to improve the external quantum efficiency (EQE) and the other is to reduce the dark current density. Using photomultiplication (PM) effect to improve the EQE is one of the most important approaches to achieve high-sensitivity photodetection.

Traditional photomultipliers, based on a complex vacuum system including photo induced electron emission, secondary electron emission, and electron optics possessing components, are bulky and of high cost, severely limiting their applications [13–16]. Avalanche photodiodes are another commonly used high-sensitivity photodetector, which are made of inorganic semiconductor materials such as silicon, germanium, indium gallium arsenide [13,17–19], and so on. Their working mechanism is that photo generated carriers are accelerated under the strong electric field induced by a large reverse bias, and then the impact ionization with the crystal lattice takes place, thereby bringing forward the avalanche multiplication effect [20–22].

Besides inorganic semiconductors, organic semiconductors have also been widely favored in the field of optoelectronics due to their advantages of a simple synthesized method and adjustable bandwidth, as well as their light weight, low cost, eco-friendliness, good flexibility, and so on. High performance organic photodetectors have been reported successively in the past few years [23–27], some of which also allow the excitation of PM effects, providing another route to realize high-sensitivity photodetectors [4,28–32]. Since the exciton binding energy of organic semiconductor materials is approximately 0.1–1.4 eV, about three orders higher than that of inorganic semiconductor materials, impact ionization cannot occur in organic PM photodetectors like in inorganic avalanche photodiodes. Instead, the working mechanism of organic PM photodetectors has been identified mainly due to the trap assisted carrier tunneling effects [33–35].

After more than 20 years of development, the active layer materials of organic PM photodetectors have transitioned from organic small molecular compounds to polymers, and their device performance has also been optimized constantly. In this review, we will firstly introduce the typical structures and working mechanisms of organic PM photodetectors along with their key performance parameters. Next, we will give a detailed review of the research progress for PM photodetectors based on organic small molecular compounds and polymers, respectively. Some important progresses in improving the quantum efficiency, dark current, response speed, and spectral performance of both types of PM photodetectors are presented. In addition, we will introduce some other working mechanisms of organic PM photodetectors. Finally, we will summarize the paper and consider prospects for the future research of organic PM photodetectors.

2. Basic Structures and Working Mechanisms of Organic PM Photodetectors

2.1. Basic Structures of Organic PM Photodetectors

The basic structures of organic PM photodetectors as shown in Figure 1 comprises of the anode, the cathode, and the active layer with a large amount of interfacial/bulk carrier traps. They can be mainly grouped into two types, the single junction type and the bulk heterojunction type, which are similar to those of organic solar cells [36–38]. The early organic PM photodetectors belonged to the single junction type with their active layers made of an *N*-type or *P*-type organic compound. In contrast, the bulk heterojunction type organic PM photodetectors possess active layers made of donor/acceptor (D/A) blend. Although the bilayer heterojunction type organic solar cells, in which the active layer consists of a stack of *N*-type and *P*-type semiconductor films, have been frequently researched, there are rare studies about applying this junction in organic PM photodetectors. In practice, extensive efforts have been made on introducing interfacial modified layers between the electrode and the active layer to improve the PM effects in organic photodetectors. In addition, doping other materials into the active layer has also been carried out for improving PM performances.

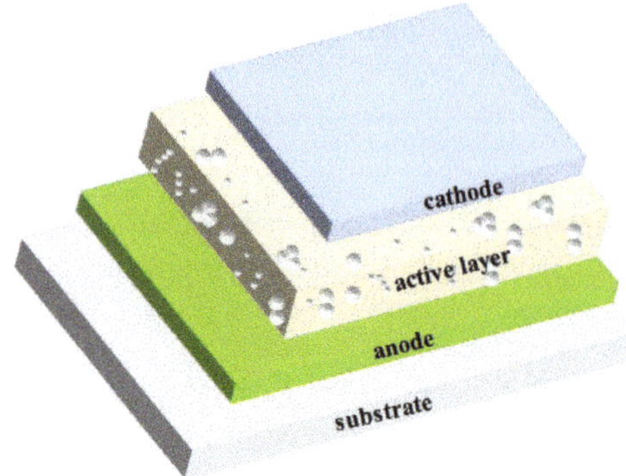

Figure 1. Structural diagram of organic PM photodetectors.

2.2. Working Mechanisms of Organic PM Photodetectors

Most of organic PM photodetectors realize EQE far exceeding 100% based on the trap assisted carrier tunneling effect. In all, the trap assisted carrier tunneling effect contains four steps which are the formation of the Schottky barrier, the capture of carriers by traps after light illuminations, the carrier transport toward the Schottky junction under applied bias, and the carrier tunneling through the narrowed Schottky barrier. Based on the type of the trapped carrier (electron or hole), the working mechanisms of organic PM photodetectors are divided into two categories. Figure 2 shows the working mechanism of the case of electron trap assisted carrier tunneling effect.

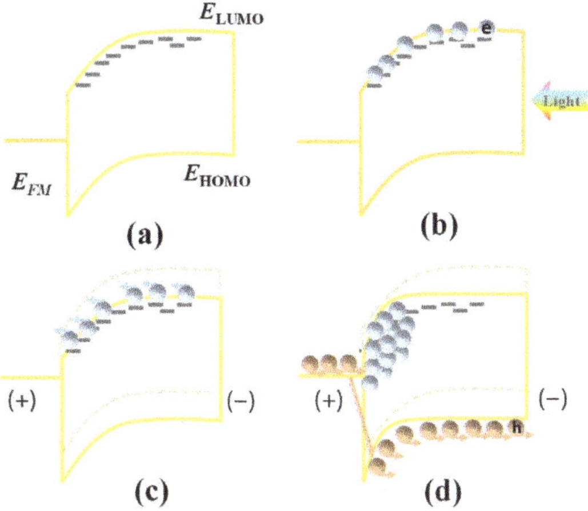

Figure 2. Working mechanism of organic photomultiplication (PM) photodetectors due to the electron trap assisted hole tunneling effect. (**a**) Energy band bending without bias; (**b**) Photo-generated electrons are captured by traps; (**c**) Trapped electrons transport toward the junction once the bias is applied; (**d**) Trapped electrons arriving at the junction cause the hole tunneling from the circuit into the semiconductor, producing the current multiplication effect. E_{FM} is the Fermi level of the metal, and E_{LUMO} and E_{HOMO} are the lowest unoccupied molecular orbital level and the highest occupied molecular orbital level of the organic semiconductor, respectively.

The premise of realizing the carrier tunneling effect is the formation of Schottky junction when the metal electrode is in contact with the semiconductor layer [39]. For the organic semiconductor with a large number of electron traps, the proper Schottky band bending without bias is shown in Figure 2a. It corresponds to the case when the Fermi energy level of the metal electrode (E_{FM}) is higher than that of the organic semiconductor material (E_{FS}). In such a case, electrons flow from metal into the semiconductor, yielding a built-in electric field toward the semiconductor with a downward energy band bending, thereby hindering the diffusion of holes in the organic semiconductor to the electrode. To achieve such a downward band bending, low work function electrodes like Ag (4.26 eV), Al (4.28 eV), and Mg (3.66 eV), are required. After light illumination, photo-generated electrons are captured by the electron traps as shown in Figure 2b. When a bias is applied with the electric field pointing from electrode to the semiconductor (reverse bias if the electrode works as cathode; forward bias if the electrode works as anode), the trapped electrons transport toward the Schottky junction as shown in Figure 2c. In the junction region, these arrived electrons narrow the Schottky junction and thus enhance the intensity of the built-in electric field, causing the holes to tunnel through the junction and inject into the organic semiconductor from the external circuit, and finally resulting in the current multiplication effect, as shown in Figure 2d.

The corresponding processes of hole trap assisted electron tunneling effect are exactly the opposite (not shown), which requires E_{FM} lower than E_{FS}. Using a high work function metal such as Au (5.1 eV),

ITO (4.7 eV), Pt (5.65 eV), and so on, can fulfill such a condition, thereby an upward band bending is constructed in the semiconductor. In addition, the applied bias for trapped hole transportation is the reverse, with the electric field pointing from the semiconductor to the electrode, facilitating the holes transport toward the Schottky junction. In practice, the surface of the metal electrode is always modified in order to adjust its work function and further regulate the band bending.

2.3. Key Performance Parameters of Organic PM Photodetectors

Key performance parameters of organic PM photodetectors include photoresponsivity, quantum efficiency, detectivity, linear dynamic range, and response time, which are listed as follows.

2.3.1. Photoresponsivity

The photoresponsivity is defined as the ratio of photocurrent to the power intensity of incident light, which characterizes the sensitivity of photodetectors to incident light. The greater photoresponsivity, the better sensitivity to incident light for PM photodetectors. Photoresponsivity can be expressed by Equation (1):

$$R = \frac{I_{ph}}{P_{in}} = \frac{I_l - I_d}{P_{in}} \tag{1}$$

where I_{ph} is photocurrent, I_l is the current under light illumination, I_d is dark current, and P_{in} is the incident light intensity.

2.3.2. External (Internal) Quantum Efficiency

The external quantum efficiency (*EQE*) is defined as the electron number detected per incident photon, as presented by Equation (2):

$$EQE = \frac{N_e}{N_p} = \frac{I_{ph}/e}{P_{in}/h\nu} \tag{2}$$

where N_e and N_p are the number of detected electrons and incident photons, respectively, h is the Planck's constant, ν is the frequency of light, and e is the electronic charge. The absorption properties of the selected materials, the structural design of the device, and the electrical properties of the materials are all key factors affecting *EQE*. In traditional photodiode type photodetectors, *EQE* is smaller than unity. However, if current gain exists, *EQE* can be greater than 1, for example, in avalanche photodiode, photoconductor, or phototransistor type photodetectors [40,41].

In organic photomultiplication photodetectors, the presence of deep traps in the organic active layer causes a long carrier recombination lifetime for one type of charge, resulting in a high photocurrent amplification (gain), similar to that happens in photoconductor type photodetectors. Gain is determined by the ratio of recombination lifetime and transit time for another type of charge to sweep across the device, as given by Equation (3):

$$Gain = \frac{\chi\tau}{T} = \frac{\chi\tau\mu V}{L^2} \tag{3}$$

where χ is the fraction of trapped electrons or holes over the total amount of the dissociated excitons, τ is the lifetime of trapped carriers, T is the transport time of the untrapped carriers flowing across the active layers, V is the applied bias, L is the active layer thickness, and μ is the field dependent mobility of the untrapped carriers. As known, in photodetectors with gain mechanisms, *EQE* is equal to *Gain* in number [42].

The internal quantum efficiency (*IQE*) is defined by the ratio of the number of carriers detected in the external circuit to the number of photons absorbed. The product of the *IQE* and the light absorption efficiency of the active layer (*abs*) is the *EQE*, as determined by Equation (4):

$$IQE = \frac{N_e}{N_p abs(\lambda)} = \frac{EQE}{abs(\lambda)} \tag{4}$$

2.3.3. Detectivity

The detectivity (*D**) is the figure of merit to characterize the capability of weak light detection for a photodetector, which can be calculated from the noise density and the photoresponsivity *R*. It is one of the most important physical parameter for photodetectors as given by Equation (5):

$$D^* = \frac{R\sqrt{Af}}{i_n} \tag{5}$$

where *A* is the active area of the detector, *f* is the electrical bandwidth, and i_n is the measured total noise current. Considering that the noise current under dark is dominated by shot noise, the detectivity can be calculated through Equation (6):

$$D^* = \frac{R}{\sqrt{2eJ_d}} \tag{6}$$

2.3.4. Linear Dynamic Range

The linear dynamic range (*LDR*) is defined as the response range of photodetector being linear over a wide range of light intensity. *LDR* can be calculated through Equation (7):

$$LDR = 20\log(\frac{P_{max}}{P_{min}}) \tag{7}$$

where and P_{min} are the maximum and minimum incident light intensities of the photocurrent density versus light intensity curve which lie within the linear response range.

2.3.5. Response Time

The response time of the detector reflect the response speed of the detector to receive incident light radiation, which includes two parts, the rise time (T_r) and the falling time (T_f). The rise (or falling) time is defined as the time for the photocurrent to rise from 10% to 90% (fall from 90% to 10%) during the on and off cycles of light illumination. The sum of the rise time and the falling time is counted as the response time of the photodetector.

3. Organic PM Photodetector Based on Small Molecular Compounds

The early organic PM photodetectors were based on organic small molecular compounds such as *N*-methyl-3,4,9,10-perylenetetracarboxyl-diimide (Me-PTC), naphthalene tetracarboxylic anhydride (NTCDA), fullerenes (C_{60}), 2,9-dimethyl quinacridone (DQ), and so on. Except for DQ, Me-PTC, NTCDA, and C_{60} are all *N*-type semiconductor materials which support the hole trap assisted electron tunneling effect. In this section, we will first introduce the progress of organic small molecular PM photodetectors based on the single junction type and the bulk heterojunction type, respectively. Then, some important progress made by researchers in improving the performance of the organic small molecular PM photodetectors will be presented.

3.1. Single Junction Type Organic Small Molecular PM Photodetectors

In 1994, Hiramoto et al. fabricated the pioneering organic PM photodetector using the Me-PTC (a *N*-type perylene pigment having methyl groups) based on a triple layer configuration of

Au/Me-PTC/Au on a glass substrate. The fabricated device produced an *IQE* of 1.0×10^4 at -16 V under 600 nm light illumination at the temperature of $-50\ ^\circ$C, as shown in Figure 3a [43]. Their later experimental work reflected that the surface of Me-PTC was very rough, leading to imperfect contact between and the metal and Me-PTC and thereby forming structural traps of holes, as indicated in Figure 3b [44]. As a result, a large number of holes were trapped by Me-PTC at the metal/Me-PTC interface, making the PM phenomenon possible. However, Me-PTC based organic PM devices cannot respond at room temperature due to too few interfacial trapped carriers. Besides Me-PTC, two other perylene pigments of PhEt-PTC and n-Bu-PTC also exhibit the PM effect due to structural traps at the imperfect metal/semiconductor interface [45,46]. Later, the same group realized the PM photodetection at room temperature based on organic small molecular material of NTCDA. The NTCDA device presented a PM effect under 400 nm light illumination and its *IQE* reached 1.3×10^5 at -16 V [47]. Subsequent studies indicate that introducing an interfacial layer of PhEt-PTC next to the NTCDA layer [26] or reducing the grain boundaries of the NTCDA film [48] can improve the response speed of the device (see the detail in Section 3.3.4).

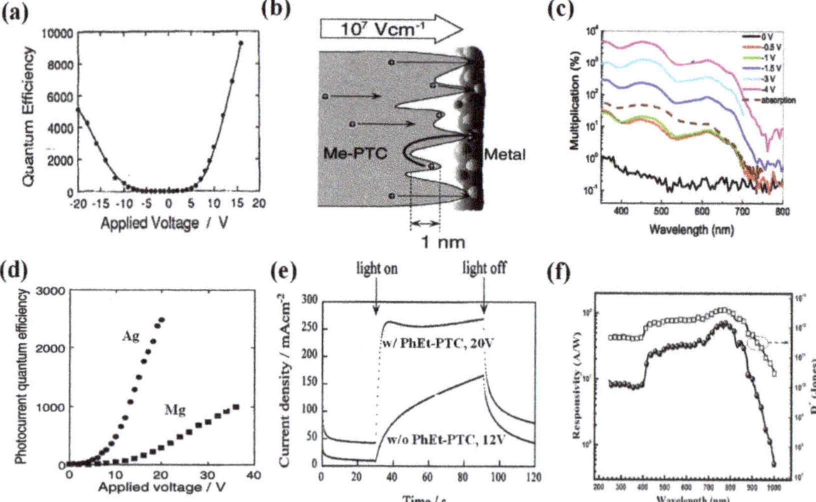

Figure 3. Device performances of various organic small molecular PM photodetectors. (**a**) Internal quantum efficiency (*IQE*) at different voltages for the Au/Me-PTC/Au device under 600 nm light illumination (Reproduced with permission from [43]. *AIP Publishing*, 1994); (**b**) Schematic view of the interfacial traps for the Au/Me-PTC/ITO device (Reproduced with permission from [44]. *AIP Publishing*, 1998); (**c**) External quantum efficiency (*EQE*) spectra under different biases and the absorption spectrum of the ITO/PEDOT:PSS/C_{60}/BCP/Al device (Reproduced with permission from [49]. *AIP Publishing*, 2007); (**d**) *IQE* at different voltages under 600 nm light illumination for the ITO/DQ/Ag and ITO/DQ/Mg devices, respectively (Reproduced with permission from [50]. *The Japan Society of Applied Physics*, 1996); (**e**) Transient current density curves of ITO/PhEt-PTC/NTCDA/Au and ITO/NTCDA/Au devices, respectively (Reproduced with permission from [26]. *AIP Publishing*, 2000); (**f**) Responsivity and detectivity spectra of the glass/ITO/TPBi/C_{70}/SnPc:C_{70}/BCP/Al incorporated with down-conversion material of 4P-NPB (Reproduced with permission from [28]. *Royal Society of Chemistry*, 2016).

The organic small molecular material C_{60} is also a commonly used material for achieving PM effects. In 2007, Huang and Yang characterized the photo current response of a device with configuration of ITO/PEDOT:PSS/C_{60}/BCP/Al, in which ITO/PEDOT:PSS was used to form a composite electrode [49]. Here, C_{60} layer formed a disordered structure, which is better than ordered structures for producing the PM effect. With the help of the composite electrode, they realize an *EQE* of 5.0×10^3% as shown in Figure 3c based on the hole trap assisted electron tunneling effect. By increasing the bias voltage the PM effect becomes more significant. However, inserting a 20 nm thick BCP layer at the PEDOT:PSS/C_{60} interface would annihilate the PM performance, reflecting that the PM behavior occurs at the PEDOT:PSS/C_{60} interface rather than within the C_{60} layer.

In contrast, the organic PM devices based on the electron trap assisted hole tunneling effect have been researched not as extensively as their hole trap counterparts. In 1996, Hiramoto et al. developed the first electron trap based PM photodetector using a *P*-type material Quinacridone (DQ) as the active layer [50]. In their device, the DQ layer was sandwiched between ITO and Ag (or Mg) electrodes. This study concluded the metal electrode plays a significant role in influencing the multiplication performance of the photodetectors. The Ag electrode with a work function a bit higher than that of the Mg electrode is much superior on producing a high multiplication factor. Specifically, the Ag electrode device has an *IQE* of 2.5×10^3 when the bias voltage is 20 V while that of the Mg electrode device is only 1.0×10^3 at a bias of 36 V; see in Figure 3d.

3.2. Bulk Heterojunction Type Organic Small Molecular PM Photodetectors

The organic small molecular PM photodetectors with bulk heterojunction type active layers were proposed later than their single junction counterparts. So far, all reported bulk heterojunction type organic small molecular PM photodetectors are made of fullerene and a *P*-type semiconductor material through co-evaporation. As early as 2002, Matsunobu et al. firstly put forward the ITO/CuPc:C_{60}/Au device for PM photodetection [51]. They found that the rise and fall times of the bulk heterojunction device are only 8 ms and 15 ms, respectively, while the response time of the single junction control device is in second time scale. This is mainly because the introduction of CuPc provides a favorable path for hole transport, therefore the accumulation time of holes at the Schottky junction can be shortened.

In 2010, Hammond et al. prepared a bulk heterojunction type PM photodetector with a structure of ITO/NTCDA/C_{60}/CuPc:C_{60}/BCP/Al [52]. They found that the insertion of a composite hole blocking layer comprising of NTCDA (2–3 nm)/C_{60} (10 nm) can facilitate the accumulation of holes at the interface between the hole blocking layer and the active layer, inducing a large amount of electrons being injected from the electrode and thereby producing a high photocurrent gain. When the composite hole blocking layer is removed, the device *EQE* increases at 0 V bias, but reduces to be less than 100% as the applied voltage increases. Such a phenomenon proves that the gain of the device occurs at the interface between the hole blocking layer and the photoactive layer, rather than within the photoactive layer.

Recently, bulk heterojunction organic small molecular PM photodetectors with a configuration of ITO/TPBi (or Bmpypb, LiF)/C_{70}/TAPC (or SnPc):C_{70}/BCP/Al were developed [28]. Here, the co-evaporated film of TAPC (or SnPc):C_{70} works as the active layer. They further elucidated that for the composite interfacial layer of TPBi (or Bmpypb, LiF)/C_{70}, TPBi (or Bmpypb, LiF) plays the role of hole blocking while C_{70} plays the role of hole accumulation. The *EQE* of their TAPC:C_{70} PM devices exceed 1.0×10^4% over the wavelength range from 350 nm to 650 nm and the response of the SnPc:C_{70} device can be extended to infrared (IR) range.

3.3. Performance Studies on Organic Small Molecular PM Photodetectors

In this subsection, we will introduce the progress of performance studies on organic small molecular PM photodetectors from the aspects of water/oxygen treatment, quantum efficiency, dark current, response speed, and spectrum adjustment.

3.3.1. Water/Oxygen Treatment

Because organic materials are sensitive to water and oxygen, organic semiconductor films and devices are usually prepared in conditions without water and oxygen. However, studies reflect that performances of the prepared organic PM photodetectors bear dramatic changes after being exposed to water and oxygen [53,54]. Sometimes, the water or oxygen treatment is beneficial to the improvement of the photo current gain. For example, Hiramoto et al. found that the adsorption of oxygen by N-type materials of Me-PTC and NTCDA suppresses the PM effect while the same treatment toward P-type material of DQ increases the PM gain factor. This is because there is more O_2^- after the organic material adsorbing O_2 molecules. The increased O_2^- act as electron traps which facilitate the carrier capture in P-type semiconductor and thereby more holes being injected, but inhibit the carrier capture in N-type semiconductor due to recombination with hole traps. They also found that after the adsorption of water, devices made of N-type material Me-PTC and P-type material DQ exhibited higher PM performance due to the increased photocarrier generation.

3.3.2. Quantum Efficiency

Devices with higher quantum efficiency have higher responsivity and the corresponding devices are more sensitive at fixed dark current. The early study showed that the hole traps within Me-PTC based PM devices are structural traps produced by non-uniform Me-PTC film interfaces. From this point, research on the effect of deposition rate and deposition method of the metal electrode on the PM performance was carried out [55]. That work concluded that, based on the thermal evaporation method, decreasing the deposition rate of metal electrode from 0.7 nm/s to 0.008 nm/s brings forward the increase of the quantum efficiency by 30 times. This is because the low rate of deposition of the Au electrode can maintain the structural traps produced by the non-uniform Me-PTC film. In addition, they found that the ion sputtering method is less effective than the thermal evaporation method for yielding high PM performance due to the same reason. For the PM photodetector made of the N-type perylene pigments, its quantum efficiency can also be improved through the solvent treatment [45]. It was demonstrated that the THF solvent treatment can change the PhEt-PTC film from amorphous to poly-crystalline, therefore the PhEt-PTC becomes coarser, producing more structural traps and thus elevating the quantum efficiency, as exhibited in Figure 3d.

3.3.3. Dark Current

The dark current determines the ability of detecting weak light for photodetectors. When the quantum efficiency of different devices are comparable, the device with lower dark current has lower noise equivalent power, which allows a weaker optical signal to be detected. Jinsong Huang and his collaborators found that the dark current of PM photodetector with configuration of ITO/PEDOT:PSS/C_{60}/BCP/Al was as high as 2 mA/cm^2 at -6 V, not suitable for the detection of weak light. They attributed the large dark current to the possible ohmic contact between PEDOT:PSS and C_{60}. In order to reduce dark current, a high carrier injection barrier is required. The same group found that by inserting a C-TPD layer between the PEDOT:PSS and C_{60} films, an electron injection barrier as high as 2.8 eV was realized, thereby significantly reducing the dark current by 3–4 orders [56]. But such a design brings barrier for electron injection from PEDOT:PSS to C_{60}, resulting in the *EQE* lower than 100%. Later, they introduced a nano-composite buffer layer C-TPD:ZnO (1:1) between PEDOT:PSS and C_{60}, through which they not only maintained the dark current at a low level but also realized the PM photodetection [57]. Compared with the device with a single C-TPD buffer layer, the composite buffer layer with ZnO nanoparticles brought additional hole traps, which could capture photo generated holes, making them recombine with the electrons in the defect state on the ZnO surface. As a result, the energy band bending between ZnO nanoparticles can be reduced, promoting the electron injection from PEDOT:PSS to C_{60}, and therefore the PM effect can be realized. Due to the reduced dark current, its LDR of PM photodetection reaches 120 dB and its detectivity at 390 nm is as high as 3.6×10^{11} Jones.

3.3.4. Response Speed

Photodetectors with high response speed are crucial for many applications. Increasing the applied bias can shorten the response time. Usually, a high quality active layer can withstand a high applied voltage with respect to a poor quality one. In 2000, Nakayama et al. compared the performances of

ITO/PhEt-PTC/NTCDA/Au and ITO/NTCDA/Au devices [26]. In their experiments, they failed to fabricate high quality NTCDA film but succeeded in making high quality PhEt-PTC film on ITO. With the help of high quality PhEt-PTC buffer film, the morphology of the NTCDA film can be improved apparently. As a result, the PhEt-PTC/NTCDA based device can withstand a high voltage. The transient measurement displays that the rise time of the PhEt-PTC/NTCDA device is only 3.7 s at a bias voltage of −20 V while that of the control is more than 60 s at −12 V, as exhibited in Figure 3e. It is emphasized that the insertion of a PhEt-PTC layer does not have negative effects on the electron injection from ITO to NTCDA because the LUMO level of PhEt-PTC is lower than that of the NTCDA. In addition, the response speed of the photodetector can be faster using a single-crystalline semiconductor, due to good carrier transport property, with respect to the polycrystalline one [48].

3.3.5. Spectrum Adjustment

Another important indicator of photodetector performances is its working wavelength range. Generally, a narrow band photodetector can be realized through a broadband photodetector integrated with a color filter. Thus, a high sensitivity photodetector that can respond at wavelength ranges of ultraviolet (UV), visible, and even IR would be quite attractive because of its wide applications.

The combination of two active materials is one approach to realize broadband response of PM devices. For example, NTCDA based PM devices can only sense UV light due to its large band gap, but in combination with Me-PTC, the composite device can respond to both UV light and the visible light [47]. High multiplication rate at the visible range is produced by the photo carrier generation taking place within the Me-PTC layer. It is noted that the triple layer device of ITO/Me-PTC/Au can only work at temperatures below zero Celsius. However, the bilayer system of NTCDA/Me-PTC allows the generation of holes by light absorption in Me-PTC and trapping of holes in the NTCDA film, resulting in the accumulation of holes at the NTCDA/Au interface and further the injection of electrons into the device.

Light transition based on down conversion materials is another effective way to broaden the spectral range of PM devices. Recently, aiming to enhance the response at the deep-UV range, Yang et al. applied a capping layer with the down conversion material of 4P-NPB on the illumination side of a bulk heterojunction PM device with configuration of glass/ITO/TPBi/C_{70}/SnPc:C_{70}/BCP/Al [28]. Their work demonstrates that when the UV light irradiates the 4P-NPB layer, the device will absorb the light and emit visible light, which will pass through into SnPc:C_{70} layer and then produce the PM response. Under illumination by light that can be absorbed by SnPc:C_{70}, the 4P-NPB layer is transparent, yielding negligible influences on the multiplication rates. Overall, the wavelength range of the PM device with detectivity exceeding 10^{11} Jones covers from deep-UV to near-IR (250–1000 nm) as shown in the Figure 3f, and the responsivity and detectivity at 780 nm are 70 A/W and 4×10^{12} Jones, respectively.

4. Organic PM Photodetectors Based on Polymers

With respect to the small molecular counterparts, although polymer PM photodetectors were developed later, they have attracted significant interests and attention of researchers attributed to their advantages of rich materials, easy process, and good compatibility with the roll-to-roll technique. Studies on single junction type polymer PM photodetectors are scarce. In 1999, Däubler et al. fabricated a single junction type polymer PM photodetector based on a *P*-type semiconductor material arylamino-PPV. Due to the large amount of electron accumulation at the arylamino-PPV/Al interface, the tunneling of holes from Al into the active layer takes place, yielding the PM photodetection with an *IQE* up to 2.0×10^3% [58]. In 2007, Campbell and Crone also observed PM in a device with configuration of ITO/PEDOT:PSS/MEH-PPV/Al [59] and its gain is around 20 under 500 nm light illumination at −20 V, but they provided an explanation that is distinct from the trapped carrier induced carrier injection.

In the following years, almost all related studies have focused on bulk heterojunction type polymer PM photodetectors, which can be categorized into three groups, depending on the heterojunction objects. The most widely studied group is the bulk heterojunction formed by organic semiconductors (e.g., polymer and fullerene derivative), and the other two are heterojunctions with inorganic materials and insulating polymers. In this section, we will first introduce the progress made in these three heterojunction type polymer PM photodetectors, respectively. Next, performance studies including reducing dark current and broad/narrow band spectrum adjustment of polymer PM photodetectors will be disclosed.

4.1. Bulk Heterojunction Based on Organic Semiconductors

We will introduce the progress made in bulk heterojunction polymer photodetectors based on organic semiconductors according to different donor/acceptor weight ratios.

4.1.1. Donor/Acceptor Weight Ratio of 1:1

In the early days, by reference to solar cells, bulk heterojunction type polymer PM devices were designed with the donor/acceptor weight ratio of 1:1. Experiences suggest that 1:1 donor/acceptor weight ratio forms a favorable interpenetrating network in solar cells which can facilitate the transport of both electrons and holes. However, with such design, one can hardly achieve current multiplication because of the short life time of both electrons and holes. To realize PM photodetection in a P3HT:PC$_{61}$BM (with the weight ratio of 1:1) device, one can incorporate inorganic nanoparticles into the active layer which will be presented in Section 4.2 [65]. Other approaches of doping organic compounds [60,66] and interface modifications [61,67,68] have been also proposed to realize the PM phenomenon in 1:1 donor/acceptor heterojunction devices as introduced in the following.

In 2010, Chen et al. incorporated an organic dye Ir-125 into the P3HT:PC$_{61}$BM (1:1) bulk heterojunction device and discovered current multiplication as well [60]. The device without dye does not have any gain. In contrast, the device with dye show PM from UV to near IR range as shown in Figure 4a. The *EQE* reaches its maximum of 7.2 × 10^3% at −1.5 V under 500 nm light illumination. The exhibited PM phenomenon is because Ir-125 dye brings forward a lot of electron traps into the active layer. With the incorporation of another organic dye Q-switch 1, the PM response of the P3HT:PC$_{61}$BM:Ir-125 device can be extended to near IR range [66].

Besides, interface modification is an alternative effective approach to induce PM in P3HT:PC$_{61}$BM (1:1) devices. In 2014, Melancon et al. introduced a semi-continuous gold (s-Au) film between the ITO electrode and the active layer, with its structural diagram displayed in Figure 4b. Their work indicated that with the help of the s-Au film, PM was successfully excited with an *EQE* of 1.5 × 10^3% at −2 V bias under 400 nm light illumination [61]. They explained this phenomenon that the s-Au film acts as a hole blocking layer which enables the accumulation of holes at the P3HT/PCBM interface and the further tunneling of electrons from ITO into the P3HT/PCBM region. In 2017, Wang et al. used PFN to modify the ITO/active layer interface but the obtained *EQE* was only slightly higher than 100% [67]. Later, the same group proposed to utilize the transparent polyethylenimine ethoxylated (PEIE) to modify the ITO surface [67]. Through this process, the work function of the electrode gets lower, bringing forward the energy barrier formed between the work function of the PEIE modified ITO and the HOMO of P3HT is 0.75 eV larger than that between the bare ITO and P3HT. The expanded energy barrier causes the enhanced interfacial accumulation of photo carrier, which is preferred to the increased photocurrent gain. The *EQE* value of the device based on PEIE modified ITO reached 3.3 × 10^3% at −1 V under 370 nm light illumination. The proposed PM device exhibits a rise time of 78 μs and a fall time of 87 μs.

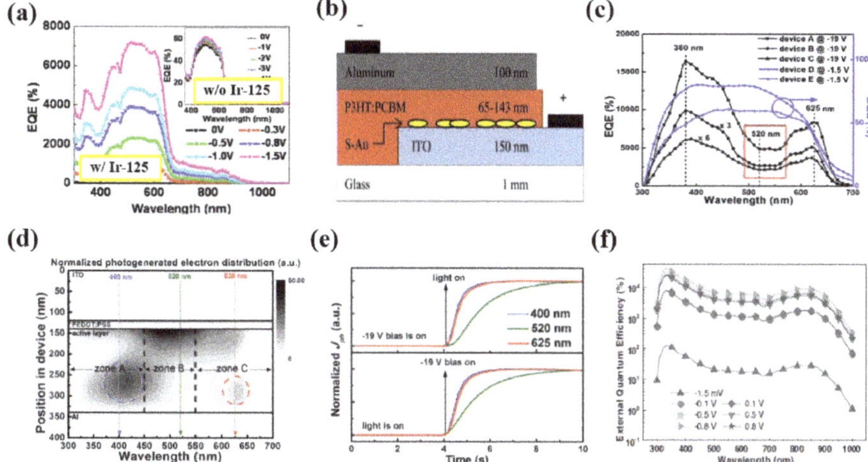

Figure 4. Device performances of various polymer PM photodetectors based on bulk heterojunctions made of organic semiconductors. (**a**) *EQE* spectra of ITO/PEDOT:PSS/P3HT:PCBM (1:1)/Ca/Al with and without Ir-125 doped in active layer (Reproduced with permission from [60]. *AIP Publishing*, 2010); (**b**) Structural diagram of ITO/s-Au/P3HT:PCBM/Al device (Reproduced with permission from [61]. *AIP Publishing*, 2014); (**c**) *EQE* spectra of ITO/PEDOT:PSS/P3HT:PC71BM/LiF/Al with different P3HT:PC71BM weight ratios (Reproduced with permission from [62]. *Springer Nature*, 2015); (**d**) Calculated wavelength dependent distribution of photogenerated electrons in the active layers of P3HT:PC71BM (100:1) without bias (Reproduced with permission from [63]. *Royal Society of Chemistry*, 2015); (**e**) Normalized transient photo current curves under light illumination at the wavelengths of 400 nm, 520 nm, and 625 nm, respectively for P3HT:PC71BM (100:1) device (Reproduced with permission from [63]. *Royal Society of Chemistry*, 2015); (**f**) *EQE* spectra measured under different bias voltages after UV light treatment for ITO/ZnO/PDPP3T:PC71BM (1:2)/Al device (Reproduced with permission from [64]. *John Wiley and Sons*, 2016).

4.1.2. Donor/Acceptor Weight Ratio Higher than 1:1

An easy means to realize PM in the donor/acceptor heterojunction photodetectors is by increasing the amount of donor (or reducing the content of acceptor) in the blend of active material, in other words, using a blend with donor/acceptor weight ration higher than 1:1. On the condition that the ratio of acceptor in the active layer is being reduced, the acceptor forms isolated islands instead of connected networks, which can trap the photo generated electrons and thereby making possible the injection of holes from the external circuit. Compared with PM devices with balanced donor/acceptor weights, the response speed of bulk heterojunction polymer PM photodetectors with less acceptor in the active layer is significantly lower, because it takes more time for trapped carriers to accumulate at the Schottky junction due to poorer carrier transport properties.

The pioneer research of this kind of PM devices was carried out by Fujun Zhang's group in 2015 [62]. They fabricated a PM photodetector with a configuration of ITO/PEDOT:PSS/P3HT:PC$_{71}$BM/LiF/Al using 100:1 P3HT/PC$_{71}$BM which exhibited an *EQE* as high as 1.7×10^4% under 380 nm light irradiation at -19 V bias, as shown in Figure 4c. The acceptor islands can trap the photo generated electrons, which will transport to and accumulate at the Schottky junction formed between the active layer and the Al electrode with applied voltage. Later, by removing the LiF buffer layer, they lowered the hole injection barrier between the active layer and the Al electrode, raising the *EQE* up to 3.8×10^4% at -19 V bias [69]. P3HT molecular arrangement with face-on is more favorable for hole transportation. Their further study indicated that rapid annealing the active layer after spin-coating can avoid atomic self-assembly, which is more helpful to form a face-on structure, bringing forward *EQE* rise up to 1.2×10^5% at -19 V under 610 nm light illumination due to the improved hole transport property [70]. Subsequently, they employed a burn-in treatment to their PM devices by applying a voltage of -25 V and -19 V for 1000 s, successively, making the device performance more stable [63]. In order to further elucidate the working mechanism of the devices, they examined the distributions of optical field within in the PM device at different wavelengths together with the transient photocurrent measurements; see Figure 4d,e. The results indicated that stronger absorption occurring closer to the Al electrode (e.g., at wavelengths of 400 nm and 625 nm) corresponds to a faster transient response due to shorter distance of electron transport. In contrast, at 520 nm wavelength light illumination, although with relatively high light absorption, its transient response is quite slow due to the absorption taking place close to the PEDOT:PSS/P3HT:PC$_{71}$BM interface. Besides PCBM, they also used IC$_{60}$BA to make the active layer with P3HT but the obtained *EQE* was 6.9×10^2% with a 100:2 donor/acceptor weight ratio [71].

Most of the active layers in bulk heterojunction polymer PM devices comprise of polymer and fullerene derivative of PCBM. Beyond fullerene and their derivatives, non-fullerene acceptor materials have also attracted attention due to their strong absorption capabilities in visible and near-IR regions, adjustable energy levels, and good stability. Through collaboration with Xiaowei Zhan's group, Zhang's group prepared the PM devices by blending P3HT with non-fullerene acceptor materials of DC-IDT2T [72] or ITIC [73], resulting in extended spectrum response and simultaneously more stable device performances. For the DC-IDT2T based PM device, its optimal donor/acceptor weight ratio was 100:1 as well and its *EQE* exceeded 1.0×10^4% over the range from 350 nm to 650 nm with a maximum *R* of 131.4 A/W and *D** of 1.43×10^{14} Jones. Moreover, compared with the PCBM based device, the DC-IDT2T based device responded much better at the near-IR wavelength range due to good absorption of DC-IDT2T. After exposing the PM devices in air for 40 h, the DC-IDT2T based device only suffered 39% degradation on *EQE* while the PCBM based device bore a degradation of 57%, indicating the DC-IDT2T acceptors are promising for developing stable PM photodetectors.

The *P*-type polymer constituting the active layer of PM photodetectors can also be regulated. In 2017, Esopi et al. used a *P*-type material F8T2 as the donor to prepare the PM devices with PCBM [74]. Their device configuration is ITO/PEDOT:PSS/F8T2:PC$_{71}$BM (100:4)/LiF/Al with the F8T2/PC$_{71}$BM weight ratio of 100:4 and its *EQE* is 5.6×10^3% at -40 V bias under 360 nm light illumination. Most importantly, F8T2 based PM devices have a very low dark current (only 2.7×10^{-7} mA/cm^2 at -1 V bias), much lower than that of the P3HT based device. This is due to the inhibited hole injection from the Al electrode to the donor, produced by the increased barrier between the HOMO level of the acceptor and Fermi level of Al (F8T2: 1.2 eV, and P3HT: 0.9 eV). But the increased barrier inevitably brought the decrease of photodetector response under light illumination. They also compared the PM performances before and after removing the LiF buffer layer, and concluded that the removal of LiF made the hole injection much easier but it also deteriorated the stability of devices.

4.1.3. Donor/Acceptor Weight Ratio Lower than 1:1

It is anticipated that decreasing the donor/acceptor weight ratio would increase the amount of hole traps in the active layer, possibly leading to PM photodetection based on the hole trap assisted electron tunneling effect. In 2016, Dongge Ma and collaborators proposed a PM photodetector based on a narrow bandgap polymer donor PDPDP3T with the configuration of ITO/ZnO/PDPP3T:PC$_{71}$BM/Al, which showed an *EQE* of 1.4×10^5% at a low bias of -0.5 V after the device was irradiated by UV light for 30 s, as displayed in Figure 4e [64]. In their active layer, the blend of PDPP3T:PC$_{71}$BM system has a weight ratio of 1:2 (lower than 1:1), which is different from the previous two situations in Sections 4.1.1 and 4.1.2. Here, they claimed that there are a lot of hole traps in their active layer, which tended to form trapped holes accumulated at the ZnO/active layer interface under reverse bias. Because large electron injection barriers were generated at both the ITO/ZnO and ZnO/active layer interfaces, the pristine device without UV treatment behaves as a photodiode without any gain. The electron blocking effect could be alleviated through UV treatment. After absorbing ultraviolet light, the ZnO nanoparticles could generate electron-hole pairs, functioning as centers to neutralize the oxygen molecules adsorbed on the surface of the ZnO nanoparticles. As a result of the desorption of oxygen molecules from the surface of the particles, a decrease of the LUMO energy differences at both the ITO/ZnO and ZnO/active layer interfaces took place, and a reduction of the electron injection barriers from ITO to $PC_{71}BM$ was realized. Such a phenomenon finally enabled the device to become a photoconductor with large gain.

Differently, the investigation carried out by Nie et al. into an inverted organic PM photodetector of ITO/lysine/PBDTT-DPP:$PC_{71}BM$ (1:2)/MoO_3/Al also included an active layer with less donor. In 2017, electron traps were identified rather than hole traps [75]. In this design, the ITO modified by lysine was the cathode while the MoO_3/Al acted as the anode. The device showed a large amount of photon to electron multiplication at room temperature with an *EQE* up to 1.6×10^5% (936.05 A/W) under 10 V bias. Inversely, no gain was found at negative bias. The gain behavior was attributed to the electron trap assisted hole tunneling from Al/MoO_3 composite electrode into the active layer. The low current under dark was due to a space charge region formed between PBDTT-DPP/MoO_3 interfaces, which could be erased after being exposed to light.

4.2. Bulk Heterojunction with Inorganic Nanoparticles or Quantum Dots

Both inorganic nanoparticles and quantum dots have been employed to form heterojunctions with polymer or polymer blend. Their incorporation provides additional carrier traps which are essential for the followed carrier tunneling at the Schottky junction.

In 2008, Chen et al. incorporated cadmium telluride (CdTe) nanoparticles into the active layer of P3HT:$PC_{61}BM$ (1:1), obtaining an *EQE* of 8.0×10^3% at -4.5 V bias under 350 nm light illumination [65]. In this study, their CdTe nanoparticles were capped with *N*-phenyl-*N*-methylthiocarbamate (PMDTC) ligands which can improve the solubility of inorganic nanoparticles in the target solution of active material. Their study implied that a solvent annealing step after film spin-coating could induce a higher concentration of CdTe nanoparticles on the top of the annealed film. Under light exposure, CdTe nanoparticles with trapped electrons lowered the energy barrier for hole injection from the top electrode to the active layer.

ZnO nanoparticles are an alternative choice of carrier trap materials which have the merits of low cost, variable synthetic strategies, and so on. In 2012, Jinsong Huang's group incorporated ZnO nanoparticles into P3HT or PVK film and fabricated the PM devices with configurations of ITO/PEDOT:PSS/PVK:TPD-Si_2/P3HT (or PVK):ZnO/BCP/Al as shown in Figure 5a [8], aiming to modify the Schottky junction for smooth hole injection into the active layer. Here, the PVK:TPD-Si_2 blend film behaved as the electron blocking layer and the BCP layer acted as the hole blocking layer, through which the dark current was controlled at an extremely low level (6.8 nA at -9 V for the PVK based device) while excellent PM performance was maintained. Specifically, the *EQE* of the P3HT:ZnO device and PVK:ZnO device are up to 13.4×10^5% and 2.4×10^5% respectively, at -9 V bias under 360 nm light illumination. Similar to the function of CdTe nanoparticles in the previous work, ZnO nanoparticles blended with the donor polymer worked as the electron traps which could modify the Schottky junction for smooth hole injection into the active layer. Later, Huang's group incorporated ZnO nanoparticles into PDTP-DFBT film which can sense light from UV to near-IR [76]. They demonstrated that the surface treatment of active layer by Ar plasma etching can effectively enhance electron trap assisted hole injection with the gain improved by 2–3 times as displayed in Figure 5b. A control sample of spin-coating ZnO nanoparticle layer on the top of active layer provided a direct evidence that excess ZnO nanoparticles created more traps.

Beyond nanoparticles, inorganic quantum dots with sizes smaller than 10 nm have also been applied in organic PM photodetectors aiming for providing carrier traps or extending the response spectrum range. In 2014, Huang's group doped PbS quantum dots and ZnO quantum dots together into the active layer comprising of P3HT:$PC_{61}BM$ (1:1) [77]. The ternary active layer of P3HT:$PC_{61}BM$:ZnO already possesses the PM photodetection ability at wavelengths shorter than 650 nm because ZnO quantum dots offer plentiful electron traps and thus enable hole tunneling. With the incorporation of PbS quantum dots, the PM performance was extended to the wavelength as long as 1000 nm, as shown in Figure 5c. The principle is that the electrons generated due to light absorption by PbS in the IR range can transfer to ZnO traps, triggering the hole injection into the active layer and further PM photodetection with extended spectrum range.

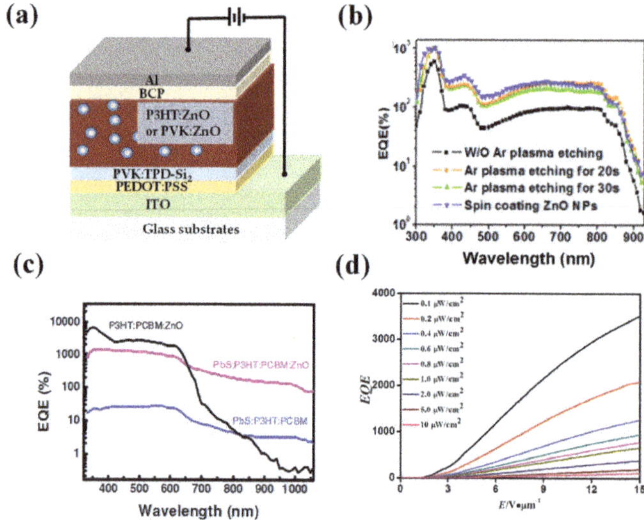

Figure 5. Device performances of various polymer PM photodetectors based on bulk heterojunctions with inorganic materials and insulating polymers. (**a**) Structural diagram of ITO/PEDOT:PSS/PVK:TPD-Si$_2$/P3HT:ZnO (PVK:ZnO)/BCP/Al device; (**b**) *EQE* spectra of the PDTP-DFBT based PM photodetectors blended with ZnO nanoparticles treated by different processes (Reproduced with permission from [76]. *AIP Publishing* 2015); (**c**) *EQE* spectra of the ITO/PEDOT:PSS/P3HT:PCBM/Al devices with the active layer doped with or without PbS and ZnO QDs (Reproduced with permission from [77]. *John Wiley and Sons*, 2014); (**d**) *EQE* spectra under different light intensities of the bulk heterojunction polymer PM photodetector realized through doping Y-TiOPc quantum dots into the insulating polymer PVB (Reproduced with permission from [27]. *Royal Society of Chemistry*, 2016).

4.3. Bulk Heterojunction with Insulating Polymers

It is an interesting finding that blending organic quantum dots into insulating polymer can also induce PM photodetection. In 2015, Peng et al. proposed a bulk heterojunction photodetector made of oxotitanium phthalocyanine in the crystal form of phase-Y (Y-TiOPc) quantum dots and an insulating polycarbonate resin PCZ-300 (shorted as Y-TiOPc@PC) in contact with two parallel metal electrodes, which had a wide spectral response from 400 nm to 940 nm and an *EQE* of 3.6 × 10^4% at 830 nm [78]. The explanation of their finding is as follows. In the device, Y-TiOPc nanoparticles were separated by PC, forming a large amount of Y-TiOPc/PC interfaces, therefore the only allowed charge transport mechanism in the active layer is based on charge tunneling. Because the energy barrier for the hole is higher by 1.4 eV than that for the electron, the tunneling probability of photo generated electrons in the conduction band is estimated to be much higher than that of photo-induced holes in the valence band of the separated Y-TiOPc quantum dots. As a result, efficient trapping of holes is produced, so that more electrons can flow through the device before the recombination of photo induced charge carriers occurs. Actually, this work did not mention the tunneling of carrier at the interface between the active layer and electrode, thus its principle should be different from that presented in Section 2.2 We will come back to this kind of mechanism for organic PM photodetection in Section 5.

Subsequently, Li et al. proposed a polymer PM photodetector based on Y-TiOPc quantum dots as well [27]. In that work, the Y-TiOPc quantum dots were dispersed into a polyvinyl butyral (PVB) based solution to make the active layer, and a blended film of polycarbonate (PC) and m-TPD were selected as the hole transport layer. The prepared photodetector exhibited an obvious PM phenomenon with the highest *EQE* of 3.5×10^5% (see in Figure 5d), an excellent photosensitivity with the maximum responsivity of 2227 A/W, and an outstanding low-light detection with the highest normalized detectivity of 3.1×10^{14} Jones under 780 nm light illumination. Different from the explanation in previous work [78], they attributed the multiplication to the enhanced external hole tunneling injection assisted by trapped electrons at the interface of active layer and ITO. They found that the hole energy barrier was only 0.2 eV at the Y-TiOPc/m-TPD interface while the electron energy barrier reached 0.7 eV at the Y-TiOPc/ITO interface, causing an unbalanced transport of electrons and holes. Such an unbalance further lead to the accumulation of electrons at the Y-TiOPc/ITO interface, enabling the following narrowing of Schottky junction and thus hole tunneling. Their study also reflected that Y-TiOPc quantum dots with smaller diameter could not only generate more photo carriers, but also contribute to the formation of a steeper band bending, promoting the injection of a large amount of holes.

4.4. Performance Studies of Polymer PM Photodetectors

In this subsection, we will introduce the progress of performance studies on polymer PM photodetectors from the aspects of dark current, broadband response, and narrowband response.

4.4.1. Dark Current

Inserting an appropriate buffer layer between the active layer and electrode can greatly reduce the dark current while maintaining the photo current [67,73]. For example, Zhang's group systemically compared the performances of ITIC based polymer PM photodetectors using PEDOT:PSS and PFN as the buffer layer with their current density-voltage (*J-V*) characteristics under dark and light respectively [73]. It was found that the PEDOT:PSS device can only work under reverse bias; in contrast, the PFN device can work effectively under both reverse and forward bias. Moreover, the dark current of the PFN device (10^{-6} mA/cm^2) was much lower with respect to the PEDOT:PSS device (10^{-4} mA/cm^2) at 0 bias; the dark current of the PFN device also decreased significantly at the bias when PM was triggered (e.g., −15 V). Such a phenomenon was explained as follows based on the energy diagrams displayed in Figure 6a,b. Under the reverse bias, the amount of holes injected from the Al electrode to the HOMO level of P3HT is relatively low because the difference between the LUMO level of P3HT and the Fermi level of Al is high (about 1.2 eV), causing the dark currents in both PEDOT:PSS and PFN devices to be low. The difference in performance between PFN and PEDOT:PSS devices is mainly due to difference between HOMO levels of these two buffer materials (PEDOT:PSS: −5.1 eV; PFN: −5.6 eV). Therefore, PFN behaves more effectively than PEDOT:PSS to prevent holes transit from active layer to the ITO electrode, yielding a reduced dark current. Under forward bias, the high HOMO level of PEDOT:PSS cannot block the hole injection from ITO to the active layer under dark, resulting in no response difference before and after light illumination. In contrast, there is a difference of 0.9 eV between the HOMO level of PFN and the Fermi level of ITO, which can effectively block holes being injected from the ITO electrode, yielding a very low dark current under forward bias.

4.4.2. Broadband Response

In order to extend the response spectrum range of organic PM devices, researcher constructed ternary bulk heterojunctions which are realized through blending two donors with different absorption spectrum range together with the acceptor, similar as in solar cells [79–85]. In 2015, Zhang's group prepared an polymer PM photodetector with the structure of ITO/PEDOT:PSS/P3HT:PTB7-Th:PC$_{71}$BM/Al [79]. The photodetector with the active layer containing only PTB7-Th possess very weak PM effect, while the ternary device shows very high multiplication rates over the wavelength range from 350 nm to 800 nm. The highest *EQE* values of the ternary device with P3HT:PTB7-Th:PC$_{71}$BM weight ratio of 50:50:1 are around 3.8×10^4% in the spectral range from 625 nm to 750 nm under -25 V bias (as shown in Figure 6c), and the corresponding *R* is 229.5 A/W and *D** is 1.91×10^{13} Jones. The broad spectral response range was due to the contribution of PTB7-Th exciton dissociation on the number of trapped electrons in PC$_{71}$BM near the Al cathode. The results also showed that, with respect to P3HT/PC$_{71}$BM junctions, PTB7-Th/PC$_{71}$BM junction induced shallower traps due to smaller LUMO differences between donor and acceptor, resulting in quick filling of the electrons as turning on the incident light and thereby a faster response.

4.4.3. Narrowband Response

Traditional photodetectors obtain narrowband response utilizing color filters, which are at the cost of light attenuation, the reduced responsivity, as well as the relatively complicated system with the increased cost [86–90]. In 2015, the Paul L. Burn and Paul Meredith's group firstly achieved narrowband organic photodetector with thick active layers through adjusting the carrier collection efficiency [88]. Later, Zhang's group applied the exact idea into polymer PM devices and implemented a filterless narrowband PM photodetector. The thicknesses of their active layers (100:1 P3HT:PC$_{71}$BM) were in micrometer scale and the highest *EQE* reached 5.3×10^4% at -60 V and the FWHM of their detection spectrum was only 28 nm [91]. Its principle is that, with thick active layer, close to the Al electrode, the amount of photo carriers decreases linearly from short wavelength to long wavelength, leading to a wavelength regime with poor hole collection efficiency at the opposite electrode under reverse bias. By using a PFN type buffer layer, the narrowband PM photodetector can work under both forward and reverse bias [92], with the reasons being present in Section 4.4.1. The PFN based PM device exhibited two narrowband response windows under forward bias (*EQE*: 7.2×10^3% or 8.2×10^3%, for 340 nm or 650 nm light illumination at 60 V) and a single narrowband response window under reverse bias (*EQE*: 1.6×10^3% for 665 nm light illumination at -60 V) as displayed in Figure 6d,e, respectively. The two narrowband response windows obtained under forward bias can be well explained based on the photo carrier generation maps as functions of position and wavelength. Close to the ITO/PFN electrode, the photo carriers in the wavelength range of 400–600 nm is much larger, corresponding to poorer hole collection efficiency at the Al electrode due to long transport length, than that outside this range. As a result, under forward bias, two wavelength regimes are left with good hole collection efficiency, that is, the bands are shorter than 400 nm and longer than 600 nm. In 2018, they also applied this idea into ternary PM photodetectors [93]. Thanks to the doped PTB7-Th, the device had a narrowband response under -50 V reverse bias, corresponding to a maximum *EQE* of 2.0×10^2% at 800 nm with an FWHM of 40 nm.

Similar works have also been carried out for bulk heterojunction polymer PM photodetectors based on inorganic quantum dots. In 2016, Huang's group developed a narrowband PM photodetector with the structure of ITO/PVK/P3HT:PC$_{60}$BM:CdTe QDs/BCP/Al with a 3.5 μm thick active layer thickness [94]. Compared with the quantum dot free photodetector, the photodetector with CdTe quantum dots maintained the low dark current under reverse bias as well as the narrow band response. Besides, the *EQE* values of the photodetector were improved significantly with the incorporation of the quantum dots which provide vast electron traps in the active layer, reaching nearly 2.0×10^2% at 660 nm light irradiation and the corresponding LDR is 110 dB. Through a device with configuration of ITO/SnO$_2$/PEIE/PDTP-DFBT:PC$_{71}$BM:PbS QDs (4 μm-thick)/MoO$_3$/Ag [95], they also obtained a filterless narrowband photodetection response with a 50 nm FWHM at near-IR range. In their design, PbS QDs behaved as hole traps, triggering the injection of electrons from Ag electrode to the active layer.

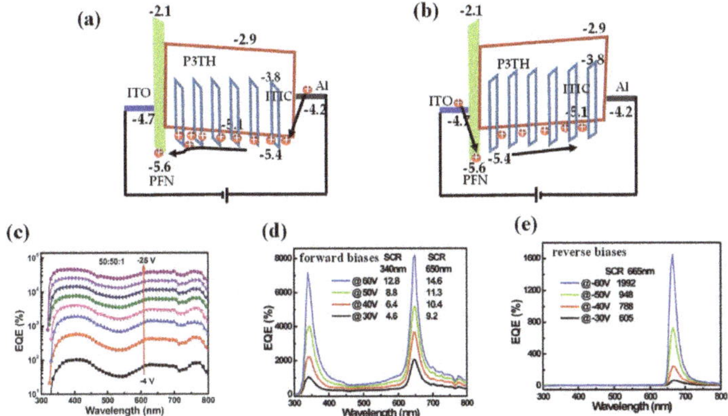

Figure 6. (a,b) Energy level diagrams of the PFN device under reverse forward biases under dark; (c) *EQE* spectra under different biases of the ITO/PEDOT:PSS/P3HT:PTB7-Th:PC$_{71}$BM (50:50:1)/Al device (Reproduced with permission from [79]. *Royal Society of Chemistry*, 2015); (d,e) Narrowband *EQE* spectra of the ITO/PFN-OX/P3HT:PC$_{61}$BM (4.0 µm)/Al device under different biases (Reproduced with permission from [92]. *Royal Society of Chemistry*, 2017).

5. Other Mechanisms of Organic PM Photodetectors

In addition to the explanations presented in Section 2.2, researchers have also put forward some other mechanisms for realizing organic PM photodetection.

In 2006, Reynaert et al. used the organic small molecular F$_{16}$CuPc as the active layer to prepare an organic PM photodetector with a structure of ITO/PEDOT:PSS/F$_{16}$CuPc/Al, yielding the *EQE* exceeding 3.0×10^3% at 633 nm [96]. They believed that the PM effect of this device was caused by the local charge induced exciton quenching in semiconductors. The reason is that F$_{16}$CuPc is a unipolar disordered organic small molecular that forms an ohmic contact with the metal electrode rather than a Schottky contact. To verify the deduction, they compared the device performances with different metal electrodes. When the Schottky contact is formed at the semiconductor/metal interface, band bending will be different for metal electrodes with different work functions, thereby the device performances would be altered. However, their experiments indicated that the devices with Au electrodes and Al electrodes have exactly the same performance, reflecting that the ohmic contact is constructed between F$_{16}$CuPc and the metal electrode. This theory is distinct from the trap assisted carrier tunneling mechanism, because the PM phenomenon of this F$_{16}$CuPc based device is a bulk effect rather than an interfacial effect.

There are also other works confirming that the gain of their organic PM detectors is not due to interfacial effects. In 2007, Campbell and Crone prepared a series of organic PM photodetectors using MEH-PPV as the active layer [59]. They found the device with bare MEH:PPV had similar gain characteristics as those of devices with active layers doped with 10 wt% PbSe quantum dots or 10 wt% C$_{60}$. Without mentioning any interfacial effects, they proposed that a small fraction of the optically excited excitons dissociate producing deeply trapped electrons and free holes. The trapped electrons can lead to photoconductive gain if the electron lifetime is longer than the hole carrier transit time. The gain per trapped electron is determined by the number of holes passing through the device during the lifetime of the trapped electron. This principle is similar to that of photoconductive type photodetectors. Later in 2009, the same researchers realized PM photodetection at near-IR range through a device with configuration of ITO/PEDOT/OSnNcCl$_2$/BCP/Ca [97] based on the circulation of free carriers in response to trapped photo carriers as well. Peng et al. clarified that their Y-TiOPc@PC based PM devices were based on the same principle of carrier circulation [78].

One can also employ inorganic quantum dots to realize PM photodetection based on the principle of multiple exciton generation. In 2005, Qi et al. prepared an organic PM photodetector by doping PbSe QDs in MEH-PPV with the structure of ITO/PEDOT:PSS/PbSe:MEH-PPV/Al [98]. According to their study, when MEH-PPV was doped with PbSe quantum dots (with an absorption peak at 1900 nm) with a diameter of 8 nm, the device's *EQE* reaches 1.5×10^2% at -8 V bias under 510 nm light illumination. In contrast, the reference device only showed an *EQE* of 40%. Their explanation was as follows. When the incident photon energy was at least three times larger than the quantum dot band gap, the PbSe QDs could absorb photons and generate multiple excitons, resulting in multiplication of carriers. Another control study was carried out by altering the size of the PbSe quantum dots to 4.5 nm which corresponds to an absorption peak at 1100 nm, but the device did not induce any PM effect. The dependence of this photodetector performance on the size of quantum dots reflects that multiple exciton generation ascribed to the strong quantum confinement effect of quantum dots is responsible for the PM effect.

6. Summary and Outlook

In this paper, we have summarized past studies on organic PM photodetectors since 1994. Performances of representative organic small molecular PM photodetectors and polymer PM photodetectors are summarized in Tables 1 and 2, respectively.

Studies on organic small molecular PM photodetectors came out early. From Table 1, one sees clearly that most of reported studies focused on *N*-type semiconductor materials which produced PM performances based on the mechanisms of hole trap assisted electron tunneling. In these devices, the carriers are captured by interfacial structural traps, which bare the limitations of low quantity and complexity to control. In addition, single junction type PM devices suffer slow transient response, that is in second time scale or even longer, due to poor carrier transport in the active layer. These limitations were alleviated by constructing a bulk heterojunction active layer. C_{60} (or C_{70}) was frequently combined with *P*-type semiconductor materials (e.g., CuPc, TAPC, SnPc) to form bulk heterojunction active layers, based on which the response time of PM photodetection can be reduced to millisecond timescale. The reason of improved response speed is that the trapped hole carriers can freely transport through the network formed by the *P*-type semiconductor material and thus the accumulation of carrier at the Schottky junction can be accelerated.

Table 2 reflects that polymer PM photodetectors have become a hot research topic since 2010, which must be closely related with the success made in the field of polymer solar cells. Compared with the single junction counterparts which have received little attention, bulk heterojunction type polymer PM photodetectors with diverse heterojunction formulas have been widely studied. For the heterojunction formula of polymer/inorganic-nanoparticles, Huang et al. realized an *EQE* of 3.4×10^5% by blending ZnO nanoparticles into the active layer P3HT [8]. For the heterojunction formula of polymer/insulator, Li et al. proposed to make the active layer by dispersing the Y-TiOPc quantum dots into a PVB solution [27], yielding an *EQE* of 3.5×10^5%. Beyond these two heterojunction formulas, the most popular researched one is the formula made of organic semiconductors comprising of donor and acceptor with different weight ratios. Introducing carrier traps is the primary task to realize current multiplication in organic bulk heterojunction type photodetectors. In 2015, Zhang et al. for the first time put forward introducing electron traps in the donor/acceptor active layer through reducing the weight ratio of acceptor, yielding an *EQE* of 1.7×10^4% [62]. To bring in carrier traps into bulk heterojunction photodetector devices with 1:1 donor/acceptor weight ratio, approaches of doping organic dyes or incorporation of a gold island film have been adopted, but the achieved *EQE*s are far below those obtained by reducing the acceptor ratio. We also noticed that using PDPP3T:PC_{71}BM blend with a weight ratio of 1:2 can also induce a PM effect with an *EQE* as high as 1.4×10^5% after the LUMO energy differences at both the ITO/ZnO and ZnO/active layer interfaces are reduced by UV irradiation [64].

Table 1. Representative organic small molecular PM photodetectors and their performances.

Mechanism	SJ/BJ	Year [Ref]	Device Structure	QE (%) at Bias	Other Performances
electrons tunneling	SJ	1994 [43]	Glass/Au/Me-PTC/Au	1.0×10^6 (*IQE*@600 nm), -16 V	Working temperature: $-50\,°C$
		1996 [47]	ITO/NTCDA/Au	1.3×10^7 (*IQE*@400 nm), -16 V	Rise time: >60 s
		2000 [26]	ITO/PhEt-PTC/NTCDA/Au	1.7×10^7 (*IQE*@400 nm), -20 V	Rise time: 3.7 s
		2007 [49]	ITO/PEDOT:PSS/C$_{60}$/BCP/Al	5.0×10^3 (*EQE*@450 nm), -4 V	-
		2014 [57]	ITO/PEDOT:PSS/C-TPD:ZnO/C$_{60}$/BCP/Al	4.0×10^2 (*EQE*@390 nm), 8 V	LDR: 120 dB; D^*: 3.6×10^{11} Jones
	BJ	2002 [51]	ITO/CuPc:C$_{60}$/Au	1.5×10^3 (*IQE*@560 nm), -14 V	Response time: ms
		2010 [52]	ITO/NTCDA/C$_{60}$/CuPc:C$_{60}$/BCP/Al	3.4×10^4 (*EQE*), -4 V	Response time: ms
		2016 [28]	ITO/TPBi/C$_{70}$/TAPC:C$_{70}$/BCP/Al	1.0×10^3 (*EQE*), -4 V	-
			4P-NPB/glass/ITO/TPBi/C$_{70}$/SnPc:C$_{70}$/BCP/Al	1.0×10^4 (*EQE*@780 nm), -10 V	R: 70 A/W; D^*: 4×10^{12} Jones
holes tunneling	SJ	1996 [50]	ITO/DQ/Ag (or Mg)	Ag: 2.5×10^5 (*IQE*@600 nm), 20 V Mg: 1.0×10^5 (*IQE*@600 nm), 36 V	Response time: 10–20 s

Table 2. Representative organic polymer PM photodetectors and their performances.

BHJ Type	Year [Ref]	Device	EQE (%) at Bias	Other Performances
Polymer/Organic 1:1	2010 [60]	ITO/PEDOT:PSS/P3HT:PCBM: Ir-125 (1:1:1)/Ca/Al	7.6×10^2@800 nm, -5 V	R: 4.9 A/W; Broadband response
	2012 [66]	ITO/PEDOT:PSS/P3HT:PCBM:Q-Switch1 (1:1:1)/Ca/Al	8.4×10^2@560 nm, -5 V	R: 4 A/W; Broadband response
		ITO/PEDOT:PSS/P3HT:PCBM: Ir-125:Q-Switch1 (1:1:0.5:0.5)/Ca/Al	5.5×10^3@560 nm, -3.7 V	R: 23 A/W; Broadband response
	2014 [61]	ITO/s-Au/P3HT:PCBM (1:1)/Al	1.5×10^3@400 nm, -2 V	-
	2017 [67]	ITO+PEIE/P3HT:PC$_{61}$BM (1:1)/Al	3.3×10^3@370 nm, -1 V	R: 14.2 A/W; D^*: 1.0×10^{12} Jones; Time: 78 µs (rise), 87 µs (decay)
Polymer/Organic Higher than 1:1	2015 [62]	ITO/PEDOT:PSS/P3HT:PC$_{71}$BM (100:1)/LiF/Al	1.7×10^4@380 nm, -19 V	-
	2015 [79]	ITO/PEDOT:PSS/P3HT:PTB7-Th: PC$_{71}$BM (50:50:1)/Al	3.8×10^4@750 nm, -25 V	R: 229.5 A/W; D^*:1.9×10^{13} Jones; Broadband response
	2016 [73]	ITO/PFN/P3HT:ITIC (100:1)/Al	2.3×10^3@625 nm, -15 V	R: 41.9 A/W; D^*: 7.1×10^{12} Jones
	2017 [91,92]	ITO/PFN-OX/P3HT:PC$_{61}$BM (100:1, 4 µm)/Al	8.2×10^3@650 nm, 60 V	D^*: 7.7×10^{11} Jones@10 V; Narrowband response
	2017 [74]	ITO/PEDOT:PSS/F8T2:PC$_{71}$BM (100:4)/LiF/Al	5.6×10^3@360 nm, -40 V	R: 15.9 A/W
	2018 [93]	ITO/PFN-OX/P3HT:PTB7-Th: PC$_{61}$BM (40:60:1,3 µm)/Al	2.0×10^2@800 nm, -50 V	D^*: >1.0×10^{11} Jones@10 V; LDR: 180 dB@550 nm, 30 V; Narrowband response

Table 2. *Cont.*

BHJ Type	Year [Ref]	Device	EQE (%) at Bias	Other Performances
Polymer/Organic Lower than 1:1	2016 [64]	ITO/ZnO/PDPP3T:PC$_{71}$BM (1:2)/Al	1.4×10^5@680 nm, 0.5 V	D^*: 6.3×10^{12} Jones; Decay time:0.27 s
	2017 [75]	ITO//Lys/ PBDTT-PP:PC$_{71}$BM (1:2)/MoO$_3$/Al	5.0×10^3@730 nm, 1 V	R: 29.5 A/W; D^*: 1.6×10^{15} Jones @735 nm; Time: 162 µs (rise), 7.9 ms (decay); LDR: 160 dB
Polymer/Inorganic	2008 [62]	ITO/PEDOT:PSS/P3HT:PCBM:CdTe (1:1)/Ca/Al	8.0×10^4@350 nm, −9 V	-
	2012 [8]	ITO/PEDOT:PSS/PVK:TPD-Si$_2$/P3HT:ZnO/BCP/Al	3.4×10^5@360 nm, −9 V	R: 1001 A/W; D^*: 3.4×10^{15} Jones; Time: 25 µs (rise), 142 µs (decay)
	2015 [76]	ITO/PEDOT:PSS/PVK:TPD-Si$_2$/PDTP-FBT:ZnONPs (1:3)/BCP/Al	2.5×10^2@800 nm, −4.5 V	R: 1.6 A/W; D^*: 7.1×10^9 Jones
	2016 [95]	ITO/SnO$_2$/PEIE/PDTP-DFBT:PC$_{71}$BM:PbS QDs (4 µm)/MoO$_3$/Ag	1.8×10^2@890 nm, −7 V	R: 1.3 A/W; D^*: 8.0×10^{11} Jones; LDR: 110 dB; Time: 318 µs; Narrowband response
	2016 [87]	ITO/PVK/P3HT:PC$_{60}$BM:CdTe QDs (1:1, 3.5 µm)/BCP/Al	2.0×10^2@660 nm, −6 V	D^*: 7.3×10^{11} Jones; LDR: 110 dB; Narrowband response
Polymer/Insulator	2015 [78]	Au/Y-TiOPc@PC/Au	3.6×10^4@830 nm, 225 kV/cm	LDR: 7.1 dB@808 nm
	2016 [27]	ITO/Y-TiOPc NPs/m-TPD/Al	3.5×10^5@780 nm, 15 V/µm	R: 2227 A/W; D^*: 3.1×10^{14} Jones; Broadband response

By comparison with organic small molecules which can only be processed through film deposition techniques, it is concluded that the solution processable polymers are the best candidates for developing low cost organic PM photodetectors because the solution process technique allows introduction of carrier traps into the bulk heterojunction active layer through blending. We also note that among the diverse methods of introducing traps into the polymer bulk heterojunction active layer, adjusting the weight ratio of acceptor/donor is the simplest one and researches on this topic might achieve greet success in the future. In order to improve the multiplication rate, it was necessary to insert a composite layer comprising of a hole blocking layer and a hole accumulation layer between the bulk heterojunction active layer and the electrode. The selection of the inserted hole blocking layer needs to be carried out delicately. On one hand, its HOMO level should be sufficiently lower than that of the *P*-type material in bulk heterojunction for reducing dark current under dark; on the other hand, its LUMO level should not be higher than that of *N*-type material in bulk heterojunction, enabling fluent electron injection from the electrode to the active layer. In addition, organic PM devices with broadband response, which can be sensitive to the light over wavelength ranges of ultraviolet (UV), visible, and even IR, can be realized through assembly of two different active materials as well as incorporation of down conversion materials, organic dyes, and so on. Especially, for polymer PM photodetectors, Zhang et al. have constructed a series of ternary bulk heterojunctions through blending two donors which respond at different spectrum ranges together with the acceptor [79,80] In order to realize filterless narrowband organic PM photodetectors, active layers with thicknesses of several micrometers were utilized to adjust the carrier collection efficiency.

Beyond the exotic achievements listed in this review for organic PM photodetectors, there are still some problems to be solved. For example, the response speeds of most reported organic PM photodetectors are slow, which cannot meet the needs of high speed photodetectors. In most of the work, the stabilities of the organic PM devices are not discussed, which is crucial to solve before putting them into practical applications. Moreover, the introduction of metal micro/nano structures into organic PM photodetectors should be consolidated with more attention as a feasible approach of manipulating the carrier generation and distribution like in organic solar cells [99–111]. It is emphasized that, recently, organic-inorganic hybrid perovskite photodetectors have also been reported with extremely high quantum efficiency [112–115] of which the principles were explained based on the trapped carrier assisted carrier tunneling or ion migration effect. And compared with organic devices, perovskite PM photodetectors require lower bias and possess faster response but their stability is inferior. Furthermore, it's worth noting that, there are also plenty of original works on organic/inorganic hybrid heterojunction photodetectors with the inorganic materials of diverse structures including bulk films [116–118], 2D materials [119–121], nanomaterials [121,122], etc. A combination of organic materials with perovskite or inorganic semiconductors might offer a promising route to realize overall high performance PM photodetectors.

Author Contributions: Conceptualization and Supervision, Y.C.; Writing—Original Draft Preparation, L.S., Q.L.; Resources, W.W., Y.Z., G.L. and T.J.; Writing—Review & Editing, Y.C. and Y.H.; and Funding Acquisition, Y.C. and Y.H.

Funding: This research was funded by National Natural Science Foundation of China (61775156, 61475109, 61605136, and U1710115), Young Talents Program of Shanxi Province, Young Sanjin Scholars Program, the Natural Science Foundation of Shanxi Province (201701D211002, and 201601D021051), Henry Fok Education Foundation Young Teachers fund, Key Research and Development (International Cooperation) Program of Shanxi Province (201603D421042), and Platform and Base Special Project of Shanxi Province (201605D131038).

Conflicts of Interest: The authors declare no conflict of interest.

References

1. Rogalski, A.; Antoszewski, J.; Faraone, L. Third-generation infrared photodetector arrays. *J. Appl. Phys.* **2009**, *105*, 091101. [CrossRef]
2. Kim, S.; Lim, Y.T.; Soltesz, E.G.; De Grand, A.M.; Lee, J.; Nakayama, A.; Parker, J.A.; Mihaljevic, T.; Laurence, R.G.; Dor, D.M.; et al. Near-infrared fluorescent type II quantum dots for sentinel lymph node mapping. *Nat. Biotechnol.* **2003**, *22*, 93–97. [CrossRef] [PubMed]
3. Sukhovatkin, V.; Hinds, S.; Brzozowski, L.; Sargent, E.H. Colloidal Quantum-Dot Photodetectors Exploiting Multiexciton Generation. *Science* **2009**, *324*, 1542–1544. [CrossRef] [PubMed]
4. Gong, X.; Tong, M.; Xia, Y.; Cai, W.; Moon, J.S.; Cao, Y.; Yu, G.; Shieh, C.-L.; Nilsson, B.; Heeger, A.J. High-Detectivity Polymer Photodetectors with Spectral Response from 300 nm to 1450 nm. *Science* **2009**, *325*, 1665–1667. [CrossRef] [PubMed]
5. Li, W.-D.; Chou, S.Y. Solar-blind deep-UV band-pass filter (250–350 nm) consisting of a metal nano-grid fabricated by nanoimprint lithography. *Opt. Express* **2010**, *18*, 931–937. [CrossRef] [PubMed]
6. McDonald, S.A.; Konstantatos, G.; Zhang, S.; Cyr, P.W.; Klem, E.J.; Levina, L.; Sargent, E.H. Solution-processed PbS quantum dot infrared photodetectors and photovoltaics. *Nat. Mater.* **2005**, *4*, 138–142. [CrossRef] [PubMed]
7. Dou, L.; Yang, Y.M.; You, J.; Hong, Z.; Chang, W.H.; Li, G.; Yang, Y. Solution-processed hybrid perovskite photodetectors with high detectivity. *Nat. Commun.* **2014**, *5*, 5404. [CrossRef] [PubMed]
8. Guo, F.; Yang, B.; Yuan, Y.; Xiao, Z.; Dong, Q.; Bi, Y.; Huang, J. A nanocomposite ultraviolet photodetector based on interfacial trap-controlled charge injection. *Nat. Nanotechnol.* **2012**, *7*, 798–802. [CrossRef] [PubMed]
9. Konstantatos, G.; Sargent, E.H. Nanostructured materials for photon detection. *Nat. Nanotechnol.* **2010**, *5*, 391–400. [CrossRef] [PubMed]
10. Wang, X.; Cheng, Z.; Xu, K.; Tsang, H.K.; Xu, J.-B. High-responsivity graphene/silicon-heterostructure waveguide photodetectors. *Nat. Photonics* **2013**, *7*, 888–891. [CrossRef]
11. Koppens, F.H.; Mueller, T.; Avouris, P.; Ferrari, A.C.; Vitiello, M.S.; Polini, M. Photodetectors based on graphene, other two-dimensional materials and hybrid systems. *Nat. Nanotechnol.* **2014**, *9*, 780–793. [CrossRef] [PubMed]
12. Liu, Z.; Parvez, K.; Li, R.; Dong, R.; Feng, X.; Mullen, K. Transparent conductive electrodes from graphene/PEDOT:PSS hybrid inks for ultrathin organic photodetectors. *Adv. Mater.* **2015**, *27*, 669–675. [CrossRef] [PubMed]
13. Kang, Y.; Liu, H.-D.; Morse, M.; Paniccia, M.J.; Zadka, M.; Litski, S.; Sarid, G.; Pauchard, A.; Kuo, Y.-H.; Chen, H.-W.; et al. Monolithic germanium/silicon avalanche photodiodes with 340 GHz gain–bandwidth product. *Nat. Photonics* **2009**, *3*, 59–63. [CrossRef]
14. Fang, Z.; Wang, Y.; Liu, Z.; Schlather, A.; Ajayan, P.M.; Koppens, F.H.L.; Nordlander, P.; Halas, N.J. Plasmon-Induced Doping of Graphene. *ACS Nano* **2012**, *6*, 10222–10228. [CrossRef] [PubMed]
15. Ju, Y.; Song, J.; Geng, Z.; Zhang, H.; Wang, W.; Xie, L.; Yao, W.; Li, Z. A microfluidics cytometer for mice anemia detection. *Lab Chip* **2012**, *12*, 4355–4362. [CrossRef] [PubMed]
16. Kabir, M.Z.; Hijazi, N. Temperature and field dependent effective hole mobility and impact ionization at extremely high fields in amorphous selenium. *Appl. Phys. Lett.* **2014**, *104*, 192103. [CrossRef]
17. Ribordy, G.; Gautier, J.-D.; Zbinden, H.; Gisin, N. Performance of InGaAs/InP avalanche photodiodes as gated-mode photon counters. *Appl. Opt.* **1998**, *37*, 2272–2277. [CrossRef] [PubMed]
18. Hayden, O.; Agarwal, R.; Lieber, C.M. Nanoscale avalanche photodiodes for highly sensitive and spatially resolved photon detection. *Nat. Mater.* **2006**, *5*, 352–356. [CrossRef] [PubMed]
19. Cova, S.; Ghioni, M.; Lacaita, A.; Samori, C.; Zappa, F. Avalanche photodiodes and quenching circuits for single-photon detection. *Appl. Opt.* **1996**, *35*, 1956–1976. [CrossRef] [PubMed]
20. Pearsall, T.P.; Temkin, H.; Bean, J.C.; Luryi, S. Avalanche gain in GexSi1-x/Si infrared waveguide detectors. *IEEE Electron Device Lett.* **1986**, *7*, 330–332. [CrossRef]
21. Renker, D. Geiger-mode avalanche photodiodes, history, properties and problems. *Nucl. Instrum. Methods Phys. Res. Sect. A* **2006**, *567*, 48–56. [CrossRef]
22. Reznik, A.; Zhao, W.; Ohkawa, Y.; Tanioka, K.; Rowlands, J.A. Applications of avalanche multiplication in amorphous selenium to flat panel detectors for medical applications. *J. Mater. Sci. Mater. Electron.* **2007**, *20*, 63–67. [CrossRef]

23. Baierl, D.; Fabel, B.; Gabos, P.; Pancheri, L.; Lugli, P.; Scarpa, G. Solution-processable inverted organic photodetectors using oxygen plasma treatment. *Org. Electron.* **2010**, *11*, 1199–1206. [CrossRef]

24. Nalwa, K.S.; Cai, Y.; Thoeming, A.L.; Shinar, J.; Shinar, R.; Chaudhary, S. Polythiophene-fullerene based photodetectors: Tuning of spectral response and application in photoluminescence based (bio)chemical sensors. *Adv. Mater.* **2010**, *22*, 4157–4161. [CrossRef] [PubMed]

25. Leem, D.-S.; Lee, K.-H.; Park, K.-B.; Lim, S.-J.; Kim, K.-S.; Wan Jin, Y.; Lee, S. Low dark current small molecule organic photodetectors with selective response to green light. *Appl. Phys. Lett.* **2013**, *103*, 043305. [CrossRef]

26. Nakayama, K.-I.; Hiramoto, M.; Yokoyama, M. A high-speed photocurrent multiplication device based on an organic double-layered structure. *Appl. Phys. Lett.* **2000**, *76*, 1194–1196. [CrossRef]

27. Li, X.; Wang, S.; Xiao, Y.; Li, X. A trap-assisted ultrasensitive near-infrared organic photomultiple photodetector based on Y-type titanylphthalocyanine nanoparticles. *J. Mater. Chem. C* **2016**, *4*, 5584–5592. [CrossRef]

28. Yang, D.; Zhou, X.; Wang, Y.; Vadim, A.; Alshehri, S.M.; Ahamad, T.; Ma, D. Deep ultraviolet-to-NIR broad spectral response organic photodetectors with large gain. *J. Mater. Chem. C* **2016**, *4*, 2160–2164. [CrossRef]

29. Wang, X.; Li, H.; Su, Z.; Fang, F.; Zhang, G.; Wang, J.; Chu, B.; Fang, X.; Wei, Z.; Li, B.; et al. Efficient organic near-infrared photodetectors based on lead phthalocyanine/C60 heterojunction. *Org. Electron.* **2014**, *15*, 2367–2371. [CrossRef]

30. Yang, D.; Zhang, L.; Yang, S.Y.; Zou, B.S. Low-voltage pentacene photodetector based on a vertical transistor configuration. *Acta Phys. Sin.* **2015**, *64*, 108503. [CrossRef]

31. Rauch, T.; Böberl, M.; Tedde, S.F.; Fürst, J.; Kovalenko, M.V.; Hesser, G.; Lemmer, U.; Heiss, W.; Hayden, O. Near-infrared imaging with quantum-dot-sensitized organic photodiodes. *Nat. Photonics* **2009**, *3*, 332–336. [CrossRef]

32. Peumans, P.; Bulović, V.; Forrest, S.R. Efficient, high-bandwidth organic multilayer photodetectors. *Appl. Phys. Lett.* **2000**, *76*, 3855–3857. [CrossRef]

33. Wu, S.-H.; Li, W.-L.; Chu, B.; Su, Z.-S.; Zhang, F.; Lee, C.S. High performance small molecule photodetector with broad spectral response range from 200 to 900 nm. *Appl. Phys. Lett.* **2011**, *99*, 023305. [CrossRef]

34. Alvarado, S.F.; Seidler, P.F.; Lidzey, D.G.; Bradley, D.D.C. Direct Determination of the Exciton Binding Energy of Conjugated Polymers Using a Scanning Tunneling Microscope. *Phys. Rev. Lett.* **1998**, *81*, 1082–1085. [CrossRef]

35. Caserta, G.; Rispoli, B.; Serra, A. Space-Charge-Limited Current and Band Structure in Amorphous Organic Films. *Phys. Status Solidi* **1969**, *35*, 237–248. [CrossRef]

36. Scharber, M.C.; Mühlbacher, D.; Koppe, M.; Denk, P.; Waldauf, C.; Heeger, A.J.; Brabec, C.J. Design Rules for Donors in Bulk-Heterojunction Solar Cells—Towards 10% Energy-Conversion Efficiency. *Adv. Mater.* **2006**, *18*, 789–794. [CrossRef]

37. Xue, J.; Uchida, S.; Rand, B.P.; Forrest, S.R. 4.2% efficient organic photovoltaic cells with low series resistances. *Appl. Phys. Lett.* **2004**, *84*, 3013–3015. [CrossRef]

38. He, Z.; Zhong, C.; Su, S.; Xu, M.; Wu, H.; Cao, Y. Enhanced power-conversion efficiency in polymer solar cells using an inverted device structure. *Nat. Photonics* **2012**, *6*, 591–595. [CrossRef]

39. Jansen-van Vuuren, R.D.; Armin, A.; Pandey, A.K.; Burn, P.L.; Meredith, P. Organic Photodiodes: The Future of Full Color Detection and Image Sensing. *Adv. Mater.* **2016**, *28*, 4766–4802. [CrossRef] [PubMed]

40. Streetman, B.G.; Banerjee, S. *Solid-State Electronic Devices*, 6th ed.; Prentice-Hall: Upper Saddle River, NJ, USA, 2005; pp. 406–409. ISBN 978-81-203-3020-7.

41. Sze, S.M.; Ng, K.K. *Physics of Semiconductor Devices*, 3rd ed.; Wiley-Interscience: Hoboken, NJ, USA, 2006; pp. 666–697. ISBN 978-0-471-14323-9.

42. Ahmadi, M.; Wu, T.; Hu, B. A Review on Organic–Inorganic Halide Perovskite Photodetectors: Device Engineering and Fundamental Physics. *Adv. Mater.* **2017**, *29*, 1605242. [CrossRef] [PubMed]

43. Hiramoto, M.; Imahigashi, T.; Yokoyama, M. Photocurrent multiplication in organic pigment films. *Appl. Phys. Lett.* **1994**, *64*, 187–189. [CrossRef]

44. Hiramoto, M.; Nakayama, K.-I.; Katsume, T.; Yokoyama, M. Field-activated structural traps at organic pigment/metal interfaces causing photocurrent multiplication phenomena. *Appl. Phys. Lett.* **1998**, *73*, 2627–2629. [CrossRef]

45. Hiramoto, M.; Nakayama, K.; Sato, I.; Kumaoka, H.; Yokoyama, M. Photocurrent multiplication phenomena at organic/metal and organic/organic interfaces. *Thin Solid Films* **1998**, *331*, 71–75. [CrossRef]

46. Nakayama, K.-I.; Hiramoto, M.; Yokoyama, M. Photocurrent multiplication at organic/metal interface and surface morphology of organic films. *J. Appl. Phys.* **2000**, *87*, 3365–3369. [CrossRef]

47. Katsume, T.; Hiramoto, M.; Yokoyama, M. Photocurrent multiplication in naphthalene tetracarboxylic anhydride film at room temperature. *Appl. Phys. Lett.* **1996**, *69*, 3722–3724. [CrossRef]

48. Hiramoto, M.; Miki, A.; Yoshida, M.; Yokoyama, M. Photocurrent multiplication in organic single crystals. *Appl. Phys. Lett.* **2002**, *81*, 1500–1502. [CrossRef]

49. Huang, J.; Yang, Y. Origin of photomultiplication in C_{60} based devices. *Appl. Phys. Lett.* **2007**, *91*, 203505. [CrossRef]

50. Hiramoto, M.; Kawase, S.; Yokoyama, M. Photoinduced Hole Injection Multiplication in p-Type Quinacridone Pigment Films. *Jpn. J. Appl. Phys.* **1996**, *35*, L349–L351. [CrossRef]

51. Matsunobu, G.; Oishi, Y.; Yokoyama, M.; Hiramoto, M. High-speed multiplication-type photodetecting device using organic codeposited films. *Appl. Phys. Lett.* **2002**, *81*, 1321–1322. [CrossRef]

52. Hammond, W.T.; Xue, J. Organic heterojunction photodiodes exhibiting low voltage, imaging-speed photocurrent gain. *Appl. Phys. Lett.* **2010**, *97*, 073302. [CrossRef]

53. Hiramoto, M.; Fujino, K.; Yoshida, M.; Yokoyama, M. Influence of Oxygen and Water on Photocurrent Multiplication in Organic Semiconductor Films. *Jpn. J. Appl. Phys.* **2003**, *42*, 672–675. [CrossRef]

54. Hiramoto, M.; Suemori, K.; Yokoyama, M. Influence of Oxygen on Photocurrent Multiplication Phenomenon at Organic/Metal Interface. *Jpn. J. Appl. Phys.* **2003**, *42*, 2495–2497. [CrossRef]

55. Hiramoto, M.; Sato, I.; Nakayama, K.-I.; Yokoyama, M. Photocurrent multiplication at organic/metal interface and morphology of metal films. *Jpn. J. Appl. Phys.* **1998**, *37*, L1184–L1186. [CrossRef]

56. Guo, F.; Xiao, Z.; Huang, J. Fullerene Photodetectors with a Linear Dynamic Range of 90 dB Enabled by a Cross-Linkable Buffer Layer. *Adv. Opt. Mater.* **2013**, *1*, 289–294. [CrossRef]

57. Fang, Y.; Guo, F.; Xiao, Z.; Huang, J. Large Gain, Low Noise Nanocomposite Ultraviolet Photodetectors with a Linear Dynamic Range of 120 dB. *Adv. Opt. Mater.* **2014**, *2*, 348–353. [CrossRef]

58. Däubler, T.K.; Neher, D.; Rost, H.; Hörhold, H.H. Efficient bulk photogeneration of charge carriers and photoconductivity gain in arylamino-PPV polymer sandwich cells. *Phys. Rev. B Condens. Matter* **1999**, *59*, 1964–1972. [CrossRef]

59. Campbell, I.H.; Crone, B.K. Bulk photoconductive gain in poly(phenylene vinylene) based diodes. *J. Appl. Phys.* **2007**, *101*, 024502. [CrossRef]

60. Chen, F.-C.; Chien, S.-C.; Cious, G.-L. Highly sensitive, low-voltage, organic photomultiple photodetectors exhibiting broadband response. *Appl. Phys. Lett.* **2010**, *97*, 103301. [CrossRef]

61. Melancon, J.M.; Živanović, S.R. Broadband gain in poly(3-hexylthiophene):phenyl-C_{61}-butyric-acid-methyl-ester photodetectors enabled by a semicontinuous gold interlayer. *Appl. Phys. Lett.* **2014**, *105*, 163301. [CrossRef]

62. Li, L.; Zhang, F.; Wang, J.; An, Q.; Sun, Q.; Wang, W.; Zhang, J.; Teng, F. Achieving EQE of 16,700% in P3HT:PC_{71}BM based photodetectors by trap-assisted photomultiplication. *Sci. Rep.* **2015**, *5*, 9181. [CrossRef] [PubMed]

63. Li, L.; Zhang, F.; Wang, W.; Fang, Y.; Huang, J. Revealing the working mechanism of polymer photodetectors with ultra-high external quantum efficiency. *Phys. Chem. Chem. Phys.* **2015**, *17*, 30712–30720. [CrossRef] [PubMed]

64. Zhou, X.; Yang, D.; Ma, D.; Vadim, A.; Ahamad, T.; Alshehri, S.M. Ultrahigh Gain Polymer Photodetectors with Spectral Response from UV to Near-Infrared Using ZnO Nanoparticles as Anode Interfacial Layer. *Adv. Funct. Mater.* **2016**, *26*, 6619–6626. [CrossRef]

65. Chen, H.Y.; Lo, M.K.; Yang, G.; Monbouquette, H.G.; Yang, Y. Nanoparticle-assisted high photoconductive gain in composites of polymer and fullerene. *Nat. Nanotechnol.* **2008**, *3*, 543–547. [CrossRef] [PubMed]

66. Chuang, S.-T.; Chien, S.-C.; Chen, F.-C. Extended spectral response in organic photomultiple photodetectors using multiple near-infrared dopants. *Appl. Phys. Lett.* **2012**, *100*, 013309. [CrossRef]

67. Wang, T.; Hu, Y.; Deng, Z.; Wang, Y.; Lv, L.; Zhu, L.; Lou, Z.; Hou, Y.; Teng, F. High sensitivity, fast response and low operating voltage organic photodetectors by incorporating a water/alcohol soluble conjugated polymer anode buffer layer. *RSC Adv.* **2017**, *7*, 1743–1748. [CrossRef]

68. Wang, Y.; Zhu, L.; Hu, Y.; Deng, Z.; Lou, Z.; Hou, Y.; Teng, F. High sensitivity and fast response solution processed polymer photodetectors with polyethylenimine ethoxylated (PEIE) modified ITO electrode. *Opt. Express* **2017**, *25*, 7719–7729. [CrossRef] [PubMed]

69. Li, L.; Zhang, F.; Wang, W.; An, Q.; Wang, J.; Sun, Q.; Zhang, M. Trap-assisted photomultiplication polymer photodetectors obtaining an external quantum efficiency of 37,500%. *ACS Appl. Mater. Interfaces* **2015**, *7*, 5890–5897. [CrossRef] [PubMed]

70. Wang, W.; Zhang, F.; Li, L.; Gao, M.; Hu, B. Improved Performance of Photomultiplication Polymer Photodetectors by Adjustment of P3HT Molecular Arrangement. *ACS Appl. Mater. Interfaces* **2015**, *7*, 22660–22668. [CrossRef] [PubMed]

71. Han, Z.; Zhang, H.; Tian, Q.; Li, L.; Zhang, F. Solution-processed polymer photodetectors with trap-assisted photomultiplication. *Sci. China Phys. Mech.* **2015**, *58*, 1–5. [CrossRef]

72. Wang, W.; Zhang, F.; Bai, H.; Li, L.; Gao, M.; Zhang, M.; Zhan, X. Photomultiplication photodetectors with P3HT:fullerene-free material as the active layers exhibiting a broad response. *Nanoscale* **2016**, *8*, 5578–5586. [CrossRef] [PubMed]

73. Miao, J.; Zhang, F.; Lin, Y.; Wang, W.; Gao, M.; Li, L.; Zhang, J.; Zhan, X. Highly Sensitive Organic Photodetectors with Tunable Spectral Response under Bi-Directional Bias. *Adv. Opt. Mater.* **2016**, *4*, 1711–1717. [CrossRef]

74. Esopi, M.R.; Calcagno, M.; Yu, Q. Organic Ultraviolet Photodetectors Exhibiting Photomultiplication, Low Dark Current, and High Stability. *Adv. Mater. Technol.* **2017**, *2*, 1700025. [CrossRef]

75. Nie, R.; Deng, X.; Feng, L.; Hu, G.; Wang, Y.; Yu, G.; Xu, J. Highly Sensitive and Broadband Organic Photodetectors with Fast Speed Gain and Large Linear Dynamic Range at Low Forward Bias. *Small* **2017**, *13*. [CrossRef] [PubMed]

76. Shen, L.; Fang, Y.; Dong, Q.; Xiao, Z.; Huang, J. Improving the sensitivity of a near-infrared nanocomposite photodetector by enhancing trap induced hole injection. *Appl. Phys. Lett.* **2015**, *106*, 023301. [CrossRef]

77. Dong, R.; Bi, C.; Dong, Q.; Guo, F.; Yuan, Y.; Fang, Y.; Xiao, Z.; Huang, J. An Ultraviolet-to-NIR Broad Spectral Nanocomposite Photodetector with Gain. *Adv. Opt. Mater.* **2014**, *2*, 549–554. [CrossRef]

78. Peng, W.; Liu, Y.; Wang, C.; Hu, R.; Zhang, J.; Xu, D.; Wang, Y. A highly sensitive near-infrared organic photodetector based on oxotitanium phthalocyanine nanocrystals and light-induced enhancement of electron tunnelling. *J. Mater. Chem. C* **2015**, *3*, 5073–5077. [CrossRef]

79. Wang, W.; Zhang, F.; Li, L.; Zhang, M.; An, Q.; Wang, J.; Sun, Q. Highly sensitive polymer photodetectors with a broad spectral response range from UV light to the near infrared region. *J. Mater. Chem. C* **2015**, *3*, 7386–7393. [CrossRef]

80. Gao, M.; Wenbin, W.; Li, L.; Miao, J.; Zhang, F. Highly sensitive polymer photodetectors with a wide spectral response range. *Chin. Phys. B* **2017**, *26*, 530–536. [CrossRef]

81. Ameri, T.; Khoram, P.; Min, J.; Brabec, C.J. Organic ternary solar cells: A review. *Adv. Mater.* **2013**, *25*, 4245–4266. [CrossRef] [PubMed]

82. An, Q.; Zhang, F.; Gao, W.; Sun, Q.; Zhang, M.; Yang, C.; Zhang, J. High-efficiency and air stable fullerene-free ternary organic solar cells. *Nano Energy* **2018**, *45*, 177–183. [CrossRef]

83. An, Q.; Zhang, F.; Li, L.; Wang, J.; Zhang, J.; Zhou, L.; Tang, W. Improved efficiency of bulk heterojunction polymer solar cells by doping low-bandgap small molecules. *ACS Appl. Mater. Interfaces* **2014**, *6*, 6537–6544. [CrossRef] [PubMed]

84. An, Q.; Zhang, F.; Sun, Q.; Zhang, M.; Zhang, J.; Tang, W.; Yin, X.; Deng, Z. Efficient organic ternary solar cells with the third component as energy acceptor. *Nano Energy* **2016**, *26*, 180–191. [CrossRef]

85. Kokil, A.; Poe, A.M.; Bae, Y.; Della Pelle, A.M.; Homnick, P.J.; Lahti, P.M.; Kumar, J.; Thayumanavan, S. Improved performances in polymer BHJ solar cells through frontier orbital tuning of small molecule additives in ternary blends. *ACS Appl. Mater. Interfaces* **2014**, *6*, 9920–9924. [CrossRef] [PubMed]

86. Dandin, M.; Abshire, P.; Smela, E. Optical filtering technologies for integrated fluorescence sensors. *Lab Chip* **2007**, *7*, 955–977. [CrossRef] [PubMed]

87. Olbright, G.R.; Peyghambarian, N.; Gibbs, H.M.; Macleod, H.A.; Van Milligen, F. Microsecond room-temperature optical bistability and crosstalk studies in ZnS and ZnSe interference filters with visible light and milliwatt powers. *Appl. Phys. Lett.* **1984**, *45*, 1031–1033. [CrossRef]

88. Armin, A.; Jansen-van Vuuren, R.D.; Kopidakis, N.; Burn, P.L.; Meredith, P. Narrowband light detection via internal quantum efficiency manipulation of organic photodiodes. *Nat. Commun.* **2015**, *6*, 6343. [CrossRef] [PubMed]

89. Li, Z.; Butun, S.; Aydin, K. Large-Area, Lithography-Free Super Absorbers and Color Filters at Visible Frequencies Using Ultrathin Metallic Films. *ACS Photonics* **2015**, *2*, 183–188. [CrossRef]

90. Xu, T.; Wu, Y.K.; Luo, X.; Guo, L.J. Plasmonic nanoresonators for high-resolution colour filtering and spectral imaging. *Nat. Commun.* **2010**, *1*, 59. [CrossRef] [PubMed]

91. Wang, W.; Zhang, F.; Du, M.; Li, L.; Zhang, M.; Wang, K.; Wang, Y.; Hu, B.; Fang, Y.; Huang, J. Highly Narrowband Photomultiplication Type Organic Photodetectors. *Nano Lett.* **2017**, *17*, 1995–2002. [CrossRef] [PubMed]

92. Miao, J.; Zhang, F.; Du, M.; Wang, W.; Fang, Y. Photomultiplication type narrowband organic photodetectors working at forward and reverse bias. *Phys. Chem. Chem. Phys.* **2017**, *19*, 14424–14430. [CrossRef] [PubMed]

93. Miao, J.; Zhang, F.; Du, M.; Wang, W.; Fang, Y. Photomultiplication Type Organic Photodetectors with Broadband and Narrowband Response Ability. *Adv. Opt. Mater.* **2018**, *6*, 1800001. [CrossRef]

94. Shen, L.; Fang, Y.; Wei, H.; Yuan, Y.; Huang, J. A Highly Sensitive Narrowband Nanocomposite Photodetector with Gain. *Adv. Mater.* **2016**, *28*, 2043–2048. [CrossRef] [PubMed]

95. Shen, L.; Zhang, Y.; Bai, Y.; Zheng, X.; Wang, Q.; Huang, J. A filterless, visible-blind, narrow-band, and near-infrared photodetector with a gain. *Nanoscale* **2016**, *8*, 12990–12997. [CrossRef] [PubMed]

96. Reynaert, J.; Arkhipov, V.I.; Heremans, P.; Poortmans, J. Photomultiplication in Disordered Unipolar Organic Materials. *Adv. Funct. Mater.* **2006**, *16*, 784–790. [CrossRef]

97. Campbell, I.H.; Crone, B.K. A near infrared organic photodiode with gain at low bias voltage. *Appl. Phys. Lett.* **2009**, *95*, 263302. [CrossRef]

98. Qi, D.; Fischbein, M.; Drndić, M.; Šelmić, S. Efficient polymer-nanocrystal quantum-dot photodetectors. *Appl. Phys. Lett.* **2005**, *86*, 093103. [CrossRef]

99. Cui, Y.; Fung, K.H.; Xu, J.; Ma, H.; Jin, Y.; He, S.; Fang, N.X. Ultrabroadband light absorption by a sawtooth anisotropic metamaterial slab. *Nano Lett.* **2012**, *12*, 1443–1447. [CrossRef] [PubMed]

100. Cui, Y.; He, Y.; Jin, Y.; Ding, F.; Yang, L.; Ye, Y.; Zhong, S.; Lin, Y.; He, S. Plasmonic and metamaterial structures as electromagnetic absorbers. *Laser Photonics Rev.* **2014**, *8*, 495–520. [CrossRef]

101. Wang, W.; Hao, Y.; Cui, Y.; Tian, X.; Zhang, Y.; Wang, H.; Shi, F.; Wei, B.; Huang, W. High-efficiency, broad-band and wide-angle optical absorption in ultra-thin organic photovoltaic devices. *Opt. Express* **2014**, *22* (Suppl. 2), A376–A385. [CrossRef] [PubMed]

102. Cui, Y.; Zhao, H.; Yang, F.; Tong, P.; Hao, Y.; Sun, Q.; Shi, F.; Zhan, Q.; Wang, H.; Zhu, F. Efficiency enhancement in organic solar cells by incorporating silica-coated gold nanorods at the buffer/active interface. *J. Mater. Chem. C* **2015**, *3*, 9859–9868. [CrossRef]

103. Hao, Y.; Song, J.; Yang, F.; Hao, Y.; Sun, Q.; Guo, J.; Cui, Y.; Wang, H.; Zhu, F. Improved performance of organic solar cells by incorporating silica-coated silver nanoparticles in the buffer layer. *J. Mater. Chem. C* **2015**, *3*, 1082–1090. [CrossRef]

104. Wang, Z.; Hao, Y.; Wang, W.; Cui, Y.; Sun, Q.; Ji, T.; Li, Z.; Wang, H.; Zhu, F. Incorporating silver-SiO_2 core-shell nanocubes for simultaneous broadband absorption and charge collection enhancements in organic solar cells. *Synth. Met.* **2016**, *220*, 612–620. [CrossRef]

105. Ji, T.; Wang, Y.; Cui, Y.; Lin, Y.; Hao, Y.; Li, D. Flexible broadband plasmonic absorber on moth-eye substrate. *Mater. Today Energy* **2017**, *5*, 181–186. [CrossRef]

106. Liu, D.; Liang, Q.; Li, G.; Gao, X.; Wang, W.; Zhan, Q.; Ji, T.; Hao, Y.; Cui, Y. Improved efficiency of organic photovoltaic cells by incorporation of auag-alloyed nanoprisms. *IEEE J. Photovolt.* **2017**, *PP*, 1–6. [CrossRef]

107. Wang, W.; Cui, Y.; Fung, K.H.; Zhang, Y.; Ji, T.; Hao, Y. Comparison of Nanohole-Type and Nanopillar-Type Patterned Metallic Electrodes Incorporated in Organic Solar Cells. *Nanoscale Res. Lett.* **2017**, *12*, 538. [CrossRef] [PubMed]

108. Peter Amalathas, A.; Alkaisi, M.M. Efficient light trapping nanopyramid structures for solar cells patterned using UV nanoimprint lithography. *Mater. Sci. Semicond. Process.* **2017**, *57*, 54–58. [CrossRef]

109. Zhang, C.; Song, Y.; Wang, M.; Yin, M.; Zhu, X.; Tian, L.; Wang, H.; Chen, X.; Fan, Z.; Lu, L.; et al. Efficient and Flexible Thin Film Amorphous Silicon Solar Cells on Nanotextured Polymer Substrate Using Sol-gel Based Nanoimprinting Method. *Adv. Funct. Mater.* **2017**, *27*, 1604720. [CrossRef]

110. Lin, Y.; Xu, Z.; Yu, D.; Lu, L.; Yin, M.; Tavakoli, M.M.; Chen, X.; Hao, Y.; Fan, Z.; Cui, Y.; et al. Dual-Layer Nanostructured Flexible Thin-Film Amorphous Silicon Solar Cells with Enhanced Light Harvesting and Photoelectric Conversion Efficiency. *ACS Appl. Mater. Interfaces* **2016**, *8*, 10929–10936. [CrossRef] [PubMed]

111. Zhang, Y.; Cui, Y.; Wang, W.; Fung, K.H.; Ji, T.; Hao, Y.; Zhu, F. Absorption Enhancement in Organic Solar Cells with a Built-In Short-Pitch Plasmonic Grating. *Plasmonics* **2014**, *10*, 773–781. [CrossRef]

112. Rui, D.; Yanjun, F.; Jungseok, C.; Jun, D.; Zhengguo, X.; Qingfeng, D.; Yongbo, Y.; Andrea, C.; Cheng, Z.X.; Jinsong, H. High-Gain and Low-Driving-Voltage Photodetectors Based on Organolead Triiodide Perovskites. *Adv. Mater.* **2015**, *27*, 1912–1918.

113. Liu, C.; Peng, H.; Wang, K.; Wei, C.; Wang, Z.; Gong, X. PbS quantum dots-induced trap-assisted charge injection in perovskite photodetectors. *Nano Energy* **2016**, *30*, 27–35. [CrossRef]

114. Chen, H.-W.; Sakai, N.; Jena, A.K.; Sanehira, Y.; Ikegami, M.; Ho, K.-C.; Miyasaka, T. A Switchable High-Sensitivity Photodetecting and Photovoltaic Device with Perovskite Absorber. *J. Phys. Chem. Lett.* **2015**, *6*, 1773–1779. [CrossRef] [PubMed]

115. Konrad, D.; Wolfgang, T.; Thomas, M.; Michael, S.; Khaja, N.M.; Michael, G. Working Principles of Perovskite Photodetectors: Analyzing the Interplay Between Photoconductivity and Voltage-Driven Energy-Level Alignment. *Adv. Funct. Mater.* **2015**, *25*, 6936–6947.

116. Goh, C.; Scully, S.R.; McGehee, M.D. Effects of molecular interface modification in hybrid organic-inorganic photovoltaic cells. *J. Appl. Phys.* **2007**, *101*, 114503. [CrossRef]

117. Levell, J.W.; Giardini, M.E.; Samuel, I.D.W. A hybrid organic semiconductor/silicon photodiode for efficient ultraviolet photodetection. *Opt. Express* **2010**, *18*, 3219–3225. [CrossRef] [PubMed]

118. Yakuphanoglu, F. Photovoltaic properties of hybrid organic/inorganic semiconductor photodiode. *Synth. Met.* **2007**, *157*, 859–862. [CrossRef]

119. Jariwala, D.; Marks, T.J.; Hersam, M.C. Mixed-dimensional van der Waals heterostructures. *Nat. Mater.* **2017**, *16*, 170–181. [CrossRef] [PubMed]

120. Jariwala, D.; Howell, S.L.; Chen, K.S.; Kang, J.; Sangwan, V.K.; Filippone, S.A.; Turrisi, R.; Marks, T.J.; Lauhon, L.J.; Hersam, M.C. Hybrid, Gate-Tunable, van der Waals p-n Heterojunctions from Pentacene and MoS2. *Nano Lett.* **2016**, *16*, 497–503. [CrossRef] [PubMed]

121. Jariwala, D.; Sangwan, V.K.; Wu, C.-C.; Prabhumirashi, P.L.; Geier, M.L.; Marks, T.J.; Lauhon, L.J.; Hersam, M.C. Gate-tunable carbon nanotube–MoS^2 heterojunction p-n diode. *Proc. Natl. Acad. Sci. USA* **2013**. [CrossRef] [PubMed]

122. Wang, J.-J.; Wang, Y.-Q.; Cao, F.-F.; Guo, Y.-G.; Wan, L.-J. Synthesis of Monodispersed Wurtzite Structure CuInSe2 Nanocrystals and Their Application in High-Performance Organic–Inorganic Hybrid Photodetectors. *J. Am. Chem. Soc.* **2010**, *132*, 12218–12221. [CrossRef] [PubMed]

nanomaterials

Article

High-Efficiency Visible Transmitting Polarizations Devices Based on the GaN Metasurface

Zhongyi Guo *, Haisheng Xu, Kai Guo, Fei Shen, Hongping Zhou, Qingfeng Zhou, Jun Gao and Zhiping Yin *

School of Computer and Information, Hefei University of Technology, Hefei 230009, China; xuhaisheng@mail.hfut.edu.cn (H.X.); kai.guo@hfut.edu.cn (K.G.); shenfei@hfut.edu.cn (F.S.); ciangela@hfut.edu.cn (H.Z.); enqfzhou@hfut.edu.cn (Q.Z.); gaojun@hfut.edu.cn (J.G.)
* Correspondence: guozhongyi@hfut.edu.cn (Z.G.); zpyin@hfut.edu.cn (Z.Y.);
   Tel.: +86-186-5515-1981 (Z.G.); +86-177-5608-163 (Z.Y.)

Received: 29 April 2018; Accepted: 10 May 2018; Published: 15 May 2018

Abstract: Metasurfaces are capable of tailoring the amplitude, phase, and polarization of incident light to design various polarization devices. Here, we propose a metasurface based on the novel dielectric material gallium nitride (GaN) to realize high-efficiency modulation for both of the orthogonal linear polarizations simultaneously in the visible range. Both modulated transmitted phases of the orthogonal linear polarizations can almost span the whole 2π range by tailoring geometric sizes of the GaN nanobricks, while maintaining high values of transmission (almost all over 90%). At the wavelength of 530 nm, we designed and realized the beam splitter and the focusing lenses successfully. To further prove that our proposed method is suitable for arbitrary orthogonal linear polarization, we also designed a three-dimensional (3D) metalens that can simultaneously focus the X-, Y-, 45°, and 135° linear polarizations on spatially symmetric positions, which can be applied to the linear polarization measurement. Our work provides a possible method to achieve high-efficiency multifunctional optical devices in visible light by extending the modulating dimensions.

Keywords: metasurfaces; orthogonal polarization; high-efficiency; polarization analyzer

1. Introduction

Optical metasurfaces, a two-dimensional planar variation of the concept of metamaterials, have been engineered to realize exotic electromagnetic properties by control of the phase, amplitude, and polarization of incident light, which are capable of being manipulated in a desirable manner [1,2]. A metasurface is usually composed of nano-antennas or nano-apertures because of introduced arbitrary abrupt phase shifts by adjusting geometric sizes [3–5]. Compared to the conventional optical devices, devices made from metasurfaces have the advantage of ultrathin, highly integrated, versatile, low-cost characteristics [6–8]. With the development of nanotechnology, many structures have been realized to demonstrate the broad applications of metasurfaces, including metalens [9–11], wave plates [12], filters [13], absorbers [14], vortex beam generation [15], and holograms [16,17].

In the visible light region, much of the previous work was based on metal nanostructures, and most of these had been proved to be low transmission as a result of Joule losses [18,19]. For semiconductor-based dielectric metasurfaces, there is no inherent loss for similar metal nanostructures; therefore these have demonstrated superior performance for providing a high transmission in many applications of metasurface devices to date [20–23]. At the beginning, a dielectric material was combined with a metal back reflector to work as the reflective metasurface, and most of these were designed on the basis of the Pancharatnam–Berry (PB) phase principle for the incidences of circularly polarized light [24]. However, there are some limitations in optical system integration for practical applications of the reflective metasurfaces, which can be overcome by transmitted dielectric metasurfaces made of high-index dielectrics, such as silicon (Si) [25–28] and titanium oxide (TiO$_2$) [29,30]. Si has evident absorption in the visible band, and the preparation process of high-cost TiO$_2$ also degrades its potential application [31]. Meanwhile, the PB phase approach is inherently limited to circularly polarized light and is unsuitable for linear polarization or for polarization-independent metasurfaces. Thus, it is important to design low-cost, high-efficiency devices for linear polarization incidences in the visible range.

In this paper, we propose a high-efficiency dielectric metasurface based on gallium nitride (GaN) nanobricks (refractive index of 2.4) for manipulating orthogonal linear polarizations simultaneously at the wavelength of 530 nm. GaN was chosen because there is no intrinsic absorption throughout the visible spectrum (GaN's band gap is about 3.4 eV) [32,33]. The polarization beam splitter (PBS) and metalens were designed successfully by tailoring the geometry of the GaN nanobricks to meet the corresponding required phases. Meanwhile, we also designed a simple polarization analyzer for analyzing the degree of linear polarization (DoLP) of the incidences, which can simultaneously focus the X-, Y-, 45°, and 135° linear polarizations in spatially symmetric positions. Our results would lead to wide applications in photonic research for visible light.

2. Design and Analysis

As is schematically shown in Figure 1, the designed GaN nanobrick, with a length, width, and height of l, w, and h, respectively, is adhered onto an Al$_2$O$_3$ substrate with a depth of d. In our designs, the phase accumulation is achieved through the waveguiding effect, which is proportional to the height (h) of the designed nanobricks; thus the aspect ratio of the designed nanobricks should be taken into consideration. For incident light with a wavelength of 530 nm, the height of the GaN nanobrick is set as h = 800 nm for obtaining the entire 2π modulating phase for the transmitting light, which was optimized and confirmed by the finite-difference time-domain (FDTD) method. For simplifying the calculation, we set the depth of the Al$_2$O$_3$ substrate as d = 300 nm. The concrete modulating phases of the transmitted linear-polarized light could be realized by changing the size of the GaN nanobricks. A square lattice with a lattice constant of p = 260 nm was chosen for the optimized and designed simulations; this was obtained from the following two standards: (i) the lattice constant must be less than half of the main working wavelength to avoid the diffraction effect; (ii) the lattice constant should be large enough for the near-field strong interactions between two neighboring nanobricks to be avoided [34]. The length and width of rectangular nanobricks affects the transmission amplitudes and phases of X- and Y-polarization, respectively.

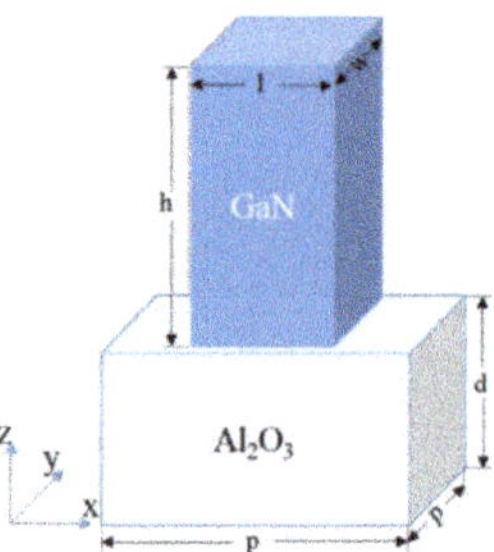

Figure 1. Schematic of the designed unit cell: p = 260 nm, d = 300 nm, and h = 800 nm.

Here, on the basis of double-phase modulation, we introduce the design of the metasurfaces with the GaN nanobricks, which can expand the modulating dimensions of linear-polarized light. In the numerical simulations, periodic boundary conditions were used in the X- and Y-directions to ensure the accuracy of the phase spectrum in varying structural dimensions. We used the transmission phase of co-polarization light, because of its rectangular structure and the small polarization conversion for the nonrotating periodic nanobricks. As shown in Figure 2a,b, with a normal incident wavelength of 530 nm for X-polarized light, the transmittance and the modulating phase of the transmitted X-polarized light could be expressed as functions of the length (l) and width (w) from 30 to 230 nm of the rectangular nanobricks, respectively. It is clear that the modulating phase of the transmitted light could almost cover the entire 2π completely, while the transmittance was close to unity. Similarly, Figure 2c,d represents the transmittance and modulating phases of transmitted Y-polarized light, respectively. It can be seen from Figure 2 that for any desired modulating phases of the transmitted X- and Y-polarized light, there always existed one rectangular GaN nanobrick that could simultaneously satisfy the needs of two orthogonally polarized modulating phases but keep the transmittance close to unity. Therefore, the method used can obtain different phase responses for orthogonal polarization states simultaneously by changing the length and width of the rectangular nanobricks, and the local modulating phases for a couple of incident orthogonal polarization states can be manipulated independently. This is the method with which we can increase the modulating dimension, which can be called the principle of double-phase modulation.

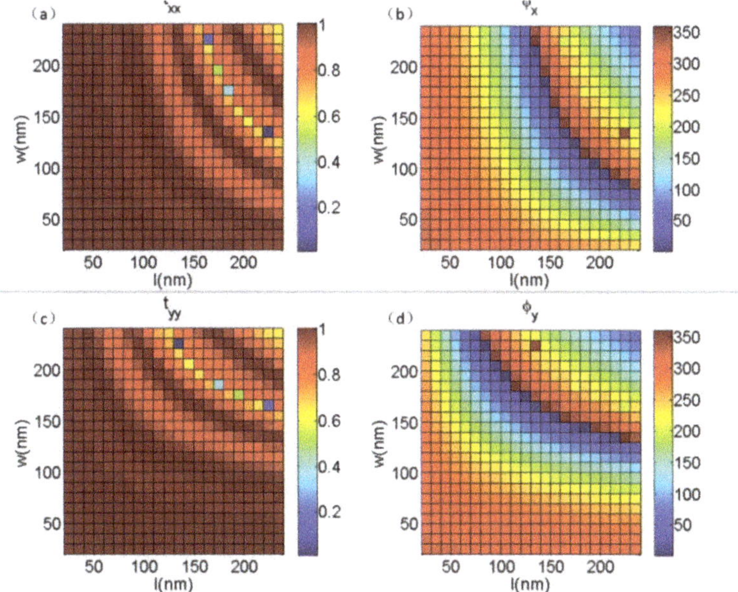

Figure 2. Transmitted light normalization: (**a**) Transmittance variation for gallium nitride (GaN) nanobricks on Al_2O_3 substrate, and (**b**) phase as a function of l and w for normal incidence of the X-polarized light. (**c**,**d**) The normalized transmittance and phase as a function of l and w for normal incidence of the Y-polarized light, respectively.

To prove that our proposed method can be applied to optical wavefront shaping, we first designed a polarization beam splitter (PBS) to deflect the X- and Y-polarized light into different directions efficiently. In order to split the orthogonal polarized light, for simplicity, we designed the device with a couple of opposite gradient phases for X- and Y-polarized light, which meant X- and Y-polarized light would be reflected into two opposite directions. As shown in Figure 2, both of the 2π transmitted modulating phases could be obtained by changing the dimensions of l and w under the X- and Y-polarized incidences, which meant that any combination of the modulating phases for the X- and

Y-polarized incidences could be realized by a GaN nanobrick with a certain length and width. On the basis of the generalized Snell's law [35]:

$$n_t \sin \theta_t - n_i \sin \theta_i = \frac{\lambda_0}{2\pi} \frac{d\varphi}{dx} \tag{1}$$

reasonable choices of parameters can control the angle of refraction, where θ_i and θ_t are the incident angle and refractive angle, respectively; n_i and n_t are the refractive indices in the incident and refracted region, respectively; λ_0 is the incident wavelength in vacuum; and $d\varphi/dx$ is the phase gradient. It is worth mentioning here that $d\varphi$ and dx are the phase difference and distance difference between two adjacent unit cells. We can utilize the last term in Equation (1) to obtain the desired angle of refraction if the wavelength of the incident light is determined. As shown in Figure 3a, eight unit cells were selected to construct a combination of the modulating opposite phases for the *X*- and *Y*-polarized incidences, and the phase differences were $\pi/4$ and $-\pi/4$ between two adjacent unit cells for covering the entire 2π, which could be determined from Figure 2. Meanwhile, the transmittance of each unit cell was close to unity to ensure that the designed device was as efficient as possible. The inset in Figure 3a shows a supercell composed of eight unit cells (GaN nanobricks). Theoretically, 45° linear-polarized light can be decomposed into *X*- and *Y*-polarized components. Therefore, if the 45° linear-polarized light is incident to the designed beam splitter, the *X*- and *Y*-polarized components will be deflected into two different directions. As shown Figure 3b,c, the concrete simulated electric fields demonstrate that the transmitted *X*- and *Y*-polarized light was refracted into two opposite directions because of the designed opposite-phase modulation difference for the transmitted *X*- and *Y*-polarized light. It should be noted that the reflected field intensity was almost zero, as depicted in Figure 3b,c, which ensured the high-efficiency transmitted characteristics of the designed GaN metasurface.

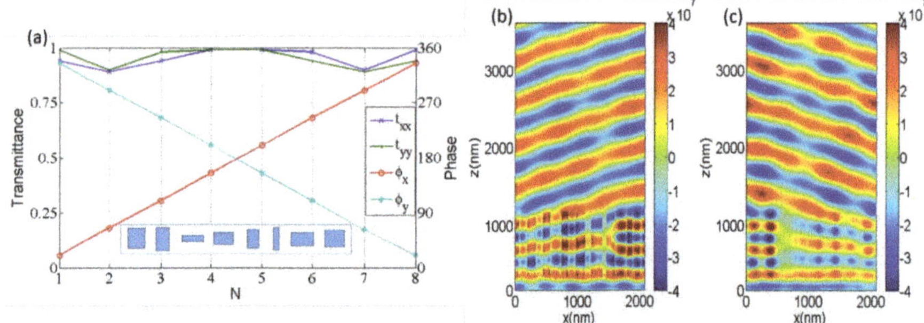

Figure 3. (**a**) The normalized transmittance and the phase of transmitted light through the eight unit cells for the *X*- and *Y*-polarized light incidences at wavelength of 530 nm. The inset in (**a**) is a supercell composed of eight unit cells. (**b**,**c**) The electric field distributions for *X*- and *Y*-polarized light, respectively; it can be clearly seen that *X*- and *Y*-polarized light is refracted into two different directions.

After calculating the results from Figure 3b,c, the refraction angle was $\pm14.7°$ for the *X*- and *Y*-polarized light, which agreed well with the theoretical value of $\pm14.76°$ from Equation (1). The transmitted efficiencies of the *X*- and *Y*-polarized light were 85% and 89.5%, respectively, when the light was incident to the designed device normally, where the deflection efficiency is defined as the ratio of transmitted power to incident power. As shown in Figure 4, when the 45° polarized light was incident to the designed device normally, the deflection efficiencies of the *X*- and *Y*-polarized light were 42.5% and 44.7%, respectively. There were some slight differences in the deflection efficiencies of the *X*- and *Y*-polarized components, which could be attributed to the non-symmetrical characteristics of the nanobrick along the 45° direction.

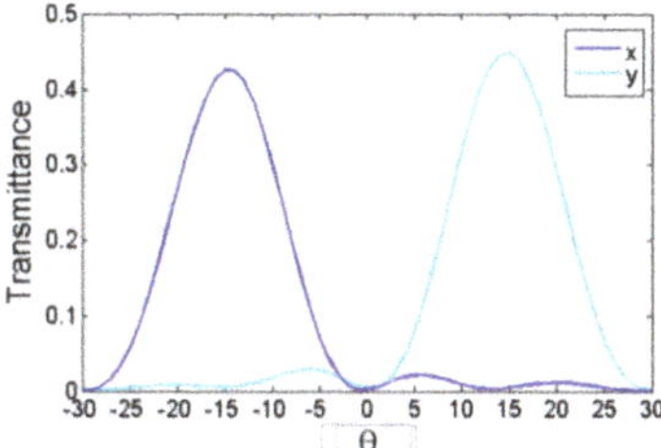

Figure 4. The transmittance of *X*- and *Y*-polarized light as a function of deflection angle (θ) under 45° polarized light incident on the bottom.

In order to demonstrate the superiority of our designs, on the basis of the principle of double-phase modulation, a difunctional metalens was also designed at the wavelength of 530 nm. Similarly to the PBS, the designed difunctional metalens can focus *X*- and *Y*-polarized light into two different places. The required phase for the focusing lens can be expressed as follows:

$$\varphi(x,y) = \frac{2\pi}{\lambda}\left(\sqrt{(x \pm x_0)^2 + (y \pm y_0)^2 + f^2} - f\right) \tag{2}$$

where λ is the incident wavelength in vacuum, x_0 is the horizontal distance between the focus and the center of the metalens, and f is the focal length. Here, we set $y = 0$, $y_0 = 0$, $x_0 = 8p = 2080$ nm, and the $f = 9\lambda = 4770$ nm. We combined φ_x and φ_y to select 56 nanobricks for the metalens in Figure 2; the locations of the focuses of the *X*- and *Y*-polarized light were $(-x_0, 0, f)$ and $(x_0, 0, f)$. With the normal incidences at the wavelength of 530 nm, Figure 5a–e demonstrates the distributions of the different transmitted intensities of the designed difunctional metalens under 0° (*X*-polarization), 30°, 42°, 45°, 60°, and 90° (*Y*-polarization) linear-polarized states of incident light, respectively. We also extracted the transmitted intensity distributions in the *X–Y* plane at the focal planes under the incidences of different linear polarizations, as shown in the lower half of each figure in Figure 5. As shown in Figure 5a,f, when 0° (*X*-polarization) and 90° (*Y*-polarization) linear-polarized light was incident to the designed metalens, there was just one focusing spot each at the positions of $(-2080$ nm, 0, 6470 nm) and $(2080$ nm, 0, 6470 nm), respectively, in the transmitted fields (the ordinates of the focuses (in Z-direction) were larger than the setup of 4770 nm because of the height of the metasurface and an additional 600 nm air layer). Because arbitrary linear polarizations can be decomposed into two orthogonal linear polarizations, thus we could approximately analyze the incident linear polarization state on the basis of the focusing intensity. For example, for the 30° linear polarization incidence, compared to the *Y*-polarized component, the *X*-polarized component was dominant; thus the focusing energy of *X*-polarized light was stronger than that of *Y*-polarized light, as shown in Figure 5b. Similarly, for the 60° linear polarization incidence, the focusing intensity of *X*-polarized light was weaker than that of *Y*-polarized light, as depicted in Figure 5e. Meanwhile, for the incidence of 45° linear polarization, as shown in Figure 5d, the focusing intensities of *X*- and *Y*-polarized components in the transmitted fields should be the same, but there was a small mismatch in the two focuses' energies, which could be attributed to the non-symmetrical characteristics of the nanobrick along the 45° direction and which may have led to a very small intensity deviation for the intensity responses of *X*- and *Y*-polarized light. After calculation, the numerical aperture (NA) of the metalens designed was 0.836 on the basis of NA $= \sin[\tan^{-1}(D/2f)]$, where D and f are the size and focal length of the lens. The NA can be made larger by increasing the size of the lens when the focal length is fixed, and the focus effect will be better.

To further confirm the versatility of the design method for analyzing arbitrary linear polarizations, a three-dimensional (3D) metalens was designed for analyzing *X*-, *Y*-, 45°, and 135° linear-polarized light simultaneously. On the basis of Equation (2), here, we set $x_0 = 8p = 2080$ nm, $y_0 = 6p = 1560$ nm, and $f = 5\lambda = 2650$ nm, and we set the focusing points as $(-x_0, -y_0)$, $(x_0, -y_0)$, $(-x_0, y_0)$, and (x_0, y_0) for the X-, Y-, 45°, and 135° linear-polarized light, respectively. The GaN nanobrick array of the designed 3D metalens is schematically shown in Figure 6, in which 768 GaN nanobricks were selected for satisfying the phase requirements on the basis of the phase distributions in Figure 2. The size of the designed metalens is 8.32 μm × 6.24 μm; there are two parts of the GaN nanobrick array (the lower part is used to focus X- and Y-linear-polarized light to the locations of $(-x_0, -y_0)$ and $(x_0, -y_0)$, respectively; the upper part is obtained by rotating the GaN nanobricks in the lower part to 45° and −45° accordingly for manipulating and focusing the 45° and 135° linear-polarized light to the locations of $(-x_0, y_0)$ and (x_0, y_0) respectively. The transmission intensity distributions in the focal plane are demonstrated in Figure 7a–f under the incidences of the X-, Y-, 45°, 135°, 30°, and 60° linear-polarized light, respectively. We can observe that the focuses were oval, which could be attributed firstly to the horizontal and vertical axes of the scale not being the same for the focus, as well as that the longitudinal structural elements provided a smaller phase distribution than that in the lateral direction. As shown in Figure 7a, under the incidence of X-polarized light, the intensity at the position of $(x_0, -y_0)$ for the Y-polarization focus was nearly zero, as there was no Y-polarized component in the incidence, but the focusing intensities for the 45° and 135° linear-polarized components were nonzero as a result of the decomposed components from the X-polarized incidence. This was similar for arbitrary linear-polarized incidences. For example, in Figure 7f, under the incidence of 60° linear-polarized light, there were four evident focusing points because of the nonzero decompositions of the X-, Y-, 45°, and 135° linear-polarized components.

Figure 8a,b illustrates the normalized transmitted intensity distribution in the focal plane $(y = -6p, x = -8p)$ in case of the 45° linear-polarized light incidence. As shown in Figure 8a, the intensity was almost undifferentiated, which agreed well with Figure 7c. The X-polarized focusing intensity was half that of the 45° linear-polarized light, as depicted in Figure 8b, because the 45° linear-polarized light could be decomposed into X- and Y-polarized components. The difference in the full width at half maximum between the X-polarized focus and 45° linear-polarized focus proves that the vertical phase profile was smaller once again, as mentioned before, which could be solved by increasing the size of the lens. The designed 3D metalens can be used as a polarization splitter to separate arbitrary linear polarizations into two pairs of orthogonal polarizations, and it can also be used as a linear polarization generator under the incidence of any type of polarized light.

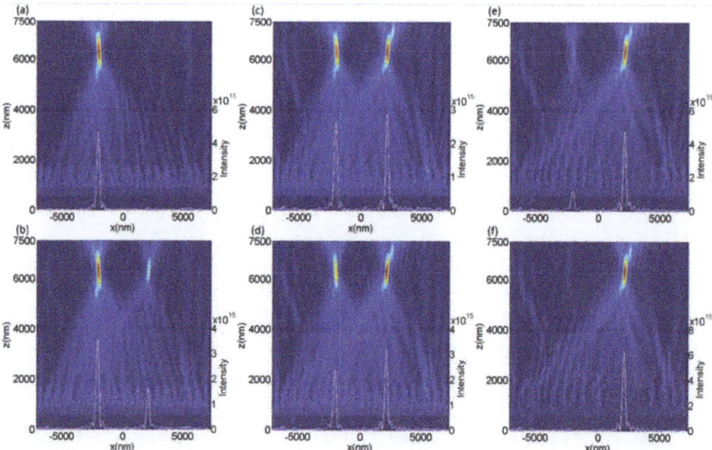

Figure 5. The distribution of transmitted intensities ($|E|^2$) under the linear polarization states of incident light are (**a**) 0°, (**b**) 30°, (**c**) 42°, (**d**) 45°, (**e**) 60°, and (**f**) 90°. The white solid and dashed lines are the intensity distribution curve and the position of focal plane.

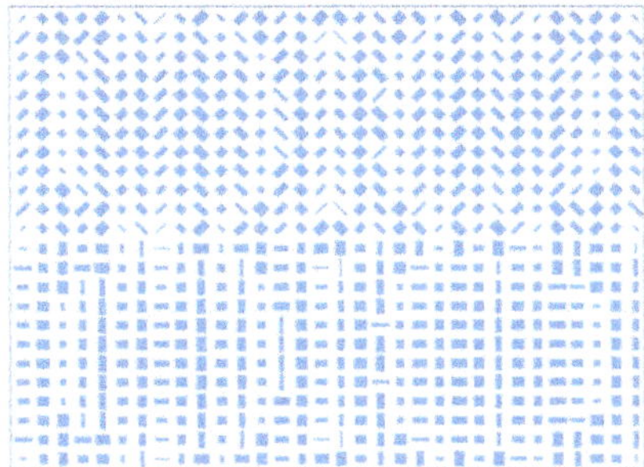

Figure 6. Schematic of the structure array of the designed 3D metalens.

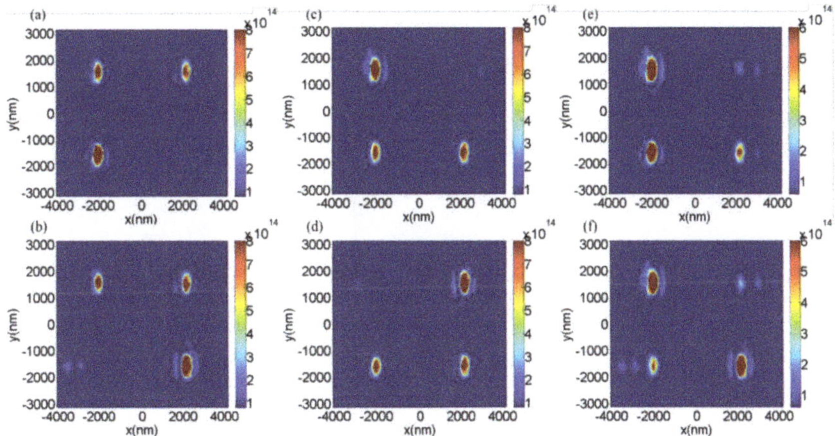

Figure 7. The distributions of transmitted intensities ($|E|^2$) in the focusing plane (X–Y) under (**a**) X-, (**b**) Y-, (**c**) 45°, (**d**) 135°, (**e**) 30°and (**f**) 60° linear-polarized incidences.

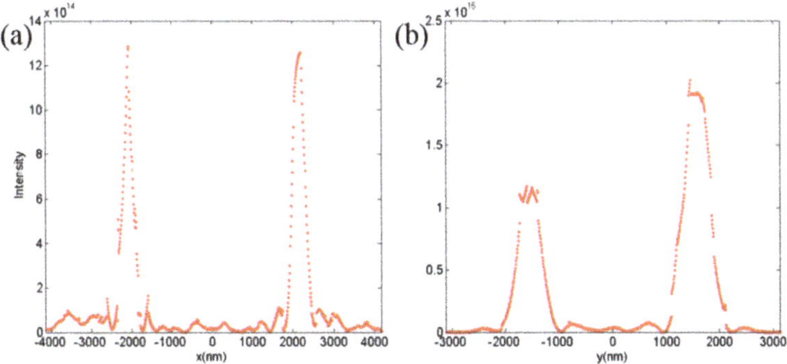

Figure 8. The transmitted intensity ($|E|^2$) profiles in focal plane at (**a**) $y = -1560$ nm and (**b**) $x = -2080$ nm under the incidence of 45° linear-polarized light.

In addition, by measuring the focusing energies at each focal point of the designed 3D metalens, we could analyze the polarization states of the arbitrary linear-polarized light. The focusing energies of X-, Y-, 45°, and 135° linear-polarized components can be expressed as I_x, I_y, I_{45}, and I_{135}, respectively, from which we could analyze the polarization direction of arbitrary linear-polarized incidences. When the incident linear polarization angle is θ, the focusing energies of the four focal points are I'_x, I'_y, I'_{45}, and I'_{135}, respectively. The Stokes parameters of I, Q, and U and the DoLP of the incidence can be obtained as follows:

$$s_1 = \frac{I'_x - aI'_y}{I'_x + aI'_y}, \quad s_2 = \frac{I'_{45} - bI'_{135}}{I'_{45} + bI'_{135}}, \quad DOLP = \frac{\sqrt{s_1^2 + s_2^2}}{I_{total}}$$

where $a = I_x/I_y$, $b = I_{45}/I_{135}$, and I_{total} is the total intensity of transmitted light. The Stokes parameters can be normalized to make the unit total light intensity:

$$Q = \frac{s_1}{\sqrt{s_1^2 + s_2^2}}, \quad U = \frac{s_2}{\sqrt{s_1^2 + s_2^2}}, \quad I = \sqrt{Q^2 + U^2},$$

and the polarization direction can be expressed as $\tan 2\theta = \frac{U}{Q}$.

Here, we calculated the power value instead of energy for analyzing the polarization direction of the incidences under simulation. When light with a polarization angle of θ was incident to the designed polarization analyzer, the powers of the X-, Y-, 45°, and 135° polarized components could be obtained at their focuses accordingly, and then we could calculate the polarization angle of the incident light. For example, when light with polarization angles of $\theta = 30°$ and $\theta = 60°$ was incident to the designed polarization analyzer, the powers of the X-, Y-, 45°, and 135° polarized components could be obtained as is depicted in Table 1. We calculated the polarization angles of the incident light as 29.87° and 59.53°, respectively, which agreed well with the true values. This means that the designed 3D metalens can be used for analyzing linear polarization states with high accuracy. If arbitrary linear-polarized light is incident, we can calculate the Stokes parameters I, Q, and U by collecting the energy at each focal point and then analyze the polarization direction of the incident linear-polarized light.

Table 1. The power values of X-, Y-, 45°, and 135° linear-polarized components at focus.

θ	P_x	P_y	P_{45}	P_{135}
30°	0.213	0.089	0.300	0.032
60°	0.079	0.245	0.295	0.034

3. Conclusions

In summary, we propose a metasurface based on the novel dielectric material GaN to realize high-efficiency modulation for both of the orthogonal linear polarizations simultaneously in the visible range. We designed a beam splitter to refract the light of orthogonal polarizations into two opposite directions, as well as a metalens to focus the X- and Y-polarized light in two spatially symmetric positions. We also designed a 3D metalens that can simultaneously focus X-, Y-, 45°, and 135° linear polarizations in spatially symmetric positions, which can be applied as a simple polarization analyzer for analyzing the polarization direction of the incidences. Our presented method can be used over the whole visible range, which gives it great potential in applications such as polarization imaging. It is promising that the designed devices can be integrated into systems to form a highly integrated polarization multiplexor.

Author Contributions: Writing-Original Draft Preparation, H.X., Z.G. and Z.Y.; Writing-Review &Editing, K.G. and F.S.; Data Curation, Q.Z., J.G., and H.Z.

Funding: This research was funded by the National Natural Science Foundation of China (61775050, 11505043); the Natural Science Foundation of Anhui Province, China (1808085MF188, 1808085QA21); the Science and Technology Planning Project of Guangdong Province, China (2016B010108002); and Fundamental Research Funds for the Central Universities (JD2017JGPY0005).

Conflicts of Interest: The authors declare that they have no competing interests.

References

1. Lin, D.; Fan, P.; Hasman, E.; Brongersma, M.L. Dielectric gradient metasurface optical elements. *Science* **2014**, *345*, 298–302. [CrossRef] [PubMed]

2. Kildishev, A.V.; Boltasseva, A.; Shalaev, V.M. Planar photonics with metasurfaces. *Science* **2013**, *339*, 1232009. [CrossRef] [PubMed]

3. Shalaev, M.I.; Sun, J.; Tsukernik, A.; Pandey, A.; Nikolskiy, K.; Litchinitser, N.M. High-efficiency all-dielectric metasurfaces for ultracompact beam manipulation in transmission mode. *Nano Lett.* **2015**, *15*, 6261–6266. [CrossRef] [PubMed]

4. Sun, S.; Yang, K.Y.; Wang, C.M.; Juan, T.-K.; Chen, W.T.; Liao, C.Y.; He, Q.; Xiao, S.; Kung, W.-T.; Guo, G.-Y.; et al. High-efficiency broadband anomalous reflection by gradient meta-surfaces. *Nano Lett.* **2012**, *12*, 6223–6229. [CrossRef] [PubMed]

5. Li, R.; Guo, Z.; Wang, W.; Zhang, J.; Zhang, A.; Liu, J.; Qu, S.; Gao, J. High-efficiency cross polarization converters by plasmonic metasurface. *Plasmonics* **2015**, *10*, 1167–1172. [CrossRef]

6. Aieta, F.; Genevet, P.; Kats, M.A.; Yu, N.; Blanchard, R.; Gaburro, Z.; Capasso, F. Aberration-free ultrathin flat lenses and axicons at telecom wavelengths based on plasmonic metasurfaces. *Nano Lett.* **2012**, *12*, 4932–4936. [CrossRef] [PubMed]

7. Wu, P.C.; Chen, J.W.; Yin, C.W.; Lai, Y.-C.; Chung, T.L.; Liao, C.Y.; Chen, B.H.; Lee, K.-W.; Chuang, C.-J.; Wang, C.-M.; et al. Visible Metasurfaces for On-chip Polarimetry. *ACS Photonics* **2017**, in press. [CrossRef]

8. Guo, Z.; Zhu, L.; Guo, K.; Shen, F.; Yin, Z. High-Order Dielectric Metasurfaces for High-Efficiency Polarization Beam Splitters and Optical Vortex Generators. *Nanoscale Res. Lett.* **2017**, *12*, 512. [CrossRef] [PubMed]

9. Khorasaninejad, M.; Chen, W.T.; Devlin, R.C.; Oh, J.; Zhu, A.Y.; Capasso, F. Metalenses at visible wavelengths: Diffraction-limited focusing and subwavelength resolution imaging. *Science* **2016**, *352*, 1190–1194. [CrossRef] [PubMed]

10. Wang, W.; Guo, Z.; Li, R.; Zhang, J.; Liu, Y.; Wang, X.; Qu, S. Ultra-thin, planar, broadband, dual-polarity plasmonic metalens. *Photonics Res.* **2015**, *3*, 68–71. [CrossRef]

11. Vo, S.; Fattal, D.; Sorin, W.V.; Peng, Z.; Tran, T.; Fiorentino, M.; Beausoleil, R.G. Sub-wavelength grating lenses with a twist. *IEEE Photonics Technol. Lett.* **2014**, *26*, 1375–1378. [CrossRef]

12. Yu, N.; Aieta, F.; Genevet, P.; Kats, M.A.; Gaburro, Z.; Capasso, F. A broadband, background-free quarter-wave plate based on plasmonic metasurfaces. *Nano Lett.* **2012**, *12*, 6328–6333. [CrossRef] [PubMed]

13. Cheng, J.; Mosallaei, H. Truly achromatic optical metasurfaces: A filter circuit theory-based design. *JOSA B* **2015**, *32*, 2115–2121. [CrossRef]

14. Azad, A.K.; Kort-Kamp, W.J.M.; Sykora, M.; Weisse-Bernstein, N.R.; Luk, T.S.; Taylor, A.J.; Dalvit, D.A.R.; Chen, H.-T. Metasurface broadband solar absorber. *Sci. Rep.* **2016**, *6*, 20347. [CrossRef] [PubMed]

15. Wang, W.; Li, Y.; Guo, Z.; Li, R.; Zhang, J.; Zhang, A.; Qu, S. Ultra-thin optical vortex phase plate based on the metasurface and the angular momentum transformation. *J. Opt.* **2015**, *17*, 045102. [CrossRef]

16. Wang, Q.; Zhang, X.; Xu, Y.; Gu, J.; Li, Y.; Tian, Z.; Singh, R.; Zhang, S.; Han, J.; Zhang, W. Broadband metasurface holograms: Toward complete phase and amplitude engineering. *Sci. Rep.* **2016**, *6*, 32867. [CrossRef] [PubMed]

17. Li, R.; Guo, Z.; Wang, W.; Zhang, J.; Zhou, K.; Liu, J.; Qu, S.; Liu, S.; Gao, J. Arbitrary focusing lens by holographic metasurface. *Photonics Res.* **2015**, *3*, 252–255. [CrossRef]

18. Qin, F.; Ding, L.; Zhang, L.; Monticone, F.; Chum, C.C.; Deng, J.; Mei, S.; Li, Y.; Teng, J.; Hong, M. Hybrid bilayer plasmonic metasurface efficiently manipulates visible light. *Sci. Adv.* **2016**, *2*, e1501168. [CrossRef] [PubMed]

19. Qi, X.; Nie, Z.; Chen, Y.; Qi, X.; Nie, Z.; Chen, Y. Directional Enhancement Analysis of All-Dielectric Optical Nanoantennas Based on SIE Formulation. *IEEE Photonics Technol. Lett.* **2018**, *30*, 123–126. [CrossRef]

20. Staude, I.; Miroshnichenko, A.E.; Decker, M.; Fofang, N.T.; Liu, S.; Gonzales, E.; Dominguez, J.; Luk, T.S.; Neshev, D.N.; Brener, I.; et al. Tailoring directional scattering through magnetic and electric resonances in subwavelength silicon nanodisks. *ACS Nano* **2013**, *7*, 7824–7832. [CrossRef] [PubMed]

21. Cheng, J.; Jafar-Zanjani, S.; Mosallaei, H. All-dielectric ultrathin conformal metasurfaces: Lensing and cloaking applications at 532 nm wavelength. *Sci. Rep.* **2016**, *6*, 38440. [CrossRef] [PubMed]

22. Evlyukhin, A.B.; Novikov, S.M.; Zywietz, U.; Eriksen, R.L.; Reinhardt, C.; Bozhevolnyi, S.I.; Chichkov, B.N. Demonstration of magnetic dipole resonances of dielectric nanospheres in the visible region. *Nano Lett.* **2012**, *12*, 3749–3755. [CrossRef] [PubMed]

23. Park, J.; Yu, K. Metasurface-based ultra-thin circular polarization analyzer integrated with semiconductor photodetectors. In Proceedings of the 22nd Microoptics Conference (MOC 2017), Tokyo, Japan, 19–22 November 2017; pp. 350–351.

24. Groever, B.; Chen, W.T.; Capasso, F. Meta-lens doublet in the visible region. *Nano Lett.* **2017**, *17*, 4902–4907. [CrossRef] [PubMed]

25. Li, Z.; Kim, I.; Zhang, L.; Mehmood, M.Q.; Anwar, M.S.; Saleem, M.; Lee, D.; Nam, K.T.; Zhang, S.; Luk'yanchuk, B. Dielectric meta-holograms enabled with dual magnetic resonances in visible light. *ACS Nano* **2017**, *11*, 9382–9389. [CrossRef] [PubMed]

26. Kamali, S.M.; Arbabi, E.; Arbabi, A.; Horie, Y.; Faraji-Dana, M.; Faraon, A. Angle-multiplexed metasurfaces: Encoding independent wavefronts in a single metasurface under different illumination angles. *Phys. Rev. X* **2017**, *7*, 041056. [CrossRef]

27. Arbabi, A.; Horie, Y.; Ball, A.J.; Bagheri, M.; Faraon, A. Subwavelength-thick lenses with high numerical apertures and large efficiency based on high-contrast transmitarrays. *Nat. Commun.* **2015**, *6*, 7069. [CrossRef] [PubMed]

28. Lin, D.; Holsteen, A.L.; Maguid, E.; Wetzstein, G.; Kik, P.G.; Hasman, E.; Brongersma, M.L. Photonic multitasking interleaved Si nanoantenna phased array. *Nano Lett.* **2016**, *16*, 7671–7676. [CrossRef] [PubMed]

29. Ma, Y.; Rui, G.; Gu, B.; Cui, Y. Trapping and manipulation of nanoparticles using multifocal optical vortex metalens. *Sci. Rep.* **2017**, *7*, 14611. [CrossRef] [PubMed]

30. Khorasaninejad, M.; Zhu, A.Y.; Roques-Carmes, C.; Chen, W.T.; Oh, J.; Mishra, I.; Devlin, R.C.; Capasso, F. Polarization-insensitive metalenses at visible wavelengths. *Nano Lett.* **2016**, *16*, 7229–7234. [CrossRef] [PubMed]

31. Chen, B.H.; Wu, P.C.; Su, V.C.; Lai, Y.-C.; Chu, C.H.; Lee, I.C.; Chen, J.-W.; Chen, Y.H.; Lan, Y.-C.; Kuan, C.-H.; et al. GaN metalens for pixel-level full-color routing at visible light. *Nano Lett.* **2017**, *17*, 6345–6352. [CrossRef] [PubMed]

32. Khodaee, M.; Banakermani, M.; Baghban, H. GaN-based metamaterial terahertz bandpass filter design: Tunability and ultra-broad passband attainment. *Appl. Opt.* **2015**, *54*, 8617–8624. [CrossRef] [PubMed]

33. Wang, Z.; He, S.; Liu, Q.; Wang, W. Visible light metasurfaces based on gallium nitride high contrast gratings. *Opt. Commun.* **2016**, *367*, 144–148. [CrossRef]

34. Wang, B.; Dong, F.; Li, Q.T.; Yang, D.; Sun, C.; Chen, J.; Song, Z.; Xu, L.; Chu, W.; Xiao, Y.-F.; et al. Visible-frequency dielectric metasurfaces for multiwavelength achromatic and highly dispersive holograms. *Nano Lett.* **2016**, *16*, 5235–5240. [CrossRef] [PubMed]

35. Ni, X.; Emani, N.K.; Kildishev, A.V.; Boltasseva, A.; Shalaev, V.M. Broadband light bending with plasmonic nanoantennas. *Science* **2012**, *335*, 427. [CrossRef] [PubMed]

 *nanomaterials*

Article

Tunable Metamaterial with Gold and Graphene Split-Ring Resonators and Plasmonically Induced Transparency

Qichang Ma, Youwei Zhan and Weiyi Hong *

Guangzhou Key Laboratory for Special Fiber Photonic Devices and Applications & Guangdong Provincial Key Laboratory of Nanophotonic Functional Materials and Devices, South China Normal University, Guangzhou 510006, China; qichangma@m.scnu.edu.cn (Q.M.); zhanyw@m.scnu.edu.cn (Y.Z.)
* Correspondence: hongwy@m.scnu.edu.cn; Tel.: +86-185-203-89309

Received: 28 November 2018; Accepted: 20 December 2018; Published: 21 December 2018

Abstract: In this paper, we propose a metamaterial structure for realizing the electromagnetically induced transparency effect in the MIR region, which consists of a gold split-ring and a graphene split-ring. The simulated results indicate that a single tunable transparency window can be realized in the structure due to the hybridization between the two rings. The transparency window can be tuned individually by the coupling distance and/or the Fermi level of the graphene split-ring via electrostatic gating. These results could find significant applications in nanoscale light control and functional devices operating such as sensors and modulators.

Keywords: metamaterials; mid infrared; graphene split-ring; gold split-ring; electromagnetically induced transparency effect

---

1. Introduction

Electromagnetically-induced transparency (EIT) is a concept originally observed in atomic physics and arises due to quantum interference, resulting in a narrowband transparency window for light propagating through an originally opaque medium [1,2]. The EIT effect extended to classical optical systems using plasmonic metamaterials leads to new opportunities for many important applications such as slow light modulator [3–6], high sensitivity sensors [7,8], quantum information processors [9], and plasmonic switches [10–12]. Generally speaking, the realization of the EIT effect is usually achieved by two kinds of schemes such as the bright-bright mode coupling [7,12–16] and the bright-dark mode coupling [17–21]. So far, many researches have obtained the EIT effect from various metamaterial structures [11,22–27]. However, most of them only consist of metallic materials and they can only be controlled by the geometry of structures, which limits their applications.

Graphene is a two-dimensional material which is composed of single layer of carbon atoms, which has been widely used in optoelectronic devices, such as optical modulators [28–30] due to its unique properties such as high electron mobility [31], high light transparency [32] and high thermal conductivity [33]. Particularly, the surface conductivity of graphene has a wide range of tunable characteristics by changing the Fermi level of the graphene via electrostatic gating, which makes it a significant potential application in high-performance tunable optical devices [34,35]. Meanwhile, graphene is widely used in metamaterial to achieve a variety of optoelectronic devices with tunable properties [36–38].

Recently, L.S. Yan et al. have investigated that a plasmonic waveguide system based on side coupled complementary split-ring resonators and the electromagnetic responses of CSRR can be flexibly handled by changing the asymmetry degree of the structure and the width of the metallic baffles [39]. W. Yu et al. have realized a dual-band EIT effect which was composed of a split-ring resonator (SRR) and a pair of metal strips. Most of the previous researches were based on metallic split-ring which lacks tunability [40]. In this paper, we propose a periodic metamaterial structure which is composed of a graphene split-ring and a gold split-ring to investigate the EIT effect in the MIR regions. A transparency window can be observed in the transmission spectrum of the structure under plane wave excitation. It is found that the coupling strength greatly depends on the separation between the two rings. Besides, the EIT window can be tuned by changing the Fermi levels of the graphene split-ring via electrostatic gating, which may offer possible applications at tunable mid-infrared functional devices, such as optical switches and modulators.

2. Materials and Methods

As shown in Figure 1a, the metamaterial structure used to demonstrate the EIT phenomenon consists of three layers. The top layer is composed of gold split-ring and graphene split ring while the middle layer plays a role as a substrate and its material is Al_2O_3 and the graphene split-ring is composed of 5 layers of graphene in our following simulations so as to enhance the modulation depth [41]. The bottom layer is doped silicon (doped Si) substrate which is used for an electrode to apply voltage to the graphene split-ring for changing its Fermi level, and another electrode is Au_1 whose material is gold as shown in Figure 1a. In the simulation of the transmission spectra, Au_1 is negligible because of the relatively small size with respect to the structure. The incident light propagates in the z direction and its polarization is along x direction. Both the gold split-ring and graphene split-ring are symmetric in y direction and their symmetry axis are coinciding. The (complex) refractive index of Al_2O_3 is obtained from Table I, page 662 in Ref. [42] and the doping concentration of doped Si is $10^{18} cm^{-3}$ and its refractive index can be obtained from Ref. [43]. Both the Al_2O_3 and doped Si have been considered having the dispersion and absorption effects in our simulations.

The optical properties of the gold material are obtained from Palik's experimental data. The surface complex conductivity of graphene $\sigma(\omega)$ is dominated by interband and intraband transitions within the random-phase approximation, which can be calculated as [44,45]:

$$\sigma(\omega) = \sigma_{\text{interband}} + \sigma_{\text{intraband}} \tag{1}$$

where

$$\sigma_{\text{interband}} = \frac{e^2}{4\hbar} \left\{ \frac{1}{2} + \frac{1}{\pi}\arctan\left(\frac{\hbar\omega - 2E_f}{2k_B T}\right) - \frac{i}{2\pi}\ln\left(\frac{\left(\hbar\omega + 2E_f\right)^2}{\left(\hbar\omega - 2E_f\right)^2 + (2k_B T)^2}\right) \right\} \tag{2}$$

$$\sigma_{\text{intraband}} = \frac{2ie^2 k_B T}{\pi\hbar^2(\omega + i\tau^{-1})}\ln\left(2\cosh\left(\frac{E_f}{2k_B T}\right)\right) \tag{3}$$

Here, ω is the optical angular frequency, T is the ambient temperature (300 K in this paper), E_f is the Fermi level of graphene, $\hbar = h/2\pi$ is the reduced Plank's constant, k_B is the Boltzmann constant, τ is the carrier relaxation time and $\tau = \hbar/2\Gamma$ [46], where Γ is the scattering rate and it is set 0.00099 eV in this paper (corresponding to the carrier relaxation time $\tau \approx 0.33$ ps). It can be seen from the above formula that the surface conductivity is greatly affected by the Fermi level and the Fermi level is determined by the carrier concentration n_g and Fermi velocity V_f. The formula $E_f = \hbar V_f(\pi n_g)^{1/2}$ with the Fermi velocity of $V_f = 10^6$ m/s can be obtained from Ref. [47] and $n_g = \varepsilon_0\varepsilon_d V_g/eH_1$, where V_g is the applied voltage and ε_d is the permittivity of Al_2O_3 and H_1 is the thickness of Al_2O_3. In the following simulation, graphene is modeled as a thin anisotropic material film and the in-plane permittivity is approximated to be [44]:

$$\varepsilon_g = 1 + \frac{i\sigma(\omega)}{\varepsilon_0\omega t_g} \tag{4}$$

where t_g is the thickness of graphene which is estimated as 1 nm [48] and ε_0 is the permittivity in vacuum. In addition, the conductivity of the N-layer graphene is $N\sigma$ [49,50].

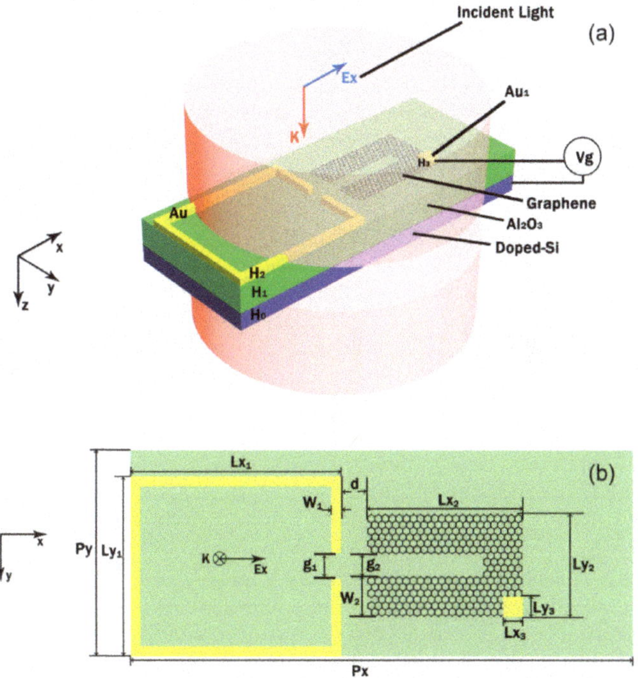

Figure 1. (**a**) Schematic diagram of the plasmonic metamaterial unit cell consists of a graphene split-ring and a gold split-ring. (**b**) Top view of metamaterial structure. The periods of the y and x direction are Py = 0.22 μm, Px = 1.34 μm, respectively. The distance between the graphene split-ring and the gold split-ring is d and the gaps of the gold split-ring and the graphene split-ring are g_1 = 8 nm, g_2 = 20 nm, respectively. The width of the gold split-ring and the graphene split-ring are W_1 = 12 nm, W_2 = 43 nm. The length of the gold split-ring in the y and x direction are Ly_1 = 200 nm, Lx_1 = 262 nm, respectively; and the length of the graphene split-ring in the y and x direction are Ly_2 = 106 nm, Lx_2 = 104 nm, respectively. The thickness of doped silicon, Al$_2$O$_3$ and Au are set as H_0 = 10 nm, H_1 = 40 nm and H_2 = 4 nm, respectively. V_g is the applied voltage and both the Au$_1$ and doped silicon are served as electrodes to apply voltages to the graphene split-ring. The thickness of the Au$_1$ is H_3 = 4 nm and its width (Ly_3) and length (Lx_3) are set as 10 nm.

In this paper, we use the finite-difference time-domain (FDTD) simulations to numerically investigate the properties of the structure. The perfectly matched layer (PML) absorbing boundary conditions are set on the z direction while the periodic boundary conditions are set on the x and y directions.

3. Results and Discussion

To demonstrate the EIT effect, we numerically calculate the simulated transmission spectra of the unit cell with the graphene split-ring and the gold split-ring are shown in Figure 1a. The plasmonic resonances of the graphene split-ring and the gold split-ring are excited at the same wavelength. In addition, the transmission spectra of the unit cell with a graphene split-ring and a gold split-ring under x-polarized and y-polarized incident light excitation, respectively. It is obviously that when using the x-polarized incident light, a transparency window is observed. For the case of y-polarized incident light, there are two plasmon resonance peaks at 3.61 µm and 7.19 µm observed and no EIT phenomenon can be found. Consequently, we choose x-polarized incident light in the following simulations. In the simulations, the E_f of the graphene split-ring is set as 0.9 eV and the coupling distance d is 20 nm. From the transmission spectra in Figure 2a, one can find that a strong plasmon resonance at 4.09 µm of both the graphene split-ring and the gold split-ring can be excited by the x-polarized incident light. The quality factor of graphene split-ring is higher than that of the gold split-ring and a transparency window with 95% transmission appears at 3.95 µm when we put the two rings together due to the interaction of them. In addition, two obvious dips can be found in the transmission spectra (orange line in Figure 2b) at 3.83 µm and 4.35 µm, respectively.

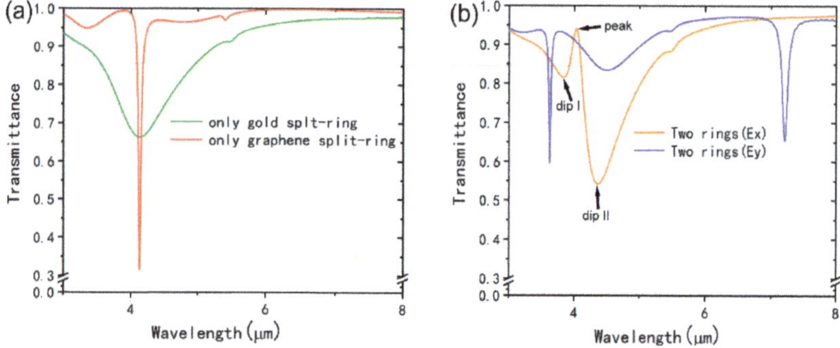

Figure 2. (**a**) Transmission spectra of the individual graphene split-ring and the individual gold split-ring. (**b**) Transmission spectra of the structure consists of graphene split-ring and gold split-ring under x-polarized and y-polarized incident light excitation, respectively.

Then, we study the response of EIT effect resulting from the variation of the layers of graphene split-ring (Figure 3a) and the periods in the x direction the total structure. The Fermi level is set as 0.9 eV and the distance of the two split-rings d is 20 nm. It can be observed that dip II is obviously blue-shifted from 7.97 µm to 4.35 µm while dip I blue-shifts slightly and its transmittance increases with the layers of the graphene split-ring N increasing from 1 to 5. In addition, it is clear shown that the transmission peak is blue-shifted and the line width of the transmission peak becomes smaller and narrower. This is because the hybridized of the gold split-ring and graphene split-ring become stronger with increasing the layer number N. It is also found that 4 layers or 5 layers of the graphene split-ring are suitable. Here we choose the graphene split-ring of 5 layers because the quality factor of transmission peak is larger than that of the 4 layers. In Figure 3b, it is shown that the period in the x direction of the plasmonic metamaterial unit cell can also influence the EIT phenomenon. With the increase of the Px from 1.34 µm to 2.34 µm, both dip I and dip II blue-shift slightly and the amplitudes of the transmittance increase. The line width of the transmission peak becomes larger and its transmittance increase with increasing the periods (Px) in the x direction.

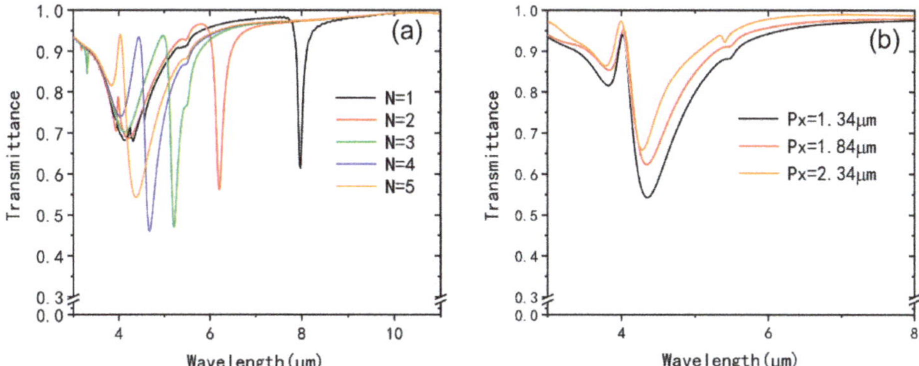

Figure 3. (**a**) Transmission spectra of the individual graphene split-ring with different layers. (**b**) Transmission spectra of the structure with different periods (Px) in the x direction.

Furthermore, we also investigate the dependence of the EIT spectra on the distance d between the two rings. As shown in Figure 4, the EIT peak gradually becomes sharper and its transmittance gradually decreases with the increasing of d. When d comes to 400 nm, the peak will disappear. It reveals that the interaction of gold split-ring and graphene split-ring becomes weaker and their interaction eventually disappears when the coupling distance is large enough. It can also find that dip I red-shifts slightly while the dip II blue-shifts with the increase of d and both their transmittance decrease. Therefore, the coupling strength of the metamaterial structure greatly depends on the distance of the gold split-ring and graphene split-ring.

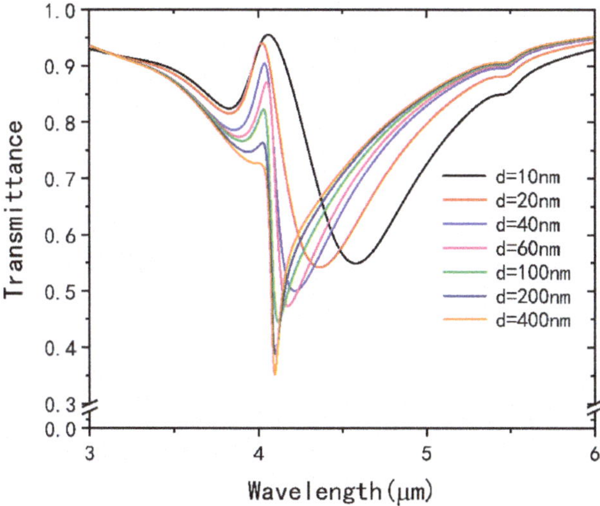

Figure 4. Transmission spectra of the metamaterial structure with different distances d.

To get further insight into the physics of the EIT phenomenon of our structure, the electric-field distributions of the z-component at wavelength I ($\lambda = 3.83$ μm) and wavelength II ($\lambda = 4.35$ μm) are shown in Figure 5a–d. The generation of dip I and dip II can be understood and predicted by the plasmon hybridization model, similar to the case for the molecular orbital theory and plasmon hybridization model [51–54]. Since the thicknesses of the graphene split-ring and the gold split-ring are different, we place two power monitors at 2 nm above the graphene split-ring and the gold split-ring, respectively. In this simulation, the distance d is 20 nm, and the Fermi level of graphene split-ring is 0.9 eV.

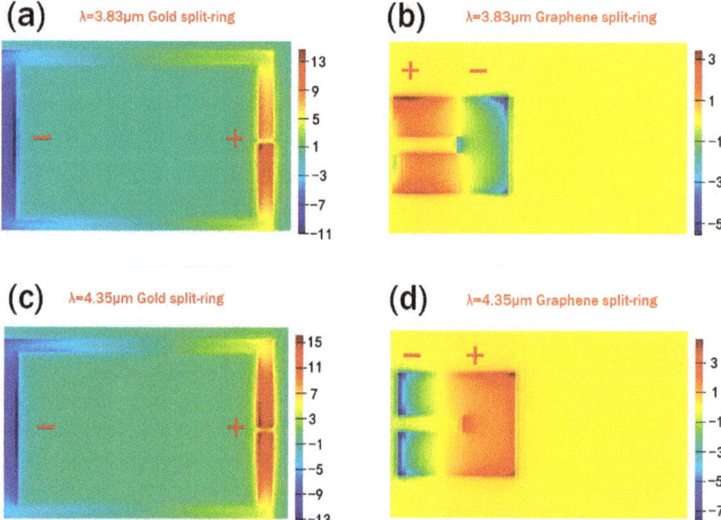

Figure 5. z-component distributions of the electric field under x-polarized continue light excitation at (**a**) λ = 3.83 μm (Au power monitor), (**b**) λ = 3.83 μm (graphene power monitor), (**c**) λ = 4.35 μm (Au power monitor), (**d**) λ = 4.35 μm (graphene power monitor).

The placement of graphene split-ring and gold split-ring can be treated as asymmetric nanoparticle dimer which has been investigated in Ref. [53]. When d is small enough, their local surface plasmons will couple to each other, which is analogous to σ hybridization of p orbital electron in the atomic outer layer. From Figure 5a,b, one can observe that the left side of the gold split-ring gathers a large amount of negative induced charge, and a large amount of positive induced charges are accumulated on the right side for the case of λ = 3.83 μm. In contrast, the positive and negative induced charges are distributed on the left and right sides of the graphene split-ring. The different charge distributions of the two rings are similar to the anti-bonding mode of the hybrid molecular system. On the other hand, one can find that the distribution of induced change in the gold split-ring remains unchanged at the wavelength λ = 4.35 μm, while the distribution of induced change in the graphene split-ring is opposite to the situation at λ = 3.83 μm from Figure 5c,d, which is similar to the bonding mode of hybrid molecular system.

As mentioned earlier, the complex surface conductivity of graphene can be changed by controlling its Fermi level, which can be realized by applying different gate-voltages to the graphene split-ring. Compared with the reported bright-bright mode coupling EIT structure fabricated by metallic material, we can realize the dynamically control of the EIT window in our metamaterial structure without reconstructing the geometries or imbedding other actively controlled materials. Figure 6a illustrates the transmission spectra with different Fermi levels, it is apparent that the EIT window can be tuned over a large range in the MIR regions by a small change of the Fermi level. When the Fermi level changes from 0.6 eV to 0.9 eV, the transmission window blue-shifts from 4.82 μm to 4.02 μm, which can be interpreted by the graphene plasmonic resonance wavelength $\lambda \sim 1/E_f^{1/2}$ [55]. At the same time, the transmittance of the dip I and dip II will decrease with increasing of the Fermi level while the transmittance of the peak almost remains unchanged. The transmission spectra of the unit cell with only graphene split-ring with different Fermi level are shown in Figure 6b. When the Fermi level changes from 0.6 eV to 0.9 eV, the resonance wavelengths blue shift from 4.99 μm to 4.12 μm and the amplitude of the transmittance are decreasing. It is easy to find that the resonance wavelength of individual graphene split-ring in Figure 6a is almost the same as Dip II in Figure 6b for the same Fermi level, which shows that the dip II is greatly affected by the graphene split-ring.

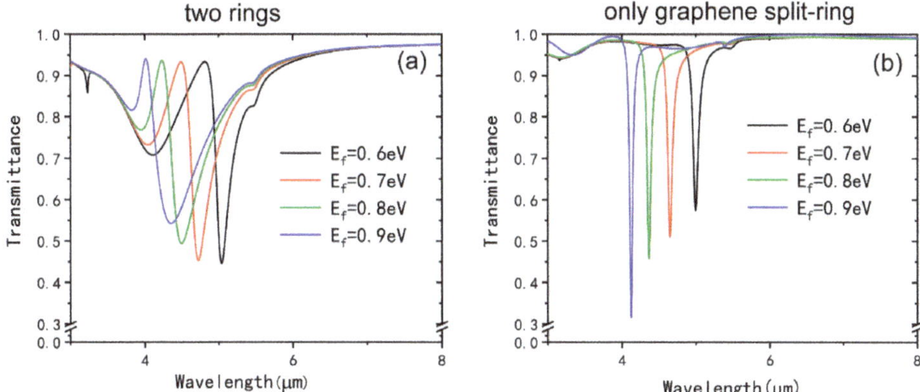

Figure 6. (**a**) Transmission spectra of the two rings with different Fermi levels of the graphene split-ring. (**b**) Transmission spectra of the individual graphene split-ring with different Fermi levels.

4. Conclusions

In this paper, we have successfully proposed that the EIT effect can be realized through bright-bright modes coupling in the MIR regions. The EIT window appears due to the hybridization between the gold split-ring and graphene split-ring and it is sensitive to the coupling distance. Moreover, we investigate the response of EIT effect resulting from the variation of the layers of graphene split-ring and periods in the x direction. Particularly, the EIT window can be dynamically tuned by adjusting the Fermi level of the graphene split-ring instead of reconstructing the structure. These properties could lead to new opportunities for many important applications, such as sensors and slow light modulators in MIR regions.

Author Contributions: Formal Analysis, Q.M. and Y.Z.; Writing-Original Draft Preparation, Q.M. and Y.Z.; Writing-Review & Editing, W.H.

Funding: This research was supported by the National Natural Science Foundation of China (Grant No. 11874019).

Acknowledgments: This research was supported by the National Natural Science Foundation of China (Grant No. 11874019) and Guangzhou Key Laboratory for Special Fiber Photonic Devices and Applications, South China Normal University, Guangzhou 510006, People's Republic of China.

References

1. Field, J.E.; Hahn, K.H.; Harris, S.E. Observation of electromagnetically induced transparency in collisionally broadened lead vapor. *Phys. Rev. Lett.* **1991**, *67*, 3062–3065. [CrossRef] [PubMed]
2. Boller, K.J.; Imamolu, A.; Harris, S.E. Observation of electromagnetically induced transparency. *Phys. Rev. Lett.* **1991**, *66*, 2593–2596. [CrossRef] [PubMed]
3. Zhu, L.; Meng, F.-Y.; Fu, J.-H.; Wu, Q.; Hua, J. Multi-band slow light metamaterial. *Opt. Express* **2012**, *20*, 4494–4502. [CrossRef] [PubMed]
4. Kravtsov, V.; Atkin, J.M.; Raschke, M.B. Group delay and dispersion in adiabatic plasmonic nanofocusing. *Opt. Lett.* **2013**, *38*, 1322–1324. [CrossRef] [PubMed]
5. Wang, G. Slow light engineering in periodic-stub-assisted plasmonic waveguide. *Appl. Opt.* **2013**, *52*, 1799–1804. [CrossRef] [PubMed]

6. Miyamaru, F.; Morita, H.; Nishiyama, Y.; Nishida, T.; Nakanishi, T.; Kitano, M.; Takeda, M.W. Ultrafast optical control of group delay of narrow-band terahertz waves. *Sci. Rep.* **2014**, *4*, 4346. [CrossRef]

7. Dong, Z.-G.; Liu, H.; Cao, J.-X.; Li, T.; Wang, S.-M.; Zhu, S.-N.; Zhang, X. Enhanced sensing performance by the plasmonic analog of electromagnetically induced transparency in active metamaterials. *Appl. Phys. Lett.* **2010**, *97*, 114101. [CrossRef]

8. Wang, J.; Fan, C.; He, J.; Ding, P.; Liang, E.; Xue, Q. Double Fano resonances due to interplay of electric and magnetic plasmon modes in planar plasmonic structure with high sensing sensitivity. *Opt. Express* **2013**, *21*, 2236–2244. [CrossRef]

9. Fleischhauer, M.; Imamoglu, A.; Marangos, J.P. Electromagnetically induced transparency: Optics in coherent media. *Rev. Mod. Phys.* **2005**, *77*, 633–673. [CrossRef]

10. Xu, H.; Li, H.; Liu, Z.; Cao, G.; Wu, C.; Peng, X. Plasmonic EIT switching in ellipsoid tripod structures. *Opt. Mater.* **2013**, *35*, 881–886. [CrossRef]

11. Chen, J.; Li, Z.; Zhang, X.; Xiao, J.; Gong, Q. Submicron bidirectional all-optical plasmonic switches. *Sci. Rep.* **2013**, *3*, 1451. [CrossRef] [PubMed]

12. Chen, J.; Wang, P.; Chen, C.; Lu, Y.; Ming, H.; Zhan, Q. Plasmonic EIT-like switching in bright-dark-bright plasmon resonators. *Opt. Express* **2011**, *19*, 5970–5978. [CrossRef] [PubMed]

13. Liu, N.; Kaiser, S.; Giessen, H. Magnetoinductive and electroinductive coupling in plasmonic metamaterial molecules. *Adv. Mater.* **2008**, *20*, 4521–4525. [CrossRef]

14. Zhang, S.; Genov, D.A.; Wang, Y.; Liu, M.; Zhang, X. Plasmon-Induced Transparency in Metamaterials. *Phys. Rev. Lett.* **2008**, *101*, 47401. [CrossRef] [PubMed]

15. Jin, X.; Lu, Y.; Zheng, H.; Lee, Y.; Rhee, J.Y.; Jang, W.H. Plasmonic electromagnetically-induced transparency in symmetric structures. *Opt. Express* **2010**, *18*, 13396–13401. [CrossRef] [PubMed]

16. Dong, Z.-G.; Liu, H.; Xu, M.-X.; Li, T.; Wang, S.-M.; Zhu, S.-N.; Zhang, X. Plasmonically induced transparent magnetic resonance in a metallic metamaterial composed of asymmetric double bars. *Opt. Express* **2010**, *18*, 18229–18234. [CrossRef] [PubMed]

17. Fedotov, V.A.; Rose, M.; Prosvirnin, S.L.; Papasimakis, N.; Zheludev, N.I. Sharp Trapped-Mode Resonances in Planar Metamaterials with a Broken Structural Symmetry. *Phys. Rev. Lett.* **2007**, *99*, 147401. [CrossRef] [PubMed]

18. Papasimakis, N.; Fedotov, V.A.; Zheludev, N.I.; Prosvirnin, S.L. Metamaterial Analog of Electromagnetically Induced Transparency. *Phys. Rev. Lett.* **2008**, *101*, 253903. [CrossRef] [PubMed]

19. Chiam, S.-Y.; Singh, R.; Rockstuhl, C.; Lederer, F.; Zhang, W.; Bettiol, A.A. Analogue of electromagnetically induced transparency in a terahertz metamaterial. *Phys. Rev. B* **2009**, *80*, 153103. [CrossRef]

20. Li, Z.; Ma, Y.; Huang, R.; Singh, R.; Gu, J.; Tian, Z.; Han, J.; Zhang, W. Manipulating the plasmon-induced transparency in terahertz metamaterials. *Opt. Express* **2011**, *19*, 8912–8919. [CrossRef]

21. Kekatpure, R.D.; Barnard, E.S.; Cai, W.; Brongersma, M.L. Phase-Coupled Plasmon-Induced Transparency. *Phys. Rev. Lett.* **2010**, *104*, 243902. [CrossRef] [PubMed]

22. Liu, N.; Weiss, T.; Mesch, M.; Langguth, L.; Eigenthaler, U.; Hirscher, M.; Sönnichsen, C.; Giessen, H. Planar Metamaterial Analogue of Electromagnetically Induced Transparency for Plasmonic Sensing. *Nano Lett.* **2010**, *10*, 1103–1107. [CrossRef] [PubMed]

23. Yang, Y.; Kravchenko, I.I.; Briggs, D.P.; Valentine, J. All-dielectric metasurface analogue of electromagnetically induced transparency. *Nat. Commun.* **2014**, *5*, 5753. [CrossRef] [PubMed]

24. Zhang, X.; Xu, N.; Qu, K.; Tian, Z.; Singh, R.; Han, J.; Agarwal, G.S.; Zhang, W. Electromagnetically induced absorption in a three-resonator metasurface system. *Sci. Rep.* **2015**, *5*, 10737. [CrossRef] [PubMed]

25. Wang, L.; Gu, Y.; Chen, H.; Zhang, J.-Y.; Cui, Y.; Gerardot, B.D.; Gong, Q. Polarized linewidth-controllable double-trapping electromagnetically induced transparency spectra in a resonant plasmon nanocavity. *Sci. Rep.* **2013**, *3*, 2879. [CrossRef] [PubMed]

26. Zhang, Y.; Li, T.; Chen, Q.; Zhang, H.; O'Hara, J.F.; Abele, E.; Taylor, A.J.; Chen, H.-T.; Azad, A.K. Independently tunable dual-band perfect absorber based on graphene at mid-infrared frequencies. *Sci. Rep.* **2015**, *5*, 18463. [CrossRef] [PubMed]

27. Wang, J.; Yuan, B.; Fan, C.; He, J.; Ding, P.; Xue, Q.; Liang, E. A novel planar metamaterial design for electromagnetically induced transparency and slow light. *Opt. Express* **2013**, *21*, 25159–25166. [CrossRef]

28. Liu, M.; Yin, X.; Ulin-Avila, E.; Geng, B.; Zentgraf, T.; Ju, L.; Wang, F.; Zhang, X. A graphene-based broadband optical modulator. *Nature* **2011**, *474*, 64–67. [CrossRef]

29. Liu, M.; Yin, X.; Zhang, X. Double-Layer Graphene Optical Modulator. *Nano Lett.* **2012**, *12*, 1482–1485. [CrossRef]

30. Liang, G.; Hu, X.; Yu, X.; Shen, Y.; Li, L.H.; Davies, A.G.; Linfield, E.H.; Liang, H.K.; Zhang, Y.; Yu, S.F.; et al. Integrated Terahertz Graphene Modulator with 100% Modulation Depth. *ACS Photonics* **2015**, *2*, 1559–1566. [CrossRef]

31. Mayorov, A.S.; Gorbachev, R.V.; Morozov, S.V.; Britnell, L.; Jalil, R.; Ponomarenko, L.A.; Blake, P.; Novoselov, K.S.; Watanabe, K.; Taniguchi, T.; et al. Micrometer-Scale Ballistic Transport in Encapsulated Graphene at Room Temperature. *Nano Lett.* **2011**, *11*, 2396–2399. [CrossRef] [PubMed]

32. Nair, R.R.; Blake, P.; Grigorenko, A.N.; Novoselov, K.S.; Booth, T.J.; Stauber, T.; Peres, N.M.R.; Geim, A.K. Fine Structure Constant Defines Visual Transparency of Graphene. *Science* **2008**, *320*, 1308. [CrossRef] [PubMed]

33. Balandin, A.A. Thermal properties of graphene and nanostructured carbon materials. *Nat. Mater.* **2011**, *10*, 569–581. [CrossRef] [PubMed]

34. Ju, L.; Geng, B.; Horng, J.; Girit, C.; Martin, M.; Hao, Z.; Bechtel, H.A.; Liang, X.; Zettl, A.; Shen, Y.R.; et al. Graphene plasmonics for tunable terahertz metamaterials. *Nat. Nanotechnol.* **2011**, *6*, 630–634. [CrossRef] [PubMed]

35. Lee, S.H.; Choi, M.; Kim, T.-T.; Lee, S.; Liu, M.; Yin, X.; Choi, H.K.; Lee, S.S.; Choi, C.-G.; Choi, S.-Y.; et al. Switching terahertz waves with gate-controlled active graphene metamaterials. *Nat. Mater.* **2012**, *11*, 936–941. [CrossRef] [PubMed]

36. Zhang, Y.; Feng, Y.; Zhu, B.; Zhao, J.; Jiang, T. Graphene based tunable metamaterial absorber and polarization modulation in terahertz frequency. *Opt. Express* **2014**, *22*, 22743–22752. [CrossRef] [PubMed]

37. Zhang, Y.; Feng, Y.; Zhu, B.; Zhao, J.; Jiang, T. Switchable quarter-wave plate with graphene based metamaterial for broadband terahertz wave manipulation. *Opt. Express* **2015**, *23*, 27230–27239. [CrossRef] [PubMed]

38. Xu, B.; Gu, C.; Li, Z.; Niu, Z. A novel structure for tunable terahertz absorber based on graphene. *Opt. Express* **2013**, *21*, 23803–23811. [CrossRef] [PubMed]

39. Guo, Y.; Yan, L.; Pan, W.; Luo, B.; Wen, K.; Guo, Z.; Luo, X. Electromagnetically induced transparency (EIT)-like transmission in side-coupled complementary split-ring resonators. *Opt. Express* **2012**, *20*, 24348–24355. [CrossRef] [PubMed]

40. Yu, W.; Meng, H.; Chen, Z.; Li, X.; Zhang, X.; Wang, F.; Wei, Z.; Tan, C.; Huang, X.; Li, S. The bright–bright and bright–dark mode coupling-based planar metamaterial for plasmonic EIT-like effect. *Opt. Commun.* **2018**, *414*, 29–33. [CrossRef]

41. Chu, H.S.; How Gan, C. Active plasmonic switching at mid-infrared wavelengths with graphene ribbon arrays. *Appl. Phys. Lett.* **2013**, *102*, 231107. [CrossRef]

42. Palik, E.D. *Handbook of Optical Constants of Solids*; Academic Press: Boston, MA, USA, 1998; Volume 3, p. 662.

43. Basu, S.; Lee, B.J.; Zhang, Z.M. Infrared Radiative Properties of Heavily Doped Silicon at Room Temperature. *J. Heat Transfer* **2009**, *132*, 23301–23308. [CrossRef]

44. Lao, J.; Tao, J.; Wang, Q.J.; Huang, X.G. Tunable graphene-based plasmonic waveguides: Nano modulators and nano attenuators. *Laser Photon. Rev.* **2014**, *8*, 569–574. [CrossRef]

45. Chen, Z.H.; Tao, J.; Gu, J.H.; Li, J.; Hu, D.; Tan, Q.L.; Zhang, F.; Huang, X.G. Tunable metamaterial-induced transparency with gate-controlled on-chip graphene metasurface. *Opt. Express* **2016**, *24*, 29216–29225. [CrossRef] [PubMed]

46. Yao, J.; Chen, Y.; Ye, L.; Liu, N.; Cai, G.; Liu, Q.H. Multiple resonant excitations of surface plasmons in a graphene stratified slab by Otto configuration and their independent tuning. *Photon. Res.* **2017**, *5*, 377–384. [CrossRef]

47. Gao, W.; Shu, J.; Qiu, C.; Xu, Q. Excitation of Plasmonic Waves in Graphene by Guided-Mode Resonances. *ACS Nano* **2012**, *6*, 7806–7813. [CrossRef] [PubMed]

48. Vakil, A.; Engheta, N. Transformation Optics Using Graphene. *Science* **2011**, *332*, 1291–1294. [CrossRef] [PubMed]

49. Casiraghi, C.; Hartschuh, A.; Lidorikis, E.; Qian, H.; Harutyunyan, H.; Gokus, T.; Novoselov, K.S.; Ferrari, A.C. Rayleigh Imaging of Graphene and Graphene Layers. *Nano Lett.* **2007**, *7*, 2711–2717. [CrossRef]

50. Yan, H.; Li, X.; Chandra, B.; Tulevski, G.; Wu, Y.; Freitag, M.; Zhu, W.; Avouris, P.; Xia, F. Tunable infrared plasmonic devices using graphene/insulator stacks. *Nat. Nanotechnol.* **2012**, *7*, 330–334. [CrossRef]

51. Jain, P.K.; Eustis, S.; El-Sayed, M.A. Plasmon Coupling in Nanorod Assemblies: Optical Absorption, Discrete Dipole Approximation Simulation, and Exciton-Coupling Model. *J. Phys. Chem. B* **2006**, *110*, 18243–18253. [CrossRef]
52. Prodan, E.; Nordlander, P. Plasmon hybridization in spherical nanoparticles. *J. Chem. Phys.* **2004**, *120*, 5444–5454. [CrossRef] [PubMed]
53. Koh, A.L.; Bao, K.; Khan, I.; Smith, W.E.; Kothleitner, G.; Nordlander, P.; Maier, S.A.; McComb, D.W. Electron Energy-Loss Spectroscopy (EELS) of Surface Plasmons in Single Silver Nanoparticles and Dimers: Influence of Beam Damage and Mapping of Dark Modes. *ACS Nano* **2009**, *3*, 3015–3022. [CrossRef] [PubMed]
54. Chen, H.; Shao, L.; Li, Q.; Wang, J. Gold nanorods and their plasmonic properties. *Chem. Soc. Rev.* **2013**, *42*, 2679–2724. [CrossRef] [PubMed]
55. Cheng, H.; Chen, S.; Yu, P.; Duan, X.; Xie, B.; Tian, J. Dynamically tunable plasmonically induced transparency in periodically patterned graphene nanostrips. *Appl. Phys. Lett.* **2013**, *103*, 203112. [CrossRef]

 *nanomaterials*

Article

Dynamically Tunable Light Absorbers as Color Filters Based on Electrowetting Technology

Jun Wu *, Yaqiong Du, Jun Xia, Tong Zhang, Wei Lei and Baoping Wang

Joint International Research Laboratory of Information Display and Visualization, School of Electronic Science and Engineering, Southeast University, Nanjing 210096, China; 220171415@seu.edu.cn (Y.D.); xiajun@seu.edu.cn (J.X.); tzhang@seu.edu.cn (T.Z.); lw@seu.edu.cn (W.L.); wbp@seu.edu.cn (B.W.)
* Correspondence: wujunseu@seu.edu.cn

Received: 21 November 2018; Accepted: 31 December 2018; Published: 6 January 2019

Abstract: A device that uses the electrowetting fluid manipulation technology to realize the reversible and dynamical modulation of the local surface plasmon resonance is invented. By varying the electrowetting voltage, the distribution of fluids media surrounding the grating structure get changed accordingly, causing the modulation of the plasmonic resonance peak. The simulation results indicated that three primary colors, that are cyan, magenta and yellow (CMY) can be respectively reflected through selecting suitable structural parameters. More importantly, for the first time, the invented fluid-based devices have exhibited fine-tuning characteristics for each primary color. Finally, the device has been proved to have a large color gamut range in the Commission International De L'E'clairage (CIE) 1931 color space.

Keywords: electrowetting; tunable absorbers; subwavelength metal grating; plasmon resonance

---

1. Introduction

Artificial micro-nano structures can produce various singular optical properties and phenomena not available in natural materials [1]. Photonics devices based on the artificial micro-nano structures have broad application prospects in many fields such as display [2], detection [3], imaging [4], remote sensing [5] and printing [6] because of the advantages such as designable optical properties, excellent process compatibility, and outstanding integration [7–9]. Benefits from the development of micromachining technology, that is the emergence of nanoprocessing methods, the feasible working band of photonic devices has extended from infrared to visible light, based on which new principles and new methods are expected to be developed for micro-nano characterization techniques and micro-nano photonic functional devices.

In recent decades, researchers have conducted extensive research on various types of micro-nano-structured photonic devices and have obtained many novel and important research results [10–13]. Color filter is one kind of common optical component that selectively reflects or transmits a specific wavelength in the visible light band and exhibits different colors. Color filters with red, green and blue primary colors are widely used in liquid crystal display, optical communication, sensor detection, imaging and a wide range of other fields [14,15]. The color filters based on periodic nanostructures have a small absorption rate of incident light, and meanwhile, the performance parameters of which, such as the central spectral position and bandwidth can be well adjusted by the structural parameters. Especially, the one based on the periodic grating structure can be easily fabricated by photolithography or nanoimprint process, which has not only low processing cost but also superior reflection and transmission properties, has attracted numerous interests [16–22]. Researchers found that the grating-based nanophotonic devices exhibit singular optical effects such as guided mode resonance and surface plasmon resonance. Furthermore, by adjusting the orientation, period, depth, etc. of the sub-wavelength grating, selective transmission (reflection/absorption) could be realized, and therefore it has attracted widespread attention in many applications [23,24]. However, it is also highly desired in some fields to make the working band of photonic device regulated dynamically and reversibly. Therefore, solid-based photonic devices with invariable structural parameters exhibit severe limitations. In recent years, researchers have made great efforts to develop adjustable plasmonic devices. An effective method is to exploit the remarkable properties of plasmons—high sensitivity to surrounding dielectric materials of plasmonic nanostructures [25–27]. For example, Wang and Chumanov have demonstrated electrochemical modulation of the LSPR intensity and frequency in Ag nanoparticles by coating with a tungsten oxide gel [28]. Zheng et al. applied a bistable, redox-controllable rotaxane molecule to a gold nanodisk array to achieve active adjustment of plasmon resonance by chemical conversion between its oxidized and reduced states. [29].

In this paper, we present an adjustable filter based on a grating structure by creatively adopting an advanced electrowetting fluid-manipulation technology in a double-layer metal–fluid–dielectric configuration. By changing the period of the grating, we have achieved the reflection of CMY (cyan, magenta, yellow) primary colors. Furthermore, the tinge of each primary color is dynamically and reversibly regulated for the first time by simply varying the electrowetting voltage applied to change the distribution of the fluids. And due to the fast speed of electrowetting regulation, the response time of the device is around 10 ms, which is more responsive and superior than many other methods. Furthermore, the benefits also include reversible regulation, miniaturization, and so forth. This tunable filter allows for a variety of tinge choices for each kind of colors, extends the color field of the display device without the demand of increasing the number of components. In addition, our fluid-based design is reusable and flexible, compared to the solid-based devices.

2. Materials and Methods

Figure 1a,b shows the principle of the adopted advanced fluid manipulation technique, which is generally named as electrowetting. Electrowetting technique is used in our research to achieve contact angle modulation. The contact angle of the conducting fluid on the solid surface is determined by the force balance at the contact point. The initial equilibrium contact angle, as shown in Figure 1a, θ_0, is given by Young's equation:

$$\gamma_{s1} + \sigma_{12} \cos \theta_0 = \gamma_{s2} \tag{1}$$

Here γ_{s1} is the surface energy per unit area between the solid surface and conducting fluid, γ_{s2} is the surface energy per unit area between the solid surface and insulating fluid, and σ_{12} is the surface tension at the interface between the two fluids.

In electrowetting devices, a voltage is usually applied between the conducting fluid and the driving electrode. In many applications, a thin hydrophobic dielectric layer is deposited onto the driving electrode, which is often referred as "electrowetting on dielectric" (EWOD). In this case, the capacitance of the dielectric layer is superior to the electrical double layer capacitance at the electrode-liquid interface. By increasing the voltage, the energy stored in the capacitor get increased, and meanwhile, the effective surface energy gets reduced. Young's equation is therefore modified as follows:

$$\gamma_{s1} - \frac{\varepsilon V^2}{2d_f} + \sigma_{12} \cos \theta_{ew} = \gamma_{s2} \tag{2}$$

Here ε is the permittivity of the dielectric, V is the applied voltage-difference, and d_f is the dielectric thickness. Combining Equations (1) and (2) yields

$$\cos \theta_{ew} = \cos \theta_0 + \frac{\varepsilon V^2}{2\sigma_{12}d_f} \tag{3}$$

Thus, electrowetting can be well used to dynamically modify the force balance at the contact point, and thereby the contact angle gets changed accordingly, as shown in Figure 1b.

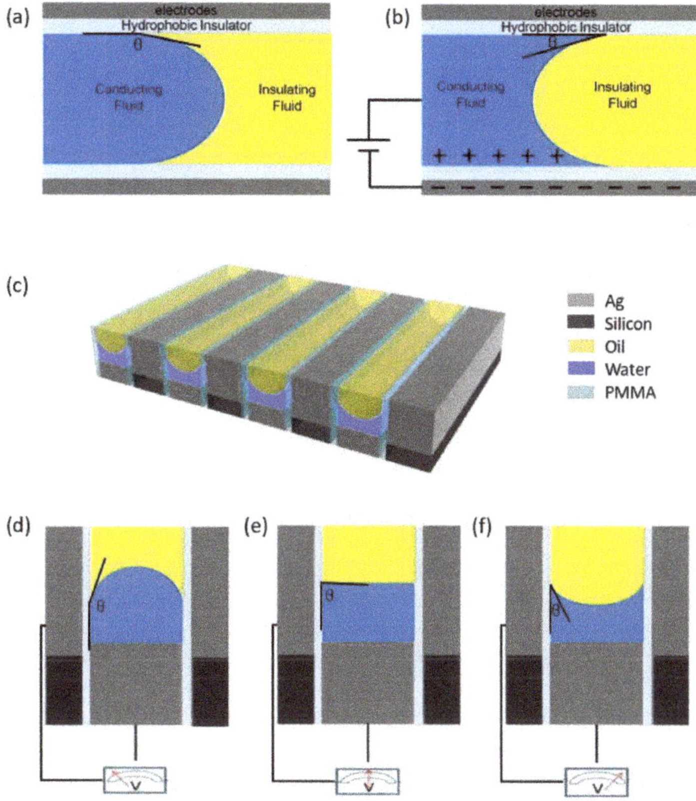

Figure 1. The principle of electrowetting is shown in (**a**,**b**). (**a**) The initial state and (**b**) indicates the situation after applying voltage. (**c**) The sketch of the model we created. In the figure, gray, black, yellow, blue and transparent areas respectively represent metallic silver, silicon elemental, oil, water and medium polymethyl methacrylate (PMMA). (**d**–**f**) The three different cases of the interface morphology after voltage application.

Based on the principle of electrowetting introduced above, we have established a two-layer metal grating structure model as shown in Figure 1c. In such design, silver and silicon are selected as the metal and dielectric materials respectively, forming the core architecture of grating filter. Compared with single-layer metal grating, the two-layer sub-wavelength metal grating has superior polarization and filtering performance. To integrate the electrowetting features in the model, a thin, hydrophobic polymethyl methacrylate (PMMA) layer is added onto the sidewalls of upper silver strips. Meanwhile, the upper and lower silver strips are acting as the driving and grounding electrodes respectively. A certain proportion of water and oil, acting as the conducting fluid and insulating fluid respectively, are then added to the space between the upper silver strips of the adjacent gratings. The interface curvature between the oil and water is initially determined by the force balance as predicted by Equation (1) and can be modulated by regulating the electrowetting voltage applied between the two electrodes as shown in Figure 1d–f.

As the surface tension of the oil is smaller than that of water, the initial state, that is, when the voltage is 0, the contact angle of water on the sidewall surface is relatively large, and the interface bulges in the direction of the oil, as shown in Figure 1d. During the process of increasing the applied voltage, the contact angle is gradually reduced, and the interface gradually flattens and then bulges toward the water, as shown in Figure 1e,f. Meanwhile, the media distribution in such device changes with the interface morphology, rendering the modulation of filtering function. If we reduce the voltage again, the centre point of oil–water interface will gradually recover to the previous height. Furthermore, the response time of electrowetting technology is particularly short, so the desired interface morphology can be quickly realized by applying a certain voltage.

To explore the relationship between the interface morphology and applying voltage, as well as the dynamical filtering function, COMSOL Multiphysics and the finite difference method are used in present study. For material selection in the simulation software FDTD, we chose oil with a refractive index of 1.7, water with a refractive index of 1.3, silicon in the material library, and silver in the palik (0–2 μm). The boundary condition is a periodic condition, and the light source is a 400–760 nm plane wave.

3. Results and Discussion

As can be predicted according to the electrowetting theory, the oil/water interface morphology between the upper silver strips can be regulated as desired in our model, by selectively varying the device parameters, e.g., the voltage applied, the spacing between strips, and the hydrophobic layer material type and thickness. During our investigation, we firstly settled our device period as 400 nm, with duty cycle as 0.5. We set the upper layer metal thickness as 100 nm, lower layer metal thickness as 140 nm, and Si layer thickness as 100 nm.

In a typical EWOD configuration, the same contact angle variation usually requires a higher applying voltage for a higher dielectric layer thickness, which is the reason why a very thin hydrophobic PMMA layer is used in present study for minimizing energy consumption as well as the component miniaturization. By slightly increasing the electrowetting voltage, the contact angle in our device can reduce significantly. Meanwhile, the interface morphology is synchronously changed from a protruding state to a concave state, as shown in Figure 2a–c. The distribution of two fluids get changed continuously, in another word, the surrounding media of the two-layer metal grating structure is changed during this whole process. Theoretically, the filtering function of the device will be also changed. It is well known that the working band of the filter can be adjusted by changing the grating period. For a specific period, changing the distribution of the liquid, that is, changing the refractive index of the environment media around the metal, can also affect the absorption of a specific wavelength.

Figure 2d–f is the corresponding electric field distribution diagram in this case. As can be observed, the electric field is mainly concentrated in the region between the ridge of the grating and the trench, that is, the transparent dielectric layer PMMA and the region where the fluid exists. From the images we can see that as the contact angle decreases, the coverage area proportion of oil which covers the upper metal surface is decreasing, and the electric field excited in the vicinity is also gradually weakened. A larger coverage area proportion of oil on the metal surface implies there is a greater dielectric constant of the surrounding environment, which will offset more free charge on the metal surface by the polarization charge generated in the surrounding medium. Then, the restoring force become smaller, and thereby the resonance energy of the surface plasmon becomes weakened. Therefore, in the process of increasing the voltage, the coverage area of oil gradually decreases, and the resonance wavelength continuously shifts blue. Figure 2g–i indicates the distribution of magnetic field, which is mainly concentrated in the region between the upper metal and the lower metal, that is, the transparent dielectric layer PMMA. It is noticed that the magnitude of the magnetic field enhancement is also related to the distribution of the fluids interface. Figure 2j shows the modulation of the resonant wavelength by different electrowetting voltages under the same structural parameters as mentioned in previous section. As can be observed, the wavelength varies from 579 nm to 542 nm continuously. The corresponding extinction tinge was also changed.

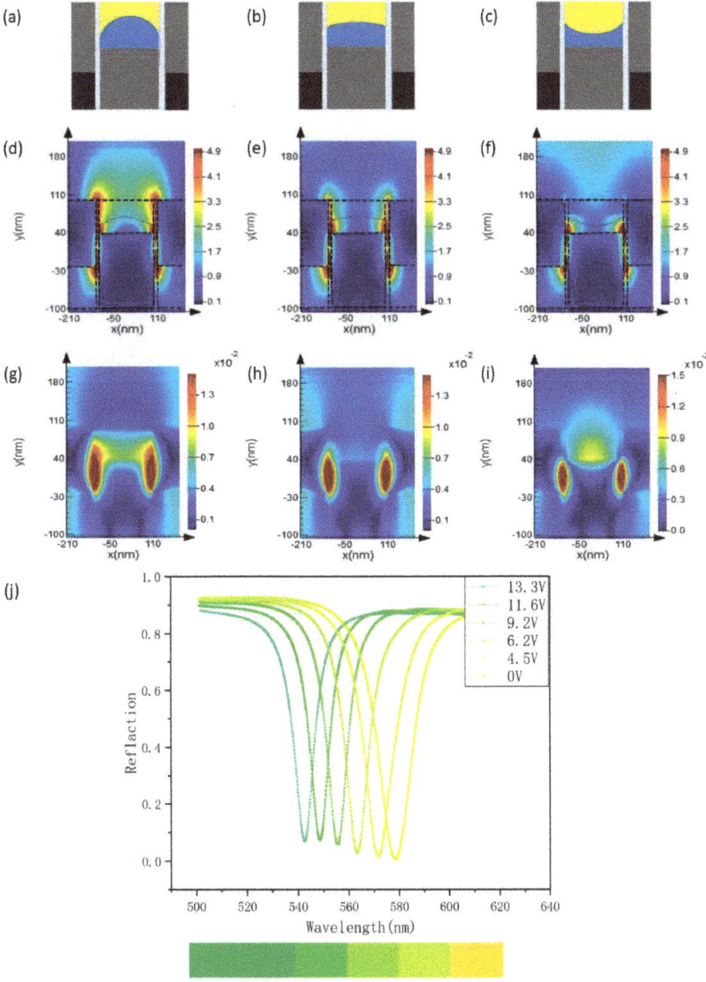

Figure 2. (a–i) The electric field and magnetic field distribution of the invented device with a period size of 400 nm working at the electrowetting voltage of 4 V, 8 V and 12 V, respectively. (j) Indicates the reflection spectrum of the device with the mentioned period size.

For a color filter based on the solid-state grating structure, the modulation in filtering function is usually realized by varying the periodic size. In current study, we also carried out the related investigation in our new designed fluid-based two-layer metal grating structure. Considering the interests of visible light, the periodic size is changed from 250 nm to 600 nm, with the increasing step of 50 nm. As can be observed in Figure 3a, the extinction spectra get adjusted significantly, to be specific, it continuously red-shifts from 462 nm to 720 nm with the periodic size increases. In another word, three primary colors, i.e., red, green and blue, can be obtained in our structures. More importantly, the extinction wavelength of each structure increases with the applying voltage, which further confirms that the invented fluid-based devices have fine-tuning characteristics for each color, including the primary one, under the role of EWOD voltage. Similar with the situation introduced in Figure 2, this novel phenomenon can be attributed to the continuous change in surrounding media of the fluid-based two-layer metal grating structure by the applying voltage.

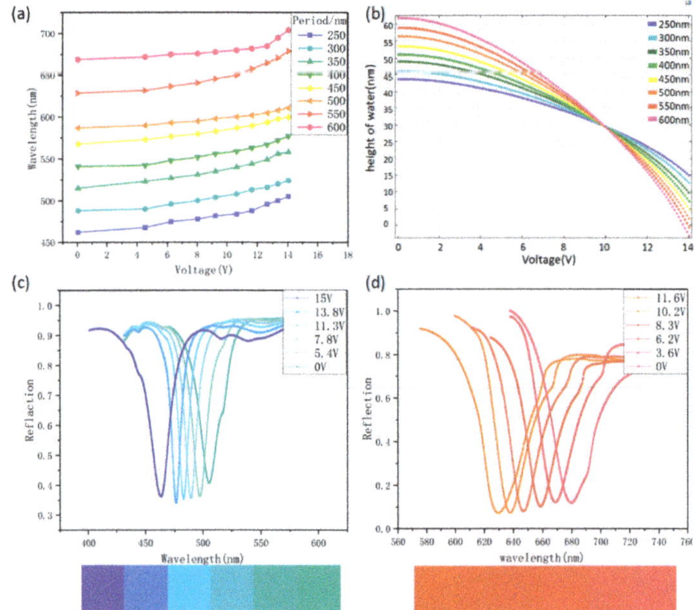

Figure 3. (**a**) The relationship of extinction wavelengths with the electrowetting voltages. (**b**) A schematic representation of the centre point height of the oil–water interface as a function of voltage. (**c**,**d**) Reflection peaks at different voltages for periods of 250 nm and 550 nm, respectively.

In Figure 3b, we present a schematic diagram of the centre point height of oil–water interface as a function of voltage for different periodic sizes. The curves are all derived from the COMSOL software, where the device parameters are accurately settled. As can be clearly observed that the height decreases nonlinearly with the increasing voltage and gets more significantly change for a larger periodic size even under the same voltage variation.

The investigation results proved that the surrounding media distribution can be well controlled by simply applying appropriate voltage and thereby the tinge of each color obtained can be regulated and selected. On this basis, we further investigated the filtering function of two cases where the periods are 250 nm and 550 nm. We can notice that these two periods respectively correspond to the extinction colors of the blue band and the red band. Furthermore, as can be seen from Figure 3c,d, they both produce a blue shift as the voltage increases. The blue color changes from 505 nm to 462 nm, and the red color changes from 680 nm to 630 nm. The corresponding tinge changes are also indicated.

By realizing the three extinction colors of red, green and blue, we can get the reflected colors of CMY. Figure 4a shows a schematic diagram of a simple color filter in which three different periods can reflect three different colors. However, unlike the conventional solid-based color filters, the device proposed here has flexible adjustability. The change of the externally applied voltage led to the changes of oil/water interface curvature inside the device cavity and causes the shift of the spectral absorption peak position. The tinge of the reflected light of the device is then changed. More importantly, this kind of device can realize the correction of performance error induced by the fabrication deviation or design defect in structural parameters, which has a significant advantage in making a device with high requirements in tinge precision without the demand of high-level processing standards.

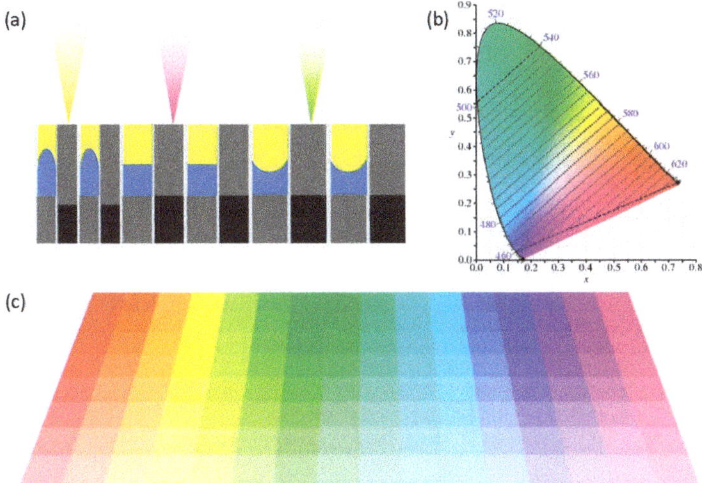

Figure 4. (**a**) The schematic diagram of the proposed integrated filtering device based on the grating structure. The periods of the three parts are 250 nm, 400 nm and 550 nm, respectively. (**b**) The range of gamuts that the filtering device can produce in CIE 1931 xy chromaticity coordinates (dashed line). (**c**) Color gradient map that can be achieved by selecting the appropriate parameters.

To provide a quantitative measure of the reflected color, we calculated the chromaticity coordinates of the spectrally reflected light and plotted it in the CIE 1931 xy color space, as shown in Figure 4b. Since the extinction color is blue, green and red, the light reflected from the device is its corresponding complementary color. Since the tinge of the extinction color varies widely, the range of tinge that can be achieved in the CIE 1931 xy color space is also quite wide. More importantly, different combinations of tinge can be achieved by applying different voltages to the gratings with different periods in the device. Integrating gratings of different periods in the same device results in a wider tinge range than conventional devices, which greatly increases the applications of the device. The colors that can be obtained are shown in Figure 4c, and each color can be realized directly or realized by mixing the colors obtained from different periods or different voltage conditions. The principle of subtractive color method is potentially useful to replace the ink and pigment to achieve various colors, avoiding the usage of the harmful substances contained in traditional inks and pigments, exhausting waste after complicated chemical processes and cumbersome processes. We can also choose different periods of grating structure to form different devices, or change the fluid material, dielectric layer thickness, grating duty cycle, metal layer height and other parameters to realize the desired color range, and thus satisfy different needs of various occasions. This further increases the flexibility of device usage. This adjustable color filter will have important application prospects in the fields of flat panel display, photovoltaic, biomimetic, green printing and security.

4. Conclusions

In summary, we have designed a new type of color filter in combination with advanced electrowetting fluid handling technology and double-layer grating structure. We fully consider the compatibility of fluid self-assembly methods with traditional micromachining methods. By changing the applied voltage between the conductive fluid and the driving electrode to change the environmental medium around the metal in the grating, the color of the extinction color can be dynamically and reversibly adjusted. The tunable device can be reused, the liquid parameters are easily changed, and the response speed is extremely fast. As such, there is a wide range of important potential applications in the fields of new flat panel display, photovoltaic, and so on.

Author Contributions: Conceptualization, J.W. and Y.D.; methodology, J.W.; investigation, Y.D.; data curation, T.Z.; writing—original draft preparation, J.W. and Y.D.; writing—review and editing, J.W. and J.X.; supervision, W.L.; project administration, B.W.; funding acquisition, J.W. and J.X.

Funding: This research was funded by National Key R&D Program of China (2017YFB1002900), National Natural Science Foundation of China (NSFC) (61775035) and Natural Science Foundation of Jiangsu Province (BK20181268).

Conflicts of Interest: The authors declare no conflict of interest.

References

1. Guo, Z.Y.; Nie, X.R.; Shen, F.; Zhou, H.P.; Zhou, Q.F.; Gao, J.; Guo, K. Actively Tunable Terahertz Switches Based on Subwavelength Graphene Waveguide. *Nanomaterials* **2018**, *8*, 665. [CrossRef] [PubMed]

2. Dong, L.; Agarwal, A.K.; Beebe, D.J.; Jiang, H.R. Adaptive liquid microlenses activated by stimuli-responsive hydrogels. *Nature* **2006**, *442*, 551–554. [CrossRef] [PubMed]

3. Subr, M.; Prochazka, M. Polarization- and Angular-Resolved Optical Response of Molecules on Anisotropic Plasmonic Nanostructures. *Nanomaterials* **2018**, *8*, 418. [CrossRef] [PubMed]

4. Liao, T.W.; Verbruggen, S.W.; Claes, N.; Yadav, A.; Grandjean, D.; Bals, S.; Lievens, P. TiO2 Films Modified with Au Nanoclusters as Self-Cleaning Surfaces under Visible Light. *Nanomaterials* **2018**, *8*, 30. [CrossRef] [PubMed]

5. Schwarz, B.; Reininger, P.; Ristanic, D. Monolithically integrated mid-infrared lab-on-a-chip using plasmonics and quantum cascade structures. *Nat. Commun.* **2014**, *5*, 4085. [CrossRef] [PubMed]

6. Elmogi, A.; Soenen, W.; Ramon, H. Aerosol-Jet Printed Interconnects for 2.5 D Electronic and Photonic Integration. *J. Lightw. Technol.* **2018**, *36*, 3528–3533. [CrossRef]

7. Bonaccorso, F.; Sun, Z.; Hasan, T. Graphene photonics and optoelectronics. *Nat. Photonics* **2010**, *4*, 611–622. [CrossRef]

8. Hutter, E.; Fendler, J.H. Exploitation of localized surface plasmon resonance. *Adv. Mater.* **2004**, *16*, 1685–1706. [CrossRef]

9. Liu, M.; Yin, X.B.; Erick, U.A. A graphene-based broadband optical modulator. *Nature* **2011**, *474*, 64–67. [CrossRef]

10. Gramotnev, D.K.; Bozhevolnyi, S.I. Plasmonics beyond the diffraction limit. *Nat. Photonics* **2010**, *4*, 83–91. [CrossRef]

11. Fan, P.; Colombo, C.; Huang, K.C.; Krogstrup, P.; Nygard, J.; Fontcuberta, I.M.A.; Brongersma, M.L. An electrically-driven GaAs nanowire surface plasmon source. *Nano Lett.* **2012**, *12*, 4943–4947. [CrossRef] [PubMed]

12. Zhu, J.; Qin, L.L.; Song, S.X.; Zhong, J.W.; Lin, S.Y. Design of a surface plasmon resonance sensor based on grating connection. *Photonic Sens.* **2015**, *5*, 159–165. [CrossRef]

13. Jules, L.H.; Nikhil, B.; Sarah, D.R.; Pedro, E. Localized surface plasmon resonance as a biosensing platform for developing countries. *Biosensors* **2014**, *4*, 172–188.

14. Song, S.C.; Ma, X.L.; Pu, M.B.; Li, X.; Guo, Y.H.; Gao, P.; Luo, X.G. Tailoring active color rendering and multiband photodetection in a vanadium-dioxide-based metamaterial absorber. *Photonics Res.* **2018**, *6*, 492–497. [CrossRef]

15. Jin, C.; Lin, C.D. Control of soft X-ray high harmonic spectrum by using two-color laser pulses. *Photonics Res.* **2018**, *6*, 434–442. [CrossRef]

16. Bibbo, L.; Khan, K.; Liu, Q.; Lin, M.; Wang, Q.; Quyang, Z.B. Tunable narrowband antireflection optical filter with a metasurface. *Photonics Res.* **2017**, *5*, 500–506. [CrossRef]

17. Li, A.; Bogaerts, W. Backcoupling manipulation in silicon ring resonators. *Photonics Res.* **2018**, *6*, 620–629. [CrossRef]

18. Kanamori, Y.; Shhimono, M.; Hane, K. Fabrication of transmission color filters using silicon subwavelength gratings on quartz substrate. *IEEE Photonics Technol. Lett.* **2006**, *18*, 2126–2128. [CrossRef]

19. Yoon, Y.; Lee, H.; Lee, S. Color filter incorporating a subwavelength Patterned grating in poly silicon. *Opt. Express* **2008**, *4*, 2374–2380. [CrossRef]

20. Lee, H.; Yoon, Y.; Lee, S. Color filter based on a subwavelength patterned metal grating. *Opt. Express* **2007**, *23*, 15457–15463. [CrossRef]

21. Kim, J.; Koo, J.; Lee, J.H. All-fiber acousto-optic modulator based on a cladding-etched optical fiber for active mode-locking. *Photonics Res.* **2017**, *5*, 391–395. [CrossRef]

22. Kanamori, Y.; Kitani, T.; Hane, K. Guided-mode resonant grating filter fabricated on silicon-on insulator Substrate. *J. Appl. Phys.* **2006**, *45*, 1883–1885. [CrossRef]

23. Ebbesen, T.W.; Lezec, H.J.; Ghaemi, H.F.; Thio, T.; Wolff, P.A. Extraordinary optical transmission through sub-wavelength hole arrays. *Nature* **1998**, *391*, 667–669. [CrossRef]

24. William, L.B.; Alain, D.; Thomas, W.E. Surface Plasmon subwavelength optics. *Nature* **2003**, *424*, 824–830.

25. Lee, K.S.; El-Sayed, M.A. Gold and silver nanoparticles in sensing and imaging: Sensitivity of plasmon response to size, shape, and metal composition. *J. Phys. Chem. B* **2006**, *110*, 19220–19225. [CrossRef] [PubMed]

26. Liu, N.; Weiss, T.; Mesch, M.; Langguth, L.; Eigenthaler, U.; Hirscher, M.; Sönnichsen, C.; Giessen, H. Planar Metamaterial Analogue of Electromagnetically Induced Transparency for Plasmonic Sensing. *Nano Lett.* **2010**, *10*, 1103–1107. [CrossRef]

27. Zhang, W.; Martin, O.J.F. A Universal Law for Plasmon Resonance Shift in Biosensing. *ACS Photonics* **2015**, *2*, 144–150. [CrossRef]

28. Wang, Z.; Chumanov, G. WO3 sol-gel modified Ag nanoparticle arrays for electrochemical modulation of surface plasmon resonance. *Adv. Mater.* **2003**, *15*, 1285–1289. [CrossRef]

29. Zheng, Y.B.; Yang, Y.W.; Jensen, L.; Fang, L.; Juluri, B.K.; Flood, A.H.; Weiss, P.S.; Stoddart, J.F.; Huang, T.J. Active Molecular Plasmonics: Controlling Plasmon Resonances with Molecular Switches. *Nano Lett.* **2009**, *9*, 819–825. [CrossRef]

 *nanomaterials*

Review

Liquid Crystal Enabled Dynamic Nanodevices

Zhenhe Ma [1], Xianghe Meng [1], Xiaodi Liu [1], Guangyuan Si [2,*] and Yan Jun Liu [3,*]

[1]   College of Information Science and Engineering, Northeastern University, Shenyang 110004, China;
      mazhenhe@163.com (Z.M.); siguang@mail.ustc.edu.cn (X.M.); youjing56789@gmail.com (X.L.)

[2]   Melbourne Centre for Nanofabrication, Victorian Node of the Australian National Fabrication Facility,
      Clayton, VIC 3168, Australia

[3]   Department of Electrical and Electronic Engineering, Southern University of Science and Technology,
      Shenzhen 518055, China

*   Correspondence: guangyuan.si@monash.edu (G.S.); yjliu@sustc.edu.cn (Y.J.L.)

Received: 26 September 2018; Accepted: 20 October 2018; Published: 23 October 2018

Abstract: Inspired by the anisotropic molecular shape and tunable alignment of liquid crystals (LCs), investigations on hybrid nanodevices which combine LCs with plasmonic metasurfaces have received great attention recently. Since LCs possess unique electro-optical properties, developing novel dynamic optical components by incorporating nematic LCs with nanostructures offers a variety of practical applications. Owing to the large birefringence of LCs, the optical properties of metamaterials can be electrically or optically modulated over a wide range. In this review article, we show different elegant designs of metasurface based nanodevices integrated into LCs and explore the tuning factors of transmittance/extinction/scattering spectra. Moreover, we review and classify substantial tunable devices enabled by LC-plasmonic interactions. These dynamically tunable optoelectronic nanodevices and components are of extreme importance, since they can enable a significant range of applications, including ultra-fast switching, modulating, sensing, imaging, and waveguiding. By integrating LCs with two dimensional metasurfaces, one can manipulate electromagnetic waves at the nanoscale with dramatically reduced sizes. Owing to their special electro-optical properties, recent efforts have demonstrated that more accurate manipulation of LC-displays can be engineered by precisely controlling the alignment of LCs inside small channels. In particular, device performance can be significantly improved by optimizing geometries and the surrounding environmental parameters.

Keywords: liquid crystals; metasurfaces; plasmonics; actively tunable nanodevices

---

1. Introduction

A surface plasmon resonance (SPR) [1–7] is formed when the incident electrons resonate with the vibration on the surface of metallic nanostructures, which has caused broad interest regarding novel applications [8–19]. Researchers have shown that plasmonic crystals [20,21] are essential for superlenses [22,23], negative refraction applications [24,25], and ultra-large nonlinearity devices [26,27]. Plasmonic metasurface-based devices [28–34] have demonstrated increasing potential practical applications citing their unique optical characteristics, which can be chosen to be more favorable than natural materials [35–37]. Since the experimental demonstration of exotic negative refraction, the concept of metamaterials has been widely favored in the scientific community, especially for enhancing polarization control [38–42], absorption [43–45], asymmetric transmission [46–48], cloaking [49–51], slow light generation [52,53], and novel sources of coherent radiation [54,55]. Similarly, metasurfaces can be easily fabricated using either a top-down or bottom-up approach, and their optical properties can be externally controlled via hybridization with a naturally available functional material, significantly expanding the range of potential practical applications [56,57].

So far, several different tuning principles [58–62] based on various types of liquid crystal (LC) molecules, such as chiral [63], nematic [64], and smectic [65] characteristics have been experimentally demonstrated to realize dynamically controllable nanodevices. Since LC-based modulation mechanism can offer an additional advantage because the LC-molecules exhibit a large optical anisotropy [66–83], investigations on LCs combined with nanostructures [84–86] or two-dimensional materials [87] have recently drawn significant attention and interest. Developing tunable optical metamaterials by incorporating nematic LCs as an electro-optic or nonlinear optical constituent has become a popular research topic [88–93]. Given the large birefringence of LCs, the optical properties of these metamaterials can be electrically or optically manipulated through a wide frequency range. Shrekenhamer and coworkers [90] have demonstrated electronically tunable metamaterial absorbers in the terahertz (THz) regime. By incorporating the active LCs into the metamaterial unit cell, the absorption was modified by 30 percent at 2.62 THz, and the resonant absorption could be tuned over 4 percent in bandwidth. A spatial modulator was achieved by Savo et al., using isothiocyanate-based LCs in the THz range [91]. Moreover, by adopting an external voltage, electro-optic switching devices have been achieved via reflection and refraction of LC-cells [92,93]. In addition, LCs are also capable of enhancing the angle reliance of a localized plasmon resonance, enabling ultra-sensitive detection of the resonance shift in the visible and near-infrared regimes [94]. Here, we summarize recent achievements of dynamically tunable nanodevices based on LCs. Given their fantastic optical properties, LCs have found extensive applications, most commonly in display techniques (high display quality, low electromagnetic radiation, large visible area, and low power consumption). On the other hand, metasurfaces offer the advantage of a significant reduction in sizes and physical dimensions and can achieve complex functionalities with simple, elegant designs. Therefore, hybrid optoelectronic devices may be enabled to realize more varied functionalities and a wider range of practical applications by combining LCs and metasurfaces.

2. Manipulating LCs and Interaction of LCs with Metasurfaces

2.1. Transmission of LCs Covered Metasurfaces

Metasurfaces have become a ubiquitous instrument for low-loss operation of phase, intensity, and polarization of light. Since they can effectively control the flow of light, many applications have been explored from microwave to visible frequencies, including lenses, beam deflectors, and hologram devices. Recent research [95] has presented dynamic manipulation on the optical response of Mie-resonant metasurfaces by controlling the alignment of LCs with an electric field.

As shown in Figure 1a,b, silicon nanodisks can be finely integrated into the LCs. The height and diameter of the nanodisks were 220 nm and 606 nm with a lattice constant of 909 nm. Note that the extraordinary and ordinary indices of nematic cells used were 1.7 and 1.51, respectively. In addition, the internal thickness of the LC-layer was fixed at 5 μm using a suitable spacer material, whilst the substrate was used as an electrode. The preferred orientation of the LC cells was induced by brushed Nylon-6. For further verification, when the voltage was "OFF", the LC alignment was parallel with the metasurface. To achieve the switching effect, the LC molecules can be redirected perpendicular to the metasurface when the voltage was turned "ON".

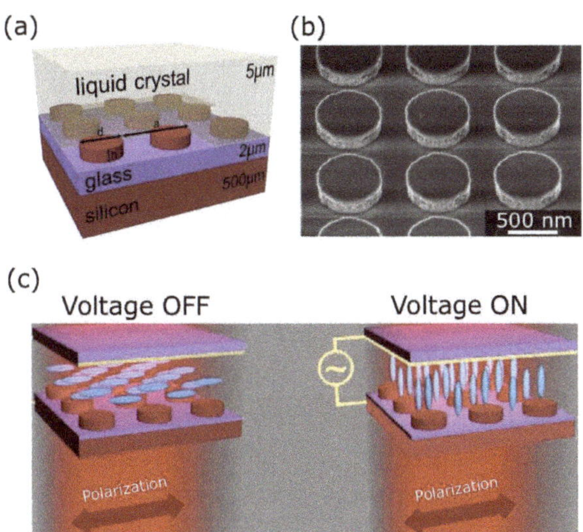

Figure 1. (**a**) Schematic drawing of hybrid structures of nanodisk metasurfaces covered with liquid crystals (LCs). (**b**) Scanning electron microscopy (SEM) image of silicon nanodisks. (**c**) Schematic diagram showing the working mechanism. Reproduced with permission from [95]. Copyright American Institute of Physics, 2017.

Thus far, most investigations have been focused on the active control of metasurfaces immersed in nematic LCs. It has been found that one can manipulate the alignment of LC-cells by applying an external electric field, which results in a significant exponential variation in a series of plasmonic devices [96–98]. Decker and coworkers have shown active control on transmission spectrum of LC-coated metasurfaces [99]. With an x-polarized incident beam, the transmittance spectrum was measured. There is only electric resonance at the wavelength of 900 nm without a bias voltage ($V = 0$ V), as shown in Figure 2a. After applying a 6 V external voltage, the device shows two magnetic resonances at wavelengths of 600 nm and 800 nm, respectively. Furthermore, the color change from yellow to almost transparent between "OFF" and "ON" states can be observed, as shown in the inset of Figure 2a. Due to the reorientation of LC-cells driven by the external electric field, the incident light had to rotate 90° to form a helical distribution when passing through the hybrid device. When the bias voltage was switched on, the LC molecules between the two electrodes were redirected (parallel with the electric field), disrupting the helical distribution. Figure 2b plots the threshold characteristic of the working principle of this active device. One can see that the normalized transmittance increases dramatically between 2 and 3 V, producing the switching effect. Once the voltage reaches 5 V, the light transmittance is saturated and the spectrum remains unchanged.

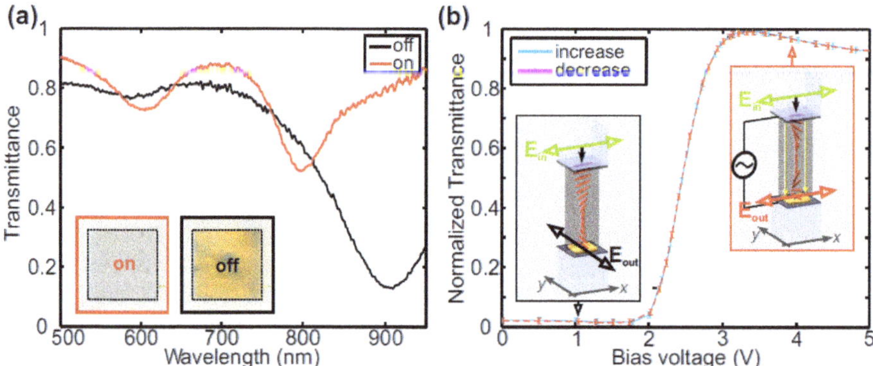

Figure 2. (**a**) Measured transmission of LC-covered metasurfaces when the external electric field is on and off. The insets demonstrate captured charge-coupled device (CCD) images showing the color difference from yellow to almost transparent. (**b**) Normalized transmittance as a function of bias voltage. The threshold behavior can be clearly observed for the switching process. Insets show schematic of LC-cells alignment without (left) and with (right) electric field. Reproduced with permission from [99]. Copyright Optical Society of America, 2013.

2.2. Transmission of PDLCs Controlled by Surface Acoustic Waves

For the past few years, polymer-dispersed liquid crystals (PDLCs) have found wide applications in a variety of fields. The orientation of LC-cells can be realigned by applying an external electric field, and therefore, the refractive index difference can be precisely modulated between cells and the polymer matrix, which means PDLCs can be shifted between transparent and opaque phases. Using surface acoustic waves (SAWs), Liu and coworkers [86] have demonstrated that one can match the refractive index between LC-cells and the polymer matrix, and therefore, achieve a switching effect. Accurate active control can be realized to drive the LC-cells to a specific orientation. Figure 3a plots the transmission of the PDLC film as a function of time with varying acoustic intensities, and the switching effect is observed clearly from the inset. Unperturbed, the PDLC film was opaque. After applying a SAW, the PDLC film became transparent and letters of "PDLC" were revealed. Note that a low SAW intensity may result in a long switch-on time with a correspondingly small optical contrast ratio. Alternatively, the switch-off time can be significantly increased by reducing the size of the droplets. For example, a PDLC-film with a uniform micron level droplet (1–3 µm) may lead to a switch-off time in the milliseconds scale. Furthermore, the driving power can be effectively decreased by adding a surfactant to PDLCs, which can act as a lubricating reagent. From Figure 3b, a clear switching effect is shown where a double exponential function is employed to fit the experimental curves, as in Reference [86]:

$$I(t) = I_{\min} + I_0 \sin^2\left\{ \frac{1}{2}\delta_1[1 - \exp(-t/\tau_1)] + \frac{1}{2}\delta_2[1 - \exp(-t/\tau_2)] \right\} \qquad (1)$$

where δ_i and τ_i ($i = 1, 2$) denote the fitting parameters. It is clear that Figure 3b shows the dynamic switching behavior which can be well-fitted by the double exponential function.

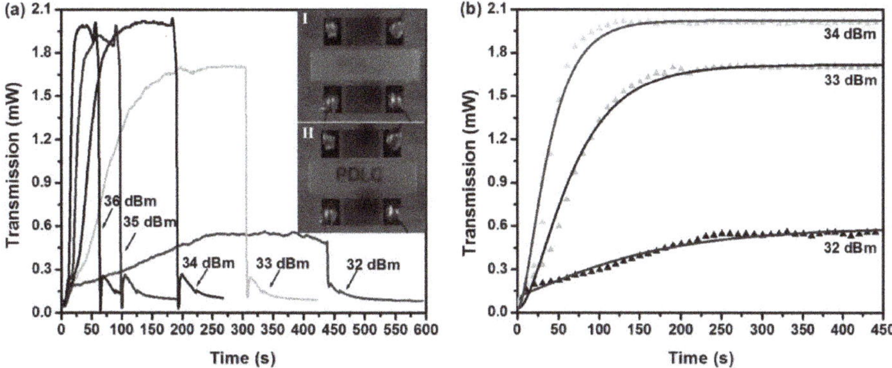

Figure 3. (**a**) Transmission as a function of time showing the switching process with varying power of surface acoustic waves (SAWs). The insets illustrate that the polymer-dispersed liquid crystal (PDLC) film is opaque without SAW and becomes transparent with SAW. (**b**) The enlarged view with time from 0 S to 450 S showing the switch-on process with theoretical fitting. Reproduced with permission from [86]. Copyright John Wiley & Sons, 2011.

2.3. Transmission of Coaxial Color Filters Covered with Photoresponsive LCs

Recently, plasmonic color filters based on annular aperture arrays (AAAs) were experimentally demonstrated by manipulating their geometric parameters [100,101]. However, the transmitted intensity was low because of large propagation losses, resulting in weak coupling between neighboring annular apertures. Most structural color filters are passive, which means the output is constant once the design is defined. Therefore, developing active color filtering devices with small pixels and high efficiency is crucial for new display techniques. By combining plasmonic nanopixels with photoresponsive LCs, all-optical color filters have been demonstrated with enhanced transmission [80]. Plasmon resonance peaks can be tuned across the whole visible band, allowing individual color outputs to be filtered out from a white light source. As shown in Figure 4a, two different resonance peaks are contained in the wavelength range of 500–800 nm, which are known as planar surface plasmons (PSPs) and cylindrical surface plasmons (CSPs), respectively. For small aperture widths (20, 40, 60, and 80 nm), one can see that CSPs are dominant. When the aperture size is 100 nm, CSPs and PSPs show similar intensity (around 7.5%). With further increased aperture widths, PSPs contribute more and grow faster than CSPs peaks. Calculations in Figure 4a, also indicate the trend of the redshift of PSPs with increasing aperture widths. Experimentally measured spectra in Figure 4b, show similar intensity change and resonance positions for PSPs peaks. However, CSPs show random positions due to fabrication imperfections. The transmission intensity of the short wavelength range is very different. Moreover, it can be seen from Figure 4c that the intensity of CSP is higher than PSP for smaller aperture sizes. Note that PSP-induced propagation gradually plays a leading role with increasing aperture sizes, and their full width at half maximum (FWHM) grows and finally reaches saturation. Figure 4c shows the peak position of the transmittance as a function of aperture size. In addition, the transmission peak induced by PSPs linearly redshifts. The transmission peak induced by CSPs linearly blueshifts with larger aperture sizes. Figure 4d illustrates CCD captured images and color changes from blue to red as aperture size grows, except for the 20 nm apertures, which show carmine because they are dominated by CSPs. Relatively high color crosstalk in measured results is because the peaks induced by PSPs and CSPs do not completely overlap with each other. Nevertheless, device performance can be further improved by carefully designing structural parameters.

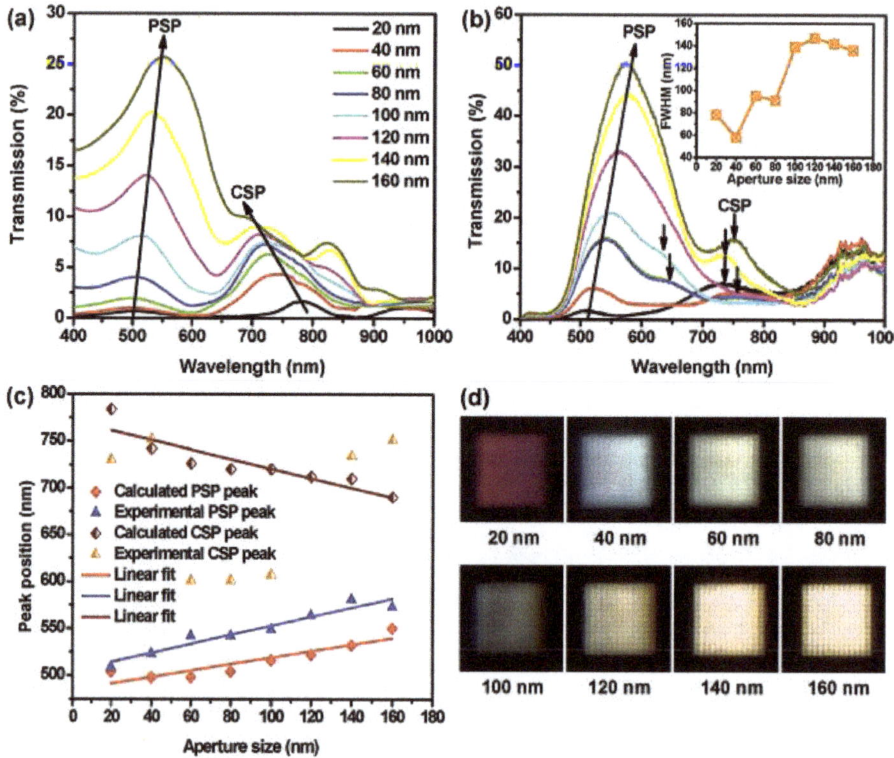

Figure 4. (**a**) Simulated and (**b**) experimental results of transmission for annular aperture arrays (AAAs) designed in a gold film. Aperture size increases from 20 nm to 160 nm in step of 20 nm. (**c**) Peak position as function of aperture size showing both cylindrical surface plasmons (CSPs) and planar surface plasmons (PSPs) peaks with linear fit. (**d**) CCD images captured by using an optical microscope. Reproduced with permission from [80]. Copyright John Wiley & Sons, 2012.

2.4. Transmission of LC Coated High Aspect Ratio Nanorods

One main obstacle to achieve high-efficiency hybrid devices using LCs and functional nanodevices is the question of how to combine them. Geometry may affect the orientation of LC cells significantly. High aspect ratio nanorods can align well with LC molecules. Using silver nanorods with large heights, it has been shown that it is possible to achieve homeotropic status [102]. In general, the transmission changes are based primarily on the excitation of the plasmonic nanorods. The transmission characteristics of the incident angle on the hybrid system have been thoroughly investigated. Experimental results demonstrate clear power dependence via transmission measurements. Figure 5 shows the effect of pumping power on transmittance at a 20° angle of incidence. As plotted in Figure 5a, one can see that the transmitted intensity is enhanced dramatically with growing pumping power until it saturates at around 10 mW. From the wavelength range of 400–500 nm, spectra show negligible changes with pumping light due to low transmitted intensities. It should be noted that more energy can pass through the hybrid system at longer wavelengths, resulting in a peak at ~650 nm. Figure 5b plots the magnified view of Figure 5a in the range of 350–450 nm, and another transmission peak between 380 nm and 390 nm can be seen as the pumping power is increased. Without pumping, the original resonance locates at around 420 nm and then shifts to 390 nm after switching on the pumping light. With further increase to pumping power, transmittance grows significantly from 6% (1 mW) to 15% (15 mW) with slight blueshift of resonance wavelength.

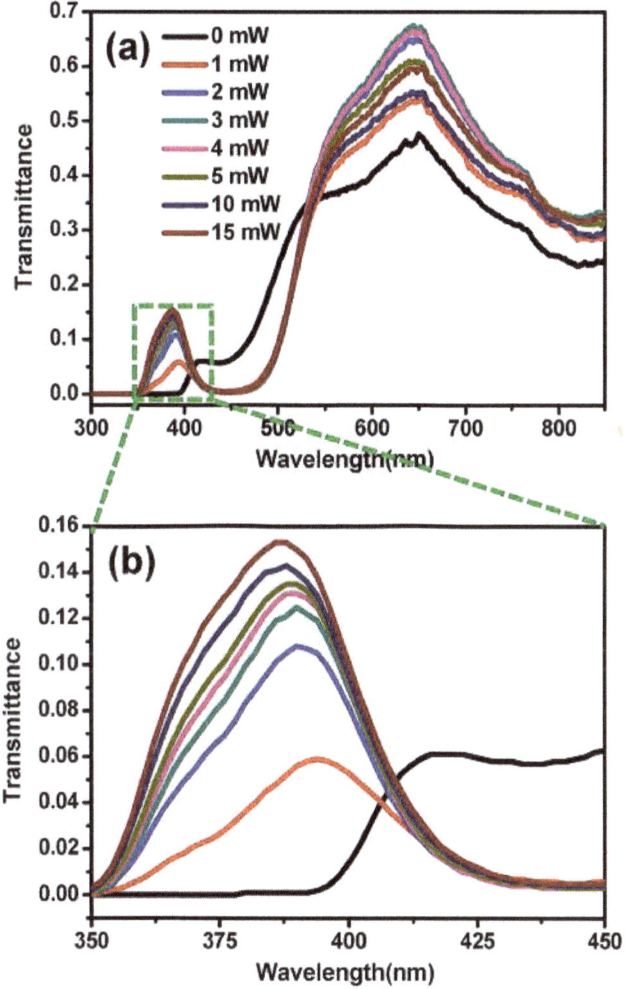

Figure 5. (**a**) Measured transmittance of silver nanorods covered by LCs as a function of wavelength for varying pumping power at 20° incidence angle and (**b**) enlarged view of 350 nm to 450 nm wavelengths showing more details of resonances located at 380–390 nm. Reproduced with permission from [102]. Copyright Royal Society of Chemistry, 2015.

2.5. Reflective Metasurface Lenses

Reflective metasurfaces [103–107] have recently attracted great interest due to their ability of manipulating electromagnetic waves at the nanoscale. Since these nanodevices demand sub-wavelength characteristics, components operating in visible frequencies are normally difficult to fabricate. Kobashi et al., [108] have shown that chiral LCs with a self-organized helical structure can lead to non-specular reflection at optical frequencies. Figure 6 shows the working principle of reflective lenses with patterned cholesteric liquid crystals (ChLCs). Figure 6a plots a reflective Fresnel lens with a parabolic phase profile wrapped to π. A functional area with circular shape was fabricated with 1 mm diameter for different phases of π and 3π, as demonstrated in Figure 6b–d. Light reflection can be produced at two different interfaces. Thus, the phase profile can be reversed in accordance with different observation direction, enabling converging or diverging effect according to the angle of incidence. As shown in Figure 6e, the laser spot shows different profiles at varying reflected surfaces compared with a mirror. Concave and convex profiles have been recorded for $m = -1$ and $m = 1$, respectively. Note that the reflective lens using ChLCs is completely reconfigurable because the convergence or divergence status can be switched freely by simply flipping the sample. The focal length is determined by Fresnel's approximation, which is related to aperture radius and phase difference. In this case, calculated focal length is 20.3 cm, which agrees well with the experimental results. This means extensive useful planar reflective optical components are enabled by using similar optoelectronic designs with ChLCs.

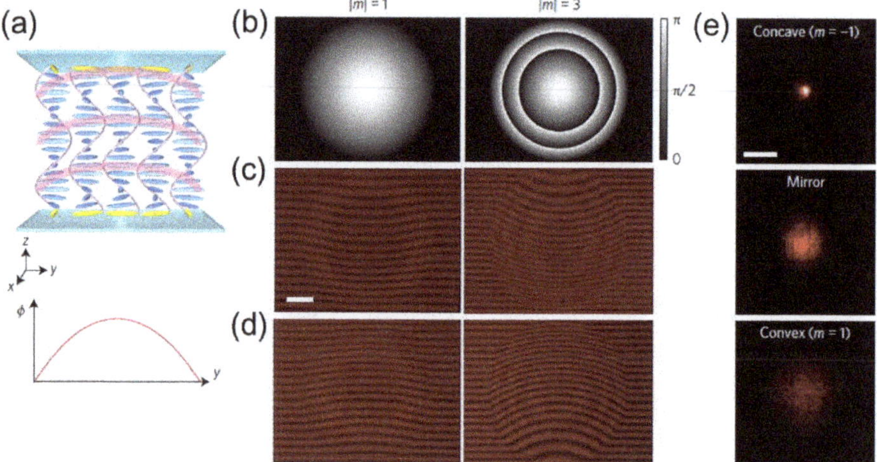

Figure 6. (a) Sketch of the working mechanism of metasurface based reflective lenses combined with patterned cholesteric liquid crystals (ChLCs) and the corresponding phase profile. (**b–d**) Device performance for different phases of π and 3π. Scale bar denotes 200 μm. (**e**) Recorded laser spot reflected from different interfaces compared with a mirror. Concave and convex profiles have been recorded for $m = -1$ and $m = 1$, respectively. Scale bar denotes 2 mm. Reproduced with permission from [108]. Copyright Springer Nature, 2016.

3. LC-Based Plasmonic Tunable Applications

In this Section, we review recent interesting tunable devices enabled by LC-plasmonic interactions, which have demonstrated practical applications ranging from resonance tuning to dynamic color switching.

3.1. Localized Surface Plasmon Resonance (LSPR) Tuning with Holographic Polymer-Dispersed Liquid Crystals (HPDLCs)

Until now, investigations on manipulating plasmon resonance have been performed and most of the achievements take advantage of tuning structural parameters and dielectric environment, with different complex designs. One main problem is that these devices normally show poor reproducibility and very limited tunable range. To further improve device performance and achieve highly reproducible results, azo-dye based holographic polymer-dispersed liquid crystals (HPDLCs) can be employed [109]. The hybrid system combines gold nanodisks with HPDLCs, which can be switched between nematic and isotropic with pumping light off and on. Figure 7a shows the principle diagram of the experimental device, which contains gold nanodisks covered with a HPDLC-layer where its orientation is controlled by an external optical field. Under light irradiation, the grating structure may suffer from thermal expansion due to localized heating of the functional area. The LC-droplets can be further squeezed, producing a shape change from oval to spherical. Scanning electron microscopy (SEM) topography of the HPDLCs transmission grating is shown in Figure 7b, with around 600 nm pitch. The nanodisk array was fabricated on a glass substrate using nanolithography, followed by reactive ion etching with ~150 nm diameter and ~320 nm periodicity.

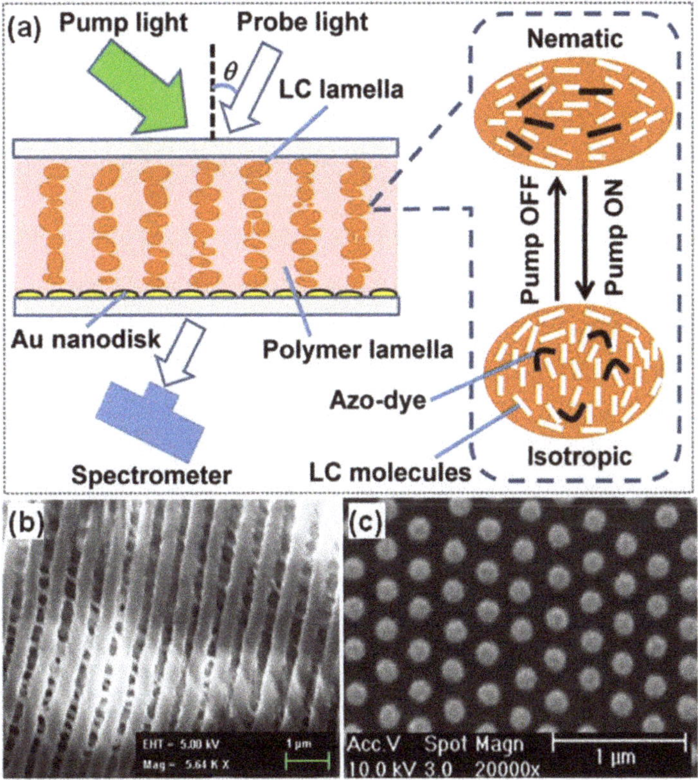

Figure 7. (a) Working principle of holographic polymer-dispersed liquid crystals (HPDLCs) transmission grating covered gold nanodisks. Modulation of resonance is realized by using a pumping light. The right part shows magnified view which demonstrates HPDLCs can be shifted between nematic and isotropic with pumping light off and on. (b) SEM topography of the HPDLCs transmission grating and (c) top-view SEM image of gold nanodisks fabricated by nanolithography followed by reactive ion etching with ~150 nm diameter and ~320 nm periodicity. Reproduced with permission from [109]. Copyright American Chemical Society, 2011.

As for the device shown in Figure 7, the measured extinction spectra at different pumping intensities are shown in Figure 8a for an incident probe angle of 42 degrees. Note that the extinction can be enhanced by decreasing pumping power with a slight redshift of resonance from around 800 nm (60 mW) to 830 nm (0 mW). However, the peak located at 580 nm remains unchanged with varying pumping power. For comparison, a control experiment was carried out which performed extinction measurement of nanodisks separated from the HPDLCs transmission gratings, as plotted in Figure 8b. It can be observed that the extinction peak significantly blueshifts to about 700 nm, compared with the hybrid system of nanodisks with the HPDLCs transmission gratings. The resonance peaks are also narrower. A lower intensity of resonance peaks compared with the coupled hybrid system is observed. However, the trend of intensity changes for peaks at longer wavelengths (~700 nm), shows similar behavior of increment extinction with smaller pumping power. For the resonance located at shorter wavelengths (~580 nm), extinction reaches saturation after the pumping power is larger than 20 mW. The coupled hybrid system also shows a periodic modulation in the absorption spectrum, forming an absorption grating, and therefore, resulting in diffraction of more light near the localized surface plasmon resonance (LSPR) peaks.

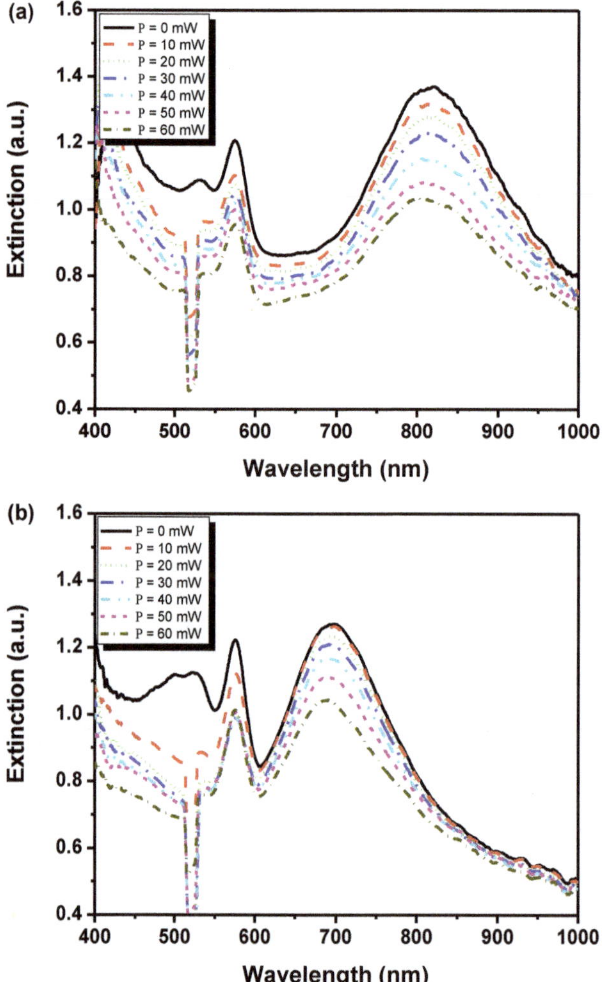

Figure 8. Measured extinction spectra as a function of wavelength with varying pumping power for (**a**) coupled system (nanodisks with HPDLCs transmission gratings), and (**b**) uncoupled nanodisks (separated from HPDLCs transmission gratings). Reproduced with permission from [109]. Copyright American Chemical Society, 2011.

To further investigate the underlying physics, nanodisks covered with dual-frequency liquid crystals (DFLCs) have been measured compared with bare gold disks, as shown in Figure 9. Note that this measurement was performed under normal incidence. The solid black and dashed red curves show different intensities and peak positions. Increased intensity with redshifted (~80 nm) resonance after combining gold nanodisks with DFLCs can be seen.

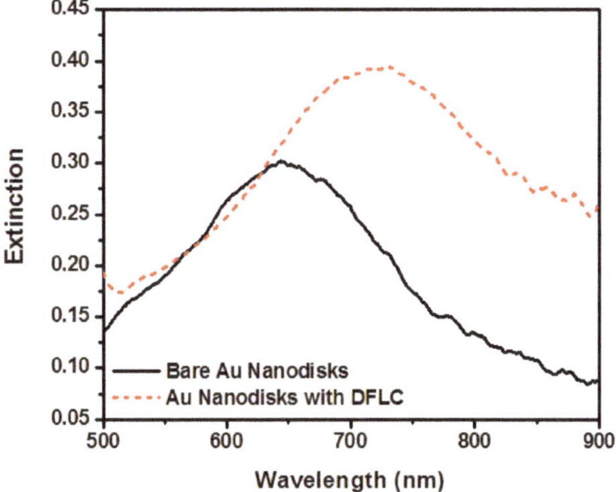

Figure 9. Comparison of measured extinction of nanodisks at normal incidence with and without dual-frequency liquid crystals (DFLCs). Reproduced with permission from [110]. Copyright American Institute of Physics, 2010.

3.2. Temperature-Dependent Tuning with Thermotropic LCs

Using thermotropic LCs, Abass and coworkers have shown an active tuning function which is temperature dependent [111]. Figure 10 shows the sketch of the experimental setup at different temperatures. It represents the diverse states of LCs. First, quantum dots (QDs) were synthesized by means of chemical processes of a consecutive ion layer absorption and reaction procedure. Beginning with an average QDs diameter of 6.5 nm, a 120 nm thick QD layer was obtained after coating on a glass substrate. A 15 nm thick silicon nitride layer was subsequently deposited on the QD layer for protecting the waveguide and smoothing the surface. Then aluminum nanoantennas were fabricated using nanoimprint lithography. Finally, thermotropic LCs were coated and a nylon alignment material was introduced to ensure that the orientation was in the direction of its lattice vectors.

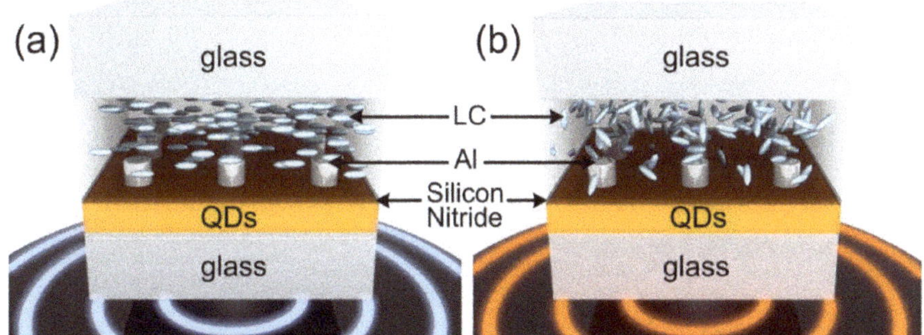

Figure 10. Schematic diagram of the temperature-dependent hybrid system using thermotropic LCs. (**a**) LC molecules are ordered at room temperature. (**b**) LC molecules are disordered at higher temperatures (>58 °C). Reproduced with permission from [111]. Copyright American Chemical Society, 2014.

Figure 11a shows the measured extinction spectra at various temperatures of the device shown in Figure 10. A halogen lamp was used as a broadband light source to illuminate the sample, thereafter the zeroth order transmittance T_0 was measured by a fiber-coupled spectrometer. There are two peaks observed in Figure 11a, which are in accordance with the hybrid plasmonic-photonic resonances. The wide resonance at long wavelengths results from the LSPRs. On the other hand, the narrow resonance at short wavelengths derives from the quasi-guided modes. Figure 11b illustrates the temperature-dependent normalized values of two modes. Interestingly, it can be seen the tendencies vary distinctly from the LSPRs to quasi-guided modes. As the transition happens, the anisotropic refractive index takes the place of the birefringent index. Figure 11c shows the $1-T_{total}$ spectra in the state of ordered and disordered LCs, respectively. The peak resonance wavelengths shown in the simulated curves are found to be in good agreement with the experimental results.

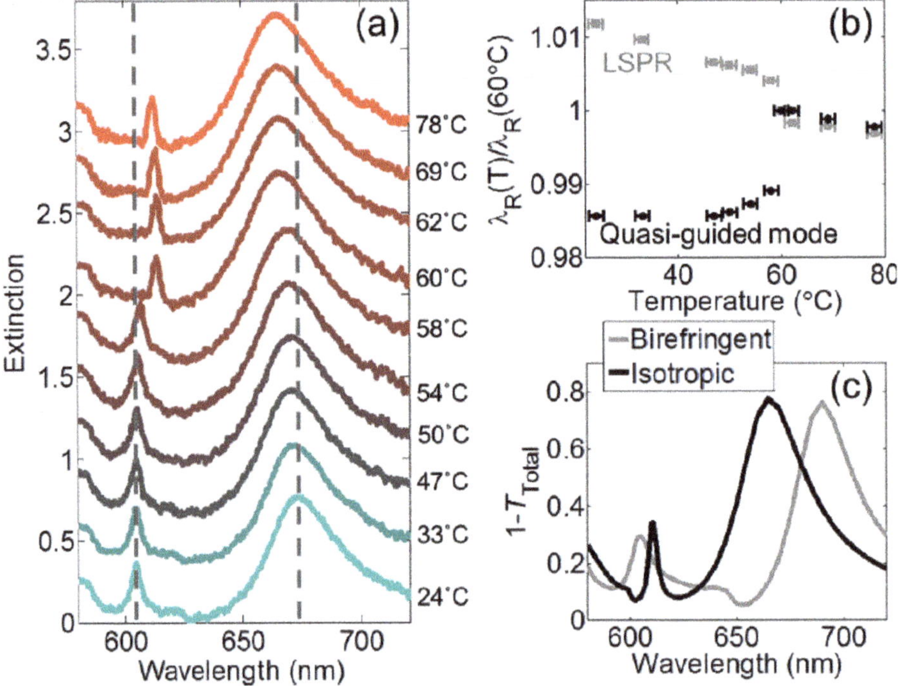

Figure 11. (a) Measured extinction as a function of wavelength at different temperatures. (b) Temperature-dependent normalization of resonance wavelength to its corresponding wavelength at 60 °C for localized surface plasmon resonance (LSPR) and quasi-guided modes. (c) The simulation spectra of two states, where the temperatures of the LCs are set to 24 °C and 60 °C, respectively. Reproduced with permission from [111]. Copyright American Chemical Society, 2014.

3.3. Active Tuning of the Fano Resonance

The collective oscillation of electrons at the interface of metal and dielectric is called surface plasmon. When neighboring nanoparticles are strongly coupled with each other, they can exhibit unique line shapes and interference, such as the famed Fano resonance. The Fano resonance is highly dependent on environmental parameters, which can be used to significantly increase the figure of merit for LSPR sensing [112–118]. Figure 12 demonstrates the working principle of a dynamic device which can control the Fano resonance using LCs [117]. In this experiment, an unpolarized white light source passed through a dark-field condenser and then focused on the sample surface. An oil-immersion objective lens was used to collect the light scattered by the nanostructures. Scattering spectra in Figure 12b were obtained at 0° polarization direction. The solid blue curve represents a Fano-like spectrum without external electric field (V_{off}). After applying a voltage to the LCs device, the Fano resonance is switched off. Similar switching behavior can be obtained at 90° polarization direction as well, as shown in Figure 12c. However, one can see that there is no Fano resonance at the V_{off} state. After switching on the voltage, a Fano-shaped profile is revealed, as shown by the red dashed line.

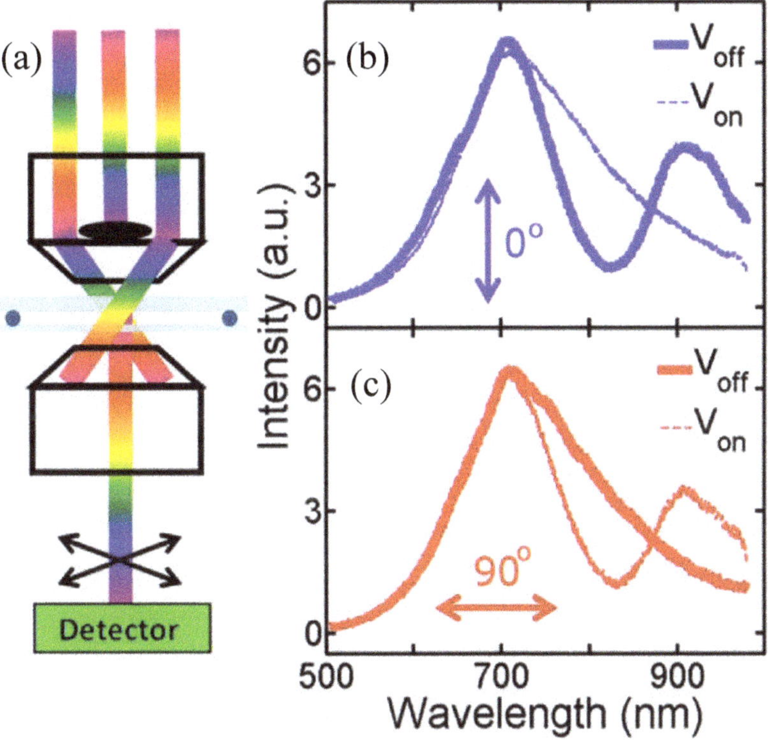

Figure 12. Dynamic manipulation of the Fano resonance using LCs. (**a**) Schematic diagram of the experimental setup of dark field microspectroscopy. (**b**,**c**) Measured scattering spectra as a function of wavelength for an individual octamer nanostructure at different polarization of (**b**) 0° and (**c**) 90°. Reproduced with permission from [117]. Copyright American Chemical Society, 2012.

3.4. Dynamic Color Tuning

Compared to traditional display technologies, there are many benefits of structural color filtering devices, including higher resolution and smaller pixels. However, their potential applications are limited because their optical characteristics remain static. By combining elaborately designed metasurfaces with highly birefringent LCs, an active reflection-type color filter has been demonstrated [119], which is not dependent on polarization. Combining the nanoimprinted structures with different geometric parameters with LCs, one can realize dynamic color tuning over the entire visible regime. As shown in Figure 13, the combination of LCs with metasurfaces can produce different color outputs [119]. Moreover, 10 μm × 10 μm pixels in the accompanying SEM images can be considered as two-dimensional gratings with different structural parameters. Such LC-plasmonic hybrid systems can be simply integrated with other display techniques. To prove this capability, the traditional transparent LCs display panel is applied, and the hybrid system can be integrated with a commercially-available thin-film-transistor (TFT) array. The aluminum metasurface was fabricated on the TFT glass plate with an 8.5 μm spacer layer, and then filled with high birefringence LCs [120].

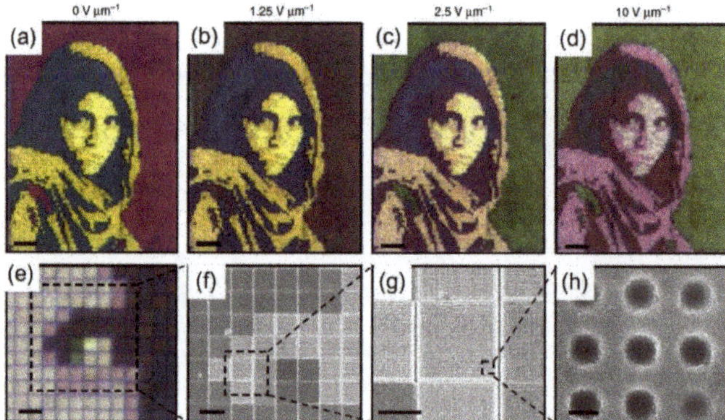

Figure 13. Dynamic color tuning using LCs. (**a–d**) Microscope images of captured portrait at varying applied voltages. (**e**) Magnified image at 10 Vμm^{-1} showing more details. (**f–h**) SEM images of fabricated structural pixels. Scale bars, (**e**) 20 μm, (**f**) 10 μm, (**g**) 5 μm, (**h**) 150 nm. Reproduced with permission from [119]. Copyright Macmillan Publishers Limited, 2015.

Figure 14a exhibits the device, and the electrical components of TFT are seen. Light can pass through the polarizer, LCs, and indium tin oxide (ITO) window to reflect the surface. Figure 14b shows color change with an applied voltage. Figure 14c illustrates an arbitrary shaped object captured with a 4× objective lens. The nanostructured surface is macroscopically patterned using UV lithography, followed by aluminum cladding. For this surface, the top ITO glass was patterned to control each letter of "UCF" individually. Moreover, a UV photo alignment was used to show how the V_{off} colors could be manipulated. Under a linearly polarized source, the azobenzene material can make the LCs uniform and perpendicular to the polarization direction.

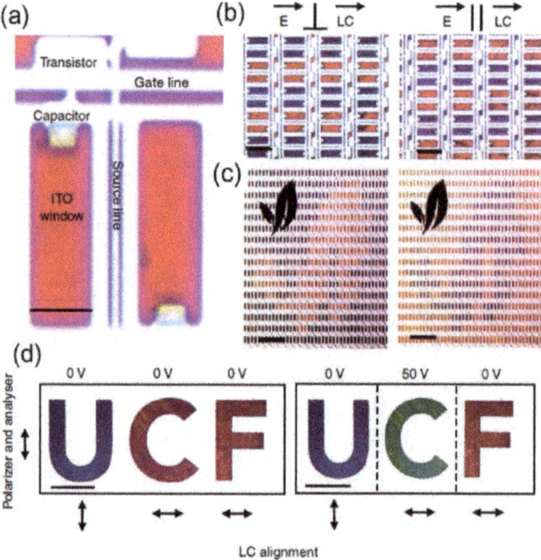

Figure 14. (**a**) Microscopic image of metasurfaces integrated with thin-film-transistor (TFT), and (**b**) shows that each row is opened every third and fourth rows in the case of parallel and vertical incident light for the top LCs orientation layer. (**c**) Arbitrary image displayed using a device photographed with a 4× objective. (**d**) Passive addressing devices with defined letters of "UCF" as a function of different parameters. Reproduced with permission from [120]. Copyright Macmillan Publishers Limited, 2017.

3.5. Narrowband Reflection Filter Controlled by LCs

Nowadays, with the development of powerful computer simulations, the design of LC-based devices can be applied even at the molecular level. More novel tunable devices can be created by combining optical components with LCs. Figure 15 shows the working principle of a narrowband resonant reflection filter which can be controlled by nematic LCs [121,122]. In this case, the coherence region was mainly probed by the evanescent field when LC cells were placed on a waveguiding layer. Transmitted and reflected spectra at varying bias voltages are plotted in Figure 15b. Owing to the strong anchoring, the resonance wavelength as a function of voltage shows S-shaped profiles (see Figure 15c for more details), which means the nanoscale coherence region cannot switch with a very fast speed.

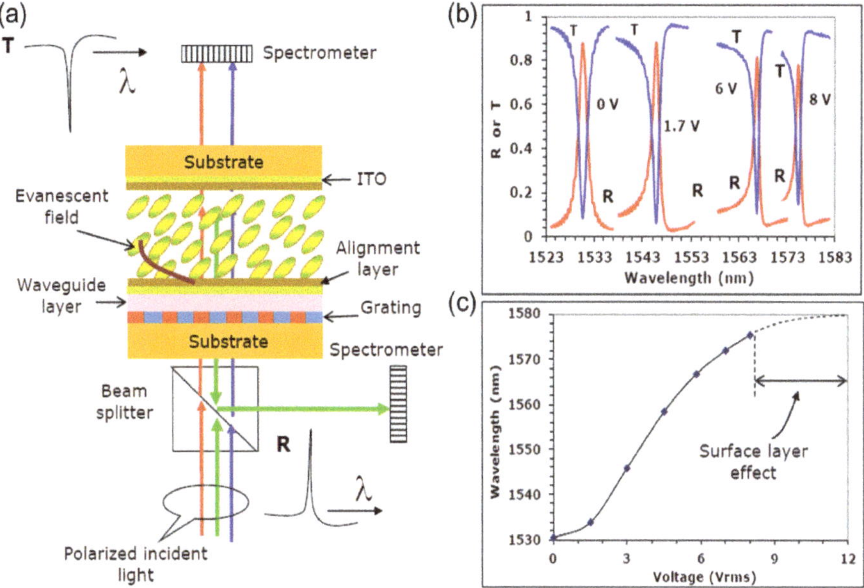

Figure 15. (**a**) Sketch of the hybrid device of reflection filter which can be controlled by nematic LCs. (**b**) Transmitted and reflected spectra at varying voltages. (**c**) Tunability curve of resonance wavelength as a function of voltage. S-shaped profiles can be observed, which means the nanoscale coherence region cannot switch with a fast speed. Reproduced with permission from [121]. Copyright Society of Photographic Instrumentation Engineers (SPIE) and Chinese Optics Letters (COL), 2012.

4. Conclusions and Outlooks

To conclude, we have reviewed the recent development of tuneable devices based on LC-plasmonic metasurfaces and their potential applications. Citing recent development of metasurfaces, as well as the exploration of new LCs functions, actively tuneable metamaterial-based devices and more practical achievements have been developed which can pave the way for new plasmonic display technologies. Nevertheless, we still need to ameliorate the sometimes less-than-ideal optical performance of these hybrid devices to achieve more varied and useful applications via consistent and reliable components. Additionally, high speed dynamic devices are of great importance for future development of multiple-functional optoelectronic nanodevices. These active nanodevices may enable more innovations due to their capability of manipulating electromagnetic waves freely at the nanoscale.

Author Contributions: Z.M. wrote Sections 1 and 2.1–2.4; X.M. wrote Sections 2.5 and 3.1; X.L. wrote Sections 3.2 and 3.3; G.S. wrote Sections 3.4 and 3.5; Y.J.L. wrote Section 4 and conceived the idea.

Funding: This research was funded by the National Natural Science Foundation of China (Grant Nos. 61771119, 31170956 and 61805113), the Natural Science Foundation of Guangdong Province (Grant Nos. 2017A030313034 and 2018A030310224), Shenzhen Science and Technology Innovation Council (Grant Nos. JCYJ20170817111349280 and KQTD2016030111203005), the Natural Science Foundation of Hebei Province (Grant No. F2018501063), Fundamental Research Funds for the Central Universities (Grant Nos. N162304007 and N162304008), Science and Technology Research Funds for Higher Education of Hebei Province (Grant No. ZD2015209).

Acknowledgments: We thank Stuart Earl from Swinburne University for helpful discussion and proofreading. This work was performed in part at the Melbourne Centre for Nanofabrication (MCN) in the Victorian Node of the Australian National Fabrication Facility (ANFF).

Conflicts of Interest: The authors declare no conflict of interest.

References

1. Halas, N.J.; Lal, S.; Chang, W.S.; Link, S.; Nordlander, P. Plasmons in strongly coupled metallic nanostructures. *Chem. Rev.* **2011**, *111*, 3913–3961. [CrossRef] [PubMed]

2. Si, G.Y.; Zhao, Y.H.; Lv, J.; Wang, F.; Liu, H.; Teng, J.; Liu, Y.J. Direct and accurate patterning of plasmonic nanostructures with ultrasmall gaps. *Nanoscale* **2013**, *5*, 4309–4313. [CrossRef] [PubMed]

3. Si, G.; Jiang, X.; Lv, J.; Gu, Q.; Wang, F. Fabrication and characterization of well-aligned plasmonic nanopillars with ultrasmall separations. *Nanoscale Res. Lett.* **2014**, *9*, 299. [CrossRef] [PubMed]

4. Jones, M.R.; Osberg, K.D.; Macfarlane, R.J.; Langille, M.R.; Mirkin, C.A. Templated techniques for the synthesis and assembly of plasmonic nanostructures. *Chem. Rev.* **2011**, *111*, 3736–3827. [CrossRef] [PubMed]

5. Si, G.Y.; Zhao, Y.H.; Lv, J.; Lu, M.; Wang, F.; Liu, H.; Xiang, N.; Huang, T.J.; Danner, A.J.; Teng, J.; et al. Reflective plasmonic color filters based on lithographically patterned silver nanorod arrays. *Nanoscale* **2013**, *5*, 6243–6248. [CrossRef] [PubMed]

6. Si, G.; Wang, Q.; Lv, J.; Miao, L.; Wang, F.; Peng, S. Interference lithography patterned large area plasmonic nanodisks for infrared detection. *Mater. Lett.* **2014**, *128*, 373–375. [CrossRef]

7. Lv, J.; Leong, E.S.P.; Jiang, X.; Kou, S.; Dai, H.; Lin, J.; Liu, Y.J.; Si, G. Plasmon-enhanced sensing: Current status and prospects. *J. Nanomater.* **2015**, *2015*, 474730. [CrossRef]

8. Yap, F.L.; Thoniyot, P.; Krishnan, S.; Krishnamoorthy, S. Nanoparticle cluster arrays for high-performance SERS through directed self-assembly on flat substrates and on optical fibers. *ACS Nano* **2012**, *6*, 2056–2070. [CrossRef] [PubMed]

9. Lv, J.; Gu, Q.; Jiang, X.; Yang, L.; Li, Z.; Ma, Z.; Si, G. Plasmonic nanoantennae with ultrasmall gaps and their application in surface enhanced Raman scattering. *Nanosci. Nanotechnol. Lett.* **2015**, *7*, 917–919. [CrossRef]

10. Tian, C.F.; Ding, C.H.; Liu, S.Y.; Yang, S.C.; Song, X.P.; Ding, B.J.; Li, Z.Y.; Fang, J.X. Nanoparticle attachment on silver corrugated-wire nanoantenna for large increases of surface-enhanced Raman scattering. *ACS Nano* **2011**, *5*, 9442–9449. [CrossRef] [PubMed]

11. Fazio, B.; D'Andrea, C.; Bonaccorso, F.; Irrera, A.; Calogero, G.; Vasi, C.; Gucciardi, P.G.; Allegrini, M.; Toma, A.; Chiappe, D.; et al. Re-radiation enhancement in polarized surface-enhanced resonant Raman scattering of randomly oriented molecules on self-organized gold nanowires. *ACS Nano* **2011**, *5*, 5945–5956. [CrossRef] [PubMed]

12. Pan, J.; Liu, S.; Yang, Y.; Lu, J. A highly sensitive resistive pressure sensor based on a carbon nanotube-liquid crystal-PDMS composite. *Nanomaterials* **2018**, *8*, 413. [CrossRef] [PubMed]

13. Pryce, I.M.; Kelaita, Y.A.; Aydin, K.; Atwater, H.A. Compliant metamaterials for resonantly enhanced infrared absorption spectroscopy and refractive index sensing. *ACS Nano* **2011**, *5*, 8167–8174. [CrossRef] [PubMed]

14. Liu, J.; He, H.; Xiao, D.; Yin, S.; Ji, W.; Jiang, S.; Luo, D.; Wang, B.; Liu, Y. Recent advances of plasmonic nanoparticles and their applications. *Materials* **2018**, *11*, 1833. [CrossRef] [PubMed]

15. Kuznetsov, A.I.; Evlyukhin, A.B.; Goncalves, M.R.; Reinhardt, C.; Koroleva, A.; Arnedillo, M.L.; Kiyan, R.; Marti, O.; Chichkov, B.N. Laser fabrication of large-scale nanoparticle arrays for sensing applications. *ACS Nano* **2011**, *5*, 4843–4849. [CrossRef] [PubMed]

16. Leong, E.S.P.; Deng, J.; Khoo, E.H.; Wu, S.; Phua, W.K.; Liu, Y.J. Fabrication of suspended, three-dimensional chiral plasmonic nanostructures with single-step electron-beam lithography. *RCS Adv.* **2015**, *5*, 96366–96371. [CrossRef]

17. Leong, E.S.P.; Liu, Y.J.; Deng, J.; Fong, Y.T.; Zhang, N.; Wu, S.J.; Teng, J.H. Fluid-enabled significant enhancement and active tuning of magnetic resonances in free-standing plasmonic metamaterials. *Nanoscale* **2014**, *6*, 11106–11111. [CrossRef] [PubMed]

18. Christensen, J.; Manjavacas, A.; Thongrattanasiri, S.; Koppens, F.H.L.; de Abajo, F.J.G. Graphene plasmon waveguiding and hybridization in individual and paired nanoribbons. *ACS Nano* **2012**, *6*, 431–440. [CrossRef] [PubMed]

19. Wang, W.H.; Yang, Q.; Fan, F.R.; Xu, H.X.; Wang, Z.L. Light propagation in curved silver nanowire plasmonic waveguides. *Nano Lett.* **2011**, *11*, 1603–1608. [CrossRef] [PubMed]

20. Si, G.; Leong, E.S.P.; Pan, W.; Chum, C.C.; Liu, Y.J. Plasmon-induced transparency in coupled triangle-rod arrays. *Nanotechnology* **2014**, *26*, 025201. [CrossRef] [PubMed]

21. Si, G.; Zhao, Y.; Leong, E.S.P.; Lv, J.; Liu, Y.J. Incident-angle dependent color tuning from a single plasmonic chip. *Nanotechnology* **2014**, *25*, 455203. [CrossRef] [PubMed]

22. Smolyaninov, I.I.; Hung, Y.J.; Davis, C.C. Magnifying superlens in the visible frequency range. *Science* **2007**, *315*, 1699–1701. [CrossRef] [PubMed]

23. Lv, J.; Jiang, X.; Ying, Y.; Si, G. Controlling light through a refined superlens. *Nanosci. Nanotechnol. Lett.* **2015**, *7*, 901–905. [CrossRef]

24. Cai, W.; Chettiar, U.K.; Kildishev, A.V.; Shalaev, V.M. Optical cloaking with metamaterials. *Nat. Photonics* **2007**, *1*, 224–227. [CrossRef]

25. Wegener, M.; Dolling, G.; Linden, S. Backward waves moving forward. *Nat. Mater.* **2007**, *6*, 475–476. [CrossRef] [PubMed]

26. Dickson, W.; Wurtz, G.A.; Evans, P.; O'Connor, D.; Atkinson, R.; Pollard, R.; Zayats, A.V. Dielectric-loaded plasmonic nanoantenna arrays: A metamaterial with tunable optical properties. *Phys. Rev. B* **2007**, *76*, 115411. [CrossRef]

27. Wurtz, G.A.; Pollard, R.; Zayats, A.V. Optical bistability in nonlinear surface-plasmon polaritonic crystals. *Phys. Rev. Lett.* **2006**, *97*, 057402. [CrossRef] [PubMed]

28. Lin, D.; Fan, P.; Hasman, E.; Brongersma, M.L. Dielectric gradient metasurface optical elements. *Science* **2014**, *345*, 298–302. [CrossRef] [PubMed]

29. Lv, J.; Yan, Y.; Zhang, W.; Liu, Y.; Jiang, Z.; Si, G. Plasmonic nanoantennae fabricated by focused ion beam milling. *Int. J. Precis. Eng. Manuf.* **2015**, *16*, 851–855. [CrossRef]

30. Chong, K.E.; Staude, I.; James, A.; Dominguez, J.; Liu, S.; Campione, S.; Subramania, G.; Luk, T.S.; Decker, M.; Neshev, D.N.; et al. Polarization-independent silicon metadevices for efficient optical wavefront control. *Nano Lett.* **2015**, *15*, 5369–5374. [CrossRef] [PubMed]

31. Lv, J.; Wang, F.; Ma, Z.; Si, G. Nanoring color filters based on Fabry-Pérot cavities. *Acta. Phys. Sin.* **2013**, *62*, 057804. [CrossRef]

32. Lv, J.; Khoo, E.H.; Leong, E.S.P.; Hu, L.; Jiang, X.; Li, Y.; Luo, D.; Si, G.; Liu, Y.J. Maskless fabrication of slanted annular aperture arrays. *Nanotechnology* **2017**, *28*, 225302. [CrossRef] [PubMed]

33. Jiang, X.; Leong, E.S.P.; Liu, Y.J.; Si, G. Tuning plasmon resonance in depth-variant plasmonic nanostructures. *Mater. Des.* **2016**, *96*, 64–67. [CrossRef]

34. Jiang, X.; Hu, S.; Li, Z.; Lv, J.; Si, G. Fabrication and characterization of plasmonic nanorods with high aspect ratios. *Opt. Mater.* **2016**, *58*, 323–326. [CrossRef]

35. Soukoulis, C.M.; Wegener, M. Past achievements and future challenges in the development of three-dimensional photonic metamaterials. *Nat. Photonics* **2011**, *5*, 523–530. [CrossRef]

36. Jiang, X.; Gu, Q.; Yang, L.; Zhao, R.; Lv, J.; Ma, Z.; Si, G. Functional plasmonic crystal nanoantennae with ultrasmall gaps and highly tunable profiles. *Opt. Laser Technol.* **2015**, *71*, 1–5. [CrossRef]

37. Xiao, D.; Liu, Y.J.; Yin, S.; Liu, J.; Ji, W.; Wang, B.; Luo, D.; Li, G.; Sun, X.W. Liquid-crystal-loaded chiral metasurfaces for reconfigurable multiband spin-selective light absorption. *Opt. Express* **2018**, *26*, 25305–25314. [CrossRef]

38. Shelby, R.A.; Smith, D.R.; Schultz, S. Experimental verification of a negative index of refraction. *Science* **2001**, *292*, 77–79. [CrossRef] [PubMed]

39. Rogacheva, A.V.; Fedotov, V.A.; Schwanecke, A.S.; Zheludev, N.I. Giant gyrotropy due to electromagnetic-field coupling in a bilayered chiral structure. *Phys. Rev. Lett.* **2006**, *97*, 177401. [CrossRef] [PubMed]

40. Earl, S.K.; James, T.D.; Gomez, D.E.; Marvel, R.E.; Haglund, R.F.; Roberts, A. Switchable polarization rotation of visible light using a plasmonic metasurface. *APL Photonics* **2016**, *2*, 016103. [CrossRef]

41. Hao, J.; Yuan, Y.; Ran, L.; Jiang, T.; Kong, A.; Chan, C.T.; Zhou, L. Manipulating electromagnetic wave polarizations by anisotropic metamaterials. *Phys. Rev. Lett.* **2007**, *99*, 063908. [CrossRef] [PubMed]

42. Plum, E.; Liu, X.X.; Fedotov, V.A.; Chen, Y.; Tsai, D.P.; Zheludev, N.I. Metamaterials: Optical activity without chirality. *Phys. Rev. Lett.* **2009**, *102*, 113902. [CrossRef] [PubMed]

43. Fedotov, V.A.; Rogacheva, A.V.; Zheludev, N.I.; Mladyonov, P.L.; Prosvirnin, S.L. Mirror that does not change the phase of reflected waves. *Appl. Phys. Lett.* **2006**, *88*, 091119. [CrossRef]

44. Landy, N.I.; Sajuyigbe, S.; Mock, J.J.; Smith, D.R.; Padilla, W.J. Perfect metamaterial absorber. *Phys. Rev. Lett.* **2008**, *100*, 207402. [CrossRef] [PubMed]

45. Wang, J.; Jiang, X.; Xia, L.; Tang, L.; Hu, S.; Lv, J.; Zhao, H.; Si, G.; Shi, R. Fabrication and optical measurement of double-overlapped annular apertures. *Opt. Mater.* **2016**, *60*, 13–16. [CrossRef]

46. Fedotov, V.A.; Mladyonov, P.L.; Prosvirnin, S.L.; Rogacheva, A.V.; Chen, Y.; Zheludev, N.I. Asymmetric propagation of electromagnetic waves through a planar chiral structure. *Phys. Rev. Lett.* **2006**, *97*, 167401. [CrossRef] [PubMed]

47. Menzel, C.; Helgert, C.; Rockstuhl, C.; Kley, E.B.; Tunnermann, A.; Pertsch, T.; Lederer, F. Asymmetric transmission of linearly polarized light at optical metamaterials. *Phys. Rev. Lett.* **2010**, *104*, 253902. [CrossRef] [PubMed]

48. Novitsky, A.V.; Galynsky, V.M.; Zhukovsky, S.V. Asymmetric transmission in planar chiral split-ring metamaterials: Microscopic Lorentz-theory approach. *Phys. Rev. B* **2012**, *86*, 075138. [CrossRef]

49. Schurig, D.; Mock, J.J.; Justice, B.J.; Cummer, S.A.; Pendry, J.B.; Starr, A.F.; Smith, D.R. Metamaterial electromagnetic cloak at microwave frequencies. *Science* **2006**, *314*, 977–980. [CrossRef] [PubMed]

50. Ergin, T.; Stenger, N.; Brenner, P.; Pendry, J.B.; Wegener, M. Three-dimensional invisibility cloak at optical wavelengths. *Science* **2010**, *328*, 337–339. [CrossRef] [PubMed]

51. Chen, H.; Chan, C.T.; Sheng, P. Transformation optics and metamaterials. *Nat. Mater.* **2010**, *9*, 387–396. [CrossRef] [PubMed]

52. Tsakmakidis, K.L.; Boardman, A.D.; Hess, O. 'Trapped rainbow' storage of light in metamaterials. *Nature* **2007**, *450*, 397–401. [CrossRef] [PubMed]

53. Papasimakis, N.; Zheludev, N.I. Metamaterial-induced transparency, sharp Fano resonances and slow light. *Opt. Photonics News* **2009**, *20*, 22–27. [CrossRef]

54. Zheludev, N.I.; Prosvirnin, S.L.; Papasimakis, N.; Fedotov, V.A. Lasing spaser. *Nat. Photonics* **2008**, *2*, 351–354. [CrossRef]

55. Hess, O.; Pendry, J.B.; Maier, S.A.; Oulton, R.F.; Hamm, J.M.; Tsakmakidis, K.L. Active nanoplasmonic metamaterials. *Nat. Mater.* **2012**, *11*, 573–584. [CrossRef] [PubMed]

56. Zheludev, N.I. The road ahead for metamaterials. *Science* **2010**, *328*, 582–583. [CrossRef] [PubMed]

57. Lv, J.; Li, Z.; Ying, Y.; Yang, L.; Yang, N.; Jiang, X.; Si, G. Recent progress on metasurface-enabled plasmonics. *Nanosci. Nanotechnol. Lett.* **2015**, *7*, 779–786. [CrossRef]

58. Driscoll, T.; Kim, H.T.; Chae, B.G.; Kim, B.J.; Lee, Y.W.; Jokerst, N.M.; Palit, S.; Smith, D.R.; Ventra, M.D.; Basov, D.N. Memory metamaterials. *Science* **2009**, *325*, 1518–1521. [CrossRef] [PubMed]

59. Samson, Z.L.; MacDonald, K.F.; De Angelis, F.; Gholipour, B.; Knight, K.; Huang, C.C.; Fabrizio, E.D.; Hewak, D.W.; Zheludev, N.I. Metamaterial electro-optic switch of nanoscale thickness. *Appl. Phys. Lett.* **2010**, *96*, 143105. [CrossRef]

60. Chen, H.T.; O'Hara, J.F.; Azad, A.K.; Taylor, A.J.; Averitt, R.D.; Shrekenhamer, D.B.; Padilla, W.J. Experimental demonstration of frequency-agile terahertz metamaterials. *Nat. Photonics* **2008**, *2*, 295–298. [CrossRef]

61. Si, G.; Zhao, Y.; Leong, E.S.P.; Liu, Y.J. Liquid-crystal-enabled active plasmonics: A review. *Materials* **2014**, *7*, 1296–1317. [CrossRef] [PubMed]

62. Kang, B.; Woo, J.H.; Choi, E.; Lee, H.H.; Kim, E.S.; Kim, J.; Hwang, T.-J.; Park, Y.-S.; Kim, D.H.; Wu, J.W. Optical switching of near infrared light transmission in metamaterial-liquid crystal cell structure. *Opt. Express* **2010**, *18*, 16492–16498. [CrossRef] [PubMed]

63. Yuan, Y.; Martinez, A.; Senyuk, B.; Tasinkevych, M.; Smalyukh, I.I. Chiral liquid crystal colloids. *Nat. Mater.* **2018**, *17*, 71–79. [CrossRef] [PubMed]

64. Zhang, F.; Zhang, W.; Sun, J.; Qiu, K.; Zhou, J.; Lippens, D. Electrically controllable fishnet metamaterial based on nematic liquid crystal. *Opt. Express* **2011**, *19*, 1563–1568. [CrossRef] [PubMed]

65. Pratibha, R.; Park, K.; Smalyukh, I.I.; Park, W. Tunable optical metamaterial based on liquid crystal-gold nanosphere composite. *Opt. Express* **2009**, *17*, 19459–19469. [CrossRef] [PubMed]

66. Urbas, A.; Klosterman, J.; Tondiglia, V.; Natarajan, L.; Sutherland, R.; Tsutsumi, O.; Ikeda, T.; Bunning, T. Optically switchable Bragg reflectors. *Adv. Mater.* **2010**, *16*, 1453–1456. [CrossRef]

67. Hsiao, V.K.S.; Li, Z.; Chen, Z.; Peng, P.C.; Tang, J.Y. Optically controllable side-polished fiber attenuator with photoresponsive liquid crystal overlay. *Opt. Express* **2009**, *17*, 19988–19995. [CrossRef] [PubMed]

68. Yin, S.; Liu, Y.J.; Xiao, D.; He, H.; Luo, D.; Jiang, S.; Dai, H.; Ji, W.; Sun, X.W. Liquid-crystal-based tunable plasmonic waveguide filters. *J. Phys. D Appl. Phys.* **2018**, *51*, 235101. [CrossRef]

69. Liu, Y.J.; Cai, Z.; Leong, E.S.P.; Zhao, X.S.; Teng, J.H. Optically switchable photonic crystals based on inverse opals partially infiltrated by photoresponsive liquid crystals. *J. Mater. Chem.* **2012**, *22*, 7609–7613. [CrossRef]

70. Hu, W.; Zhao, H.Y.; Song, L.; Yang, Z.; Cao, H.; Cheng, Z.H.; Liu, Q.; Yang, H. Electrically controllable selective reflection of chiral nematic liquid crystal/chiral ionic liquid composites. *Adv. Mater.* **2010**, *22*, 468–472. [CrossRef] [PubMed]

71. Sutherland, R.L.; Tondiglia, V.P.; Natarajan, L.V.; Bunning, T.J.; Adams, W.W. Electrically switchable volume gratings in polymer-dispersed liquid crystals. *Appl. Phys. Lett.* **1994**, *64*, 1074–1076. [CrossRef]

72. Humar, M.; Ravnik, M.; Pajk, S.; Musevic, I. Electrically tunable liquid crystal optical microresonators. *Nat. Photonics* **2009**, *3*, 595–600. [CrossRef]

73. Choi, S.S.; Morris, S.M.; Huck, W.T.S.; Coles, H.J. Electrically tuneable liquid crystal photonic bandgaps. *Adv. Mater.* **2009**, *21*, 3915–3918. [CrossRef]

74. Liu, Y.J.; Leong, E.S.P.; Wang, B.; Teng, J.H. Optical transmission enhancement and tuning by overlaying liquid crystals on a gold film with patterned nanoholes. *Plasmonics* **2011**, *6*, 659–664. [CrossRef]

75. Zhang, F.; Zhao, Q.; Kang, L.; Gaillot, D.P.; Zhao, X.; Zhou, J.; Lippens, D. Magnetic control of negative permeability metamaterials based on liquid crystals. *Appl. Phys. Lett.* **2008**, *92*, 193104. [CrossRef]

76. Su, H.; Wang, H.; Zhao, H.; Xue, T.; Zhang, J. Liquid-crystal-based electrically tuned electromagnetically induced transparency metasurface switch. *Sci. Rep.* **2017**, *7*, 17378. [CrossRef] [PubMed]

77. Mehrzad, H.; Mohajerani, E. Liquid crystal mediated active nano-plasmonic based on the formation of hybrid plasmonic-photonic modes. *Appl. Phys. Lett.* **2018**, *112*, 061101. [CrossRef]

78. Xiao, S.; Chettiar, U.K.; Kildishev, A.V.; Drachev, V.; Khoo, I.C.; Shalaev, V.M. Tunable magnetic response of metamaterials. *Appl. Phys. Lett.* **2009**, *95*, 033115. [CrossRef]

79. Lin, Y.; Yang, Y.; Shan, Y.; Gong, L.; Chen, J.; Li, S.; Chen, L. Magnetic nanoparticle-assisted tunable optical patterns from spherical cholesteric liquid crystal Bragg reflectors. *Nanomaterials* **2017**, *7*, 376. [CrossRef] [PubMed]

80. Liu, Y.J.; Si, G.Y.; Leong, E.S.P.; Xiang, N.; Danner, A.J.; Teng, J.H. Light-driven plasmonic color filters by overlaying photoresponsive liquid crystals on gold annular aperture arrays. *Adv. Mater.* **2012**, *24*, OP131–OP135. [CrossRef] [PubMed]

81. Jiang, X.; Gu, Q.; Wang, F.; Lv, J.; Ma, Z.; Si, G. Fabrication of coaxial plasmonic crystals by focused ion beam milling and electron-beam lithography. *Mater. Lett.* **2013**, *100*, 192–194. [CrossRef]

82. Evans, P.R.; Wurtz, G.A.; Hendren, W.R.; Atkinson, R.; Dickson, W.; Zayats, A.V.; Pollard, R. Electrically switchable nonreciprocal transmission of plasmonic nanorods with liquid crystal. *Appl. Phys. Lett.* **2007**, *91*, 043101. [CrossRef]

83. Kossyrev, P.A.; Yin, A.J.; Cloutier, S.G.; Cardimona, D.A.; Huang, D.H.; Alsing, P.M.; Xu, J.M. Electric field tuning of plasmonic response of nanodot array in liquid crystal matrix. *Nano Lett.* **2005**, *5*, 1978–1981. [CrossRef] [PubMed]

84. Kasyanova, I.V.; Gorkunov, M.V.; Artemov, V.V.; Geivandov, A.R.; Mamonova, A.V.; Palto, S.P. Liquid crystal metasurfaces on micropatterned polymer substrates. *Opt. Express* **2018**, *26*, 20258–20269. [CrossRef] [PubMed]

85. Reznikov, Y.; Buchnev, O.; Tereshchenko, O.; Reshetnyak, V.; Glushchenko, A.; West, J. Ferroelectric nematic suspension. *Appl. Phys. Lett.* **2003**, *82*, 1917–1919. [CrossRef]

86. Liu, Y.J.; Ding, X.; Lin, S.C.S.; Shi, J.; Chiang, I.K.; Huang, T.J. Surface acoustic wave driven light shutters using polymer-dispersed liquid crystals. *Adv. Mater.* **2011**, *23*, 1656–1659. [CrossRef] [PubMed]

87. Wang, L.; Ge, S.; Hu, W.; Nakajima, M.; Lu, Y. Graphene-assisted high-efficiency liquid crystal tunable terahertz metamaterial absorber. *Opt. Express* **2017**, *25*, 23873–23879. [CrossRef] [PubMed]

88. Komar, A.; Paniagua-Dominguez, R.; Microshnichenko, A.; Yu, Y.F.; Kivshar, Y.S.; Kuznetsov, A.I.; Neshev, D. Dynamic beam switching by liquid crystal tunable dielectric metasurfaces. *ACS Photonics* **2018**, *5*, 1742–1748. [CrossRef]

89. Dai, H.; Chen, L.; Zhang, B.; Si, G.; Liu, Y.J. Optically isotropic, electrically tunable liquid crystal droplet arrays formed by photopolymerization-induced phase separation. *Opt. Lett.* **2015**, *40*, 2723–2726. [CrossRef] [PubMed]

90. Shrekenhamer, D.; Chen, W.C.; Padilla, W.J. Liquid crystal tunable metamaterial absorber. *Phys. Rev. Lett.* **2013**, *110*, 177403. [CrossRef] [PubMed]

91. Savo, S.; Shrekenhamer, D.; Padilla, W.J. Liquid crystal metamaterial absorber spatial light modulator for THz applications. *Adv. Opt. Mater.* **2014**, *2*, 275–279. [CrossRef]

92. Buchnev, O.; Ou, J.Y.; Kaczmarek, M.; Zheludev, N.I.; Fedotov, V.A. Electro-optical control in a plasmonic metamaterial hybridised with a liquid-crystal cell. *Opt. Express* **2013**, *21*, 1633–1638. [CrossRef] [PubMed]

93. Werner, D.H.; Kwon, D.H.; Khoo, I.C.; Kildishev, A.V.; Shalaev, V.M. Liquid crystal clad near-infrared metamaterials with tunable negative-zero-positive refractive indices. *Opt. Express* **2007**, *15*, 3342–3347. [CrossRef] [PubMed]

94. Xie, J.; Zhang, X.; Peng, Z.; Wang, Z.; Wang, T.; Zhu, Z.S.; Wang, Z.; Zhang, L.; Zhang, J.; Yang, B. Low electric field intensity and thermotropic tuning surface plasmon band shift of gold island film by liquid crystals. *J. Phys. Chem. C* **2012**, *116*, 2720–2727. [CrossRef]

95. Komar, A.; Fang, Z.; Bohn, J.; Sautter, J.; Decker, M.; Miroshnichenko, A.; Pertsch, T.; Brener, L.; Kivshar, Y.S.; Staude, I.; et al. Electrically tunable all-dielectric optical metasurfaces based on liquid crystals. *Appl. Phys. Lett.* **2017**, *110*, 071109. [CrossRef]

96. Zhao, Q.; Kang, L.; Du, B.; Li, B.; Zhou, J.; Tang, H.; Liang, X.; Zhang, B. Electrically tunable negative permeability metamaterials based on nematic liquid crystals. *Appl. Phys. Lett.* **2007**, *90*, 011112. [CrossRef]

97. Linden, S.; Enkrich, C.; Wegener, M.; Zhou, J.; Koschny, T.; Soukoulis, C.M. Magnetic response of metamaterials at 100 Terahertz. *Science* **2004**, *306*, 1351–1353. [CrossRef] [PubMed]

98. Decker, M.; Feth, N.; Soukoulis, C.M.; Linden, S.; Wegener, M. Retarded long-range interaction in split-ring-resonator square arrays. *Phys. Rev. B* **2011**, *84*, 085416. [CrossRef]

99. Decker, M.; Kremers, C.; Minovich, A.; Staude, I.; Miroshnichenko, A.E.; Chigrin, D.; Neshev, D.N.; Jagadish, C.; Kivshar, Y.S. Electro-optical switching by liquid-crystal controlled metasurfaces. *Opt. Express* **2013**, *21*, 8879–8885. [CrossRef] [PubMed]

100. Si, G.; Zhao, Y.; Liu, H.; Teo, S.L.; Zhang, M.; Huang, T.J.; Danner, A.J.; Teng, J. Annular aperture array based color filter. *Appl. Phys. Lett.* **2011**, *99*, 033105. [CrossRef]

101. Zhao, Y.; Zhao, Y.; Hu, S.; Lv, J.; Ying, Y.; Gervinskas, G.; Si, G. Artificial structural color pixels: A review. *Materials* **2017**, *10*, 944. [CrossRef] [PubMed]

102. Si, G.; Leong, E.S.P.; Jiang, X.; Lv, J.; Lin, J.; Dai, H.; Liu, Y.J. All-optical, polarization-insensitive light tuning properties in silver nanorod arrays covered with photoresponsive liquid crystals. *Phys. Chem. Chem. Phys.* **2015**, *17*, 13223–13227. [CrossRef] [PubMed]

103. Yu, N.; Genevet, P.; Kats, M.A.; Aieta, F.; Tetienne, J.P.; Capasso, F.; Gaburro, Z. Light propagation with phase discontinuities: Generalized laws of reflection and refraction. *Science* **2011**, *334*, 333–337. [CrossRef] [PubMed]

104. Meinzer, N.; Barnes, W.L.; Hooper, I.R. Plasmonic meta-atoms and metasurfaces. *Nat. Photonics* **2014**, *8*, 889–898. [CrossRef]

105. Kildishev, A.V.; Boltasseva, A.; Shalaev, V.M. Planar photonics with metasurfaces. *Science* **2013**, *339*, 1232009. [CrossRef] [PubMed]

106. Yu, N.; Capasso, F. Flat optics with designer metasurfaces. *Nat. Mater.* **2014**, *13*, 139–150. [CrossRef] [PubMed]

107. Fattal, D.; Li, J.; Peng, Z.; Fiorentino, M.; Beausoleil, R.G. Flat dielectric grating reflectors with focusing abilities. *Nat. Photonics* **2010**, *4*, 466–470. [CrossRef]

108. Kobashi, J.; Yoshida, H.; Ozaki, M. Planar optics with patterned chiral liquid crystals. *Nat. Photonics* **2016**, *10*, 389–392. [CrossRef]

109. Liu, Y.J.; Zheng, Y.B.; Liou, J.; Chiang, I.K.; Khoo, I.C.; Huang, T.J. All-optical modulation of localized surface plasmon coupling in a hybrid system composed of photoswitchable gratings and Au nanodisk arrays. *J. Phys. Chem. C* **2011**, *115*, 7717–7722. [CrossRef] [PubMed]

110. Liu, Y.J.; Hao, Q.; Smalley, J.S.T.; Liou, J.; Khoo, L.H.; Huang, T.J. A frequency-addressed plasmonic switch based on dual-frequency liquid crystals. *Appl. Phys. Lett.* **2010**, *97*, 091101. [CrossRef]

111. Abass, A.; Rodriguez, S.R.; Ako, T.; Aubert, T.; Verschuuren, M.; Thourhout, D.V.; Beeckman, J.; Hens, Z.; Rivas, J.G.; Maes, B. Active liquid crystal tuning of metallic nanoantenna enhanced light emission from colloidal quantum dots. *Nano Lett.* **2014**, *14*, 5555–5560. [CrossRef] [PubMed]

112. Lassiter, J.B.; Sobhani, H.; Fan, J.A.; Kundu, J.; Capasso, F.; Nordlander, P.; Halas, N.J. Fano resonances in plasmonic nanoclusters: Geometrical and chemical tunability. *Nano Lett.* **2010**, *10*, 3184–3189. [CrossRef] [PubMed]

113. Liu, N.; Weiss, T.; Mesch, M.; Langguth, L.; Eigenthaler, U.; Hirscher, M.; Sönnichsen, C.; Giessen, H. Planar metamaterial analogue of electromagnetically induced transparency for plasmonic sensing. *Nano Lett.* **2010**, *10*, 1103–1107. [CrossRef] [PubMed]

114. Verellen, N.; Dorpe, P.V.; Huang, C.; Lodewijks, K.; Vandenbosch, G.E.; Lagae, L.; Moshchalkov, V.V. Plasmon line shaping using nanocrosses for high sensitivity localized surface plasmon resonance sensing. *Nano Lett.* **2011**, *11*, 391–397. [CrossRef] [PubMed]

115. Ye, J.; Wen, F.; Sobhani, H.; Lassiter, J.B.; Dorpe, P.V.; Nordlander, P.; Halas, N.J. Plasmonic nanoclusters: Near field properties of the Fano resonance interrogated with SERS. *Nano Lett.* **2012**, *12*, 1660–1667. [CrossRef] [PubMed]

116. Wu, C.H.; Khanikaev, A.B.; Adato, R.; Arju, N.; Yanik, A.A.; Altug, H.; Shvets, G. Fano-resonant asymmetric metamaterials for ultrasensitive spectroscopy and identification of molecular monolayers. *Nat. Mater.* **2012**, *11*, 69–75. [CrossRef] [PubMed]

117. Chang, W.S.; Lassiter, J.B.; Swanglap, P.; Sobhani, H.; Khatua, S.; Nordlander, P.; Halas, N.J.; Link, S. A plasmonic Fano switch. *Nano Lett.* **2012**, *12*, 4977–4982. [CrossRef] [PubMed]

118. Buchnev, O.; Podoliak, N.; Kaczmarek, M.; Zheludev, N.I.; Fedotov, V.A. Electrically controlled nanostructured metasurface loaded with liquid crystal: Toward multifunctional photonic switch. *Adv. Opt. Mater.* **2015**, *3*, 674–679. [CrossRef]

119. Franklin, D.; Chen, Y.; Vazquez-Guardado, A.; Modak, S.; Boroumand, J.; Xu, D.; Wu, S.; Chanda, D. Polarization-independent actively tunable colour generation on imprinted plasmonic surfaces. *Nat. Commun.* **2015**, *6*, 7337. [CrossRef] [PubMed]

120. Franklin, D.; Frank, R.; Wu, S.T.; Chanda, D. Actively addressed single pixel full-colour plasmonic display. *Nat. Commun.* **2017**, *8*, 15209. [CrossRef] [PubMed]

121. Abdulhalim, I. Liquid crystal active nanophotonics and plasmonics: From science to devices. *J. Nanophotonics* **2012**, *6*, 061001. [CrossRef]

122. Abdulhalim, I. Optimized guided mode resonant structure as thermooptic sensor and liquid crystal tunable filter. *Chin. Opt. Lett.* **2009**, *7*, 667–670. [CrossRef]

 *nanomaterials*

Article

Updating the Role of Reduced Graphene Oxide Ink on Field Emission Devices in Synergy with Charge Transfer Materials

Minas M. Stylianakis [1,*,†], George Viskadouros [1,2,*,†], Christos Polyzoidis [1], George Veisakis [1], George Kenanakis [3], Nikolaos Kornilios [1], Konstantinos Petridis [1,4] and Emmanuel Kymakis [1]

[1]   Center of Materials Technology and Photonics & Electrical Engineering Department, Technological Educational Institute (TEI) of Crete, Heraklion 71004 Crete, Greece; polyzoidis@staff.teicrete.gr (C.P.); gveisakis@staff.teicrete.gr (G.V.); kornil@staff.teicrete.gr (N.K.); c.petridischania@gmail.com (K.P.); kymakis@staff.teicrete.gr (E.K.)
[2]   Department of Mineral Resources Engineering, Technical University of Crete, Chania, 73100 Crete, Greece
[3]   Institute of Electronic Structure and Laser, Foundation for Research and Technology-Hellas, N. Plastira 100, Heraklion, 70013 Crete, Greece; gkenanak@iesl.forth.gr
[4]   Department of Electronic Engineering Technological Educational Institute (TEI) of Crete, Chania, 73132 Crete, Greece
*   Correspondence: stylianakis@staff.teicrete.gr (M.M.S.); viskadouros@staff.teicrete.gr (G.V.); Tel.: +30-2810-379775 (M.M.S.)
†   These authors contributed equally to this work.

Received: 10 December 2018; Accepted: 14 January 2019; Published: 22 January 2019

Abstract: Hydroiodic acid (HI)-treated reduced graphene oxide (rGO) ink/conductive polymeric composites are considered as promising cold cathodes in terms of high geometrical aspect ratio and low field emission (FE) threshold devices. In this study, four simple, cost-effective, solution-processed approaches for rGO-based field effect emitters were developed, optimized, and compared; rGO layers were coated on (a) n+ doped Si substrate, (b) n^+-Si/P3HT:rGO, (c) n^+-Si/PCDTBT:rGO, and (d) n^+-Si/PCDTBT:PC$_{71}$BM:rGO composites, respectively. The fabricated emitters were optimized by tailoring the concentration ratios of their preparation and field emission characteristics. In a critical composite ratio, FE performance was remarkably improved compared to the pristine Si, as well as n^+-Si/rGO field emitter. In this context, the impact of various materials, such as polymers, fullerene derivatives, as well as different solvents on rGO function reinforcement and consequently on FE performance upon rGO-based composites preparation was investigated. The field emitter consisted of n^+-Si/PCDTBT:PC$_{71}$BM(80%):rGO(20%)/rGO displayed a field enhancement factor of ~2850, with remarkable stability over 20 h and low turn-on field in 0.6 V/μm. High-efficiency graphene-based FE devices realization paves the way towards low-cost, large-scale electron sources development. Finally, the contribution of this hierarchical, composite film morphology was evaluated and discussed.

Keywords: field emission; graphene; reduced graphene oxide; polymer composites; graphene ink; cold cathode; Fowler–Nordheim

---

1. Introduction

Upon the existence of a strong electric field that passes through a barrier, electrons are emitted by a sharpened cathode tip. This quantum mechanical tunneling phenomenon is well known as field emission (FE) or "cold cathode" emission. Since higher aspect ratios (height/tip radius) are pursued for the generation of higher FE currents at lower applied electric fields; material properties and cathode configuration are essential for FE characteristics [1]. The wide expansion of FE in various applications including electron guns [2], microwave power amplifiers [3], X-ray tubes [4], neutralizers used in space propulsion devices [5], electron beam lithography [6], and large area field emission sources like flat panel field emission displays (FEDs) [7] stipulates intensive research efforts towards the design and production of electron emitting cold cathodes with improved performance (Figure 1). Aside from that, due to the high-performance requirements of FE materials such as high FE current as well as electrical and mechanical durability in highly distorted geometries, recent FE potential for the realization of integrated flexible devices still constitutes an additional engineering and research challenge. The field enhancement factor β is considered as crucial parameter for the performance of a FE cathode, which can be boosted by increasing the aspect ratio, corresponding to the emitter height over its tip radius. In this context, nanotubes and nanosheets have been considered as emerging materials for FE applications due to their unique 1D and 2D nanostructured geometry, respectively [8,9]. Indeed, the fact that the aspect ratio becomes increasingly high is attributed to their nanoscale tip curvatures and relatively long microscale lengths. Therefore, such nanostructures promote the efficient electron emission at weak applied fields through the generation of a large electric field enhancement.

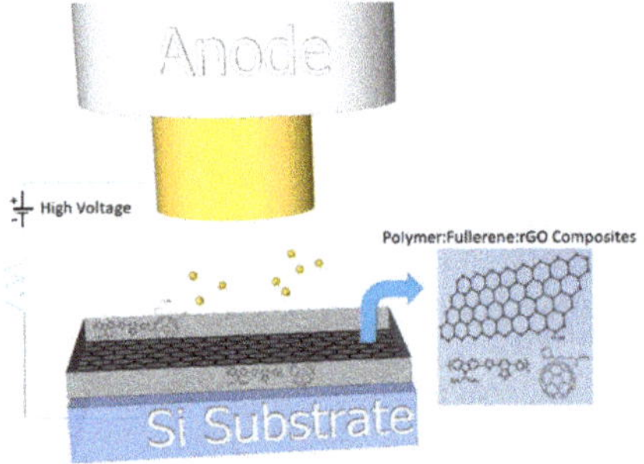

Figure 1. Schematic representation of a field emitter cathode based on hydroiodic acid (HI)-treated reduced graphene oxide (rGO)-charge transfer materials composites.

Although various 1D nanostructures have been placed under the research scope, especially carbon nanotubes (CNTs) have attracted industry's early interest as emerging FE [8–10] for CNT-based FEDs commercialization [11] yet with questionable survival in market. One of the prime concerns is the absence of long lasting FE stability [12] by virtue of the rapid degradation witnessed in CNTs-based emitters, although their electron emission density is tremendous [13]. In addition, in oxygen-rich conditions, CNTs display a perpetual decrease in the FE current density while on the contrary, V_{th} increases [14]. Due to this fact a bottleneck occurs for CNT-based applications, especially in FEDs, under low vacuum or gas purged conditions.

The research rush for graphene and 2D single layer semiconducting materials as inorganic analogues of graphene that demonstrate exceptional physical, optical, and electrical properties has to do mainly with its 2D atomic layer structure [15,16]. However, a low number of studies on the FE properties from such materials, like MoS_2 sheets, have been undertaken [17]. Moreover, WS_2 nanotubes also exhibit good FE performance and stability comparable to carbon nanostructure-based field emitters by virtue of their relatively small bandgap, limited number of dangling bonds, mechanical stability, and nontoxicity. Therefore, they pose a promise for a spectrum of technological applications [18]. Yet, WS_2 nanotubes have to be complemented with a semiconductive polymer, normally P3HT [19], in order to prevent aggregation of the NTs thus enhancing performance of cathodes without polymer. What is more, a polymer matrix is often needed in order to observe protruding nanotubes onto microstructured substrates. This may be ascribed to the low surface tension of NTs/P3HT solution and the large substrate roughness [18]. Alternatively, since the conducting polymer composite serves as an intermediate layer between the WS_2 nanotube or CNT emitters and a conducting substrate, the electrical contact is improved through the reduction of contact resistance, thereby enhancing overall FE performance [20]. Also, the low electron affinity, wide bandgap excellent transport properties, and flexibility of such polymer-based cathodes have rendered composites of semiconductive polymers with 1D or 2D nanomaterials vital to final related FE optimization [21]. Furthermore, recently, the way to exploit transition metal dichalcogenide (TMD) in FE was cleared with geometrically modulated, CVD-grown MoS_2, and $MoSe_2$ monolayer semiconductors that were suspended with 1D nanoarrays of ZnO was demonstrated, thereby enabling the geometrical modulation and tuning charge transfer [22].

An extended use of graphene nanosheets [15,23] for FE cathodes development is present, thanks to their extraordinary physical, mechanical and chemical properties [18,19,23]. Moreover, the planar 2D structure of graphene renders it to an ideal material for large-area FE devices realization. Since graphene consists of flakes with high aspect ratio as well as sub-nanometer edges, turning graphene into a superior FE is easy, hence allowing the electrons' extraction at low threshold electric fields with high geometric field enhancement. Although CVD is one of the most successful methods to produce a graphene-based FE, since edge states dominantly contribute to emitting electrons, controlling the orientation of edges perpendicular to substrates is indispensable. In the CVD case, the resulting graphene must go through complex processes such as laser writing [1], plasma treatment [24], or substrate modification [25] so as to create vertical edges, thereby engaging expensive fabrication equipment and long-term procedures.

Other graphene-based FE cathode fabrication routes include liquid phase exfoliation from graphite or exfoliation of graphene from a carbon cloth's fibers followed by coating graphene solution on a conductive substrate (e.g., Cu) [26,27], electrophoretic deposition (EPD) [27,28], screen printing accompanied with selective photoetching techniques [29], or even blade and ultrasonic spray coating [30] to name just few of them.

Apart from the previous techniques, chemical oxidation of bulk graphite by Hummers is a strong candidate for resulting to the high yield production of graphene oxide (GO). For instance, a low-cost and successful chemical method to synthesize FE-applicable reduced GO, the annual-ring graphene (ARG) method [31], which demonstrates the advantage of facile synthesis and yields abundant vertical edges on the cross-section of ARG, hence equaling or even surpassing FE performances of CNT [13], cannon-structured graphene film [32] and rGO/MnO_2 composite [33], not to mention the ultralight weight and flexible nature of the cathode. In general, albeit GO has a similar atomic single layer structure to graphene, the existence of oxygen functional groups attached onto sp^3 hybridized carbon atoms impacts on its conductivity loss, mainly due to the presence of epoxy and hydroxide groups on its basal plane [34], as illustrated in Figure 2. In order to come up against this side effect and to partially recover the conductivity, various reduction methods have been developed to reduce GO in the form of rGO towards the realization of efficient electronic functionalities.

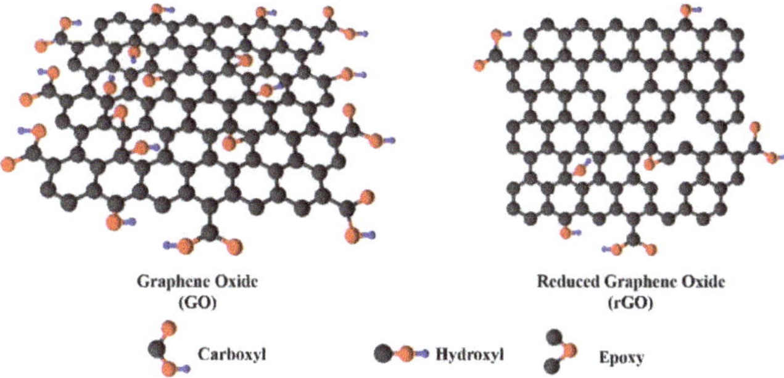

Figure 2. Depictions of GO and rGO structures.

Chemical reduction can in general be achieved in both liquid and gas phases [35]. Chemical reducing methods include agents that range from hydrazine [36–39] and Hydrogen Iodide gas [40] until even ascorbic acid [41], KOH, or NaOH [42]. Aside from chemical, thermal reduction at high temperature under inert atmosphere [34,43], and application of either electric [44], or electromagnetic field [45] are also successful reduction routes.

Before moving into plain comparison of FE parameters, considering the way of GO reduction is essential to deeper understanding the correlation between FE and GO, while taking into account only the geometric aspect ratio is insufficient on its own. The oxygen/carbon atomic ratio of the rGO lattice, determined by X-ray photoelectron spectroscopy (XPS), indicates that GO lattice parameters vary depending on the applied reduction technique. In the meanwhile, the Work Function (WF) among different rGO samples, validated by UPS measurements, gave similar trends and thus resulted in different FE characteristics. In general, lower content of oxygen-containing groups exhibits an enhanced FE performance mainly thanks to the WF increase [46]. Besides, the electron transfer at the interfaces between the substrate and the cathode material is expected to be highly promoted, since the states' density in the cases of oxygen elimination (e.g., HI-assisted rGOs) is also anticipated to be higher.

The role of GO concentration, namely flake density, impacts on the degree of reduction and morphology. Proper partial reduction leaves acidic groups on rGO surface and the polar nature of GO remains and accounts for stable, homogeneous dispersions in solvents, more notably in DMF and NMP [47]. In the case of low rGO-flake density films, the number of emitters is diminished, and the geometrical characteristics of the emitters are resultingly of minor importance are diminished, thus leading to a sharp decline in their performance as cold cathodes. On the one hand, the enhanced roughness at low rGO concentration suggests random sheets orientation onto the planar substrate, while on the other hand, screening effects are observed in the case of high-density films [48,49] followed by an understated emission performance.

On the other hand, polymer concentration affects morphology in a significant manner, while a polymer:rGO concentration ratio exists for optimized FE performance. For high polymer contents fewer rGO sheets are exposed to vacuum, while the thermal conductivity may be four times higher than ambient air for the case of P3HT:rGO, hence easing heat dissipation and achieving higher stability contrary to bare rGO [21]. For low polymer content, rGO flakes may become more preferentially oriented parallel to the substrate [36] and fewer emitting edges may be exposed to vacuum; as a result, the emission performance would be decreased again.

Edge density was also increased upon the deposition of an additional rGO layer onto the composite films, thanks to the smoother surface of the top rGO layer in the case of higher rGO content. In other cases, where the polymer concentration was higher, rGO arrays protruded from the polymer bulk, favoring the synergy of two conduction systems: conduction between rGO flakes of the same array, as well as between rGO flakes and the polymer [50]. The relevant charge transfer mechanisms are illustrated in Figure 3. Upon the additional rGO layer deposition onto the polymer:rGO composite layer, the DMF solvent of the rGO dispersion (or any other suitable solvent that might be applied) may dilute some polymer of the underlying layer, thereby a part of the rGO flakes and/or arrays penetrate the polymer. Therefore, since the polymer:rGO ratio increases together with the polymer content, rGO flake density into the polymeric scaffold gets consequently minimized, thus leading to a relevant deterioration in the edge density of rGO exposed to vacuum.

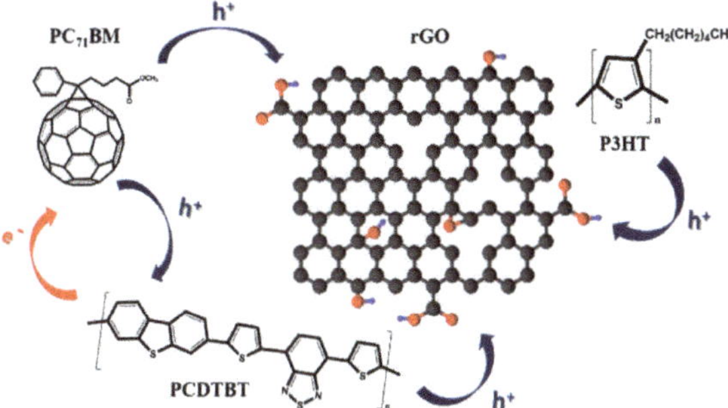

Figure 3. Comprehensive depiction of current conduction routes encountered in the studied composite cases.

Laser processing plays a crucial role in fabricating 2D-based field emitters. To begin with, structuring Si substrate to create conical, well-separated spikes that are perpendicular to substrate is crucial to good GO deposition and edge protrusion. The microspike arrays are reportedly fabricated by femtosecond (fs) laser texturing of n-type Si, usually followed by consequent removal of grown oxides by HF on spikes' surface [36]. Secondly, direct laser writing (DLW) is an additional promising technique for the rapid and facile fabrication of graphene for various applications [1,51]. Proper laser treatment gives rise to preferential protrusion of rGO sheets from the substrate and in series to FE characteristics superior to those of pristine rGO. rGO bundles align themselves perpendicular to the substrate, while at the same time sharp graphene edges are protruding out of the bundles. Epidermal treatment of rGO prevents thermal damaging of substrate. As a side advantage, DLW retains the performance of the developed flexible cathodes upon extensive bending conditions, thereby having good potential for graphene-flexible FE cathodes. Finally, selective laser reduction of GO has received attention due to its high potential [52,53] in a diversified field of applications, including FE among them.

This study aims to offer a deeper insight on how three optimized FE enhancement routes contribute to rGO-based FE enhancement. Section 2 discusses the realization of HI-assisted reduction of GO and subsequent rGO ink formation, as well as the optimization process of P3HT:rGO, PCDTBT:rGO and rGO-PCDTBT:PC$_{71}$BM composites by diversifying rGO relative concentrations, whereas final FE cathode fabrication steps are presented. Results are presented in Section 3, and compared between each other in Section 4, while subjecting to discussion concerning FE enhancement in terms of morphology, solvent, reduction agent, or material type (plain rGO ink, polymer:rGO, and polymer:fullerene blend:rGO composite ink). Introducing a PCDTBT:PC$_{71}$BM blend to FE cathodes has to the best of our knowledge never been suggested before. For the first time a HI-assisted rGO has been introduced into FE cathodes in synergy with charge transfer materials' blends; fullerene-based materials with high electron affinity and polymers are studied. The obtained results are superior to those of previous studies with rGO and polymers. [21] Moreover, the different impacts of fullerene and polymer constituents according to their relative concentrations, physical and chemical contribution are interesting subjects for further investigation.

2. Materials and Methods

2.1. Preparation of Starting GO Powder

GO was prepared from purified natural graphite powder (Alfa Aesar, ~200 mesh) according to Hummers' method [54,55]. H_2SO_4 (40 mL) was added to 500 mg graphite and stirred for several minutes, followed by the addition of $NaNO_3$ (375 mg). Further stirring took place inside a cooled beaker for 2 h, during which 3 g of $KMnO_4$ was added. Further stirring step lasted for other 4 h when the reaction mixture was left to reach room temperature before being heated at 35 °C for 30 min.

It was then poured into a flask containing deionized water (50 mL) and the stirring temperature was further raised to 70 °C for 15 min. The mixture was then decanted into 250 mL of deionized water; the unreacted $KMnO_4$ was removed by subsequent addition of 3% H_2O_2 (2 mL). The obtained graphite oxide was purified by repeated centrifugation, which initially yielded a sediment of acidic nature; thus, deionized water and HCl were added to capture sulfate ions so that sediment reaches pH 7. The final product was oven-dried and sieved.

2.2. Production of rGO Ink

A highly efficient HI/AcOH assisted reduction route was chosen to be implemented, as it is described in our previous publication [56]. In a similar manner, our GO was reduced using a mixture of hydriodic acid (HI, 55%)/acetic acid (AcOH). More specifically, the as-prepared GO powder (0.1 g) was sonicated in AcOH (37 mL) for 2 h. Then HI (2 mL) was added and the mixture was stirred at 40 °C for 40 h. Afterwards, the product was isolated by filtration and purified through a three-step washing procedure: (1) aqueous solution of saturated sodium bicarbonate ($NaHCO_3$, 3 × 2.5 mL); (2) DI water (3 × 2.5 mL); and (3) acetone (2 × 2.5 mL). Finally, the resulting rGO was dried at 60 °C in a vacuum oven overnight. In the aftermath, the reduced GO powder underwent pulverization procedure with mortar and pestle. Next, rGO powder (20 mg) was added in three vials containing THF (20 mL), DCB (4 mL), and DCB:CB (3:1 *v/v*, 20 mL), respectively, in order to predetermine the initial concentration to 1 mg/mL. Then, they were sonicated using an ultrasonic probe (Hielscher UP200Ht) for 1 h and centrifuged at 4000 rpm for 45 min (Allegra X-22) to remove the large aggregates. Afterwards, the supernatants were carefully collected with a pipette and surfactant-free viscous rGO inks were formed (~0.5 mg/mL in THF, ~0.8 mg/mL in DCB, and ~0.85 mg/mL in DCB:CB). The resulting rGO inks are depicted in Figure 4a.

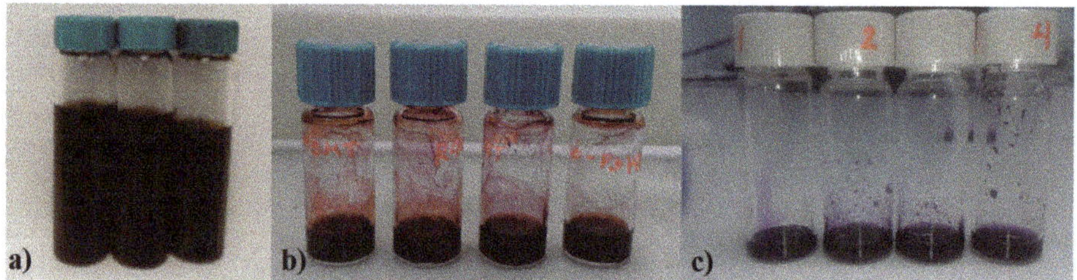

Figure 4. (**a**) Unmixed rGO ink, (**b**) P3HT:rGO, and (**c**) PCDTBT:rGO blends in controlled volume ratios.

2.3. Preparation of rGO-Polymers Composite Inks

P3HT (10 mg) and PCDTBT (4 mg) were dissolved in THF (1 mL) and DCB (1 mL), respectively. Polymers:rGO composite inks were prepared by adding rGO ink to the P3HT and PCDTBT solutions, respectively into different combinations of relative volume ratios (0:100%, 20:80%, 40:60%, 60:40%, and 80:20%). In order to prepare homogeneous polymers:rGO composite inks, the mixtures were further sonicated in an ultrasonic bath (Elmasonic S30H) for another 1.5 h and finally were left undisturbed for 15 min, in order to settle down any agglomerates. Polymers:rGO composite inks are depicted in Figure 4b,c, respectively.

2.4. Preparation of rGO-PCDTBT:PC$_{71}$BM Composite Ink

On the other hand, the polymer–fullerene powder was unable to dissolve adequately 4 mg/mL PCDTBT:PC_{71}BM solution in THF; therefore, a mixture of DCB:CB (3:1 in volume ratio) was finally applied to dissolve successfully the solid components with sufficient magnetic stirring. PCDTBT:PC_{71}BM were dissolved in DCB:CB (3:1 *v/v*, 1 mL) with 1:4 (4 mg:16 mg) ratio and stirred overnight at 70 °C. At next, PCDTBT:PC_{71}BM:rGO composites were realized by adding rGO ink, prepared in DCB:CB, to the PCDTBT:PC$_{71}$BM blend in different combinations with relative volume ratios (0:100%, 20:80%, 40:60%, 60:40%, and 80:20%). In order to prepare homogeneous PCDTBT:PC$_{71}$BM:rGO composite inks, the above described procedure was also followed.

2.5. FE Cathodes Realization

Silicon wafers (n$^+$ Si) were subsequently rinsed with acetone and isopropyl alcohol. Composite inks were then realized and coated on top of substrate surface with the drop casting method. Subsequent rGO layer was later on coated in the same manner. All films were placed in an oven for an adequate time until the full removal of the solvents. Final FE cathode structure is illustrated in Figure 1.

2.6. Fowler–Nordheim Theory

The FE properties of graphene can be determined using the Fowler–Nordheim law [26]. In brief, the relationship of FE current density (J) with the applied electric field E is demonstrated below as Equation (1):

$$J = \eta\,\alpha\,\frac{(\beta E)^2}{\varphi}e^{-b\frac{\varphi^{3/2}}{\beta E}} \tag{1}$$

where constants alpha (α) and b correspond to $\alpha = 1.54 \times 10^{-6}$A eV V^{-2} and $b = 6.83 \times 10^3$ eV$^{-3/2}$V^{-1}, φ represents the work function of the material, g is related with the emitters' geometrical efficiency, beta (β) is the field enhancement factor, and E corresponds to the macroscopic electrical field in V/μm. E is calculated from E = V/d, in which V denotes the applied potential and d the interelectrode distance. According to Fowler–Nordheim equation, the part of the ln(J/E2) − (1/E) plot that is linear gives proof of electron movement through the tunneling barrier of the emitter in the specific area.

The slope of the previous equation (kFN) derives the field enhancement factor β according to Equation (2):

$$\beta = b\,\frac{\varphi^{3/2}}{kFN} \tag{2}$$

Additional parameters for assessing FE quality include the E$_{to}$ (turn-on electric field), E$_{th}$ (threshold of electric field), the intensity of luminescence, and the stability of emission. What is sought is minimum values of E$_{to}$ and E$_{th}$, maximum values of the field enhancement factor and emission current density, as well as maximum stability so as to yield an ideal field emitter.

3. Results

3.1. Field Emission Measurements

FE measurements were executed under vacuum of less than 10^{-6} Torr. Samples were used as cold cathodes in a short circuit-protected planar diode system, as depicted in Figure 1. More details regarding the experimental FE setup can be found elsewhere [48]. Current density−voltage (J−V) curves were obtained while the interelectrode distance, d = 200 μm in our case, was controlled by a stepper motor. It is pointed out that FE characteristics remained unaffected by the anode location. In order to confirm the stability of the devices, as well as J–V reproducibility, continuous emission cycles were performed. A high voltage (HV) source (PS350-SRS) supplied a voltage with variable sweep step between electrodes. A digital picoammeter (Keithley 485) was used for the sake of measuring FE current. The emission current stability versus time was investigated by monitoring the emitted current density rate over a long time period of consecutive operation.

3.1.1. First Case: HI-Reduced GO Emitter

A set of FE cathodes with different rGO ink ratios was realized and subsequently characterized. Figures 5a,b and 6 display the FE response of the best FE cathodes per each rGO ink ratio. Indeed, the beta factor stated at a value of 660, while the E_{th} value was 1.6 V/μm. Numerical results are shown in Table 1.

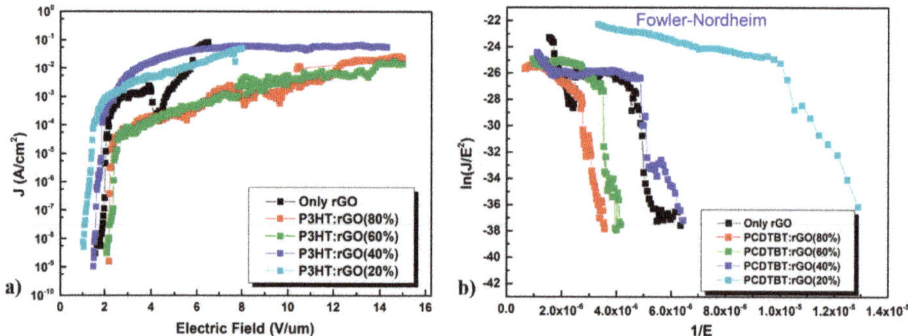

Figure 5. (**a**) Logarithmic plot of the current density, measured as a function of the electric field E (*J–E*), obtained by different concentration ratios of rGO ink in composite n^+-Si/P3HT:rGO field emitters. Lowest electrical field threshold appears for the P3HT:rGO(20%) case. (**b**) Fowler–Nordheim curves of the *J–E* plots of FE with different concentration ratios of rGO ink in composites n^+-Si/PCDTBT:rGO.

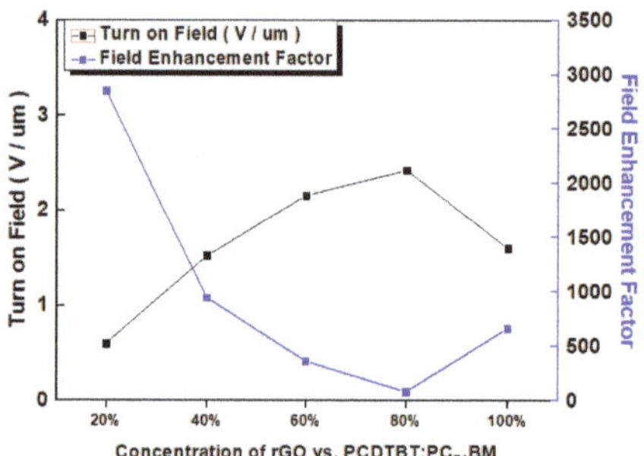

Figure 6. Variation of the turn on field (black line) and the enhancement factor (blue line) in different concentrations of $PCDTBT:PC_{71}BM:rGO$ and P3HT:rGO composite inks.

Table 1. Field enhancement factor and turn-on field figures classified per composite type (columns) and for HI/AcOH-reduced rGO ink with diversified ratios.

rGO Ratio (%)	n^+-Si/P3HT:rGO		n^+-Si/PCDTBT:rGO		n^+-Si/PCDTBT:PC$_{71}$BM:rGO	
	Field Enhancement β	Turn-on Field E_{to} (V/μm)	Field Enhancement β	Turn-on Field E_{to} (V/μm)	Field Enhancement β	Turn-on Field E_{to} (V/μm)
100%	660 ± 10	1.60 ± 0.1	660 ± 10	1.60 ± 0.1	660 ± 10	1.60 ± 0.1
80%	300 ± 10	2.23 ± 0.1	170 ± 10	2.80 ± 0.1	80 ± 10	2.43 ± 0.1
60%	420 ± 10	2.05 ± 0.1	625 ± 10	2.40 ± 0.1	360 ± 10	2.16 ± 0.1
40%	915 ± 10	1.53 ± 0.1	1090 ± 10	1.58 ± 0.1	950 ± 10	1.53 ± 0.1
20%	2500 ± 10	1.03 ± 0.1	2050 ± 10	0.80 ± 0.1	2850 ± 10	0.60 ± 0.1

The ± values denote the standard deviation of each measured or estimated quantity.

3.1.2. Second Case: Polymer:rGO Composite Emitter

Among all different volume ratio variants, that of 80% polymer (P3HT or PCDTBT):20% rGO gave the best results. Here, a set of five different ratio batches have been fabricated and tested, while each case includes five fabricated cathode samples. Accordingly, Figure 5a depicts a comparative chart of the best samples per each one of all five ratios. Also, as demonstrated in Table 1, huge deviation in FE values for certain concentration scenarios indicates inability in perfectly controlling process parameters. Yet, repeatability of bad results for these concentration ratios implies that only the optimized parameters of the ratio 80:20% are of practical importance.

In the *J–E* plots, three different functions of an emitter can be observed: (1) current density saturation; (2) field emission; and (3) no emission. FE performance was enhanced in the case that graphene flakes were vertically oriented with respect to the substrate, when the emission was weakened, due to screening effects, over a critical rGO proportion.

Besides, FE performance can be further improved, in terms of low turn on field and high enhancement factor, by adopting three different approaches, according to the Fowler–Nordheim theory: (1) by shaping organized emission arrays, thus increasing the emitting area; (2) by selecting low work function φ materials, such as graphene-based materials; and (3) by increasing the geometrical aspect ratio of the emitter and therefore to improve the enhancement factor. On the one hand, as the polymer concentration in the composite emitter is increased, the relative concentration of graphene flakes in the polymeric matrix is decreased. Consequently, the density of the graphene flakes in the vacuum is also diminished. On the other hand, when the polymer content is very low, rGO flakes exhibit random protrusions, with reference to the substrate due to the Van der Waals forces, also resulting in limited emitting potential.

3.1.3. Third Case: PCDTBT:PC$_{71}$BM:rGO Composite Emitter

In a similar manner to the polymer:rGO case, a batch with 80% blend:20% rGO ink volume ratio gave the best results. Again, per each case, a set of five different ratio batches has been fabricated and tested, while each batch includes five fabricated cathode samples and only the best cathode results per batch are presented in Figure 5b. Accordingly, Figures 6 and 7 depict a comparative chart of the best batches per each ratio with respect to the Fowler–Nordheim characterization and the overall FE performance parameters, respectively. All FE values per volume ratio set are included in Table 1.

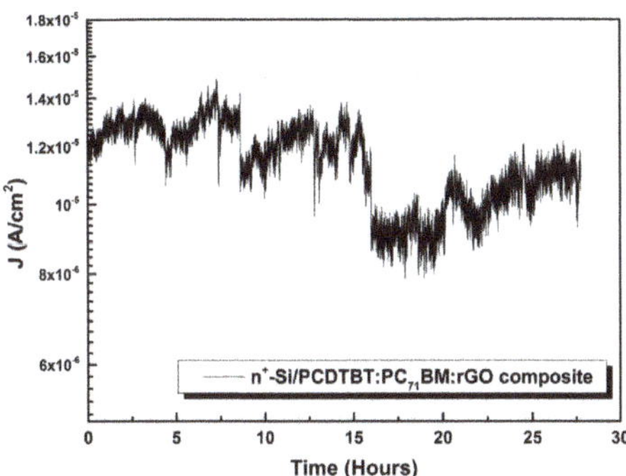

Figure 7. The evolution of the emission current density at a constant bias voltage of 1500 volts over a long period of continuous operation for the best rGO cathodes measured.

3.1.4. Field Emitter Stability

The evolution of the emission current density at a constant bias voltage of 1500 volts was tracked over a long period of over 25 h continuous operation for the best rGO-containing cathode that had been chosen among all samples. Figure 7 demonstrates the respectable stability of the emitter's current density over time, resulting in a final drop of approximately 16% with regard to the initial current. These results explain why PCDTBT:PC$_{71}$BM:rGO composite emitter has a higher level of long-term stability under the field emission experimental conditions related other rGO composite emitters that have been tested in the literature [34].

3.2. Morphological Characterization

Figures 8 and 9 demonstrate scanning electron microscopy ((FESEM JEOL-JSM7000F) images which were extracted in order to verify the creation and morphology of extruding edges both prior and after FE.

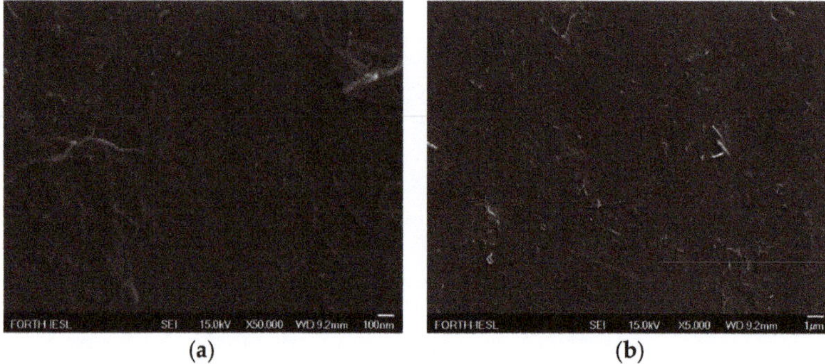

(a)   (b)

Figure 8. Top views of FE cathode surface with clearly visible edges in higher (**a**) and lower magnification scale (**b**). Distance among neighboring edges is remarkably close in the order of few microns or less.

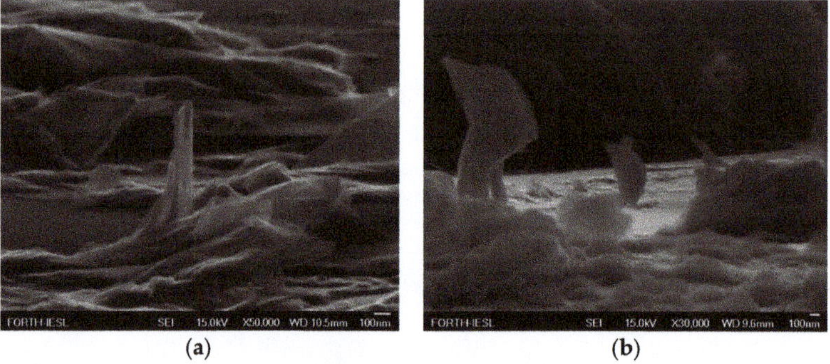

(a)   (b)

Figure 9. SEM images of a FE cathode edges in the cases of rGO bundles alone (**a**) and P3HT(80% v/v):rGO blend(20% v/v) (**b**). Thickness inhomogeneity due to the applied drop casting method has to be taken into account together with the formation of protruding edges. Impact of rGO sheets is clearly visible especially in (**a**).

4. Discussion

In general, HI/AcOH is by far more efficient by other reducing agents like hydrazine: optimum reduction dictates the compromise of yielding a partially reduced rGO. Although being of low occurrence, acid groups are present on rGO surface and result to a remaining polar nature of rGO, hence enabling rGO to form stable and homogeneous dispersions in THF [21].

As already mentioned in Section 2.4, the polymer–fullerene powder (PCDTBT:PC$_{71}$BM) was unable to dissolve adequately in THF, therefore a mixture of DCB:CB was finally applied to dissolve successfully the solid components with the assistance of magnetic stirring. A possible explanation to this fact is that blend residues which are insoluble may probably contain the highest molecular weight fraction which could not be solubilized [57]. It can be said that high MW values adversely affect polymer solubility. Low solubility of polymers may direct to a strong interchain bonding or even chain aggregation. Seemingly, orthodichlorobenzene did not result to any further aggregation of PCDTBT chain, while it could simultaneously decelerate drying of the wet film [58]. Furthermore, the effect of THF additive in polymer and blend morphology mostly has to do with the solvent's higher affinity for polymers than for PCBM. Diversifying the dipole moment by employing solvents similar to THF may control this affinity. Moreover, it has to be mentioned that THF, which has low boiling point, dissolves P3HT very easily, while such an easy dissolution for PCBM appears for the higher boiling point o-DCB, let alone the good solubility of PCDTBT [59].

With respect to the case of P3HT:rGO composite-based FE cathodes, among all different concentration variants, the volume ratio of 80% polymer:20% rGO ink gave the best results. FE trends are in accordance with relevant previous literature [21]. Indeed, component concentration significantly affects composite morphology and the dependent FE performance; Lower rGO flakes' concentration inside the composite blend, reportedly, may favor the orientation of sheets at angles different to the planar substrate, hence inducing higher roughnesses [21]. As the P3HT:rGO ratio decreases, i.e., the polymer content increases, the rGO density into the polymer matrix decreases, giving rise to a corresponding decrease in the density of the rGO edges exposed to vacuum. On much higher rGO flakes' concentration; however, screening of coated surface takes presumably place that deteriorates FE according to various studies [21,48,49]. Higher polymer content means fewer rGO sheets that are exposed to vacuum, while it has been demonstrated that thermal conductivity favors heat dissipation thus prolonging stability, in contrast with the case of bare rGO. On the other hand, lower polymer content allows rGO sheets to orientate themselves parallel to the substrate, hence resulting to fewer emitting edges and decreased FE performance. In general, at higher rGO ink concentrations, the base of the protruding bundles is a part of the rGO layer, whereas in the case of samples with high polymer content, bundles are sticking out of a polymer bulk. This observation suggests that there are two conduction processes present, conduction between rGO within flakes of a bundle and between polymer and sheets [50]. A final remark is that lower emitter edge density has been observed in high and low concentrations [21]. We conclude presuming that 80:20% analogy of P3HT:rGO better approaches the trade-off between screening and the density of protruding rGO edges, thus leading to better FE characteristics witnessed in Table 1, Figures 5 and 6.

The differences observed in field enhancement factor and in turn-on field between the composite emitters with different polymers (P3HT and PCDTBT) are directly related to their morphology instability and their different ionization state [60,61]. In this work, the highest field enhancement factor observed on semicrystalline p-type polymer P3HT and the lowest electrical threshold for amorphous p-type polymer of PCDTBT.

Besides, considering the third FE case of PCDTBT:PC$_{71}$BM blends complementing rGO, it is already known that for blend containing OSCs a low temperature annealing at approximately 70 °C suffices for optimizing PCE, thus also blend morphology and phase separation [62,63]. The same logic is valid in the FE case as well, since charge carrier transfer is thereby optimized. It has been observed that the upper part of PCDTBT:PC$_{71}$BM films is relatively enriched in PC$_{71}$BM with negative gradient of fullerene concentration when examining deeper inside the annealed layer, yet in no case exceeds upper surface concentration of PC$_{71}$BM the one of PCDTBT. Of course, PC$_{71}$BM tends to reach the upper interface with air demonstrating a mild trend for phase separation, but PC$_{71}$BM accumulation on the upper surface does not at all increase roughness due to PC$_{71}$BM crystallites. Furthermore, a mild annealing at 70 °C neither impacts on PC$_{71}$BM distribution towards depth, nor does it significantly affect blend crystallization, but enables the PCDTBT:PC$_{71}$BM blend to self-organize into an optimal morphology for charge generation and extraction and, most importantly, eases the removal of trapped blend solvent [64]. Similar to the previous case of HI-reduced GO, the 20% rGO ink:80% polymer–fullerene blend ratio appears to yield optimized FE characteristics. The fact that rGO contributes best with the same concentration as in polymer:rGO, may indicate a nonsignificant impact of fullerene on final FE contrary to rGO. Yet, in accordance with not directly FE-relevant studies that include $PC_{71}BM$ [65] or $PC_{61}BM$ [56], the fullerene–rGO interaction is highlighted in terms of further improving charge transfer from the blend to emitting edges and a significant contact resistance reduction. Finally, it has to be noticed that in all cases research excluded rGO-free implementation of FE cathodes, since this research, tries to explain the electron emission from composites emitters with rGO flakes and polymer or charge transfer materials; such research would be out of the scope of current study.

5. Conclusions

In this study, four rGO-based field effect emitters were developed, optimized, and compared. rGO layers prepared by HI/AcOH reduction method and were coated on (a) n^+ doped Si substrate, (b) n^+-Si/P3HT:rGO, (c) n^+-Si/PCDTBT:rGO, and (d) n^+-Si/PCDTBT:$PC_{71}BM$:rGO composites, respectively. We investigated the FE properties of different concentrations of polymer composite solutions to control the structural end electrical properties of the substrate. It is found that the cathodes based on PCDTBT:$PC_{71}BM$:rGO displayed a field enhancement factor of ~2850, with remarkable stability over 25 h and low turn-on field in 0.6 V/μm.

The threshold field, enhancement factor and the remarkable stability of FE current were remarkably improved compared to the pristine Si demonstrating that is a promising FE cathode with potential applications in vacuum microelectronics and FEDs.

Finally, taking into account the high potential of PCDTBT:$PC_{71}BM$ as successful blend candidate in conventional OSC devices, its optimized coating on a TCO or TCO/HTL substrate, either rigid or flexible, might result in significant number of photogenerated electrons and a further decrease in contact resistance that might further boost FE figures of merit in future, therefore possibly promising a new alternative type of photodetectors.

Author Contributions: Conceptualization, M.M.S. and G.V. (George Viskadouros); Methodology, M.M.S. and G.V. (George Viskadouros); Formal Analysis, G.V. (George Viskadouros); Investigation, M.M.S. and G.V. (George Viskadouros); Data Curation, G.V. (George Viskadouros), C.P., N.K., G.K., and G.V. (George Veisakis); Writing–Original Draft Preparation, M.M.S., G.V. (George Viskadouros), and C.P.; Writing–Review & Editing, M.M.S., G.V. (George Viskadouros), and K.P.; Supervision, E.K.; Project Administration, M.M.S. and G.V. (George Viskadouros); Funding Acquisition, M.M.S.

Funding: This research was funded by State Scholarships Foundation (SSF) and cofinanced by the European Union (European Social Fund—ESF) and Greek national funds through the action entitled "Reinforcement of Postdoctoral Researchers", in the framework of the Operational Programme "Human Resources Development Program, Education and Lifelong Learning" of the National Strategic Reference Framework (NSRF) 2014–2020, Grant No. 13992.

Acknowledgments: The authors thank IESL/FORTH for providing facilities.

Conflicts of Interest: The authors declare no conflicts of interest.

References

1. Viskadouros, G.; Konios, D.; Kymakis, E.; Stratakis, E. Direct laser writing of flexible graphene field emitters. *Appl. Phys. Lett.* **2014**, *105*, 203104. [CrossRef]
2. De Jonge, N.; Lamy, Y.; Schoots, K.; Oosterkamp, T.H. High brightness electron beam from a multi-walled carbon nanotube. *Nature* **2002**, *420*, 393–395. [CrossRef] [PubMed]
3. Teo, K.B.K.; Minoux, E.; Hudanski, L.; Peauger, F.; Schnell, J.-P.; Gangloff, L.; Legagneux, P.; Dieumegard, D.; Amaratunga, G.A.J.; Milne, W.I. Carbon nanotubes as cold cathodes. *Nature* **2005**, *437*, 968. [CrossRef] [PubMed]
4. Yue, G.Z.; Qiu, Q.; Gao, B.; Cheng, Y.; Zhang, J.; Shimoda, H.; Chang, S.; Lu, J.P.; Zhou, O. Generation of continuous and pulsed diagnostic imaging x-ray radiation using a carbon-nanotube-based field-emission cathode. *Appl. Phys. Lett.* **2002**, *81*, 355–357. [CrossRef]

5. Velasquez-Garcia, L.F.; Akinwande, A.I.; Martinez-Sanchez, M. A Planar Array of Micro-Fabricated Electrospray Emitters for Thruster Applications. *J. Microelectromech. Syst.* **2006**, *15*, 1272–1280. [CrossRef]

6. Fomani, A.A.; Akinwande, A.I.; Velasquez-Garcia, L.F. Resilient, Nanostructured, High-Current, and Low-Voltage Neutralizers for Electric Propulsion of Small Spacecraft in Low Earth Orbit. *J. Phys. Conf. Ser.* **2013**, *476*, 012014. [CrossRef]

7. Engelsen, D.E. The temptation of field emission displays. *J. Phys. Proc.* **2008**, *1*, 355–365. [CrossRef]

8. Rinzler, A.G.; Hafner, J.H.; Nikolaev, P.; Lou, L.; Kim, S.G.; Tomanek, D.; Nordlander, P.; Colbert, D.T.; Smalley, R.E. Unraveling Nanotubes: Field Emission from an Atomic Wire. *Science* **1995**, *269*, 1550–1553. [CrossRef]

9. Fan, S.; Chapline, M.G.; Franklin, N.R.; Tombler, T.W.; Cassell, A.M.; Dai, H. Self-Oriented Regular Arrays of Carbon Nanotubes and Their Field Emission Properties. *Science* **1999**, *283*, 512–514. [CrossRef]

10. Giubileo, F.; Di Bartolomeo, A.; Iemmo, L.; Luongo, G.; Urban, F. Field Emission from Carbon Nanostructures. *Appl. Sci.* **2018**, *8*, 526. [CrossRef]

11. Jiang, K. Carbon Nanotubes for Displaying. In *Micro and Nano Technologies, Industrial Applications of Carbon Nanotubes*, 1st ed.; Peng, H., Li, Q., Chen, T., Eds.; Elsevier: London, UK, 2017; pp. 101–127, ISBN 9780323414814.

12. Kayastha, V.K.; Ulmen, B.; Yap, Y.K. Effect of graphitic order on field emission stability of carbon nanotubes. *Nanotechnology* **2007**, *18*, 035206. [CrossRef] [PubMed]

13. Bonard, J.M.; Salvetat, J.P.; Stockli, T.; Forro, L.; Chatelain, A. Field emission from carbon nanotubes: Perspectives for applications and clues to the emission mechanism. *Appl. Phys. A* **1999**, *69*, 245–254. [CrossRef]

14. Wadhawan, A.; Stallcup, R.E.; Stephens, K.F.; Perez, J.M.; Akwani, I.A. Effects of O_2, Ar, and H_2 gases on the field-emission properties of single-walled and multiwalled carbon nanotubes. *Appl. Phys. Lett.* **2001**, *79*, 1867. [CrossRef]

15. Nicolosi, V.; Chhowalla, M.; Kanatzidis, M.G.; Strano, M.S.; Coleman, J.N. Liquid Exfoliation of Layered Materials. *Science* **2013**, *340*, 1420. [CrossRef]

16. Chhowalla, M.; Shin, H.S.; Goki, E.; Li, L.-J.; Loh, K.P.; Zhang, H. The chemistry of two-dimensional layered transition metal dichalcogenide nanosheets. *Nat. Chem.* **2013**, *5*, 263–275. [CrossRef] [PubMed]

17. Kashid, R.V.; Late, D.J.; Chou, S.S.; Huang, Y.-K.; De, M.; Joag, D.S.; More, M.A.; Dravid, V.P. Enhanced Field-Emission Behavior of Layered MoS_2 Sheets. *Small* **2013**, *9*, 2730–2734. [CrossRef] [PubMed]

18. Stratakis, E.; Eda, G.; Yamaguchi, H.; Kymakis, E.; Fotakis, C.; Chhowalla, M. Free-standing graphene on microstructured silicon vertices for enhanced field emission properties. *Nanoscale* **2012**, *4*, 3069–3074. [CrossRef]

19. Viskadouros, G.; Zak, A.; Stylianakis, M.; Kymakis, E.; Tenne, R.; Stratakis, E. Enhanced field emission of WS_2 Nanotubes. *Small* **2014**, *10*, 2398–2403. [CrossRef]

20. Latham, R.V. *High Voltage Vacuum Insulation*, 1st ed.; Elsevier: London, UK, 1981; ISBN 9780080533971.

21. Viskadouros, G.M.; Stylianakis, M.M.; Kymakis, E.; Stratakis, E. Enhanced field emission from reduced graphene oxide polymer composites. *ACS Appl. Mater. Interfaces* **2014**, *6*, 388–393. [CrossRef]

22. Yang, T.H.; Chiu, K.C.; Harn, Y.W.; Chen, H.Y.; Cai, R.F.; Shyue, J.J.; Lee, Y.H. Electron Field Emission of Geometrically Modulated Monolayer Semiconductors. *Adv. Funct. Mater.* **2018**, *28*, 1706113. [CrossRef]

23. Yamaguchi, H.; Murakami, K.; Eda, G.; Fujita, T.; Guan, P.; Wang, W.; Gong, C.; Boisse, J.; Miller, S.; Acik, M.; et al. Field emission from atomically thin edges of reduced graphene oxide. *ACS Nano* **2011**, *5*, 4945. [CrossRef]

24. Khare, R.T.; Gelamo, R.V.; More, M.A.; Late, D.J.; Rout, C.S. Enhanced field emission of plasma treated multilayer graphene. *Appl. Phys. Lett.* **2015**, *107*, 123503. [CrossRef]

25. Chen, L.; Qu, L.; Deng, J.H. High-efficiency field emission from pressed nickel foam–flat graphene–vertical graphene hybrids. *Mater. Lett.* **2016**, *176*, 165–168. [CrossRef]

26. Chen, L.; Yu, H.; Zhong, J.; Song, L.; Wu, J.; Su, W. Graphene field emitters: A review of fabrication, characterization and properties. *Mater. Sci. Eng. B* **2017**, *220*, 44–58. [CrossRef]

27. Liu, J.L.; Zeng, B.Q.; Wang, X.R.; Zhu, J.F.; Fan, Y. Ultra-low field electron emission of graphene exfoliated from carbon cloth. *Appl. Phys. Lett.* **2012**, *101*, 153104. [CrossRef]

28. Diba, M.; Fam, D.W.H.; Boccaccini, A.R.; Shaffer, M.S.P. Electrophoretic deposition of graphene-related materials: A review of the fundamentals. *Prog. Mater. Sci.* **2016**, *82*, 83–117. [CrossRef]

29. Wu, C.X.; Li, F.S.; Zhang, Y.A.; Guo, T.L. Field emission from vertical graphene sheets formed by screen-printing technique. *Vacuum* **2013**, *94*, 48–52. [CrossRef]

30. Wang, Q.; Zhang, Z.; Liao, Q.; Kang, Z.; Zhang, Y. Enhanced field emission properties of graphene-based cathodes fabricated by ultrasonic atomization spray. *RSC Adv.* **2018**, *8*, 16207–16213. [CrossRef]

31. Ren, X.; Hou, X.; Yu, M.; Ma, J.; Qiu, H. Annual-ring shaped graphene as a free-standing field emitter with high performance. *Mater. Lett.* **2018**, *210*, 133–135. [CrossRef]

32. Liu, J.L.; Zeng, B.Q.; Wang, W.Z.; Li, N.N.; Guo, J.; Fang, Y.; Deng, J.; Li, J.; Hao, C. Graphene electron cannon: High-current edge emission from aligned graphene sheets. *Appl. Phys. Lett.* **2014**, *104*, 023101. [CrossRef]

33. Hareesh, K.; Suryawanshi, S.R.; Phatangare, A.B.; More, M.A.; Bhoraskar, V.N.; Dhole, S.D. Facile single pot synthesis of MnO_2 nanorods-reduced graphene oxide (rGO) nanocomposite: Structural, chemical and field emission investigations. *Mater. Lett.* **2016**, *185*, 472–475. [CrossRef]

34. Sygellou, L.; Viskadouros, G.; Petridis, C.; Kymakis, E.; Galiotis, C.; Tasis, D.; Stratakis, E. Effect of the reduction process on the field emission performance of reduced graphene oxide cathodes. *RSC Adv.* **2015**, *5*, 53604–53610. [CrossRef]

35. Yoo, B.M.; Shin, H.J.; Yoon, H.W.; Park, H.B. Graphene and graphene oxide and their uses in barrier polymers. *J. Appl. Polym. Sci.* **2014**, *131*, 39628. [CrossRef]

36. Viskadouros, G.; Konios, D.; Kymakis, E.; Stratakis, E. Electron field emission from graphene oxide wrinkles. *RSC Adv.* **2016**, *6*, 2768–2773. [CrossRef]

37. Stankovich, S.; Dikin, D.A.; Piner, R.D.; Kohlhaas, K.A.; Kleinhammes, A.; Jia, Y.; Wu, Y.; Nguyen, S.B.T.; Ruoff, R.S. Synthesis of graphene-based nanosheets via chemical reduction of exfoliated graphite oxide. *Carbon* **2007**, *45*, 1558–1565. [CrossRef]

38. Li, X.; Wang, X.; Zhang, L.; Lee, S.; Dai, H. Chemically derived, ultrasmooth graphene nanoribbon semiconductors. *Science* **2008**, *319*, 1229–1232. [CrossRef]

39. Gomez-Navarro, C.; Weitz, R.T.; Bittner, A.M.; Scolari, M.; Mews, A.; Burghard, M.; Kern, K. Electronic Transport Properties of Individual Chemically Reduced Graphene Oxide Sheets. *Nano Lett.* **2007**, *7*, 3499–3503. [CrossRef]

40. Moon, I.K.; Lee, J.; Ruoff, R.S.; Lee, H. Reduced graphene oxide by chemical graphitization. *Nat. Commun.* **2010**, *1*, 73. [CrossRef]

41. Zhang, J.; Yang, H.; Shen, G.; Cheng, P.; Guo, S. Reduction of graphene oxide via L-ascorbic acid. *Chem. Commun.* **2010**, *46*, 1112–1114. [CrossRef]

42. Fan, X.; Peng, W.; Li, Y.; Li, X.; Wang, S.; Zhang, G.; Zhang, F. Deoxygenation of Exfoliated Graphite Oxide under Alkaline Conditions: A Green Route to Graphene Preparation. *Adv. Mater.* **2008**, *20*, 4490–4493. [CrossRef]

43. Akhavan, O. The effect of heat treatment on formation of graphene thin films from graphene oxide nanosheets. *Carbon* **2010**, *48*, 509–519. [CrossRef]

44. Guo, Y.; Wu, B.; Liu, H.; Ma, Y.; Yang, Y.; Yu, J.G.; Liu, Y. Electrical Assembly and Reduction of Graphene Oxide in a Single Solution Step for Use in Flexible Sensors. *Adv. Mater.* **2011**, *23*, 4626–4630. [CrossRef]

45. Wu, Z.S.; Pei, S.; Ren, W.; Tang, D.; Gao, L.; Liu, B.; Li, F.; Liu, C.; Cheng, H.M. Field Emission of Single-Layer Graphene Films Prepared by Electrophoretic Deposition. *Adv. Mater.* **2009**, *21*, 1756–1760. [CrossRef]

46. Munoz-Marquez, M.A.; Zarrabeitia, M.; Castillo-Martínez, E.; Eguía-Barrio, A.; Rojo, T.; Casas-Cabanas, M. Composition and Evolution of the Solid-Electrolyte Interphase in $Na_2Ti_3O_7$ Electrodes for Na-Ion Batteries: XPS and Auger Parameter Analysis. *ACS Appl. Mater. Interfaces* **2015**, *7*, 7801–7808. [CrossRef]

47. Konios, D.; Stylianakis, M.M.; Stratakis, E.; Kymakis, E. Dispersion behaviour of graphene oxide and reduced graphene oxide. *J. Colloid Interfaces Sci.* **2014**, *430*, 108–112. [CrossRef]

48. Zorba, V.; Tzanetakis, P.; Fotakis, C.; Spanakis, E.; Stratakis, E.; Papazoglou, D.G.; Zergioti, I. Silicon electron emitters fabricated by ultraviolet laser pulses. *Appl. Phys. Lett.* **2006**, *88*, 081103. [CrossRef]

49. Nillson, L.; Groening, O.; Emmenegger, C.; Kuettel, O.; Schaller, E.; Schlapbach, L.; Kind, H.; Bonard, J.M.; Kern, K. Scanning field emission from patterned carbon nanotube films. *Appl. Phys. Lett.* **2000**, *76*, 2071. [CrossRef]

50. Kymakis, E.; Amaratunga, G.A.J. Electrical properties of single-wall carbon nanotube-polymer composite films. *J. Appl. Phys.* **2006**, *99*, 084302. [CrossRef]

51. El-Kady, M.F.; Kaner, R.B. Direct laser writing of graphene electronics. *ACS Nano* **2014**, *8*, 8725–8729. [CrossRef]

52. Papazoglou, S.; Petridis, C.; Kymakis, E.; Kennou, S.; Raptis, Y.S.; Chatzandroulis, S.; Zergioti, I. In-situ sequential laser transfer and laser reduction of graphene oxide films. *Appl. Phys. Lett.* **2018**, *112*, 183301. [CrossRef]

53. Kymakis, E.; Savva, K.; Stylianakis, M.M.; Fotakis, C.; Stratakis, E. Flexible Organic Photovoltaic Cells with In Situ Nonthermal Photoreduction of Spin-Coated Graphene Oxide Electrodes. *Adv. Funct. Mater.* **2013**, *23*, 2742–2749. [CrossRef]

54. Stylianakis, M.M.; Stratakis, E.; Koudoumas, E.; Kymakis, E.; Anastasiadis, S.H. Organic Bulk Heterojunction Photovoltaic Devices Based on Polythiophene–Graphene Composites. *ACS Appl. Mater. Interfaces* **2012**, *4*, 4864–4870. [CrossRef]

55. Luo, Z.; Lu, Y.; Somers, L.A.; Johnson, A.T.C. High Yield Preparation of Macroscopic Graphene Oxide Membranes. *J. Am. Chem. Soc.* **2009**, *131*, 898–899. [CrossRef]

56. Kakavelakis, G.; Maksudov, T.; Konios, D.; Paradisanos, I.; Kioseoglou, G.; Stratakis, E.; Kymakis, E. Efficient and Highly Air Stable Planar Inverted Perovskite Solar Cells with Reduced Graphene Oxide Doped PCBM Electron Transporting Layer. *Adv. Energy Mater.* **2017**, *7*, 1602120. [CrossRef]

57. Berton, N.; Ottone, C.; Labet, V.; de Bettignies, R.; Bailly, S.; Grand, A.; Morell, C.; Sadki, S.; Chandezon, F. New Alternating Copolymers of 3,6-Carbazoles and Dithienylbenzothiadiazoles: Synthesis, Characterization, and Application in Photovoltaics. *Macromol. Chem. Phys.* **2011**, *212*, 2127–2141. [CrossRef]

58. Hoven, C.V.; Dang, X.D.; Coffin, R.C.; Peet, J.; Nguyen, T.Q.; Bazan, G.C. Improved performance of polymer bulk heterojunction solar cells through the reduction of phase separation via solvent additives. *Adv. Mater.* **2010**, *22*, E63–E66. [CrossRef]

59. Mahadevapuram, R.C.; Carr, J.A.; Chen, Y.; Bose, S.; Nalwa, K.S.; Petrich, J.W.; Chaudhary, S. Low-boiling-point solvent additives can also enable morphological control in polymer solar cells. *Synth. Met.* **2013**, *185–186*, 115–119. [CrossRef]

60. Cai, W.; Zhong, C.; Duan, C.; Hu, Z.; Dong, S.; Cao, D.; Lei, M.; Huang, F.; Cao, Y. The influence of amino group on PCDTBT-based and P3HT-based polymer solar cells: Hole trapping processes. *Appl. Phys. Lett.* **2015**, *106*, 233302. [CrossRef]

61. Wang, D.H.; Kim, J.K.; Seo, J.H.; Park, O.O.; Park, J.H. Stability comparison: A PCDTBT/PC$_{71}$BM bulk-heterojunction versus a P3HT/PC$_{71}$BM bulk-heterojunction. *Sol. Energy Mater. Sol. Cells* **2012**, *101*, 249–255. [CrossRef]

62. Staniec, P.A.; Parnell, A.J.; Dunbar, A.D.; Yi, H.; Pearson, A.J.; Wang, T.; Hopkinson, P.E.; Kinane, C.; Dalgliesh, R.M.; Donald, A.M.; et al. The Nanoscale Morphology of a PCDTBT:PCBM Photovoltaic Blend. *Adv. Energy Mater.* **2011**, *1*, 499–504. [CrossRef]

63. Park, S.H.; Roy, A.; Beaupré, S.; Cho, S.; Coates, N.; Moon, J.S.; Moses, D.; Leclerc, M.; Lee, K.; Heeger, A.J. Bulk heterojunction solar cells with internal quantum efficiency approaching 100%. *Nat. Photon.* **2009**, *3*, 297–302. [CrossRef]

64. Wang, T.; Pearson, A.J.; Dunbar, A.D.; Staniec, P.A.; Watters, D.C.; Yi, H.; Ryan, A.J.; Jones, R.A.; Iraqi, A.; Lidzey, D.G. Correlating Structure with Function in Thermally Annealed PCDTBT:PC$_{70}$BM Photovoltaic Blends. *Adv. Funct. Mater.* **2012**, *22*, 1399–1408. [CrossRef]

65. Bonaccorso, F.; Balis, N.; Stylianakis, M.M.; Savarese, M.; Adamo, C.; Gemmi, M.; Pellegrini, V.; Stratakis, E.; Kymakis, E. Functionalized Graphene as an Electron-Cascade Acceptor for Air-Processed Organic Ternary Solar Cells. *Adv. Funct. Mater.* **2015**, *25*, 3870–3880. [CrossRef]

 *nanomaterials*

Article

Room Temperature Resonant Photocurrent in an Erbium Low-Doped Silicon Transistor at Telecom Wavelength

Michele Celebrano [1], **Lavinia Ghirardini** [1], **Marco Finazzi** [1], **Giorgio Ferrari** [2], **Yuki Chiba** [3], **Ayman Abdelghafar** [3], **Maasa Yano** [3], **Takahiro Shinada** [4] and **Takashi Tanii** [3] and **Enrico Prati** [5,*]

[1]   Dipartimento di Fisica, Politecnico di Milano, Piazza Leonardo da Vinci 32, I-20133 Milano, Italy;
      michele.celebrano@polimi.it (M.C.); lavinia.ghirardini@polimi.it (L.G.); marco.finazzi@polimi.it (M.F.)

[2]   Dipartimento di Elettronica, Informazione e Bioingegneria, Politecnico di Milano, Via Colombo 81,
      I-20133 Milano, Italy; giorgio.ferrari@polimi.it

[3]   School of Science and Engineering, Waseda University, 3-4-1 Ohkubo, Shinjuku, Tokyo 169-8555, Japan;
      chiba@tanii.nano.waseda.ac.jp (Y.C.); aymanabdelghafar@toki.waseda.jp (A.A.);
      m.ntr0504@ruri.waseda.jp (M.Y.); tanii@waseda.jp (T.T.)

[4]   Center for Innovative Integrated Electronic Systems, Tohoku University, Sendai 980-8572, Japan;
      shinada@cies.tohoku.ac.jp

[5]   Istituto di Fotonica e Nanotecnologie, Consiglio Nazionale delle Ricerche, Piazza Leonardo da Vinci 32,
      I-20133 Milano, Italy

*   Correspondence: enrico.prati@cnr.it

Received: 1 February 2019; Accepted: 5 March 2019 ; Published: 11 March 2019

Abstract: An erbium-doped silicon transistor prepared by ion implantation and co-doped with oxygen is investigated by photocurrent generation in the telecommunication range. The photocurrent is explored at room temperature as a function of the wavelength by using a supercontinuum laser source working in the µW range. The 1-µm^2 transistor is tuned to involve in the transport only those electrons lying in the Er-O states. The spectrally resolved photocurrent is characterized by the typical absorption line of erbium and the linear dependence of the signal over the impinging power demonstrates that the Er-doped transistor is operating far from saturation. The relatively small number of estimated photoexcited atoms ($\approx 4 \times 10^4$) makes Er-dpoed silicon potentially suitable for designing resonance-based frequency selective single photon detectors at 1550 nm.

Keywords: erbium; silicon transistor; photocurrent

1. Introduction

Er implanted in silicon has received a renewed interest after the advent of single-photon-based quantum communications, because of its capability to transmit [1,2] and receive [3] photons at a wavelength compatible with commercial optical fibers, and because of its compatibility with silicon photonics [4–6]. Er-doped silicon junctions [7,8] and transistors [3,9] have been explored in the past. The photocurrent effect observed by an individual Er atom has been reported at 4 K, therefore, far from practical operation temperatures [3]. There, the $^4I_{13/2} \rightarrow {}^4I_{15/2}$ transition of the Er^{3+} ion in the silicon transistor is stimulated by a laser tuned at the resonance wavelength. We already explored single atom and impurity band effects in single ion implanted transistors [10–13] by including As [14], P [15] and Ge [16] atoms in view of nanoelectronic application, and the individual photon emission regime by photoluminescence of ErO$_x$ dots in silicon [6]. Several near-infrared detectors have been demonstrated in the past starting from Reference [17], including some based on CMOS technology with high sensitivity [18,19], but they are not frequency-selective, which would be relevant to reduce dark counts without the use of filters when detection of single monochromatic photons is targeted in an integrated device. In this article, the photocurrent of an ErO_x doped transistor is modulated by telecom wavelength irradiation as a function of both irradiation power and frequency. We investigate the room temperature photocurrent regime, where applications can be designed, by exploiting a relatively small number of Er atoms, of the order of 4×10^4, fed by a laser operated in the μW regime. In the following, the process to co-implant Er and O in back-gated transistors is outlined, and the photocurrent characterization is described at telecom wavelengths around 1550 nm. The photocurrent is proportional to the power of the laser and it reveals an absorption frequency dependence at coincidence with the Er absorption line. ErO_x co-doped transistor can operate at room temperature as a frequency selective light sensor, tuned at the same wavelength range of Er emission, towards nearby room temperature single photon emitter [20] and receiver [21] resonators based on few Er atoms.

2. Materials and Methods

2.1. Device Fabrication and Transport Characterization

Back-gated transistors were fabricated on a silicon-on-insulator (SOI) wafer by conventional complementary metal-oxide-silicon (CMOS) processes. The n^+ source and drain regions were doped with phosphorous at a concentration of 10^{21} cm^{-3}, and the n^- channel region was left undoped with 10^{15} cm^{-3}, thus, the transistors operate in the accumulation mode. Each transistor consists of 150 nm-buried oxide layer and 90 nm-SOI layer capped with 15 nm-silicon dioxide layer. The transistor size ranges between W = 1–10 μm, with $L = 1$ μm. The photocurrent effects reported in this work refer to the minimum size of W = 1 μm. The channel was doped with erbium at a dose of 3×10^{13} cm^{-2} at 20 keV. Oxygen was then implanted at a dose of 2.5×10^{13} cm^{-2} at 25 keV with a sacrificial layer deposited on the wafer to adjust the implantation depth of the two dopants. After removing the sacrificial layer, the wafer was thermally annealed at 900 °C for 30 min to promote the association of erbium and oxygen. Because of the out-diffusion of Er from the surface consequent to the annealing process, which reduces the Er of a factor of ≈ 7 [6,22], we estimate that the Er atoms in the channel of the transistor are of the order of 4×10^4. According to the literature, the annealing of Si:ErO_x at 900 °C determines the formation of a defect which lies at $E_C \approx -150$ meV where E_C is the conduction band edge, and it is partially ionized by thermal excitations at 300 K [8]. Such defect is dominant at high annealing temperature with respect to other Er-O related defects, while other Er defects disappear when O is present [23]. Such defects are not electrically explored in diode structures, while they can be explored by a transistor device, thanks to the control by the back gate voltage (Figure 1a). The formation of such defects is observed by the sub-threshold transport of the devices. Compared to the transfer characteristics curves of a control transistor not implanted with Er (Figure 1b), the Er-related defects at −150 meV determine a negative shift of the threshold voltage (data refer to L = 1 μm, W = 2 μm devices at $V_{DS} = 100$ mV). During a voltage sweep from negative to positive values the first accessed energy levels are those provided by the ErO defects, and only at about 150 meV higher in energy the conduction band is directly accessed.

2.2. Optical Setup for Photocurrent Characterization

The setup employed for the optical and photocurrent characterization is the modified commercial confocal microscope (WITec GmbH, Ulm, Germany) sketched in Figure 2a. The photoluminescence (PL) of the sample is excited by a continuous wave (CW) fiber-coupled laser diode at 782 nm wavelength. Light from such laser source, collimated and reflected by a beam splitter cube (Thorlabs Inc., Newton, NJ, USA), is focused by a 0.85 NA objective (Nikon Instruments Europe BV, Amsterdam, The Netherlands) to a diffraction-limited spot of about 1 μm in diameter on the sample. The emission from Er ions is collected through the same objective in epireflection geometry, and it is sent to a set of long-pass filters (1300 and 1350 nm cut-off wavelengths, Thorlabs Inc.) to reject any residual reflected pump light for the PL detection. The filtered emission is then imaged with an InP/InGaAs single photon avalanche diode (SPAD). To excite the photocurrent in the device, instead, we switch our light source to a CW fiber-coupled laser diode with a wavelength of 1550 nm.

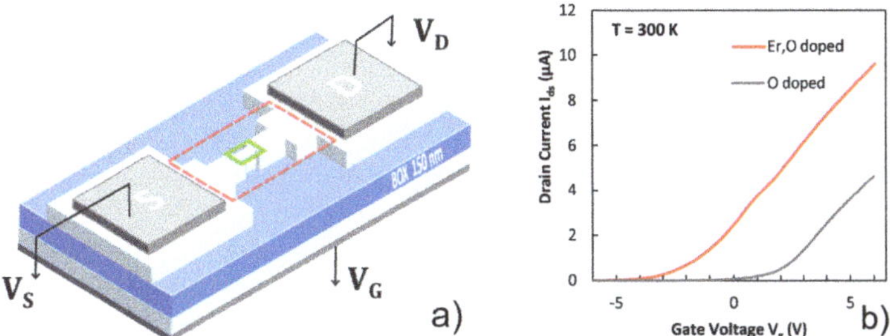

Figure 1. (**a**) Schematic of a transistor fabricated in an n-type (100) silicon on insulator (SOI) wafer. The transistor is based on a phosphorous-doped (1×10^{21} cm^{-3}) n^+ source and drain regions and a phosphorous-doped (1×10^{15} cm^{-3}) n^- channel region (size of L = 1 µm, W = 1 µm) and operates in the accumulation mode. The red dashed line encloses the optically investigated area. (**b**) I-V curve of two transistors differing only by the Er implantation measured at room temperature with $V_{DS} = 100$ mV. The shift of the threshold voltage is caused by the defect formation in the silicon band gap after 900 °C annealing .

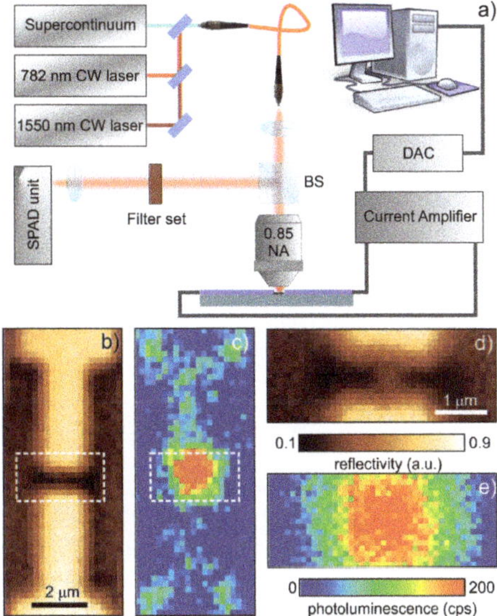

Figure 2. (**a**) Schematic of the setup used for the optical and electrical characterization of our sample. BS: beam splitter cube, DAC: data acquisition card. (**b**) Reflectivity and (**c**) photoluminescence (PL) maps of the whole transistor recorded at the telecom band while exciting it with the continuous wave (CW) laser at 782 nm. (**d**,**e**) Zooms of the maps shown in (**b**,**c**), respectively (see dashed lines), which correspond to the active region of the transistor. The PL is mostly coming from the area of the Er-implanted channel.

3. Results and Discussion

3.1. Photoluminescence Characterization

We now turn to the photonic characterization of the device. First, we performed PL and photocurrent measurements with the modified commercial confocal microscope introduced in the previous section and sketched in Figure 2a. We record emission PL and reflectivity maps form areas up to 80×80 μm^2 from the sample, by raster scanning it through the excitation and collection process with a piezoelectric stage. Figure 2b,c shows the reflectivity and emission maps respectively recorded from our Er-doped transistor. From a comparison of the sample topography and PL maps, we can confirm that the majority of the emission takes place in the channel area (highlighted region in Figure 2c and its zoom in Figure 2e), thus, verifying that the Er ions implantation is confined to this region. The lower intensity signal coming from the adjacent regions is attributed to a small PL contribution from the transistor contacts, while emission from the substrate in the spectral region is considered to be negligible.

3.2. Photocurrent Characterization

To perform the photocurrent characterization, we switch to a laser source with a wavelength of 1550 nm, and we couple it to the sample through the same setup described above. The drain-source current of the ErO_x doped transistor is converted into a voltage signal using a custom transimpedance amplifier with a conversion factor of 10 nA/V and a bandwidth of 16 Hz. The output voltage of the amplifier is connected to the input channel of the microscope and converted to the digital domain with a resolution of 3 pA simultaneously to the optical signal (see Figure 2a). In this way, optical reflectivity and photocurrent maps are simultaneously recorded, as exemplified in Figure 3a,b respectively. By comparing the sample topography and the photocurrent maps, the main current contribution originates when the Er-implanted channel region is optically excited (red dot in Figure 3b, ON state), while some background electrical current is generated from the other areas of the sample (green dot in Figure 3b, OFF state). By subtracting this background current signal recorded on the substrate to the current coming from the optical excitation of the transistor channel, we can extract the pure photocurrent generated with a $V_{DS} = 5$ mV by telecom wavelength irradiation, as plotted in Figure 3c (blue diamonds). Here, the two previously mentioned regimes, namely the photogenerated current and the conduction band transport respectively, are distinguishable. In particular, the photocurrent generated for $V_G \lesssim -2$ V is the one related to the presence of the defects.

It is worth mentioning here that by performing measurements with a non-zero gate voltage applied and zero bias between source and drain, we could observe a spurious current signal with two opposite signs at coincidence with the electrode edges, as shown in Figure 3b. Moreover, we observe that when the channel is almost completely depleted at gate voltages V_G below -6 V, a presumably different effect induces a photocurrent of the order of pA with reversed sign at the source and the drain sides respectively. This artifact has been already reported in the literature [24] and it is attributed to trapped photo-generated electrons that are effective in the suppression of the electric field near the electrodes. Although the rejection of such spurious signal is rather challenging, we were able to avoid such an artifact by collecting the signal in the middle of the channel (see red spot in Figure 3b), where the signal level for the photocurrent induced by the trapped electrons at the electrodes is negligible. A more detailed characterization of such effect is reported in the Figure S1 in the Supplementary Material.

We then performed wavelength dependent photocurrent measurements using the same acquisition setup mentioned above, by switching to a supercontinuum light source. In order to observe effects related to the defects only, the gate voltage is set to -4 V (see dashed red line in Figure 3c), at which value the defects are partially filled while the conduction band is almost empty at room temperature. To separate the different excitation wavelengths within the wide band of the exciting supercontinuum source, the light was chromatically dispersed by a Pellin-Broca prism and then selected by a slit before being coupled to a single mode fiber. In this way we could attain a relatively broadband monochromator, which allowed us to vary the wavelength of the excitation between 1400 nm and 1700 nm in steps of 20 nm with an input power level of about 1 mW for each spectral window. This experimental arrangement is the exact reciprocal of that described in detail elsewhere [25]. We also determined the optical throughput of our system by measuring the laser power impinging on the sample at each operating wavelength, as a normalization reference for the photocurrent measurement. The pure photocurrent generated by the Er-doped transistor is extrapolated from the maps recorded for each exciting wavelength, as described above. The photocurrent spectrum thus, obtained is then normalized by the power spectrum of the supercontinuum source through the optical setup to account for the achromatism of the setup. The result of this operation is shown in Figure 4a. The normalized photocurrent spectrum thus, obtained shows a main peak around 1526 nm, with a shoulder around 1448 nm. Such spectral features are in excellent agreement with the wavelength dependent absorption cross-section of commercial Er-doped optical fibers. Additionally, as shown by the linearity of the photocurrent dependence on the optical pump power at $\lambda = 1550$ nm (see the inset of Figure 4), the Er-doped transistor is operating far from its electrical saturation regime.

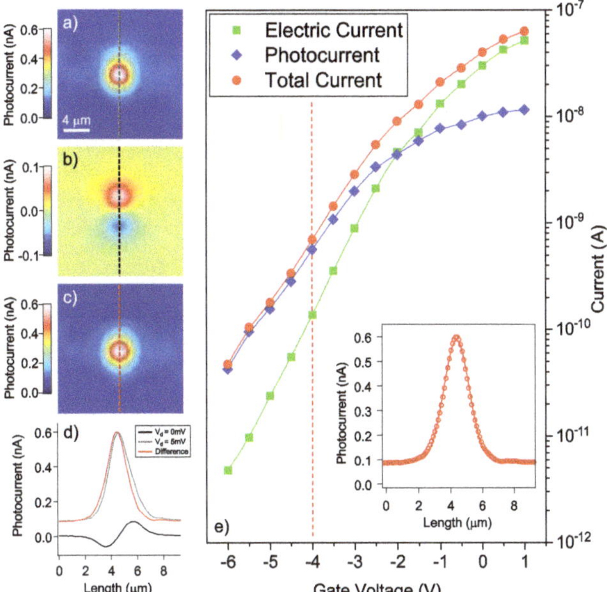

Figure 3. (**a**) Overall photocurrent map recorded on the transistor when exciting the sample at 1550 nm while applying $V_G = -4$ V and $V_{DS} = 5$ mV . (**b**) Photocurrent recorded when no net VDS is applied. (**c**) Pure photocurrent map obtained by subtracting the overall photocurrent in (**a**) by the background photocurrent in (**b**). (**d**) Vertical line profiles drawn on (**a–c**). (**e**) Drain photocurrent in ohmic regime ($V_{DS} = 5$ mV) as a function of the gate voltage in dark condition (no light, green squares) and the peak photocurrent recorded on maps such as (B) while illuminating the device with an optical power of about 1 mW (red circles). The photocurrent (blue diamonds) calculated by taking the difference between the current in ON (signal) and OFF (back-ground) conditions, i.e. the spots where the photocurrent was recorded while photoexciting on the substrate and on the Er-doped channel, respectively.

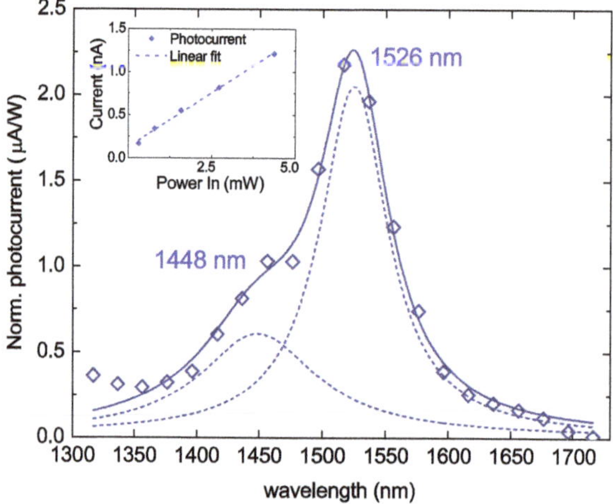

Figure 4. Frequency dependent photoconductivity measurements observed from the photocurrent, realized by exciting the Er-doped transistor with a supercontinuum light source, normalized by the optical power throughput of the source through our setup (empty diamonds). Double peak Gaussian fit to highlight the position of the spectral features of the photoconductivity measurement (dashed lines), which reproduce the wavelength dependent absorption cross-section of Erbium [26]. Inset: linear dependence of the photoconductivity on the input optical power, measured at $\lambda = 1550$ nm.

4. Conclusions

To conclude, we demonstrated that an ErO_x co-doped silicon transistor exhibits a tunable response at room temperature according to a linewidth profile typical of erbium absorption. Electrons are pumped from the ErO_x complex activated by the annealing at 900 °C. The linear dependence of the signal over the impinging power demonstrates that the Er-doped transistor is operating far from electrical saturation. The relatively small number of photoexcited atoms estimated ($\approx 4 \times 10^4$), combined with a frequency-selective response, points towards a viable employment of ErO_x defects in resonator-based single photon detectors at telecom wavelength at room temperature.

Supplementary Materials: The Supplementary Materials are available online at www.mdpi.com/2079-4991/9/3/416/s1.

Author Contributions: Conceptualization, E.P. and G.F.; methodology, Y.C., A.A., M.Y., T.T. and T.S.; validation, M.C., L.G., E.P. and G.F.; formal analysis, M.C. and E.P.; investigation, M.C., L.G., E.P. and G.F.; data curation, M.C.; writing—original draft preparation, M.C. and E.P.; writing—review and editing, L.G. and E.P.; supervision, M.F., E.P. and T.T.

Funding: This work was supported by a Grant-in-Aid for Basic Research (B) and Young Scientists (A) from MEXT.

Acknowledgments: E.P. acknowledges JSPS Fellowship 2014, and the CNR Short-Term Mobility Programs 2015 and 2016.

Conflicts of Interest: The authors declare no conflict of interest. The funders had no role in the design of the study; in the collection, analyses, or interpretation of data; in the writing of the manuscript, or in the decision to publish the results.

References

1. Simon, C.; Afzelius, M.; Appel, J.; Boyer de la Giroday, A.; Dewhurst, S.J.; Gisin, N.; Hu, C.Y.; Jelezko, F.; Kröll, S. Quantum memories—A review based on the European integrated project "Qubit Applications (QAP)". *Eur. Phys. J. B* **2010**, *58*, 1–22.

2. Dibos, A.M.; Raha, M.; Phenicie, C.M.; Thompson, J.D. Atomic Source of Single Photons in the Telecom Band. *Phys. Rev. Lett.* **2018**, *120*, 243601. [CrossRef] [PubMed]

3. Yin, C.; Rancic, M.; de Boo, G.G.; Stavrias, N.; McCallum, J.C.; Sellars, M.J.; Rogge, S. Optical addressing of an individual erbium ion in silicon. *Nature* **2013**, *497*, 91–94. [CrossRef] [PubMed]

4. Priolo, F.; Gregorkiewicz, T.; Galli, M.; Krauss, T.F. Silicon nanostructures for photonics and photovoltaics. *Nat. Nanotechnol.* **2014**, *9*, 19–32. [CrossRef] [PubMed]

5. Prati, E.; Celebrano, M.; Ghirardini, L.; Biagioni, P.; Finazzi, M.; Shimizu, Y.; Tu, Y.; Inoue, K.; Nagai, Y.; Shinada, T.; et al. Revisiting room-temperature 1.54 μm photoluminescence of ErOx centers in silicon at extremely low concentration. In Proceedings of the 2017 Silicon Nanoelectronics Workshop, Kyoto, Japan, 4–5 June 2017.

6. Celebrano, M.; Ghirardini, L.; Finazzi, M.; Shimizu, Y.; Tu, Y.; Inoue, K.; Nagai, Y.; Shinada, T.; Chiba, Y.; Abdelghafar, A.; et al. 1.54 μm photoluminescence from Er:Ox centers at extremely low concentration in silicon at 300 K. *Opt. Lett.* **2017**, *42*, 3311–3314. [CrossRef] [PubMed]

7. Franzò, G.; Coffa, S.; Priolo, F.; Spinella, C. Mechanism and performance of forward and reverse bias electroluminescence at 1.54 μm from Er-doped Si diodes. *J. Appl. Phys.* **1997**, *81*, 2784–2793. [CrossRef]

8. Priolo, F.; Franzò, G.; Coffa, S.; Polman, A.; Libertino, S.; Barklie, R.; Carey, D. The erbium-impurity interaction and its effects on the 1.54 μm luminescence of Er^{3+} in crystalline silicon. *J. Appl. Phys.* **1995**, *78*, 3874–3882. [CrossRef]

9. Zhang, Q.; Hu, G.; de Boo, G.G.; Rancic, M.; Johnson, B.C.; McCallum, J.C.; Du, J.; Sellars, M.; Yin, C.; Rogge, S. Single rare-earth ions as atomic-scale probes in ultra-scaled transistors. *arXiv* **2018**, arXiv:1803.01573.

10. Shinada, T.; Okamoto, S.; Kobayashi, T.; Ohdomari, I. Enhancing semiconductor device performance using ordered dopant arrays. *Nature* **2005**, *437*, 1128–1131. [CrossRef] [PubMed]

11. Prati, E.; Shinada, T. Atomic scale devices: Advancements and directions. In Proceedings of the 2014 IEEE International Electron Devices Meeting, San Francisco, CA, USA, 15–17 December 2014; pp. 1.2.1–1.2.4.

12. Prati, E.; Shinada, T. (Eds.) *Single-Atom Nanoelectronics*; Pan Stanford Publishing: Singapore, 2013.

13. Shinada, T.; Prati, E.; Tamura, S.; Tanii, T.; Teraji, T.; Onoda, S.; Ohshima, T.; McGuinness, L.P.; Rogers, L.; Naydenov, B.; et al. Opportunity of single atom control for quantum processing in silicon and diamond. In Proceedings of the 2014 Silicon Nanoelectronics Workshop, Honolulu, HI, USA, 8–9 June 2014.

14. Prati, E.; Hori, M.; Guagliardo, F.; Ferrari, G.; Shinada, T. Anderson–Mott transition in arrays of a few dopant atoms in a silicon transistor. *Nat. Nanotechnol.* **2012**, *7*, 443—447. [CrossRef] [PubMed]

15. Prati, E.; Kumagai, K.; Hori, M.; Shinada, T. Band transport across a chain of dopant sites in silicon over micron distances and high temperatures. *Sci. Rep.* **2016**, *6*, 19704. [CrossRef] [PubMed]

16. Prati, E.; Chiba, Y.; Yano, M.; Kumagai, K.; Hori, M.; Ferrari, G.; Shinada, T.; Tanii, T. Single ion implantation of Ge donor impurity in silicon transistors. In Proceedings of the 2015 Silicon Nanoelectronics Workshop, Kyoto, Japan, 14–15 June 2015.

17. Fan, H.Y.; Ramdas, A.K. Infrared Absorption and Photoconductivity in Irradiated Silicon. *J. Appl. Phys.* **1959**, *30*, 1127–1134. [CrossRef]

18. Geis, M.W.; Spector, S.J.; Grein, M.E.; Schulein, R.T.; Yoon, J.U.; Lennon, D.M.; Deneault, S.; Gan, F.; Kaertner, F.X.; Lyszczarz, T.M. CMOS-Compatible All-Si High-Speed Waveguide Photodiodes with High Responsivity in Near-Infrared Communication Band. *IEEE Photonics Technol. Lett.* **2007**, *19*, 152–154. [CrossRef]

19. Bradley, J.D.B.; Jessop, P.E.; Knights, A.P. Silicon waveguide-integrated optical power monitor with enhanced sensitivity at 1550 nm. *Appl. Phys. Lett.* **2005**, *86*, 241103. [CrossRef]

20. McKeever, J.; Boca, A.; Boozer, A.D.; Miller, R.; Buck, J.R.; Kuzmich, A.; Kimble, H.J. Deterministic Generation of Single Photons from One Atom Trapped in a Cavity. *Science* **2004**, *303*, 1992–1994. [CrossRef] [PubMed]

21. Ritter, S.; Nölleke, C.; Hahn, C.; Reiserer, A.; Neuzner, A.; Uphoff, M.; Mücke, M.; Figueroa, E.; Bochmann, J.; Rempe, G. An elementary quantum network of single atoms in optical cavities. *Nature* **2012**, *484*, 195–200. [CrossRef] [PubMed]

22. Shimizu, Y.; Tu, Y.; Abdelghafar, A.; Yano, M.; Suzuki, Y.; Tanii, T.; Shinada, T.; Prati, E.; Celebrano, M.; Finazzi, M.; et al. Atom probe study of erbium and oxygen co-implanted silicon. In Proceedings of the 2017 Silicon Nanoelectronics Workshop, Kyoto, Japan, 4–5 June 2017.

23. Kenyon, A.J. Erbium in silicon. *Semicond. Sci. Technol.* **2005**, *20*, R65–R84. [CrossRef]

24. Agostinelli, T.; Caironi, M.; Natali, D.; Sampietro, M.; Biagioni, P.; Finazzi, M.; Duò, L. Space charge effects on the active region of a planar organic photodetector. *J. Appl. Phys.* **2007**, *101*, 114504. [CrossRef]

25. Celebrano, M.; Baselli, M.; Bollani, M.; Frigerio, J.; Bahgat Shehata, A.; Della Frera, A.; Tosi, A.; Farina, A.; Pezzoli, F.; Osmond, J.; et al. Emission Engineering in Germanium Nanoresonators. *ACS Photonics* **2015**, *2*, 53–59. [CrossRef]

26. Miniscalco, W.J.; Quimby, R.S. General procedure for the analysis of Er^{3+} cross sections. *Opt. Lett.* **1991**, *16*, 258–260. [CrossRef] [PubMed]

MDPI
St. Alban-Anlage 66
4052 Basel
Switzerland
Tel. +41 61 683 77 34
Fax +41 61 302 89 18
www.mdpi.com

Nanomaterials Editorial Office
E-mail: nanomaterials@mdpi.com
www.mdpi.com/journal/nanomaterials